ENVIRONMENTAL CHEMISTRY

ENVIRONMENTAL CHEMISTRY

An Analytical Approach

KENNETH S. OVERWAY

WILEY

Published by John Wiley & Sons, Inc., Hoboken, New Jersey
Published simultaneously in Canada

Library of Congress Cataloging-in-Publication Data

Names: Overway, Kenneth S., 1971- author.
Title: Environmental chemistry : an analytical approach / Kenneth S. Overway.
Description: Hoboken : John Wiley & Sons, Inc., [2017] | Includes
bibliographical references and index.
Identifiers: LCCN 2016034813 (print) | LCCN 2016036066 (ebook) | ISBN
9781118756973 (hardback) | ISBN 9781119085508 (pdf) | ISBN 9781119085492
(epub)
Subjects: LCSH: Environmental chemistry.
Classification: LCC TD193 .O94 2017 (print) | LCC TD193 (ebook) | DDC
577/.14–dc23
LC record available at https://lccn.loc.gov/2016034813

Cover Design: Wiley
Cover Images: Earth © NASA;
Graphs courtesy of author

Typeset in 10/12pt TimesLTStd-Roman by SPi Global, Chennai, India

Printed in the United States of America

10 9 8 7 6 5 4 3 2 1

Periodic Table of the Elements

1 H 1.00794*																1 H 1.00794*	2 He 4.002602
3 Li 6.941*	4 Be 9.012182											5 B 10.811*	6 C 12.0107*	7 N 14.00674*	8 O 15.9994*	9 F 18.9984032	10 Ne 20.1797
11 Na 22.98976928	12 Mg 24.3050											13 Al 26.9815386	14 Si 28.0855*	15 P 30.973762	16 S 32.066*	17 Cl 35.4527*	18 Ar 39.948
19 K 39.0983	20 Ca 40.078	21 Sc 44.955912	22 Ti 47.867	23 V 50.9415	24 Cr 51.9961	25 Mn 54.938045	26 Fe 55.845	27 Co 58.933195	28 Ni 58.6934	29 Cu 63.546	30 Zn 65.38	31 Ga 69.723	32 Ge 72.63	33 As 74.92160	34 Se 78.96	35 Br 79.904	36 Kr 83.798
37 Rb 85.4678	38 Sr 87.62	39 Y 88.90585	40 Zr 91.224	41 Nb 92.90638	42 Mo 95.96	43 Tc (97.9072)	44 Ru 101.07	45 Rh 102.90550	46 Pd 106.42	47 Ag 107.8682	48 Cd 112.411	49 In 114.818	50 Sn 118.710	51 Sb 121.760	52 Te 127.60	53 I 126.90447	54 Xe 131.293
55 Cs 132.9054519	56 Ba 137.327	57 La 138.90547	72 Hf 178.49	73 Ta 180.94788	74 W 183.84	75 Re 186.207	76 Os 190.23	77 Ir 192.217	78 Pt 195.084	79 Au 196.966569	80 Hg 200.59	81 Tl 204.3833*	82 Pb 207.2	83 Bi 208.98040	84 Po (208.9824)	85 At (209.9871)	86 Rn (222.0176)
87 Fr (223.0197)	88 Ra (226.0254)	89 Ac (227.0278)	104 Rf (265.1167)	105 Db (268.125)	106 Sg (271.133)	107 Bh (270)	108 Hs (277.150)	109 Mt (276.151)	110 Ds (281.162)	111 Rg (280.164)	112 Cn (285.174)	113 Uut (284.178)	114 Fl** (289.187)	115 Uup (288.192)	116 Lv** (293)	117 Uus (294)	118 UUo (294)

58 Ce 140.116	59 Pr 140.90765	60 Nd 144.242	61 Pm (144.9127)	62 Sm 150.36	63 Eu 151.964	64 Gd 157.25	65 Tb 158.92535	66 Dy 162.500	67 Ho 164.93032	68 Er 167.259	69 Tm 168.93421	70 Yb 173.054	71 Lu 174.9668
90 Th 232.03806	91 Pa 231.03588	92 U 238.02891	93 Np (237.0482)	94 Pu (244.0642)	95 Am (243.0614)	96 Cm (247.0704)	97 Bk (247.0703)	98 Cf (251.0796)	99 Es (252.0830)	100 Fm (257.0951)	101 Md (258.0984)	102 No (259.1010)	103 Lr (262.1096)

S.E. Van Bramer 8/29/2012

* 1995 IUPAC Values from Pure Appl. Chem., Vol 68, No. 12, pp. 2339-2359, 1996. doi: 10.1351/pac199668122339, http://pac.iupac.org/publications/pac/pdf/1996/pdf/6812x2339.pdf

**Names for elements 114 and 116 are from Pure Appl. Chem., Vol. 84, No. 7, pp. 1669–1672, 2012. doi: 10.1351/PAC-REC-11-12-03

All other values from: 2009 IUPAC Values from Pure Appl. Chem., Vol. 83, No. 2, pp. 359–396, 2011. doi:10.1351/PAC-REP-10-09-14, http://pac.iupac.org/publications/pac/pdf/2011/pdf/8302x0359.pdf

-Elements with one weight have uncertainty in the last digit.

-Elements with the weight in parenthesis, weight is given for the longest lived isotope.

Electronegativity Table of the *p*-block Elements

				1 **H** 2.20	2 **He**
5 **B** 2.04	6 **C** 2.55	7 **N** 3.04	8 **O** 3.44	9 **F** 3.98	10 **Ne**
13 **Al** 1.61	14 **Si** 1.90	15 **P** 2.19	16 **S** 2.58	17 **Cl** 3.16	18 **Ar**
31 **Ga** 1.81	32 **Ge** 2.01	33 **As** 2.18	34 **Se** 2.55	35 **Br** 2.96	36 **Kr**
49 **In** 1.78	50 **Sn** 1.96	51 **Sb** 2.05	52 **Te** 2.10	53 **I** 2.66	54 **Xe** 2.60
49 **Ti** 1.8	50 **Pb** 1.8	51 **Bi** 1.9	52 **Po** 2.0	53 **At** 2.2	54 **Rn**

CONTENTS

PREFACE

Careful readers of this textbook will find it difficult to avoid the conclusion that the author is a cheerleader for collegiate General Chemistry. I have taught General Chemistry at various schools for over a decade and still enjoy the annual journey that takes me and the students through a wide array of topics that explain some of the microscopic and macroscopic observations that we all make on a daily basis. The typical topics found in an introductory sequence of chemistry courses really do provide a solid foundation for understanding most of the environmental issues facing the world's denizens. After teaching Environmental Chemistry for a few years, I felt that the textbooks available were missing some key features.

Similar to a movie about a fascinating character, an origin story is needed. In order to appreciate the condition and dynamism of our current environment, it is important to have at least a general sense of the vast history of our planet and of the dramatic changes that have occurred since its birth. The evolution of the Earth would not be complete without an understanding of the origin of the elements that compose the Earth and all of its inhabitants. To this end, I use Chapter 1 to develop an abridged, but hopefully coherent, evolution of our universe and solar system. It is pertinent that this origin story is also a convenient occasion to review some basic chemical principles that should have been learned in the previous courses and will be important for understanding the content of this book.

As a practical matter when teaching Environmental Chemistry, I was required to supplement other textbooks with a primer on measurement statistics. My students and I are making environmental measurements soon after the course begins, so knowing how to design an analysis and process the results is essential. In Chapter 2, I provide a minimal introduction to the nature of measurements and the quantitative methods and tools used in the process of testing environmental samples. This analysis relies heavily on the use of spreadsheets, a skill that is important for any quantitative scientist to master. This introduction to measurements is supplemented by an appendix that describes several of the instruments one is likely to encounter in an environmental laboratory.

Finally, the interdependence of a certain part of the environment with many others becomes obvious after even a casual study. A recursive study of environmental principles, where the complete description of an environmental system requires one to back up to study the underlying principles and the exhaustive connections between other systems followed by a restudy of the original system, is the natural way that many of us have learned about the environment. It does not, however, lend itself to the encapsulated study that a single semester represents. Therefore, I have divided the environment into the three interacting domains of The Atmosphere (Chapter 3), The Lithosphere (Chapter 4), and The Hydrosphere (Chapter 5). In each chapter, it is clear that the principles of each of these domains affect the others. Studies of the environment beyond a semester will require a great deal of recursion and following tangential topics in order to understand the whole, complicated picture. Such is the nature of most deep studies, and this textbook will hopefully provide the first steps in what may be a career-long journey.

Shall we begin?

Ken Overway
Bridgewater, Virginia
December, 2015

ABOUT THE COMPANION WEBSITE

This book is accompanied by a companion website:

www.wiley.com/go/overway/environmental_chemistry

The website includes:

- Powerpoint Slides of Figures
- PDF of Tables
- Regression Spreadsheet Template

INTRODUCTION

You are "greener" than you think you are. What I mean is that you have been twice recycled. You probably are aware that all of the molecules that make up your body have been recycled from the previous organisms, which is similar to the chemical cycles you will read about later in this book, such as the carbon cycle and the nitrogen cycle. The Earth is nearly a closed system, and it receives very little additional matter from extraterrestrial sources, except for the occasional meteor that crashes to the Earth. So, life must make use of the remains of other organism and inanimate sources in order to build organism bodies.

What you may not have been aware of is that the Earth and the entire solar system in which it resides were formed from the discarded remains of a previous solar system. This must be the case since elements beyond helium form only in the nuclear furnace of stars. Further, only in the core of a giant star do elements beyond carbon form, and only during the supernova explosion of a giant star do elements beyond iron form. Since the Earth contains all of these elements, it must be the result of at least a previous solar system. This revelation should not be entirely unexpected when you examine the vast difference between the age of the universe (13.8 billion years old) and the age of our solar system (4.6 billion years old). What happened during the 9.2 billion year gap? How did our solar system form? How did the Earth form? What are the origins of life? To answer these questions, the story of the chemical history of the universe since the Big Bang is required. Much of what you learned in General Chemistry will help you understand the origin of our home planet. It may seem like it has been 13.8 billion years since your last chemistry course, so a review is warranted. Ready?

1

ORIGINS: A CHEMICAL HISTORY OF THE EARTH FROM THE BIG BANG UNTIL NOW – 13.8 BILLION YEARS OF REVIEW

> Not only is the Universe stranger than we imagine, it is stranger than we can imagine.
> —Sir Arthur Eddington

> I'm astounded by people who want to 'know' the universe when it's hard enough to find your way around Chinatown.
> —Woody Allen

1.1 INTRODUCTION

Georges-Henri Lemaître (1894–1966), a Jesuit priest and physicist at Université Catholique de Louvain, was the first person to propose the idea of the Big Bang. This theory describes the birth of our universe as starting from a massive, single point in space at the beginning of time (literally, $t = 0$ s!), which began to expand in a manner that could loosely be called an explosion. Another famous astrophysicist and skeptic of Lemaître's hypothesis, Sir Fred Hoyle (1915–2001), jeeringly called this the "Big Bang" hypothesis. Years later, with several key experimental predictions having been observed, the Big Bang is now a theory. Lemaître developed his hypothesis from solutions to Albert Einstein's (1879–1955) theory of general relativity. Since this is not a mathematics book, and I suspect you are not interested in tackling the derivation of these equations (neither am I), so let us examine the origin of our environment and the conditions that led to the Earth that we inhabit. This chapter is not meant to be a rigorous and exhaustive explication of the Big Bang and the evidence for the evolution of the universe, which would require a deep background in atomic particle physics and cosmology. Since this is an environmental chemistry text, I will only describe items that are relevant for the environment in the context of a review of general chemistry.

1.2 THE BIG BANG

1.2.1 The Microwave Background

The first confirmation of the Big Bang comes from the prediction and measurement of what is known as the microwave background. Imagine you are in your kitchen and you turn on an electric stove. If you placed your hand over the burner element, you would feel it heat up. This feeling of heat is a combination of the convection of hot air touching your skin

---

Environmental Chemistry: An Analytical Approach, First Edition. Kenneth S. Overway.

Companion website: www.wiley.com/go/overway/environmental_chemistry

and infrared radiation. As the heating element warms up, you would notice the color of it changes from a dull red to a bright orange color. If it could get hotter, it would eventually look whitish because it is emitting most of the colors of the visible spectrum. What you have observed is Wien's Displacement Law, which describes blackbody radiation.

$$\lambda_{max} = \frac{2.8977685 \times 10^{-3}\,\frac{\mathrm{m}}{\mathrm{K}}}{T} \tag{1.1}$$

This equation shows how the temperature (T) of some black object (black so that the color of the object is not mistaken as the reflected light that gives an apple, e.g., its red or green color) affects the radiation (λ_{max}) the object emits. On a microscopic level, the emission of radiation is caused by electrons absorbing the heat of the object and converting this energy to light.

Gamma rays: Excites energy levels of the nucleus; sterilizing medical equipment
X-Rays: Refracts from the spaces between atoms and excites core e^-; provides information on crystal structure; used in medical testing
Ultraviolet: Excites and breaks molecular bonds; polymerizing dental fillings
Visible: Excites atomic and molecular electronic transitions; our vision
Infrared: Excites molecular vibrations; night vision goggles
Microwave: Excites molecular rotations; microwave ovens
Radio waves: Excites nuclear spins; MRI imaging and radio transmission

Table 1.1 Certain regions of the electromagnetic (EM) spectrum provide particular information about matter when absorbed or emitted.

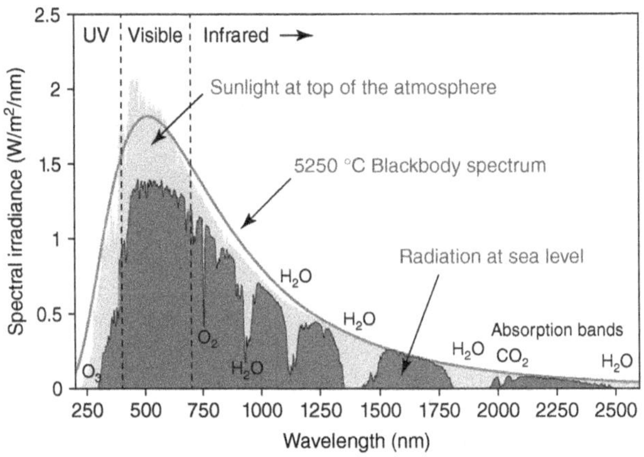

Figure 1.1 Another view of the solar radiation spectrum showing the difference between the radiation at the top of the atmosphere and at the surface. Source: Robert A. Rhode http://en.wikipedia.org/wiki/File:Solar_Spectrum.png. Used under BY-SA 3.0 //creative commons.org/licenses/by-sa/3.0/deed.en.

The λ_{max} in Wien's equation represents, roughly, the average wavelength of a spectrum, such as in Figure 1.1, which shows the emission spectrum of the Sun. Wien's Law also lets us predict the temperature of different objects, such as stars, by calculating T from λ_{max}.

Robert Dicke (1916–1977), a physicist at Princeton University, predicted that if the universe started out as a very small, very hot ball of matter (as described by the Big Bang) it would cool as it expanded. As it cooled, the radiation it would emit would change according to Wien's Law. He predicted that the temperature at which the developing universe would become transparent to light would be when the temperature dropped below 3000 K. Given that the universe has expanded a 1000 times since then, the radiation would appear red-shifted by a factor of 1000, so it should appear to be 3 K. How well does this compare to the observed temperature of the universe?

☞ For a review of the EM spectrum, see Review Example 1.1 on page 22.

When looking into the night sky, we are actually looking at the leftovers of the Big Bang, so we should be seeing the color of the universe as a result of its temperature. Since the night sky is black except for the light from stars, the background radiation from the Big Bang must not be in the visible region of the spectrum but in lower regions such as the infrared or the microwave region. When scientists at Bell Laboratories in New Jersey used a large ground-based antenna to study emission from our Milky Way galaxy in 1962, they observed a background noise that they could not eliminate no matter which direction they pointed the antenna. They also found a lot of bird poop on the equipment, but clearing that out did not eliminate the "noise." They finally determined that the noise was the background emission from the Big Bang, and it was in the microwave region of the EM spectrum (Table 1.1), just as Dicke predicted. The spectral temperature was measured to be 2.725 K. This experimental result was a major confirmation of the Big Bang Theory.

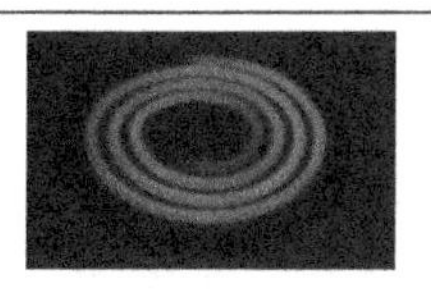

Figure 1.2 A glowing electric stove element. Courtesy K. Overway.

Blackbody Radiation

The electric heater element (Figure 1.2) demonstrates blackbody radiation. Any object that has a temperature above 0 K will express its temperature by emitting radiation that is proportional to its temperature. Wien's Displacement Law gives the relationship between the average wavelength of the radiation and the temperature. The Earth emits infrared radiation as a result of its temperature, and this leads to the greenhouse effect, which is discussed later. The person in the photos in Figure 1.3 also emits radiation in the infrared, allowing an image of his arm and hand to be seen despite the visible opacity of the plastic bag.

Figure 1.3 While visible radiation cannot penetrate the plastic bag, the infrared radiation, generated by the blackbody radiation of the man's body, can. Source: NASA.

Infrared Thermography

Infrared thermography is an application of Wien's Law and is a key component of a home energy audit. One of the most cost-effective ways to conserve energy is to improve the insulation envelope of one's house. Handheld infrared cameras, seen in Figure 1.4, allow homeowners or audit professionals to see air leaks around windows and doors. On a cold day, an uninsulated electrical outlet or poorly insulated exterior wall could be 5–8 °F colder than the surroundings. When the handheld thermal camera is pointed at a leak, the image that appears on the screen will clearly identify it by a color contrast comparison with the area around it.

Figure 1.4 A thermal camera used to find cold spots in a leaky house. Source: Passivhaus Institut "http://en.wikipedia.org/wiki/File:SONEL_KT-384.jpg." Used under BY-SA 3.0 //creativecommons.org/licenses/by-sa/3.0/deed.en.

Example 1.1: Blackbody Radiation

Wien's Displacement Law is an important tool for determining the temperature of objects based on the EM radiation that they emit and predicting the emission profile based on the temperature of an object.

1. Using Wien's Displacement Law (Eq. (1.1)), calculate the λ_{max} for a blackbody at 3000 K.
2. Using Wien's Displacement Law, calculate the λ_{max} for the Earth, which has an average surface temperature of 60 °F.
3. In which portion of the EM spectrum is the λ_{max} for the Earth?

Solution: See Section A.1 on page 231.

After the development of modern land-based and satellite telescopes, scientists observed that there were other galaxies in the universe besides our own Milky Way. Since this is true, the universe did not expand uniformly – with some clustering of matter in some places and very little matter in others. Given what we know of gravity, the clusters of matter would not expand at the same rate as matter that is more diffuse. Therefore, there must be some hot and cold spots in the universe, and the microwave background should show this. In 1989, an advanced microwave antenna was launched into space to measure

Pickering's "Harem" (1913) Edward Charles Pickering, who was a director of the Harvard College Observatory in the late 19th century, had a workforce of young men acting as "computers" – doing the very tedious work of calculating and categorizing stars using the astrophotographs that the observatory produced. In 1879, Pickering hired Williamina Fleming, an immigrant who was a former teacher in Scotland but recently struck by misfortune (abandoned by her husband while pregnant) as a domestic servant. Sometime after having noticed Fleming's intelligence, Pickering was reported to have said to one of his male computers that his housekeeper could do a better job. He hired her and went on to hire many more women because they were better at computers than their male counterparts, and they were paid about half the wages of the men (meaning Pickering could hire twice as many of them!). This group came to be known as "Pickering's Harem" and produced several world-renowned female astronomers that revolutionized the way we understand stars and their composition. Source: Harvard-Smithsonian Center for Astrophysics (https://www.cfa.harvard.edu/jshaw/pick.html). See Kass-Simon and Farnes (1990, p. 92).

this predicted heterogeneity of temperature. Further studies and better satellites produced an even finer measurement of the microwave background. The observation of the heterogeneity of the microwave background is further direct and substantial evidence of the Big Bang theory.

1.2.2 Stars and Elements

In the late 19th century, the Harvard College observatory was the center of astrophotography (see Pickering's "Harem" featurette). Astronomers from this lab were among the first to see that the colors of stars could be used to determine their temperatures and their compositions. When the EM radiation from these stars is passed through a prism, the light is dispersed into its component wavelengths, much like visible light forms a rainbow after it passes through a prism. An example of such a spectrum can be seen in Figure 1.5. Blackbody radiation (described by Wien's Law) predicts that the spectrum of the Sun should be continuous – meaning it should contain the full, unbroken spectrum – but Figure 1.5 shows that this is obviously not the case. The black lines in the spectrum indicate the presence of certain atoms and molecules in the outer atmosphere of the Sun that are absorbing some very specific wavelengths of light out of the solar spectrum. The position of the lines is a function of the energy levels of the electrons in the atoms and can be treated as an atomic fingerprint. The same sort of phenomenon happens to sunlight that reaches the surface of the Earth and is related to two very important functions of the Earth's atmosphere (see Figure 1.1), the ozone layer and the greenhouse effect, which you will learn about in Chapter 3.

☞ For a review of the interactions between light and matter, see Review Example 1.2 on page 23.

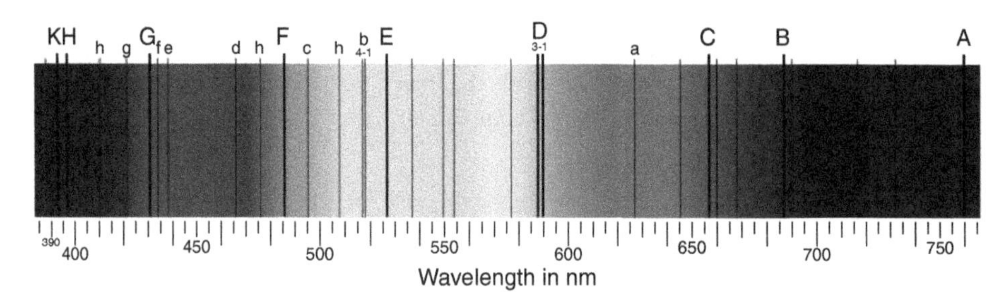

Figure 1.5 A solar spectrum showing the absorption lines from elements that compose the outer atmosphere of the Sun. Notice the sodium "D" lines, the hydrogen "C" line, and the "A" and "B" lines associated with O_2. Source: https://en.wikipedia.org/wiki/File:Fraunhofer_lines.svg.

Prefix Name	Symbol	Value
tera	T	10^{12}
giga	G	10^{9}
mega	M	10^{6}
kilo	k	10^{3}
hecto	h	10^{2}
deka	d	10^{1}
centi	c	10^{-2}
milli	m	10^{-3}
micro	μ	10^{-6}
nano	n	10^{-9}
pico	p	10^{-12}
femto	f	10^{-15}

Table 1.2 Common metric prefixes and their numerical values.

The result of all of the early astrophotography was the realization that the Sun was made mostly out of hydrogen, an unknown element, and trace amounts of other elements such as carbon and sodium. The spectroscopic fingerprint of this unknown element was so strong that scientists named it after the Greek Sun god Helios. Helium was eventually discovered on the Earth in 1895 as a by-product of radioactive decay processes in geologic formations. Early astronomers realized that the Sun and other stars contained a variety of different elements, other than hydrogen and helium, in their photospheres. Some of these elements were the result of fusion processes in the core of the stars, and other elements are the result of a star's formation from the remains of a previous generation of stars. The study of the life cycle of stars and nuclear fusion processes continued through the 20th century with the use of increasingly more powerful particle accelerators and telescopes. These studies have allowed physicists to understand the formation of the universe and the deep chasm of time between the Big Bang and the present day. In order to understand the origin of matter and the chemical principles that allow us to understand environmental chemistry, we need to take a closer look at the time line of our universe.

1.2.3 Primordial Nucleosynthesis

All of the evidence for the Big Bang may be interesting, but as an environmental chemist, you are probably wondering where all of the elements in the periodic table came from and why the Earth is the way it is (iron/nickel core, silicate crust, oceans, an atmosphere containing mostly nitrogen and oxygen). All of these elements were made in a process known as nucleosynthesis, which happened in three stages and at different points in the history of the universe.

In the initial seconds after the Big Bang, temperatures were so high that elements did not exist. As the universe cooled, the subatomic particles that chemists would recognize, protons, neutrons, and electrons, began to form. Most of the matter was in the form of hydrogen (around 75%) and helium (25%), with a little bit of lithium and other heavier elements. For a long time, temperatures were too high to allow the formation of neutral atoms, so matter existed as a plasma in much of the first half million years of the universe, with electrons separated from nuclei. Electrostatic repulsion was still present, and it prevented nucleons from combining into heavier elements. Eventually, the universe cooled enough and neutral atoms formed. The matter of the universe, at this point, was locked into mostly the elements of hydrogen and helium. It would take about 100 Myrs before heavier elements would form as a result of the birth of stars.

☞ For a review of atomic structure, see Review Example 1.4 on page 24.

☞ For a review of metric prefixes, see Review Example 1.3 on page 24.

1.2.4 Nucleosynthesis in Massive Stars

When the clouds of hydrogen and helium coalesced into the first stars, they began to heat up. The lowest energy fusion reactions are not possible until the temperature reaches about 3×10^6 Kelvin, so these protostars would only have been visible once they were hotter than about 3000 K when their blackbody radiation would have shifted into the visible spectrum. Synthesis of heavier elements, such as iron, requires temperatures around 4×10^9 Kelvin. Not all stars can reach this temperature. In fact, the surface temperature of our Sun is around 5800 K, and the core temperature is about 15×10^6 K, which is not even hot enough to produce elements such as carbon in significant amounts. Given that our Earth has a core made of mostly iron, a crust made from silicon oxides, and the asteroid belt in our solar system is composed of meteors that contain mostly iron-based rocks, our solar system must be the recycled remnants of a much larger and hotter star. High-mass stars are the element factories of the universe and develop an onion-like structure over time, where each layer has a different average temperature and is dominated by a different set of nuclear fusion reactions. As you can see from Figure 1.6, the two most abundant elements (H and He) were the result of primordial nucleosynthesis. The remaining peaks in the graph come from favored products of nuclear reactions, which occur in the various layers in a high-mass star. The layers are successively hotter than the next as a result of the increased density and pressure that occur as the star evolves. These layers develop over the life of the star as it burns through each set of nuclear fuel in an accelerating rate.

First-generation high-mass stars began their life containing the composition of the universe just after the Big Bang with about a 75:25 ratio of hydrogen and helium. Their lifetime was highly dependent on their mass, with heavier stars having shorter life cycles, thus the times provided in the following description are approximate. During the first 10 Myrs of the life of a high-mass star, it fuses hydrogen into helium. These reactions generate a lot of energy since the helium nucleus has a high binding energy. The release of energy produces the light and the heat that are necessary to keep the star from collapsing under the intense gravity (think of the Ideal Gas Law: $PV = nRT$ and the increase in volume that comes with an increase in temperature). The helium that is produced from the hydrogen fusion reactions sinks to the core since it is more dense. This generates a stratification as the core is enriched in nonreactive helium and hydrogen continues to fuse outside the core.

Once most of the hydrogen fuel is exhausted, the star starts to lose heat, and the core begins to collapse under the immense gravitational attraction of the star's mass. The helium nuclei cannot fuse until the electrostatic repulsion between the +2 nuclear charges of the

H Fusion Layer ($T \approx 3 \times 10^6$ K)
Hydrogen fusion involves several processes, the most important of which is the proton–proton chain reaction or P–P Chain. The P–P chain reactions occur in all stars, and they are the primary source of energy produced by the Sun. Hydrogen nuclei are fused together in a complicated chain process that eventually results in a stable He-4 nucleus.

$${}^{1}_{1}\mathrm{H} + {}^{1}_{1}\mathrm{H} \rightarrow {}^{2}_{1}\mathrm{H} + {}^{0}_{+1}\beta^{+} + \nu_e \quad \text{(R1.1)}$$

$${}^{2}_{1}\mathrm{H} + {}^{1}_{1}\mathrm{H} \rightarrow {}^{3}_{2}\mathrm{He} + {}^{0}_{0}\gamma \quad \text{(R1.2)}$$

$${}^{3}_{2}\mathrm{He} + {}^{3}_{2}\mathrm{He} \rightarrow {}^{4}_{2}\mathrm{He} + 2{}^{1}_{1}\mathrm{H} \quad \text{(R1.3)}$$

For your convenience, here is a summary of nuclear particles you will see.
alpha particle (${}^{4}_{2}\alpha$ **or** ${}^{4}_{2}$He) a helium-4 nucleus
beta particle (${}^{0}_{-1}\beta$) an electron, negligible mass
positron (${}^{0}_{+1}\beta^{+}$) antimatter electron, negligible mass
gamma particle (${}^{0}_{0}\gamma$) high-energy photon
neutrino (ν_e) very rare particle, negligible mass

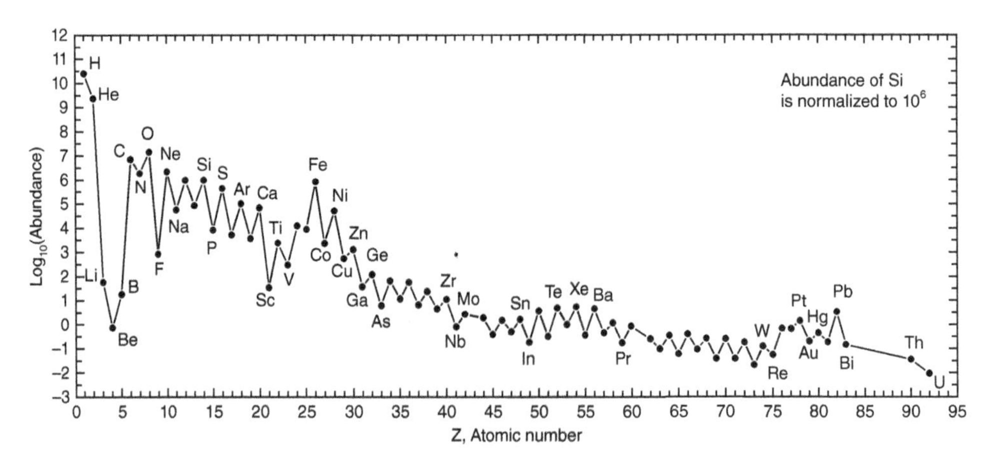

Figure 1.6 Relative abundances of the elements in the universe. Note that the *y*-axis is a logarithmic scale. Source: http://en.wikipedia.org/wiki/Abundance_of_the_chemical_elements. Used under BY-SA 3.0 //creativecommons.org/licenses/by-sa/3.0/deed.en.

☞ For a review of writing and balancing nuclear reactions, see Review Example 1.5 on page 25.

He Fusion Layer ($T \approx 1.8 \times 10^8$ K)
The fusion reaction that begins with helium is often referred to as the triple-alpha reaction, because it is a stepwise fusion of three nuclei.

$${}^{4}_{2}\mathrm{He} + {}^{4}_{2}\mathrm{He} \rightleftharpoons {}^{8}_{4}\mathrm{Be} \quad \text{(R1.4)}$$

$${}^{4}_{2}\mathrm{He} + {}^{8}_{4}\mathrm{Be} \rightarrow {}^{12}_{6}\mathrm{C} + {}^{0}_{0}\gamma \quad \text{(R1.5)}$$

To a small but significant extent, O-16 is also produced by the addition of another alpha particle.

C Fusion Layer ($T \approx 7.2 \times 10^8$ K)
Several different elements heavier than carbon are synthesized here.

$${}^{12}_{6}\mathrm{C} + {}^{12}_{6}\mathrm{C} \rightarrow {}^{24}_{12}\mathrm{Mg} + {}^{0}_{0}\gamma \quad \text{(R1.6)}$$

$${}^{12}_{6}\mathrm{C} + {}^{12}_{6}\mathrm{C} \rightarrow {}^{23}_{11}\mathrm{Na} + {}^{1}_{1}\mathrm{H} \quad \text{(R1.7)}$$

$${}^{12}_{6}\mathrm{C} + {}^{12}_{6}\mathrm{C} \rightarrow {}^{20}_{10}\mathrm{Ne} + {}^{4}_{2}\mathrm{He} \quad \text{(R1.8)}$$

$${}^{12}_{6}\mathrm{C} + {}^{12}_{6}\mathrm{C} \rightarrow {}^{16}_{8}\mathrm{O} + 2{}^{4}_{2}\mathrm{He} \quad \text{(R1.9)}$$

O Fusion Layer ($T \approx 1.8 \times 10^9$ K)
Some example reactions involving oxygen fusion.

$${}^{16}_{8}\mathrm{O} + {}^{16}_{8}\mathrm{O} \rightarrow {}^{32}_{16}\mathrm{S} + {}^{0}_{0}\gamma \quad \text{(R1.10)}$$

$${}^{16}_{8}\mathrm{O} + {}^{16}_{8}\mathrm{O} \rightarrow {}^{31}_{15}\mathrm{P} + {}^{1}_{1}\mathrm{H} \quad \text{(R1.11)}$$

$${}^{16}_{8}\mathrm{O} + {}^{16}_{8}\mathrm{O} \rightarrow {}^{31}_{16}\mathrm{S} + {}^{1}_{0}\mathrm{n} \quad \text{(R1.12)}$$

$${}^{16}_{8}\mathrm{O} + {}^{16}_{8}\mathrm{O} \rightarrow {}^{30}_{14}\mathrm{Si} + 2{}^{1}_{1}\mathrm{H} \quad \text{(R1.13)}$$

Ne Fusion Layer ($T \approx 1.2 \times 10^9$ K)
Some representative reactions involving neon.

$${}^{20}_{10}\mathrm{Ne} + {}^{0}_{0}\gamma \rightarrow {}^{16}_{8}\mathrm{O} + {}^{4}_{2}\mathrm{He} \quad \text{(R1.14)}$$

$${}^{20}_{10}\mathrm{Ne} + {}^{4}_{2}\mathrm{He} \rightarrow {}^{24}_{12}\mathrm{Mg} + {}^{0}_{0}\gamma \quad \text{(R1.15)}$$

nuclei is overcome, so no He fusion proceeds at the current temperature. As the pressure on the core increases, the temperature increases (think about the Ideal Gas Law again). Eventually, the core temperature increases to about 1.8×10^8 K, which is the ignition temperature of fusion reactions involving helium. As helium begins to fuse, the core stabilizes, and now the star has a helium fusion core and a layer outside of this where the remaining hydrogen fuses. The helium fusion core produces mostly carbon nuclei (along with other light nuclei), which are nonreactive at the core temperature, and thus, the carbon begins to sink to the center of the star forming a new core with a helium layer beyond the core and a hydrogen layer beyond that. The helium fusion process is much faster than hydrogen fusion, because helium fusion produces much less heat than hydrogen fusion so the star must fuse it faster in order to maintain a stable core (it would collapse if enough heat was not produced to balance gravity). Helium fusion lasts for about 1 Myrs.

The process described earlier repeats for the carbon core – collapse of the core, ignition of carbon fusion, pushing the remaining helium fusion out to a new layer, and a newfound stability. Carbon fusion produces a mixture of heavier elements such as magnesium, sodium, neon, and oxygen and lasts for about 1000 years because the binding energy difference between carbon and these other elements is even smaller, requiring a faster rate of reaction to produce the same heat as before. The new core eventually ignites neon, producing more oxygen and magnesium, pushing the remaining carbon fusion out to a new layer, and exhausting the neon supply after a few years. Next comes oxygen fusion, lasting only a year due to the diminishing heat production. The final major stage involves the ignition of silicon to form even heavier elements such as cobalt, iron, and nickel – lasting just seconds and forming the final core. At this point, the star resembles the onion-like structure seen in Figure 1.7 and has reached a catastrophic stage in its life cycle because iron and nickel are the most stable nuclei and fusing them with other nuclei consumes energy instead of generating it. The star has run out of fuel.

A dying star first cools and begins to collapse under its enormous mass. As it collapses, the pressure and temperature of the core rise, but there is no other fuel to ignite. Eventually, the temperature in the core becomes so immense that the binding energy holding protons and neutrons together in the atomic nuclei is exceeded. The result is a massive release of neutrons and neutrinos, and a supernova explosion results. Johannes Kepler (1571–1630) observed a supernova star in 1604. It was so bright that it was visible during the daytime.

The shock wave of neutrons that arises from the core moves through the other layers and causes the final stage of nucleosynthesis – neutron capture. All of the elements synthesized thus far undergo neutron capture and a subsequent beta emission reaction to produce an

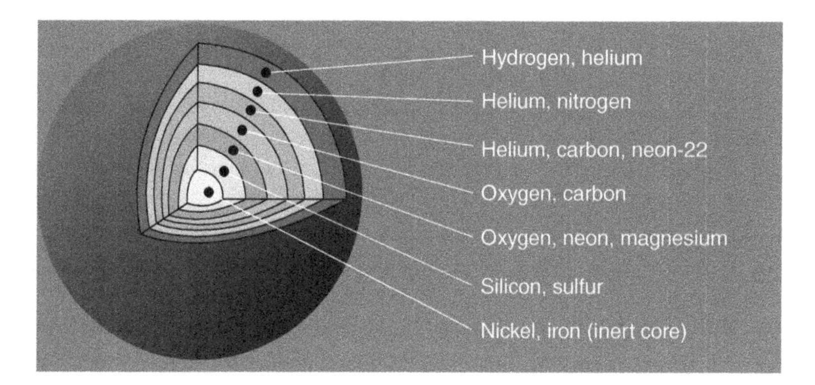

Figure 1.7 The onion-like layers of a giant star develop as it ages and approaches a supernova explosion. Source: http://commons.wikimedia.org/wiki/File:Massive_star_cutaway_pre-collapse_%28pinned%29.png. Used under CC0 1.0 //creativecommons.org/publicdomain/zero/1.0/deed.en.

element with a larger atomic number. This produces elements heavier than Fe-56.

$$^{56}_{26}\text{Fe} + {}^{1}_{0}n \rightarrow {}^{57}_{26}\text{Fe} \quad \text{(R1.16)}$$

$$^{57}_{26}\text{Fe} \rightarrow {}^{57}_{27}\text{Co} + {}^{0}_{-1}\beta \quad \text{(R1.17)}$$

The cobalt nucleus goes on to absorb another neutron and then beta-decays to form copper. This process continues until uranium is formed, which is the heaviest stable element in the periodic table. The couplet of Reactions R1.16 and R1.17 is just one example of an array of reactions where a single nucleus could absorb several neutrons (most probable since the supernova shock wave is neutron rich) and then undergoes beta decay. Eventually, the explosion blows off the outer layers of the star and forms an interstellar cloud of particles rich in elements from hydrogen to uranium. The core of the giant star either becomes a white dwarf star or, if the original star was very large, forms a black hole.

These massive stars have relatively short lives that are in the millions to hundreds of millions of years. Our Sun, which is a smaller star, will have a life span of about 10 Gyrs. The shock wave from a supernova explosion often causes the larger interstellar cloud from which the star formed to coalesce to form the next-generation star. Given the short life span of a high-mass star and given that they are the only stars massive enough to form the elements heavier than carbon, there may have been a few generations of stars that formed and exploded before our Sun formed some 9 Gyrs after the Big Bang.

Si Fusion Layer ($T \approx 3.4 \times 10^9$ K)
Silicon fusion involves a complicated series of alpha capture reactions that produce several of the elements between Si and Fe on the periodic table.

$$^{28}_{14}\text{Si} + {}^{4}_{2}\text{He} \rightarrow {}^{32}_{16}\text{S} \quad \text{(R1.18)}$$

$$^{32}_{16}\text{S} + {}^{4}_{2}\text{He} \rightarrow {}^{36}_{18}\text{Ar} \quad \text{(R1.19)}$$

$$^{36}_{18}\text{Ar} + {}^{4}_{2}\text{He} \rightarrow {}^{40}_{20}\text{Ca} \quad \text{(R1.20)}$$

This alpha capture continues until the nucleus of Ni-56 is produced, which is radioactive and has a half-life of 6 days. Ni-56 undergoes electron capture to form Co-56, which is also radioactive ($t_{1/2}$=77 days) and undergoes electron capture to form Fe-56. Thus, the final result of the very last fusion process in these giant stars is iron. This will be an important fact to remember, because it explains why iron is so abundant on the Earth and in the other terrestrial planets and asteroids.

1.2.5 Nucleosynthesis Summary

A short time after the Big Bang, the nuclei of hydrogen and helium formed, and as they cooled from a plasma to form neutral elements, the process of primordial nucleosynthesis ceased. Over a 100 million years later, the first stars began to form. The temperatures and pressures in the cores of these stars were enough to begin the process of fusion, converting hydrogen and helium into heavier nuclei. Small and medium stars, such as our Sun, usually do not produce elements heavier than carbon, whereas high-mass stars develop core temperatures high enough to synthesize elements through the first row of the transition metals (nickel and iron). Supernova explosions of high-mass stars complete the nucleosynthesis cycle by producing elements from iron to uranium through neutron capture and beta emission.

It is important to review the onion-like structure in these massive stars. As outlined earlier, there are stages where one of the products of one layer becomes the fuel for another layer. The accumulation of these elements (C, O, Si, and Fe) explains their relatively high abundance in second-, third-, and later-generation solar systems such as in the composition of the Earth.

The story of the Big Bang is a monumental achievement of brilliant scientists and powerful instrumentation that began when scientists observed differences between the spectra of stars, and now scientists use powerful telescopes and nuclear accelerators to study the processes of nucleosynthesis. It is still a field with a few mysteries left to uncover, such as the nature of Dark Matter. The most recent discovery came in July of 2012 with the announcement of the discovery of the Higgs Boson particle by the science team at the Large Hadron Collider.

Now that we know how the elements in the periodic table were produced and how the universe formed, the story of our home planet comes next. We need to zoom way in from the vastness of the universe to a single solar system in a galaxy that contains 100 billion stars. Carl Sagan summarized it best.

> We find that we live on an insignificant planet of a humdrum star lost in a galaxy tucked away in some forgotten corner of a universe in which there are far more galaxies than people.
> Sagan (1980, p. 193)

Yet this planet is very dear to us and represents the bounds of almost all we know. If humility is a virtue, then cosmology offers us plenty of that.

1.3 SOLAR NEBULAR MODEL: THE BIRTH OF OUR SOLAR SYSTEM

Pierre-Simon Laplace (1749–1827) proposed the precursor to the current solar nebular model at the end of the 18th century, describing the origin of our solar system consistent with the methodological naturalism that is a cornerstone of modern science. Legend has it (possibly apocryphal, but still instructive) that when Laplace explained his hypothesis to Napoleon Bonaparte, Napoleon asked, "How can this be! You made the system of the world, you explain the laws of all creation, but in all your book you speak not once of the existence of God!" Laplace responded with, "I did not need to make such an assumption." For Laplace, the solar system's formation could be explained by physical laws – there was no need to insert the "a miracle happened here" assumption. Scientists in pursuit of the chemical origins of life are increasingly coming to the same conclusion. Source: Pierre-Simon Laplace http://en.wikipedia.org/wiki/File:Pierre-Simon_Laplace.jpg. Used under CC0 (http://creativecommons.org/publicdomain/mark/1.0/deed.en). See Ball (1960, p. 343).

After one or more generations of high-mass stars lived their lives and ejected a soup of elements into the interstellar cloud from which they formed, the stage was set for the formation of our own solar system. The whole process probably took about 50 Myrs, but about 9 Gyrs after the Big Bang, an interstellar cloud on the outer edge of the Milky Way galaxy began to coalesce under the force of gravity. What started out as a collection of atoms, dust, rocks, and other debris began to collapse inward from all directions and adopt orbits around the center of mass. All of the orbits from all of the particles probably appeared nearly spherical from the outside, but as the cloud shrunk in size under the force of gravity, it started to flatten out. Much like an ice-skater that starts a spin with his arms out, as he pulls his arms inward, his rotational speed increases due to the conservation of angular momentum. This rotational speed increase eventually caused all of the random orbits of particles to spin around the center in the same direction and the cloud flattened into a disc.

At the center of the disc was most of the material from the cloud, but as the rotational speed increased, some of the particles of the cloud gained enough speed to establish a stable orbit around the center without continuing to collapse inward. Particles with similar stable orbits began to condense to local clusters and formed planetesimals – large chunks of material that would eventually form planets. The center of this rotating disc continued to gain mass to form a protostar – not hot enough to start fusion. As the particles and gas collided, they converted much of their momentum into heat. Within a certain radius from the center of the disc, called the frost line, temperatures were warm enough that the collapsing material remained gaseous and eventually became molten as terrestrial planets formed. These objects typically collected only the "rocky" material since the more volatile material did not condense and thus remained relatively small. Outside of the frost line, temperatures were low and gaseous, and icy planets formed. Planets beyond the frost line condensed very quickly because of the low temperatures, which caused the dust and gas to form larger and larger planetesimals, which had increasing gravitational force as the mass accumulated. This explains why the planets beyond the frost line (approximately the asteroid belt) are massive compared to the smaller, inner planets. The two largest known objects in the asteroid belt are Vesta, a rocky asteroid inside the frost line, and Ceres, a spherical icy asteroid outside of the frost line.

Eventually, the protostar at the center of this rotating disc reached 3×10^6 K and the process of hydrogen fusion began. When the Sun went nuclear, it began producing enough energy to eject charged particles, such as electrons and protons, known as the solar wind. As this wind swept through the newly formed solar system, it cleared away most of the interplanetary dust and gases that had not formed into planetesimals. What remained was the four terrestrial planets (Mercury, Venus, Earth, and Mars), still molten, and the large gaseous planets and planetoids outside of the frost line (Jupiter, Saturn, Uranus, Neptune, and the Kuiper Belt objects such as Pluto and Eris). This distinction between the inner and outer planets is a confirmation of the Solar Nebular Model, developed in the 18th century by Pierre-Simon Laplace (see featurette).

Another confirmation of the Solar Nebular Model is the orbits and orbital axes of the individual planets. All of the planets orbit the Sun in the same counterclockwise direction and in the same plane (except some objects in the Kuiper belt). The Sun and all of the planets, except two, also spin about their own axes in the counterclockwise direction – give or take a few degrees. The two exceptions are Venus, which rotates clockwise, and Uranus, which rotates on its side. While scientists are still trying to determine what happened to these two exceptions, the majority of the other planets give a cautious confirmation of the Solar Nebular Model. The Earth is a "well-behaved" planet with its counterclockwise rotation around its north pole. It is the most studied of the planets for a good reason – we live on it! Its formation is the next part of the story.

Geological Dates

Geologists and cosmologists use the unit *annum* to refer to time in the past. So, for example, 1 billion years ago would be 1 Ga (*giga annum*). In some textbooks, you might see this listed equivalently as 1 Gyrs, but the international standard method is to use annum. So, Ga, Ma, and ka in this text refer to 10^9, 10^6, and 10^3 years ago. You will see the Gyrs, Myrs, and kyrs used when a *duration* is used, which does not place the event in a geological or cosmic time line.

1.3.1 The Ages of the Earth

Over its 4.6 Gyr history,[1] the Earth has changed quite significantly. Most of this change has occurred over the first 3.8 Gyr – for nearly a billion years, the Earth has been relatively stable. In a snapshot overview, the Earth started out as molten and then cooled to form an atmosphere composed mostly of carbon dioxide and molecular nitrogen with acidic oceans rich in dissolved metals. The modern Earth has an oxidizing atmosphere with very little carbon dioxide, rich in molecular oxygen, a significant ozone layer, and basic oceans. These drastic changes have occurred over geologic eons. There is evidence to suggest that the changes described next were the result of abiotic and biotic forces. This statement bears repeating – the early forms of life on the Earth (archaebacteria, eubacteria, and others) have significantly contributed to the dramatic evolution of the entire planet.

1.3.1.1 Hadean Eon (4.6 to 4.0 Ga) During the earliest stages of development, the Earth was transformed from a molten planet to one with a solid crust in a process termed *accretion*. Earth's accretion process took about 10–100 million years, during which it was being bombarded by dust, debris, and other planetesimals in and near its orbit. One important collision between the proto-Earth and a planetesimal the size of Mars, approximately 45 Myrs after the formation of the solar system, led to the formation of the Moon and added the last

☞ *Accretion* is the process by which an object grows by acquiring additional mass. In the accretion of Earth, debris in the orbit was accumulated along with bombardment by asteroids from outside the Earth's orbit.

[1]Geologic time intervals vary from source to source, but the ranges presented in this chapter come from the 2010 designations set by the International Commission on Stratigraphy. Extensive charts can be viewed on their website (http://www.stratigraphy.org).

10% of the Earth's mass, finalized the spin velocity, and set the final tilt angle at about 23° from the vertical.

A combination of heating from radioactive decay and impact heat from collisions kept the Earth molten or caused frequent remelting throughout the early part of this eon. It is even likely that soon after some of the early oceans formed, meteor impacts caused the oceans to revaporize. Because the Earth was spinning while it was molten, its angular momentum caused the Earth to form an oblate spheroid (it is fatter around the middle than from pole to pole) and not a perfect sphere.

☞ *Differentiation* is the process by which the molten material of the early Earth separated, according to density, with the lighter material on the surface and the denser material in the core.

The various elements in the molten Earth eventually began to separate according to their density and melting point in a process called *differentiation*. If you examine Table 1.3, you will see that the densities and the melting points of the common metals and metal oxides indicate that iron and iron-based compounds have higher densities and would remain molten at lower temperatures, and they therefore sunk to the core of the planet as the less dense metals and silicates floated toward what was to become the crust. This differentiation resulted in a core that is about 85% Fe and 5% Ni and is nearly 1/3 of the Earth's total mass. The mantle and crust are dominated by silicate minerals containing a variety of alkali, alkaline earth, and transition metals.

Compounds	Melting Point (°C)	Density (g/mL)
SiO_2	1688–1700	2.65
Si	1410	2.32
CaO	2614	3.25–3.38
MgO	2852	3.58
Al_2O_3	2072	3.97
Fe_2O_3	1565	5.24
FeO	1369	5.72
Fe	1535	7.86
Ni	1455	8.9
NiO	1984	6.67

Values from CRC Handbook, 72nd ed.

Table 1.3 Melting points and densities of the major constituents of the mantle and crust.

What evidence is there for this assertion? Careful studies by seismologists and geologists have confirmed these details. Whenever there is an earthquake or nuclear weapon test, seismologists can observe seismic waves traveling through the Earth, and they observe a gradual increase in the density of the mantle until the waves reach what must be the core, where the density increases significantly. Compression waves (called *p*-waves) can travel through solids and liquids, whereas shearing waves (called *s*-waves) can only travel through solids coherently; thus, the inability of *s*-waves to travel through the core strongly suggests that it is liquid (at least the outer core). Further, the magnetic field of the Earth is consistent with convective flows of a liquid outer core. The inner core is solid, not because it is cooler than the outer core but because of the tremendous pressure the rest of the mantle and outer core exert. Temperature and pressure estimates in the core are 5400 °C and 330 GPa, which place the core in the solid region of its phase diagram.

As the surface of the Earth cooled and the higher melting point compounds rose to the surface, the crust solidified. The Earth and Venus are large enough to contain sufficient radioisotopes in the mantle and core to slow the cooling process down significantly; thus, the Earth still retains a molten outer core and a semimolten mantle. Smaller planets also delayed their complete solidification, but have long since become solid, such as Mercury, the Moon, and Mars. The Earth's surface remains active and young due to the residual trapped heat that keeps portions of the mantle in a semiliquid state. Further, the tectonic plates that form the Earth's surface are constantly shifting and recycling the crust into the molten interior. As a result, most of the lithospheric surface of the Earth is no older than 200 Ma. The oldest parts of the Earth's crust are found at the center of tectonic plates, such as in Australia and in northern Quebec near Hudson Bay, which date to 4.03 and 4.28 Ga, respectively. Rocks from the Moon and meteors have been dated to 4.5 Ga, which would corroborate the theory that the smaller bodies, such as the Moon and asteroids, would have cooled much quicker than the Earth and would have not had the tectonic activity that caused the crust of the Earth to remelt.

Isotope	Half-Life (years)
U-235	7.03×10^8
K-40	1.248×10^9
U-238	4.47×10^9
Th-232	1.40×10^{10}
Pt-190	4.5×10^{11}
Cd-113	8.04×10^{15}
Se-82	$> 9.5 \times 10^{19}$
Te-130	8×10^{20}
Te-128	2.2×10^{24}

Values from CRC Handbook, 93rd ed.

Table 1.4 Selected radioisotopes and their half-lives.

Table 1.4 contains a list of common radioactive isotopes and their measured half-lives. Most of these radioisotopes are heavy atoms formed during the last few seconds of the supernova explosions that formed the stellar cloud from which our solar system was born. The reliable decay of some of these isotopes can be used to date the age of the Earth. While a rock is molten, the contents are ill-defined as elements move around in the liquid. Once a rock solidifies, then its constituent elements are locked into place. If the molten rock is allowed to cool slowly, then crystals form. Crystals have very regular atomic arrangements and unusual purity since the lowest energy configuration of a crystal pushes out atomic impurities while it is forming. It is the same reason that icebergs are mostly freshwater even though they form in very salty oceans. Zircon ($ZrSiO_4$) is a silicate mineral that regularly includes uranium and thorium but stringently rejects lead atoms when it crystallizes, allowing it to be an atomic clock for measuring the age of a rock. The two most common isotopes of uranium are U-235 and U-238 and make up 0.7% and 99.3%, respectively, of the naturally occurring uranium. U-235 has a half-life of 0.7 Gyr, and U-238 has a half-life of 4.5 Gyr. Both isotopes naturally decay to different isotopes of lead.

The U-Pb dating system is typically used to date objects that are at least 1 Ma. The beauty of the U-Pb dating system is that any lead found in the zircon crystals is the result of the uranium decay, and since both uranium isotopes are present when uranium is present, there are two different clocks that can provide the age of the crystal. This dual dating system and others confirm the deep age of the Earth and is in agreement with estimates for the age of the Sun and other celestial bodies in our solar system.

☞ For a review of the half-life kinetics, see Review Example 1.6 on page 25.

During the Hadean Eon, the Earth's atmosphere changed drastically. The primordial atmosphere was almost certainly lost before the planet cooled. A combination of a relentless solar wind, very high temperatures when the Earth was still molten, and an atmosphere composed of light constituents, such as molecular hydrogen and helium, caused the primordial atmosphere to escape into space. It should seem strange to us that while the Sun is mostly hydrogen and helium, and since the planets formed from the same stellar cloud that formed the Sun, Earth has virtually no hydrogen and helium in its atmosphere. Other planets such as Jupiter and Saturn are exactly as predicted – made of mostly hydrogen and helium – but not the inner, rocky planets. They all were small enough and hot enough to lose most of their molecular hydrogen and helium during the initial stages of formation.

Once the surface of the Earth began to cool, the gases dissolved in the molten mantle and crust released as a result of volcanism, as well as volatile compounds delivered to the Earth by comets and asteroids, formed the second atmosphere. These gases (some combination of CO_2, H_2O, N_2, NH_3, SO_2, CO, CH_4, and HCl) were heavy enough and cool enough to be unable to escape the gravity of the cooling planet. Over the next 100 million years it took the planet to cool and sweep clean the other debris and planetesimals in its orbit, the atmosphere increased in density and volume.

At some point in the Hadean Eon, the temperature cooled sufficiently that water in the atmosphere began to condense and fall to the surface. Evidence of this was derived from zircon crystals dated to around 4.4 Ga and having O–18 isotope levels that could only have formed in the presence of liquid water (Wilde *et al.*, 2001). The hydrosphere that formed took much of the carbon dioxide with it since carbon dioxide forms an equilibrium in water with carbonic acid.

$$CO_2(g) + H_2O(l) \rightleftharpoons H_2CO_3(aq) \qquad K_H = 3.5 \times 10^{-2} \qquad \text{(R1.21)}$$

This set the stage for a tremendous reduction in the amount of carbon dioxide in the atmosphere, for not only did it dissolve into the hydrosphere, but many dissolved metals in the hydrosphere formed insoluble compounds with the carbonate ion. This locked tremendous amounts of carbon dioxide into geologic formations, such as limestone, and out of the atmosphere. Carbon dioxide is a potent greenhouse gas, which will be discussed in Chapter 3, and its loss to our atmosphere from a majority constituent to its current level at less than 1% allowed the Earth to become a moderately warm planet, unlike the fiery hell that Venus became. Venus has an atmosphere that has about 90 times more pressure compared to Earth, and it is mostly carbon dioxide with 3% molecular nitrogen. It is likely that the Earth's atmosphere was initially similar, but because Earth cooled enough to condense water, most of the carbon dioxide and water were locked into the oceans and rocks, leaving a much thinner atmosphere with a smaller greenhouse effect.

☞ For a review of the solubility rules for ionic compounds, see Review Example 1.7 on page 26.

☞ For a review of naming ionic compounds, see Review Example 1.8 on page 27.

So what might have happened to the Earth if the atmospheric water had not condensed? Modern Venus, which is about 18% less massive than the Earth, probably holds the answer. The primordial atmospheres of Venus and Earth after accretion contained large amounts of carbon dioxide, water, molecular nitrogen, and a few minor components such as sulfur dioxide. The current evidence from atmospheric probes and spectroscopy shows that there is virtually no water in the atmosphere of Venus. There are clouds of sulfuric acid, which were formed from the combination of water and sulfur trioxide, but all of the free water vapor is gone. If Earth and Venus are so similar in other compositional ways, what happened to the oceans' worth of water that Venus presumably once had? The answer may seem strange, but the water was lost to space.

☞ For a review of bond photolysis, see Review Example 1.10 on page 28.

Carbon dioxide contains two C=O double bonds, water contains two H−O single bonds, and molecular nitrogen contains one N≡N triple bond. Examining Table I.7 (Appendix I on page 325) shows clearly that carbon dioxide and molecular nitrogen have very strong bonds (804 and 946 kJ/mol, respectively). Water, on the other hand, is assembled from relatively weak bonds by comparison (465 kJ/mol). These gases would have been present in the upper layers of the atmosphere as well as in the lower layers. In the upper layers, all of these gases would have been exposed to radiation from the Sun strong enough to break bonds. The O–H bond in water can be broken by radiation with a maximum wavelength of 257 nm. An equilibrium can be established between photolyzed water and its reformation, but two of the hydrogen atom fragments can react to form molecular hydrogen ($h\nu$ represents the energy of a photon).

$$H_2O(g) + h\nu \rightleftharpoons H(g) + OH(g) \quad (R1.22)$$

$$2H(g) \rightarrow H_2(g) \quad (R1.23)$$

☞ For a review of the solubility rules for gases in liquids, see Review Example 1.11 on page 29.

The fate of the highly reactive hydroxyl radical will be discussed thoroughly in Chapter 3, but atomic and molecular hydrogen, even on a planet with a relatively cool atmosphere, has enough energy to escape into space. After a few billion years, it appears that the entire mass of hydrogen in the water that currently forms our oceans had done something very different on Venus – it photolyzed, and the hydrogen diffused into space leaving behind a very dry atmosphere.

The formation of the hydrosphere in the late Hadean initially resulted in acidic oceans due to the tremendous amount of carbonic acid formed from the dissolved carbon dioxide. Acidic solutions tend to dissolve reduced metals; thus, the first oceans were rich in metals such as iron, calcium, sodium, and magnesium. As the early oceans were forming, there was likely a tremendous amount of rain on the lithosphere. The acidic rain would have reacted with reduced metals on the surface, such as iron, in the following manner.

$$2H^+(aq) + Fe(s) \rightleftharpoons H_2(g) + Fe^{2+}(aq) \quad (R1.24)$$

The H^+ represents an acidic proton in the aqueous solution. Note the relationship between the amount of acid and the amount of iron that is dissolved as a result of the reaction. This 2:1 relationship is referred to as the stoichiometric relationship between the reactants. Stoichiometry is the study of these relationships and the link between the reactants consumed and the products produced. Reactions, such as the aforementioned one, were responsible for adding much of the dissolved metals to the watershed and into the early oceans.

☞ For a review of the process of balancing reactions, see Review Example 1.12 on page 29.

As mentioned earlier, the carbonate ion precipitated some of these cations to form sedimentary rocks such as limestone. These sedimentary rocks then are a record of the local conditions of the oceans when the precipitation occurred. The ions contained in the rocks are a direct result of the ions present when the sediment was formed. This is one of the tools that geologists use to draw conclusions about events that occurred millions and billions of years ago.

☞ For a review of stoichiometry, see Review Example 1.16 on page 34.

It is possible that the earliest form of life (archaebacteria) could have developed near the end of the Hadean Eon (4.1 Ga) (Nisbet and Sleep, 2001). Carbonate rocks found with some of the oldest rocks in Greenland show a distinct enrichment of the heavier isotope of carbon, C-13, which is indirect evidence that early bacteria, which prefer the lighter C-12, would have been absorbing C-12 into the developing biosphere and thereby enriching the inorganic carbon with C-13. The presence of significant quantities of methane in the atmosphere, although not exclusively of biological origin, also provides indirect evidence that methanogenic bacteria evolved as early as (3.8–4.1 Ga) (Battistuzzi, Feijao, and Hedges,

2004). No indisputable proof of life's emergence during the late Hadean Eon has been discovered, but the evidence provides a cautious indication. Stronger evidence for life in the Archean comes from carbonaceous microstructures and filaments found in rocks that are around 3.4 Ga. The methanogenic bacteria would play an important role in enhancing the greenhouse effect, moderating the surface temperatures, and acting as an atmospheric UV filter.

1.3.1.2 Archean Eon (4.0 to 2.5 Ga) Arising out of the Hadean Eon, the Earth had a solid crust, an acidic hydrosphere, and an atmosphere that was mostly molecular nitrogen and carbon dioxide – significantly less CO_2 than it had before the water vapor condensed to form the oceans. It should be noted that the young Sun was also evolving and less luminous than it is presently. During the Hadean Eon, the Sun was about 70% as bright as it is presently, and during the Archean Eon, it increased its luminosity from about 75 to 85% of its present state. This combination of a young Sun and a reduced greenhouse effect eventually plunged the Earth into a severe ice age, cold enough to bring some level of glaciation near the equator. The rising methane levels as a result of methanogenic bacteria delayed this deep freeze until the end of the Archean Eon.

Once methane concentrations reached about 10% of the carbon dioxide levels in the early atmosphere, an organic haze would have developed. This haze has been observed in laboratories and on Titan, one of the moons of Saturn. The development of this haze was important for the eventual colonization of land-based life since the absence of ozone in the atmosphere would have meant harsh levels of UV radiation from the Sun. This would have made life outside of the oceans nearly impossible. The organic haze would have provided some protection from the most damaging UV radiation, but falls well short of the current protection the surface of the Earth receives from the modern ozone layer.

As the early life forms continued to evolve in the Archean Eon, different metabolic mechanisms began to emerge. The first photosynthetic organisms used the methane and sunlight in an anoxygenic type of photosynthesis. This eventually led to oxygenetic photosynthesis soon after the microorganisms colonized the land masses. This set the stage for another dramatic event in the evolution of the planet known as the Great Oxidation, which took place in the Proterozoic Eon (Battistuzzi, Feijao, and Hedges, 2004).

After the evolution of oxygenic photosynthetic organisms, the level of molecular oxygen in the atmosphere began to increase. Since molecular oxygen is such a reactive molecule, a combination of factors would keep the levels low for a long time: methane in the atmosphere reacted with molecular oxygen to form carbon dioxide, a large reservoir of dissolved metals in the hydrosphere formed metal oxides and precipitated from solution, and reduced minerals in the lithosphere were oxidized. One of the primary pieces of evidence of the rise of atmospheric oxygen comes from sedimentary rocks known as banded-iron formations (BIFs). Episodic increases in atmospheric or hydrospheric oxygen levels resulted in the precipitation of iron(III) oxide from the oceans, which formed a sediment. Iron(III) oxide has an intense color and is the reason for the rusty color of iron-rich soils and rocks. During the times when oxygen levels were not very high, hydrothermal vents spewing alkaline water as a result of the carbonate-rich ocean bedrock of the Archean (Shibuya *et al.*, 2010) caused dissolved silicates to precipitate and formed layers of chert (ranging in color from whitish to dark grayish) that fell on top of the iron(III) oxide and formed the bands of the BIFs.

Stable isotope studies using chromium-53 show that BIFs, at certain times in geologic history, are a result of atmospheric increases in molecular oxygen instead of hydrospheric increases (Frei *et al.*, 2009). This is significant because it indicates that the oxygen increases were very likely caused by land-based life and not microbial life in the oceans. The evidence for this assertion comes from redox chemistry involving dissolved iron cations in the hydrosphere and chemically weathered chromium minerals from terrestrial sources. As BIFs were being formed, some redox chemistry involving Cr(VI) ions resulted in the inclusion of

Redox reactions are a class of reactions where electrons are transferred between reactants. There must be an element that is *oxidized*, losing its electrons to an element that is *reduced*. Many important reactions involve redox chemistry, including metabolism and batteries.

Cr in the BIFs along with the iron. The essential reactions are given as follows (notice that the first reaction is balanced for charge, and then see the review box on redox reactions).

$$Cr^{6+} + 3Fe^{2+} \rightarrow Cr^{3+} + 3Fe^{3+} \quad (R1.25)$$

$$8H^{+}(aq) + CrO_4^{2-}(aq) + 3Fe^{2+}(aq) \rightarrow 4H_2O(l) + Cr^{3+}(aq) + 3Fe^{3+} \quad (R1.26)$$

The first reaction in the aforementioned couplet shows the elemental *redox* chemistry and the transfer of electrons from iron to chromium. The second reaction more accurately shows the *reduction* of chromium in the chromate ion while the iron(II) ion is being *oxidized*. A chemical analysis of this reaction shows that under low pH conditions from acidic rain (high concentrations of H^+ ions), terrestrial manganese oxide oxidizes chromium(III) minerals to chromate ions (see Reaction R1.27), which are highly soluble and enter the oceans by way of the watershed.

$$2Cr(OH)_3(s) + 2H^{+}(aq) + 3MnO_2(s) \rightarrow 2CrO_4^{2-}(aq) + 3Mn^{2+}(aq) + 4H_2O(l) \quad (R1.27)$$

This terrestrial source of chromium is enriched in Cr-53 compared to the background concentration of chromium in the ocean. The chromate ions in the oceans then undergo reduction by iron(II) (see Reaction R1.25) and the Cr(III) and Fe(III) ions are precipitated as oxides, forming the reddish-colored layers of the BIFs. Other sources of Cr(III) could come from hydrothermal vents and microbial sources, but these inputs of chromium produce Cr-53 poor BIFs. What this means is that if scientists study the amount of Cr-53 in various BIF samples, they can determine which ones were caused by atmospheric increases in oxygen due to land-based plants and which were caused by aquatic photosynthesis. Studies of BIFs show that there was one major atmospheric oxygen event in the late Archean Eon at around 2.8–2.3 Ga and another in the Proterozoic Eon at about 0.8 Ga. These events are sometimes referred to as Great Oxidation Events (or GOEs).

For a review of redox reactions, see Review Example 1.13 on page 30.

Regardless of the source of BIFs, they can be used to place major hydrospheric and atmospheric chemistry events along a geologic time line. The oldest of these BIFs date to around 3.5 Ga, but the largest occurrence appears at 2.7 Ga after which occurred the first GOE. After the formation of BIFs at 2.7 Ga, oceanic concentrations of dissolved iron were low enough to allow molecular oxygen levels in the atmosphere to rise to levels above 1 part per million and set in motion the next evolutionary stage in the development of the Earth (Canfield, 2005).

1.3.1.3 Proterozoic Eon (2.5 to 0.5 Ga) A combination of the aforementioned factors resulted in the first deep ice age of the Earth. A rapid increase in oxygen levels as a result of the first GOE resulted in a concomitant decrease in carbon dioxide and methane levels, two very effective greenhouse gases. Photosynthetic organisms proliferated and absorbed large quantities of carbon dioxide, molecular oxygen was oxidizing the methane haze and converting it to carbon dioxide, and the young Sun was still only about 85% of its current luminosity. This combination of factors was responsible for plunging the Earth into a deep freeze, covering most of the planet and sending glaciers to regions near the equator. Evidence for two other similar events occurred at 800 and 600 Ma. Volcanism and the release of significant quantities of carbon dioxide, combined with a warming Sun, were likely responsible for enhancing the greenhouse effect and thawing the planet.

Molecular oxygen levels continued to rise above 1% of current levels, and they would reach 10% by the end of the eon. As you will learn in Chapter 3, atmospheric oxygen strongly absorbs UV-C radiation near 150 nm and, as a result, forms ozone. As the methane haze was thinning due to increased oxygen levels, it was being replaced by the oxygen/ozone UV filtration system that we rely on in modern times. The presence of photosynthetic life was making it possible for terrestrial organisms to thrive in the absence of harsh UV radiation.

1.3.1.4 Phanerozoic Eon (0.5 Ga to Present) This final and current eon saw molecular oxygen levels continue to increase, a rapid expansion of biodiversity known as the Cambrian Explosion, and a stabilization of the global climate. The development of land-based vascular plants in this eon further accelerated the change from the reducing atmosphere of the early Earth to the oxidizing atmosphere that exists presently. While levels of molecular oxygen are currently around 21% and molecular nitrogen levels are around 78%, there is evidence to suggest that oxygen levels may have peaked at 35% during the Carboniferous period about 300 Ma. These elevated levels of atmospheric oxygen may have significantly impacted the evolution of some species. The most controversial of the suggestions has been the loose link between oxygen levels and the gigantism of insects and amphibians present in the fossil record (Butterfield, 2009).

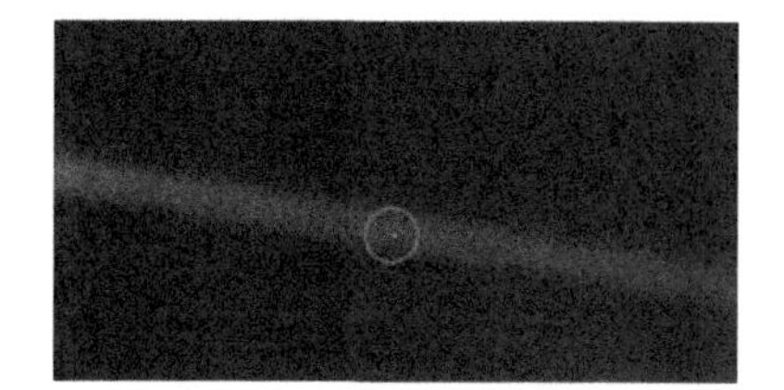

Figure 1.8 This photograph of the Earth, titled the Pale Blue Dot, was taken by the Voyager spacecraft on February 14, 1990, from somewhere near the edge of our solar system, about 6 billion kilometer from Earth. The Earth appears as a pixel against the vast background of empty space. Source: NASA.

1.3.1.5 Summary At the end of this section, you should have gained an appreciation for the thoroughly comprehensive change that the Earth has experienced over the 4.6 Gyr of its history – from a hellish, lifeless mass of chemicals to an exquisitely inhabitable "pale blue dot" (The Pale Blue Dot. photograph, 1990) (see Figure 1.8) that contains life in regions where humans thought life was not possible.

Some of the early changes in the Earth were due to chemical and physical forces such as density, temperature, and chemical reactivity. These forces turned the Earth from a molten mass to a very heterogeneous, rocky planet with a distinct atmosphere, hydrosphere, and lithosphere. The atmosphere has been transformed from a very thick, reducing layer with a large greenhouse effect to a thin, oxidizing layer with a temperate greenhouse effect – having lost much of its carbon dioxide to the hydrosphere and lithosphere and gained 21% of its volume in molecular oxygen from the biosphere. The hydrosphere has been transformed from a hot, acidic layer to a temperate, alkaline layer and the purported birthplace of life. The lithosphere has been transformed from a somewhat homogeneous mass of molten material to a highly differentiated layer with a complex structure and rejuvenating tectonic properties. In sum, the Earth is a marvel of complexity and wonder.

The next section will introduce some of the characteristics of life on Earth. While life began sometime in the late Hadean or early Archean Eon, its real heydays began in the Phanerozoic eon, beginning some 540 Ma with the Cambrian explosion (Figure 1.9). Even before then, very simple forms of life have had dramatic impacts on the planet, of which the most important contribution has been molecular oxygen. The most recent newcomer to the planet, *Homo Sapiens*, has the unique characteristic of being the only single species to have a measurable impact on the global environment. In 2008, the Geological Society of London took up a proposal to add a new epoch to the Phanerozoic eon called the *Anthropocene* to mark the end of the Holocene epoch, which likely ended at the beginning of the Industrial Revolution in the late 19th century. The human activities of tremendous population growth (it has increased by ×7 since 1800 when the human population was about 1 billion), unprecedented consumption of resources, environmentally significant accumulation of waste products, and the geologically instantaneous spread of invasive species around the planet have had an impact on the Earth that no other single species seems to have had. The unique form of consciousness that our species has exploited during our development has allowed us to be capable of realizing our impact on the planet and to simultaneously deny that our species is bound to the biological constraints that we clearly observe in other species (Birkey, 2011; Horowitz, 2012; Kapur, 2012). This situation, so rich in irony, has led one biologist to claim that of all of the amazing technological advances our species has developed (medicine, energy harvesting, agriculture, etc.), the only one that is not contributing to the problems associated with the Anthropocene is contraception (Grogan, 2013). Given the current debate in the United States over contraception in health-care insurance and domestic/international funding of reproductive health technologies and education, the use of this technology to curb the changes wrought by the Anthropocene seems dim at best.

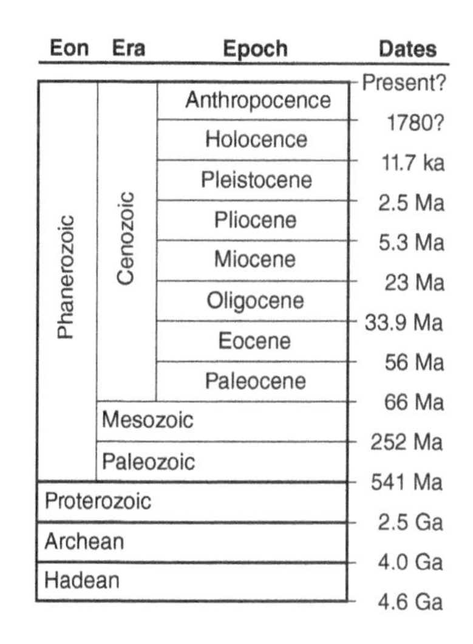

Figure 1.9 This stratigraphic chart shows the chronology of the history of the Earth, with the latest eon, the Phanerozoic eon, broken into eras and epochs. Our ancestors began using tools (2.6 Ma) in the Pliocene, mastered fire (800 ka) and invented agriculture (12 ka) in the Pleistocene, and emerged from the last ice age (10 ka) in the Holocene. Some scientists have proposed a new epoch to recognize the global impact humans have had on the Earth since the Industrial Revolution. Courtesy K. Overway, 2013.

I'm starting to paint an entirely too gloomy picture here. Let us lighten things up a bit by examining the amazing properties that life has harnessed to produce effects that animate scientists in the fields of biology, chemistry, physics, psychology, sociology, the humanities, ...

1.4 LIFE EMERGES

The earliest evidence for life comes in the form of fossilized bacteria in sedimentary rocks found in western Australia and South Africa that have been dated to around 3.5 Ga (Fitz, Reiner, and Rode, 2007). Indirect evidence of an earlier emergence of life is found in C-12 enriched carbon deposits and C-13-enriched carbonate in Greenland rocks that date to 4.1–3.8 Ga. Thus, while the date of the emergence of life remains uncertain, it developed near the end of the Hadean Eon or near the beginning of the Archean Eon.

While the examination of the origins of life is usually considered the domain of evolutionary biology, the emergence of life on the Earth had such a major impact on the state of the planet that environmental scientists would say life and the Earth *coevolved.* Along with the physical forces of temperature, radiation from the Sun, and acid/base chemistry, biological organisms consumed and produced chemicals on such a scale that they affected the surface of the planet so indisputably that most of the current environmental conditions that we live under are wholly attributable to biotic forces.

Biological organisms are found in many different forms and in many different environments. Our modern understanding of organisms shows that they are amazingly complex but follow several fundamental rules. Life is so complex that some have claimed that it was not possible for life to emerge from the chaos of the Hadean Eon without the guiding intervention of a supernatural or superintelligent force. The "irreducible complexity" is at the center of the discredited intelligent design (ID) movement, which sought to add ID explanations for life alongside evolutionary theory in the public school system, much as the Creationism movement tried to do in the 20th century. While organisms are complex, they exhibit a combination of four features that, individually, are understood and can be explained by modern biologists based on a modern understanding of the Theory of Evolution.

According to Dr Robert Hazen, a scientist at the Carnegie Institution of Washington's Geophysical Laboratory, the emergence of life would have required the formation of four essential ingredients: biomolecules, such as amino acids and nucleotides; macromolecular constructions of the biomolecules; self-replication of the macromolecules; and finally, molecular evolution. While a complete understanding of the origin of life on the Earth still evades scientists, it is an active field of research. Each of the requirements for life is becoming better understood, and eventually, all of the pieces of the puzzle will likely come together to allow scientists to produce the first synthetic organism.

1.4.1 Biomolecules

In 1953, Stanley Miller (1930–2007), a chemistry graduate student studying at the University of Chicago, stunned the world with the results of his efforts to synthesize biomolecules by using only the simple chemicals and conditions that scientists at the time believed to represent the early Earth. Miller combined water, methane, ammonia, and hydrogen in a glass reactor and exposed the mixture to heat and an electrical discharge to simulate lightning (see Figure 1.10). Within 2 weeks, the reactor had produced several amino acids, the building blocks of proteins. Miller reported that 11 amino acids had been synthesized, but a modern analysis of the residue of his experiment showed that 20 amino acids had been formed. Since then, other scientists have modified the conditions of the reaction and have been able to produce other biomolecules in addition to amino acids. These conditions are modeled on the current understanding of the Hadean and Archean atmosphere and hydrosphere, hydrothermal vents, surface and shallow marine environments, and interstellar

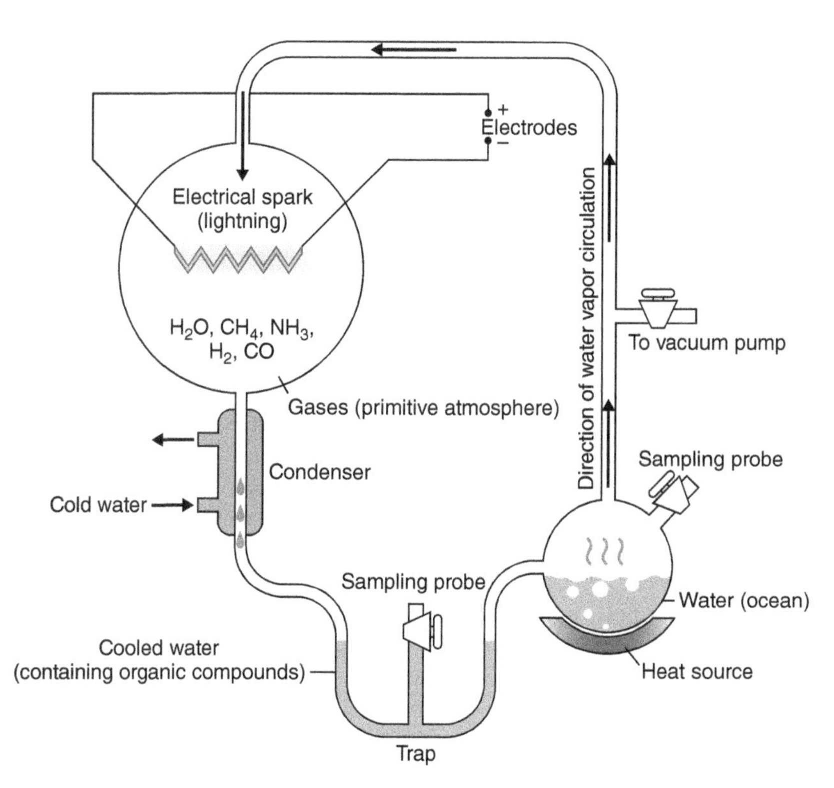

Figure 1.10 A diagram of the reaction vessel that Stanley Miller used to produce organic molecules from simple inorganic reactants. The Miller–Urey experiment proved conclusively that organic molecules could be made with very simple ingredients. Source: Yassine Mrabet http://en.wikipedia.org/wiki/Miller-Urey_experiment. Used under BY-SA 3.0 //creativecommons.org/licenses/by-sa/3.0/deed.en.

environments. Life seems to survive in a variety of environments, from the super-hot and acidic environment of hydrothermal vents on the ocean floor to dust particles high in the atmosphere, so the expansion of conditions under which biomolecules can form was necessary and has consistently yielded positive results. It is clear from the research that most of the organic molecules necessary for life are the inevitable result of mixing simple inorganic molecules in a variety of prebiotic environments.

1.4.2 Macromolecules

Macromolecules are polymers of the simpler, biomolecular monomeric units. For example, sucrose is a disaccharide or polymer of glucose and fructose. When each of these two monomers is connected via a dehydration reaction, a molecule of sucrose results (see Figure 1.11). Proteins are polymers of amino acids, starches are polymers of sugars, and RNA and DNA are polymers of nucleotides. Early life would have required the formation of macromolecules before it could begin. The emergence of life becomes difficult to explain in some cases since some macromolecules have a tendency to self-assemble under certain conditions, but others would spontaneously disassemble. Amino acids, for example, can condense to form an amide bond and a dipeptide, which is the process that forms proteins. The following is an example showing the formation of the dipeptide from two glycine amino acids.

$$H_2NCH_2COOH + H_2NCH_2COOH \rightleftharpoons H_2NCH_2CONHCH_2COOH + H_2O \qquad \text{(R1.28)}$$

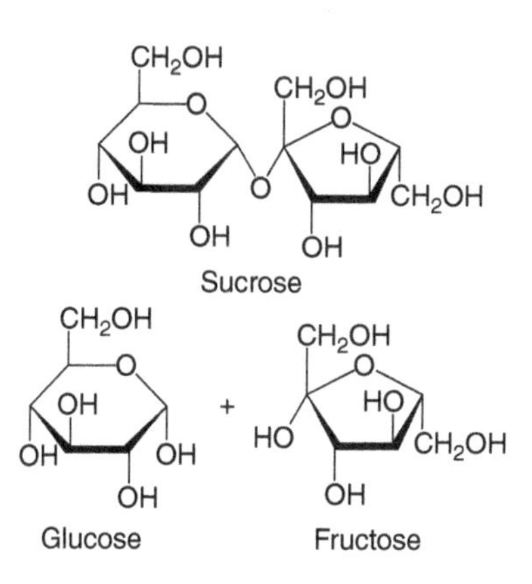

Figure 1.11 Common table sugar, sucrose, is a polymer of two simpler sugars. The monomers glucose and fructose link together to form sucrose. Courtesy NEUROtiker, 2007.

This reaction is not spontaneous at room temperature and requires an input of energy. The Gibbs free energy of the reaction demonstrates this.

$$\begin{aligned}\Delta G^\circ_{rxn} &= \Sigma\Delta G^\circ_{f\text{ products}} - \Sigma\Delta G^\circ_{f\text{ reactants}} \\ &= \left(\Delta G^\circ_{f\text{ gly-gly}} + \Delta G^\circ_{f\text{ H}_2\text{O}}\right) - 2\left(\Delta G^\circ_{f\text{ glycine}}\right) \\ &= ((1\text{ mol})(-483.796\text{ kJ/mol}) + (1\text{ mol})(-237.1\text{ kJ/mol})) \\ &\quad - ((2\text{ mol})(-366.979\text{ kJ/mol})) \\ &= +13.1\text{ kJ}\end{aligned} \tag{1.2}$$

☞ For a review of thermodynamics, see Review Example 1.14 on page 32.

Reactions with $\Delta G > 0$ are nonspontaneous, meaning that the reactants are favored and the products are unlikely to form unless energy is input into the reaction. The aforementioned calculation only considers peptide formation under standard conditions (1 M concentrations and 25 °C). Under elevated temperatures, such as those that occur around deep sea hydrothermal vents, and high salt concentrations, such as in tidal pools, peptide formation is spontaneous (Imai *et al.*, 1999; Cleaves, Aubrey, and Bada, 2009). Another study showed that under dry conditions and high pressures, such as in deep crustal sediments, glycine formed polymer chains of up to 10 units long, which were stabilized by the high pressures (Ohara, Kakegawa, and Nakazawa, 2007). Yet another study showed that in the presence of carbonyl sulfide, a volcanic gas that is present at deep sea hydrothermal vents, assists the formation of peptide bonds with 80% efficiency at room temperature (Leman, Orgel, and Ghadiri, 2004). The lesson learned here is that the tabular values that are listed in Appendix I need to be used only under the conditions to which they are measured. For most laboratory experiments in an undergraduate laboratory, they are appropriate, but conditions vary greatly across the planet, and life may have arisen under conditions that we would find extreme or in the presence of chemical species that are unfamiliar to the generalist.

☞ *Amphiphilic* molecules possess both a hydrophobic and a hydrophilic region in the same molecule. This dual solubility leads to the spontaneous formation of macromolecules in order to minimize the energy of solvation.

Fatty acids and phospholipids, on the other hand, are *amphiphilic* and are often barely soluble in water due to the poor solubility of the hydrophobic portion of the molecule. When in high enough concentration, they will self-organize into macromolecular structures such as micelles, vesicles, and bilayer sheets as seen in Figure 1.12. These structures lower the energy of the molecules since having the hydrophobic chains (represented by the stringy tails of the structure on the right in Figure 1.12) in contact with highly polar water molecules interrupts the hydrogen bonding between water molecules. Interrupting these H-bonds requires energy, so these amphiphilic molecules are forced to self-assemble into these complex structures by the sheer energetics of the environment, absent any biological or supernatural agency. The bilayers and vesicles that form spontaneously are analogous to primitive cell membranes. A further interesting observation made by some scientists studying these vesicles is that they can be autocatalytic - the formation of one structure can stimulate the formation of others that are identical. Thermodynamic predictions are not the whole picture, however. While the formation of polypeptides ostensibly seems to be unfavored, thermodynamically unfavorable chemical products are quite common because the reactions and the environment under which they form can result in kinetic inhibition as a result of a high activation energy. If a thermodynamically unfavorable product is formed under conditions of high heat or energy, more than enough to overcome the activation energy, and then is quickly placed in an environment that has a lower temperature, then the thermodynamically unstable products are "frozen in place" without enough energy to get over the activation energy barrier and allow the thermodynamically favored products to form. Take, for example, the formation of glycine ($C_3H_7NO_2$), a simple amino acid. Its production and thermodynamics can be summarized in the reaction and calculation as follows.

$$3CH_4(g) + NH_3(g) + 2H_2O(l) \rightarrow C_3H_7NO_2(aq) + 6H_2(g) \tag{R1.29}$$

$$\begin{aligned}\Delta G^\circ_{rxn} &= \Sigma \Delta G^\circ_{f\ \mathrm{products}} - \Sigma \Delta G^\circ_{f\ \mathrm{reactants}}\\ &= ((1\ \mathrm{mol})(-88.23\ \mathrm{kJ/mol}) + (6\ \mathrm{mol})(0\ \mathrm{kJ/mol}))\\ &\quad - ((3\ \mathrm{mol})(-50.8\ \mathrm{kJ/mol}) + (1\ \mathrm{mol})(-16.4\ \mathrm{kJ/mol}) + (2\ \mathrm{mol})(-237.1\ \mathrm{kJ/mol}))\\ &= +554.77\ \mathrm{kJ}\end{aligned}$$

According to these calculations, the formation of glycine is not thermodynamically favored, yet we know that many scientists were able to produce glycine and other amino acids under various conditions. The Miller–Urey experiment, for example, used electrical discharge to simulate lightning. This lightning provided the activation energy to allow some glycine to be formed. Once it is formed, it is likely to remain intact. If there was no kinetic barrier, then the products of the reaction would quickly return to the more stable starting materials.

While the energetics of condensation reactions of essential biomolecules such as amino acids, sugars, and nucleic acids prohibit self-assembly in the aqueous phase, such assemblies have been observed on the surface of minerals. Self-assembly of these macromolecules is often enhanced by episodic evaporation – a condition that would be mimicked at the edge of tidal pools. RNA chains of 50 base pairs have been recovered from the mixture of RNA nucleotides and clay. One experimental mixture of clay, RNA nucleotides, and lipids resulted in the formation of a vesicle around clay particles that had adsorbed the RNA – a situation strikingly close to a primitive cell (Hanczyc, Fujikawa, and Szostak, 2003). It appears that under certain conditions, the spontaneous formation of biomolecules is energetically favored.

A trickier problem to solve is the *biochirality* that dominates life on Earth. Chiral centers in biomolecules are predominately made up of L-amino acids and D-sugars (except a very few organisms and bacteria (Harada, 1970)). Why? At this point, there is no conclusive answer. Clay minerals have highly charged surfaces that attract polar molecules, such as amino acids, and bring them in close proximity to each other. It has been known since the late 1970s that clays promote the formation of small polypeptides. Certain mineral surfaces, such as quartz, fracture in such a way that the surface is chiral and attracts organic enantiomers. While there is an equal distribution of "left-handed" and "right-handed" quartz surfaces around the globe, it could be that the first successful, self-replicating biomolecule happened to form on a "left-handed" surface. Its early emergence meant that it autocatalytically dictated the formation of other biomolecules – meaning that if the clock was rewound, it could have been "right-handed" biomolecules.

Another mechanism for preferentially producing peptides from only L-amino acids is salt-induced peptide formation (SIPF). SIPF reactions involve octahedral complexes with copper centers, which show a stereospecific preference for L-amino acid ligands to link to other L-amino acid ligands. If one or more of the amino acids are of the D-isomer, then a peptide bond is not favored to form. This mechanism is especially enhanced by high salt concentrations and episodic evaporation cycles, much like the edges of tidal pools. In some cases, the product preference between L-L and D-D peptides is 400:1 (Plankensteiner, Reiner, and Rode, 2005).

It seems, therefore, that there are a few plausible ways for stereospecific macromolecules to emerge from the biomolecules that seem to inevitably form from abiotic conditions on Earth. Being composed of stereospecific macromolecules could also describe nonliving entities, such as crystals. Life, however, is qualitatively different from minerals and crystals. Life replicates itself and adapts to environmental conditions. The next two features of life will help distinguish it from other very interesting inanimate objects.

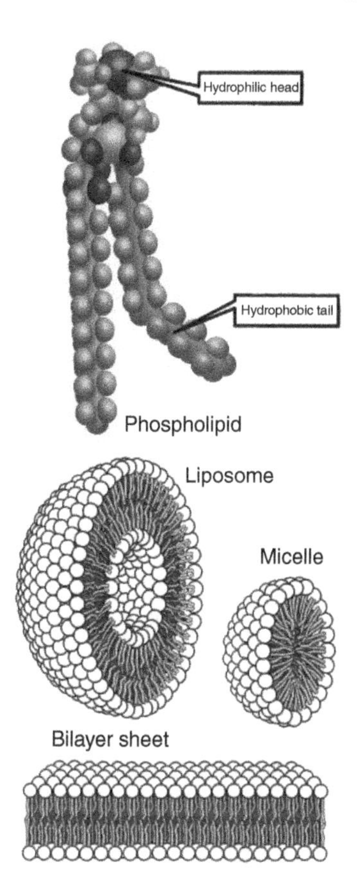

Figure 1.12 Phospholipids, which can be characterized as having separate hydrophilic and hydrophobic regions, spontaneously form bilayers, micelles, and vesicle in aqueous solutions. Courtesy Ties van Brussel and Mariana Ruiz Villarreal, 2010.

☞ *Biochirality* is a property of biological molecules that manifests itself in the "handedness" or optical activity of the monomeric units of the molecules. This means that compounds generated by biotic processes preferentially produce one of the optical isomers while abiotic chemical processes usually produce a 50/50 mixture of the isomers.

1.4.3 Self-Replication

The next ingredient of life is its ability to self-replicate. When cells divide, they form nearly identical copies of the parent cell, with replicated genetic information and molecular machinery. This replication process is a very complex interdependence of the biomolecules. DNA is considered the primary director of current biological activity because it stores the genetic information for the construction of other biomolecules, but it requires intricate protein and RNA machinery in order to carry out these instructions and to self-replicate. RNA, similar to DNA in that it can encode genetic information, has catalytic properties and can self-replicate unlike DNA. Thus, early hypotheses assumed that life emerged from an "RNA world" where the first replicators were segments of RNA. While this seems plausible in freshwater systems, in a marine-based primordial soup, RNA self-assembly is not favored due to the high salt content (Fitz, Reiner, and Rode, 2007). If life emerged from a marine environment, then what seems more likely is that life emerged from protein-based self-replicators because of the SIPF mechanism discussed earlier.

Replication is a curious and wonderful thing. In the case of minerals, replication is manifested in the process commonly called crystallization. Crystallization is a delicate balance between the driving forces of entropy and enthalpy. According to Eq. (1.16) (see page 32), the enthalpy (ΔH) and entropy terms ($T\Delta S$) must add together to make the free energy term (ΔG) negative in order for the crystallization process to be spontaneous. Since crystals are highly ordered solids, the entropy term for the formation of a crystal is negative, indicating that the crystal is more ordered than the liquid. This is also true for any amorphous solid compared to a liquid – both crystals and amorphous solids have less entropy than liquids – but a crystal has much less entropy than an amorphous solid. The structure of a crystal leads each atom or molecule to participate in more bonds than a similar atom or molecule in an amorphous solid; thus, there are more bonds per atom and therefore a lower enthalpy is achieved. This leads to a more exothermic transformation, and the enthalpy term is large enough to compensate for the entropy term.

Crystal growth only occurs under a limited range of conditions where the temperature and pressure promote solidification, but only barely. If the temperature were significantly below the melting point, then the energetics favor very fast solidification and a tremendous number of very tiny crystals form quickly, which makes the solid look very noncrystalline to us. If the temperature is just barely below the melting point and it is cooled very slowly, then a small crystal that starts to form is allowed to grow larger and larger. The appearance of the first tiny crystal becomes a site of nucleation, causing the crystal to replicate itself and form a macroscopic crystal. You have probably seen this happen on a car windshield during the winter – when the conditions are just right, the first ice that forms causes the crystal to grow across the windshield and forms a distinctive pattern. This nucleation process is a form of catalysis, the essential property of the first replicators.

☞ A *catalyst* is a chemical that lowers the activation energy of a reaction, allowing the reaction to proceed much faster than an uncatalyzed reaction.

Current experiments (Bada, 2004; Joyce, 2002) done *in vitro* (in a test tube) with RNA self-replication suggest that polymeric chains of RNA must be between 20 and 100 units long in order to achieve *catalytic* activity. Catalysis is important for self-replication because the formation and breaking of chemical bonds often requires a tremendous amount of energy. A C–H bond, for example, is 414 kJ/mol (see Table I.7 on page 325). If that amount of energy is added to a mole of water, the resulting temperature increase would be staggering.

$$
\begin{aligned}
q &= mC\Delta T \\
\Delta T &= \frac{q}{mC} \\
&= \frac{414,000\ \text{J}}{\left(18.02\ \text{g H}_2\text{O}\right)\left(4.184\ \frac{\text{J}}{\text{K}\cdot\text{g}}\right)} \\
&= 5491\ ^\circ\text{C}
\end{aligned}
$$

Catalysts are able to dramatically reduce these high energy barriers by stabilizing the molecule while a bond is breaking.

Catalysts are a feature of chemical kinetics since they influence the rate of a reaction. As you can see in Figure 1.13, a catalyzed reaction changes the reaction mechanism and results in a smaller activation energy (E_a). In this example, the uncatalyzed reaction involves the collision of ozone and an oxygen atom, which yields two units of molecular oxygen. The transition state between the reactants and products, sometimes called the *activated complex*, is the breaking of one of the bonds that holds ozone together. You know from drawing the Lewis structure of ozone that O_3 is held together by a double bond and a single bond. Remember that the number of bonds is equal to half of the difference between the electrons needed to form octets minus electrons the structure has. In this case, bonds $= \frac{1}{2}(24 - 18) = 3$. Since ozone has two resonance structures, the net bond strength is 1.5 bonds. The transition state represents the energy required to shift the resonance electrons to form a double bond and the breaking of the remaining single bond. This energy demand results in the activation energy.

In the catalyzed reaction, a chlorine atom plays an intermediary role by changing the mechanism and forming a more stable activated complex (ClO). This dramatically smaller activation energy has a significant impact on the rate of the reaction.

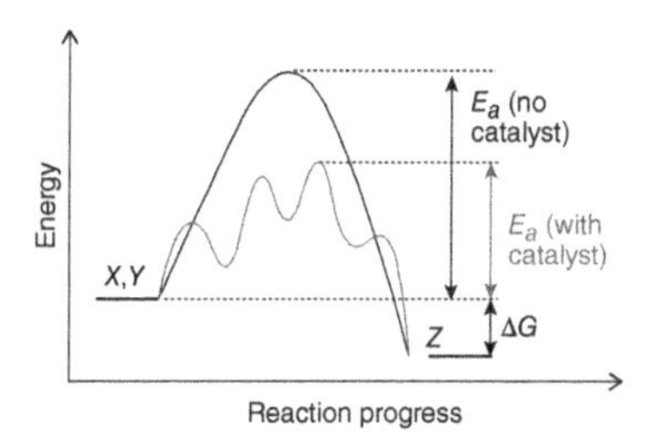

Figure 1.13 This reaction diagram shows the energetic difference between the uncatalyzed and catalyzed conversions of a generalized reaction. Notice the height of the activation energy (E_a) for each mechanism. The catalyzed reaction is much faster because the activation energy is much lower. Source: Smokefoot. http://en.wikipedia.org/wiki/Catalysis. Used under CC0 1.0 // creativecommons.org/publicdomain/zero/1.0/deed.en.

1.4.4 Molecular Evolution

Molecular evolution is the final property of life and one of the bridges between the fields of chemistry and biology. On a very simplistic level, the process of evolution involves inheritance of genetic information from a parent, the unavoidable mutation rate due to loss of fidelity during the copying process or as a result of some external factor (such as exposure to radiation), and the survival or elimination of these mutants as a result of environmental factors.

☞ For a review of drawing Lewis structures, see Review Example 1.18 on page 38.

Once the first replicator (or replicators) emerged from the prebiotic soup, each new generation of replicator molecules contained some mutations that were deleterious for the environment in which they existed (and led to elimination) or were advantageous (and led to survival). While the path from the first replicator to the present biology, which is exclusively DNA-based, has not been demonstrated conclusively by any researcher, the fields of synthetic biology and molecular evolution are actively pursuing mechanisms and pathways that examine simple replicators in environments that might have been present in the late Hadean and early Archean eons.

The algorithm of evolution shares many of the same features as the scientific method, where the ideas (hypotheses) of scientists are tested against experimental observations. The ideas survive if they are able to explain experimental results, but are more often refined (mutated) or rejected (eliminated). While science is sometimes a blind process, it is more often a directed process in which the scientist refines a hypothesis using intuition and information gathered from experimental observations. Sometimes, several hypotheses can explain a single experimental observation, but more often a single hypothesis will emerge as the best (characterized as more parsimonious or more powerful). The analogy between the scientific method and biological evolution breaks down here because evolution is a blind process driven by random mutation instead of the directional control imposed by the scientists. Further, evolution often produces several mutants that can survive in a given environment. While the process of science aims toward refinement in the hypotheses that eventually lead to theories, biological evolution requires that each mutant be only sufficient enough to survive, not the best of the mutants that survive. So in very rough terms, the scientific process is cyclical where refinement or elimination of a hypothesis is driven by empirical observations and theoretical logic in a guided process, whereas the process of evolution is a linear process where the survival or elimination of random variations in inherited genetics is driven by environmental filtration (Rosenberg and Shea, 2008).

☞ The *transition state* or *activated complex* is the high energy, transient chemical structure that is the bridge between the structures of the reactants and products. It was once impossible to study, but with the invention of femtosecond lasers, these structures have been observed.

Once Charles Darwin described biological evolution as a blind algorithm that very slowly accumulates adaptations and complexity, some claim that the philosophy of science crossed a point of no return. No longer could any sort of complexity in the world, be it abiotic or biotic, be soundly described as irreducibly complex or in need of an intelligent designer. All complexity could be explained by infinitesimally small changes over vast stretches of time. This is another distraction into philosophy, but if you are interested in reading more about this philosophical claim, read *Darwin's Dangerous Idea* by Daniel Dennett (1996).

As this chapter closes, I hope you are convinced that in order to study the present state of the environment, it is important to understand the processes that led to it. Before life emerged on Earth, the evolution of the planet could be explained entirely as a result of chemical principles acting on matter that was produced by the Big Bang, synthesized into the known elements by the fiery cores and supernova explosions of massive stars, and aggregated into the planet at the birth of our solar system. Physical energetics such as density led to the formation of the core and mantle, cooling temperatures led to the formation of different heterogeneous segments of the planet (atmosphere, hydrosphere, and lithosphere), and interactions of light from the Sun continues to drive chemical and physical processes. Once life emerged, the Earth evolved under the concomitant forces of chemistry and biology to become the hospitable place that we call home. If we are to understand how our environment is changing as a result of natural and *anthropogenic* forces, we must understand the chemistry that underlies these forces. That is not a simple thing, as testified by the list of important terms at the end of this chapter. But, hey, you are a science major, so you knew that explaining and predicting complicated systems requires a lot of background knowledge. It would not be interesting if it were any other way.

☞ *Anthropogenic* effects are those caused by human civilization. While humans are animals, their impact on the environment has been qualitiatively different than other *biotic* or *abiotic* effects since the Industrial Revolution.

1.5 REVIEW MATERIAL

The following review examples explain and demonstrate some of the important chemical principles that are required in order to understand environmental chemistry. These examples should be a review of what you learned in previous chemistry courses.

Review Example 1.1: Electromagnetic Radiation

The stuff we commonly call light is a thin slice of the EM spectrum. It is called electromagnetic because the radiation is a fast-moving wave that contains energy. Much like the ripples on the surface of a smooth pond after a rock has been tossed into it, the waves carry the energy. In the pond analogy, it requires energy to lift the water molecules up and then push them down; thus, the amplitude of the wave expresses the energy of the disturbance (the rock impacting the surface). Instead of a wave of water, however, EM radiation is a disturbance in the electric and magnetic fields that exist in space. It is a wave that oscillates an electric field perpendicular to an oscillating magnetic field.

These waves can be characterized by their wavelength, frequency, and amplitude. The wavelength (represented by λ, the Greek character lambda) is measured from peak to peak and is an expression of distance because the radiation is traveling in a certain direction. Visible light, for example, ranges in wavelength from 400 to 750 nm. The unit "nm" refers to nanometers or 10^{-9} m (see Table 1.2 on page 4). The frequency of a wave (represented by ν, the Greek character nu) is the speed at which it oscillates. Red light from a stop light oscillates at a frequency of 4×10^{14} times per second (4×10^{14} Hz). This implies that the time from one peak to another takes only 2×10^{-15} s because light travels at a fixed speed referred to as c or the speed of light. These three properties of radiation are related by the following equation.

$$\lambda \nu = c \tag{1.3}$$

If you think about using this equation to describe how you walk it should make perfect sense. Think of ν as the frequency of your strides, and think of λ as the length of each stride. If your legs move at a certain frequency and cover a specific distance per stride, then the product of these two terms determines how fast you move when you walk. For EM radiation, the speed is fixed at 2.99792458×10^{8} m/s in a vacuum – this is the value of c.

One more property of EM radiation that is related to the wavelength and frequency of light is its energy. The energy is related by the following equation.

$$E = h\nu = \frac{hc}{\lambda} \tag{1.4}$$

The energy (E) is proportional to the rate at which the wave oscillates (ν). Think about moving your hands and arms up and down. If you move them up and down slowly, it takes less energy than if you move them very quickly. The constant of proportionality in this equation is Planck's constant (h), which is $6.62606957 \times 10^{-34}$ J·s.

Exercises

1. Rank the following regions of the EM spectrum in order from the shortest to the longest wavelength: infrared, ultraviolet, visible, X-rays, radio waves.
2. What is the frequency of yellow light with a wavelength of 604 nm?
3. What is the energy of this yellow light?
4. My favorite local radio station has a frequency of 90.7 MHz. What is its wavelength?

Solution: See Section A.1 on page 231.

Review Example 1.2: Interactions between Light and Matter

Light and matter interact in some very important ways. In many cases, it would be impossible to bridge the gap between the large-scale observations we make about objects in our everyday life and the very small scale of atoms and electrons. Light can bridge the gap for us. Remember that photons contain electric and magnetic fields. This means that light can interact with particles that produce electric or magnetic fields. Electrons are the outermost part of an atom, so they commonly interact with light. A photon of light can be absorbed by an electron, giving it more energy and promoting it to a higher energy level, and a moving electron can emit a photon of light when it loses energy as it falls from a higher quantum level to a lower level. A third option is that there is no interaction, a process called transmission, because the energy of the photon does not match a possible energy jump for the electron from one level to another. These interactions mean that light can reveal a lot of the properties of matter that cannot be seen or observed on the macroscopic scale. This area of science is called spectroscopy – the study of the interaction between matter and light.

When light gets absorbed by matter, the light passing through an object gets less intense, as seen in Figure 1.14. Imagine a beam of monochromatic light passing through a cuvette containing an aqueous solution of a certain chemical – call it a dye. If you were to put the light source, with a certain intensity (I_0), on one end of the cuvette and look through the glass from the other end, you would see the light after it comes through the cuvette (with an intensity of I). If there were a lot of the dye dissolved in the water, you would see a very dim light (small I, maybe even close to zero). If the solution were very dilute with little of the dye, you would see a bright light (a large I, close to I_0). The intensity of the transmitted light is inversely proportional to the concentration (C) of the dye. This is known as Beer's Law. Here are a few features of the mathematics.

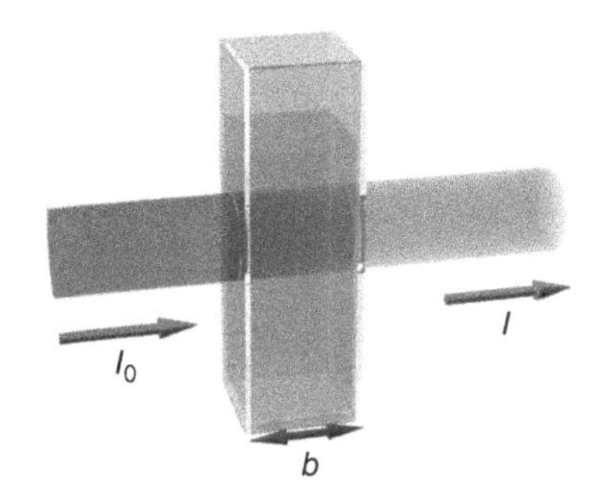

Figure 1.14 The Beer–Lambert Law, or just Beer's Law, relates the absorption of electromagnetic radiation by matter. Courtesy CarlosRC, 2008.

$$T = \frac{I}{I_0} \tag{1.5}$$

$$\%T = \frac{I}{I_0} \times 100\% \tag{1.6}$$

$$Abs = -\log\left(\frac{I}{I_0}\right) = -\log T \tag{1.7}$$

$$Abs = \varepsilon bC \tag{1.8}$$

In Eqs (1.5) and (1.6), transmittance and percent transmittance are defined using the ratio of the transmitted intensity and initial intensity. In Eq. (1.7), absorbance is defined. Equation (1.8) is a form of Beer's Law where the Greek character ε is called the molar absorptivity and describes how strongly a chemical absorbs. Dyes absorb very strongly and have large ε's.

Other compounds absorb poorly and have small ε's. Finally, the term b is the length of the cuvette, as shown in Figure 1.14.

Exercises

1. If 15% of the light is absorbed by a sample in a cuvette, calculate the absorbance.
2. Convert an absorbance of 0.157 to %T.
3. Given an absorbance of 0.088 and an ε term equal to 54.2 M^{-1} cm^{-1}, calculate the concentration of a dye in a 1-cm cuvette.

Solution: See Section A.1 on page 232.

Review Example 1.3: Metric Prefixes

Metric prefixes are convenient symbols that represent very large and very small numbers without the use of scientific notation. There are many more prefixes than are represented in Table 1.2 (see page 4), but these are the prefixes that you should memorize as they will be commonly used in this book and in your field of study. These prefixes can be treated as a symbolic constant. For example, 12.5 µm = 12.5×10^{-6} m. Notice how the μ symbol can be replaced by its numerical value.

Exercises

1. Convert 1.05 mm to µm.
2. Convert 46.2 ng/µL to g/L.
3. Express 0.154×10^{-6} L as nL.

Solution: See Section A.1 on page 232.

Review Example 1.4: The Structure of the Atom

Elements are composed of protons and neutrons (both are nucleons) that reside in the dense core of an atom that is surrounded by electrons. Each element is designated by its atomic number (Z), which represents the number of protons in the nucleus. Most elements naturally occur with varying numbers of neutrons. These are known as isotopes and are distinguished by the mass number (A), which is the sum of the number of protons and neutrons. Atoms can also form ions, which result from the addition or loss of one or more electrons and the gain of a proportional charge. All of these variations can be symbolized as follows.

$$^{A}_{Z}X^{charge}$$

An example would be $^{235}_{92}U^{3+}$, which is the isotope of uranium used in nuclear power plants. This symbol indicates that the atom has lost 3 electrons (thus the +3 charge) and contains 143 neutrons (143 n + 92 p = 235). The careful observer will note that stating U and stating 92 is redundant since uranium is the element with 92 protons. Sometimes, isotopes are also written as U-235 or uranium-235 when the charge state is not relevant.

Exercises

Complete Table 1.5 using your knowledge of the structure of an atom.

Symbol	# of protons	# of neutrons	# of electrons
	35	45	36
$^{23}_{11}Na^{+1}$			

Table 1.5 Atomic structure example with answers.

Solution: See Section A.1 on page 233.

Review Example 1.5: Balancing Nuclear Reactions

Nuclear reactions need to be balanced just like chemical reactions, except the focus is on balancing nucleons. If you look closely at the nuclear reactions in Section 1.2.4, you will see that adding all of the mass numbers (superscripts) on both sides of the reaction arrow gives an equality. The same is true for the atomic numbers (subscripts). For example, the following reaction is one of the possible mechanisms in the oxygen fusion layer of a supergiant star.

$$^{16}_{8}O + ^{16}_{8}O \rightarrow ^{A}_{Z}?? + ^{4}_{2}He \tag{R1.30}$$

Two oxygen nuclei combine to give a total mass number of 32 and a total atomic number of 16 on the left. Since the right side contains an alpha particle, the unknown product must have an atomic number of $16 = Z + 2$, where $Z = 14$. Similarly, the mass number of the unknown particle must be $32 = A + 4$, where $A = 28$. What element has an atomic number of 14? Silicon! The unknown particle must be Si-28.

$$^{16}_{8}O + ^{16}_{8}O \rightarrow ^{28}_{14}Si + ^{4}_{2}He \tag{R1.31}$$

Exercises

1. Balance the following nuclear reactions, and identify the unknown reactant or product.
   (a) $^{12}_{6}C + ^{12}_{6}C \rightarrow ^{A}_{Z}?? + 2^{4}_{2}He$
   (b) $^{12}_{6}C + ^{A}_{Z}?? \rightarrow ^{56}_{26}Fe$
   (c) $^{201}_{80}Hg + ^{0}_{-1}\beta \rightarrow ^{A}_{Z}??$
   (d) $^{19}_{9}F \rightarrow ^{A}_{Z}?? + ^{19}_{8}O$
   (e) Label the reactions in the previous problem as alpha decay, beta capture, beta emission, neutron capture, positron emission, or fusion. Use the label to figure out what the reaction does.
   (f) Write the two-step electron capture reactions that start with Ni-56 and end with Fe-56. See the description in the Si Fusion box on page 7.

Solution: See Section A.1 on page 233.

Review Example 1.6: Half-Life

The half-life is a convenient measure for gauging the lifetime of radioisotopes. Half-life is defined as the time needed for a radioisotope to lose half of its mass.

For a first-order reaction, a half-life can be determined from the rate constant and by rearranging Eq. (1.9).

$$\ln([A]_t) = \ln([A]_0) - kt \tag{1.9}$$

By definition, the initial concentration of a species is always twice as much as the current concentration ($[A]_t = \frac{1}{2}[A]_0$)

$$\begin{aligned} t_{1/2} &= \frac{\ln\left(\frac{[A]_t}{[A]_0}\right)}{-k} \\ &= \frac{\ln\left(\frac{\frac{1}{2}[A]_0}{[A]_0}\right)}{-k} \\ &= \frac{\ln\left(\frac{1}{2}\right)}{-k} \end{aligned} \tag{1.10}$$

For a second-order reaction, a half-life can be determined from second-order integrated rate equation.

$$\frac{1}{[A]_t} = \frac{1}{[A]_0} + kt \quad (1.11)$$

Using the half-life definition ($[A]_t = \frac{1}{2}[A]_0$), a second-order half-life equation can be developed.

$$t_{1/2} = \frac{\frac{1}{[A]_t} - \frac{1}{[A]_0}}{k}$$
$$= \frac{\frac{1}{\frac{1}{2}[A]_0} - \frac{1}{[A]_0}}{k}$$
$$= \frac{\frac{2}{[A]_0} - \frac{1}{[A]_0}}{k}$$
$$t_{1/2} = \frac{1}{k[A]_0} \quad (1.12)$$

Exercises

1. Ni-56 has a first-order rate constant (k) of 0.115 day^{-1}. Determine its half-life.
2. Determine the second-order half-life of NO in the following reaction if it has a rate constant of 245 $atm^{-1}s^{-1}$ when the NO concentration (partial pressure in this case) is 0.00172 atm.

$$2NO(g) \rightarrow N_2(g) + O_2(g) \quad (R1.32)$$

Solution: See Section A.1 on page 233.

Review Example 1.7: Solubility Rules

The solubility of certain ionic compounds plays an incredibly important role in solution chemistry, mineral weathering, and geologic formations, such as the stalactites and stalagmites in caves. Common anions found in freshwater and marine environments, such as sulfate, nitrate, carbonate, and halides, interact with common metal cations to form soluble and insoluble ionic compounds. Insoluble compounds form minerals such as calcite ($CaCO_3$), gypsum ($CaSO_4$), and starkeyite ($MgSO_4 \cdot 4\,H_2O$). The stalactites and stalagmites form as a result of the precipitation of limestone (another form of $CaCO_3$) as saturated water drips from the cavern ceiling and evaporates. The solubility rules that describe this are listed as follows.

Solubility Rules

- All ionic compounds that include alkali metals and ammonium are very soluble.
- All nitrates and acetates are very soluble.
- All halides are soluble except halides of Ag^+, Cu^+, Hg_2^{2+}, and Pb^{2+}.
- All sulfates are soluble except sulfates of alkaline earths below magnesium (Ba^{2+}, Ca^{2+}, Sr^{2+}), Hg_2^{2+} and Pb^{2+}.
- All carbonates are insoluble except those of alkali metals and ammonium.
- All hydroxides are insoluble except those of alkali metals (Ba^{2+} and Sr^{2+} hydroxides are slightly soluble).

Exercises

Label each of the following compounds as soluble (S) or insoluble (I) in water.

____	$CaCO_3$	____	Na_2CO_3	____	$(NH_4)_3PO_4$
____	$MgSO_4$	____	Hg_2Cl_2	____	$HgNO_3$
____	Na_2CO_3	____	NaCl	____	$Pb(CH_3COO)_2$
____	K_3PO_4	____	NaOH	____	CuCl
____	$CaSO_4$	____	$Mg(OH)_2$	____	$CuCl_2$

Solution: See Section A.1 on page 234.

Review Example 1.8: Naming Ionic Compounds

Compound names follow certain rules depending on the type of compound in question. The compounds in Table 1.3 (describing the density of material in the Earth's crust) are ionic compounds because they consist of metals and nonmetals, which, because of their very different electronegativities, fully ionize when forming stable compounds. Thus, they follow the ionic compound naming schema that can be summarized as:

1. Name the cation first and the anion last.
2. Name the cation as the element (unless it is a polyatomic cation, such as ammonium, NH_4^+).
3. Name the anion with its root name but replace the suffix with *-ide* (unless it is a polyatomic anion, such as sulfate, SO_4^{2-}) (Table 1.6).
4. If the cation is a transition metal or one of the heavier metals in the lower p-block of the periodic table (such as Pb or Sn), then add parentheses and Roman numerals to express the oxidation state of the metal, with exceptions for common, single oxidation state metals: Ag, Zn, Cd, Zr.
5. If the compound is a hydrate (with a dot H_2O after it), then follow the name of the compound with a) a Greek cardinal prefix (see Table 1.9) specifying the number of water molecules and b) the word "hydrate". Table 1.7 gives a few examples of how the ionic compound naming scheme works.

Formula	Name
Al_2O_3	Aluminum oxide
$CaSO_4 \cdot 4\,H_2O$	Calcium sulfate tetrahydrate
Fe_2O_3	Iron(III) oxide
NaCl	Sodium chloride

Table 1.7 Names of some ionic compounds.

Exercises

1. Complete the Table 1.8.

Formula	Name
CoO	
Co_2O_3	
	Magnesium nitrate
	Sodium sulfate hexahydrate

Table 1.8 Name the following ionic compounds.

Solution: See Section A.1 on page 234.

Formula	Name
CH_3COO^-	Acetate ion
NH_4^+	Ammonium ion
CO_3^{2-}	Carbonate ion
CrO_4^{2-}	Chromate ion
CN^-	Cyanide ion
$Cr_2O_7^{2-}$	Dichromate ion
OH^-	Hydroxide ion
NO_3^-	Nitrate ion
NO_2^-	Nitrite ion
$C_2O_4^{2-}$	Oxalate ion
MnO_4^-	Permanganate ion
PO_4^{3-}	Phosphate ion
SO_4^{2-}	Sulfate ion
S^{2-}	Sulfide ion
SO_3^{2-}	Sulfite ion

In order to do this, you need to know the formula and charge of the common polyatomic ions.

Table 1.6 You will be reading about many different ionic compounds throughout this textbook, and you need to be able to derive the formula from the name and vice versa.

Review Example 1.9: Naming Covalent Compounds

Unlike ionic compounds, molecular compounds are held together with covalent bonds and can be combined in a variety of ways. There is, for example, only one way to combine Mg^{2+} and OH^- such that the result is a neutral compound: $Mg(OH)_2$. Since there is only one combination, it would be redundant to name the compound magnesium dihydroxide because there is no other form (the compound magnesium trihydroxide does not exist). Thus, the name is simply magnesium hydroxide. Molecular compounds can have many more combinations because they are composed of atoms that have similar electronegativities, and therefore, they share the electrons in covalent bonds. For example, nitrogen and oxygen can be combined in several ways: NO, N_2O, NO_2, N_2O_4, and N_2O_5. Each of these compounds exists and therefore must have a unique name – nitrogen oxide is not sufficient to describe all of the variations. Here are the general rules to follow in naming molecular compounds.

1. The least electronegative element is usually written first.
2. The first element is given its elemental name.

Number	Prefix
1	Mono
2	Di
3	Tri
4	Tetra
5	Penta
6	Hexa
7	Hepta
8	Octa
9	Nona
10	Deca

Table 1.9 Greek cardinal prefixes.

3. The second element is given its root name with the suffix changed to *-ide*.
4. Prefixes are given to each element to specify the number of times they appear in the formula. Use the Greek cardinal prefixes listed in the Table 1.9. Note the following exceptions.
   (a) If the first element has a subscript of 1, do not use the mono prefix.
   (b) If a prefix results in a double vowel, drop the first vowel in the combination (*tetraoxide* becomes *tetroxide* and *monooxide* becomes *monoxide*).

Table 1.10 gives a few examples of how the molecular compound naming scheme works.

Formula	Name
NO	Nitrogen monoxide
N_2O	Dinitrogen monoxide
NO_2	Nitrogen dioxide
N_2O_4	Dinitrogen tetroxide
N_2O_5	Dinitrogen pentoxide
SF_6	Sulfur hexafluoride

Table 1.10 Common covalent compound names.

Note the difference in the names of NO_2 and NO_2^- from the polyatomic ion table. The charge is important as it changes the name and vastly changes the chemical and physical properties.

Exercises

1. Complete Table 1.11.

Formula	Name
SO_2	
	Sulfur trioxide
P_2O_4	
	Carbon tetrachloride

Table 1.11 Name the following covalent compound.

Solution: See Section A.1 on page 234.

Review Example 1.10: Photolysis

Chemical bonds can be broken in a process called photolysis. The electrons in the bonds can absorb sufficient energy to overcome the energy that holds the bonding electrons in their orbital. For example, the O–H single bond in water can be broken if a photon of 257 nm or shorter is absorbed. Here is the calculation:
O–H bond = 465 kJ/mol (from Table I.7 in Appendix I on page 325)

$$\text{total energy per photon} = (465\ \tfrac{\text{kJ}}{\text{mol}})\left(\frac{1\text{ mol}}{6.02214129\times10^{23}\text{ photons}}\right)\left(\frac{1000\text{ J}}{1\text{ kJ}}\right)$$
$$= 7.7215\times10^{-19}\text{ J per photon} \quad (1.13)$$

Now use the equation relating photon energy and wavelength.

$$E = \frac{hc}{\lambda} \quad (1.14)$$

Solve for λ

$$\lambda = \frac{hc}{E}$$

Substitute in the available numbers.

$$\lambda = \frac{(6.62606957 \times 10^{-34}\ \text{J} \cdot \text{s})(2.99792458 \times 10^{8}\ \text{m/s})}{7.7215 \times 10^{-19}\ \text{J/photon}}$$
$$= 2.5726 \times 10^{-7}\ \text{m} = 257\ \text{nm} \quad (1.15)$$

Since photon energy and wavelength are inversely proportional, photon wavelengths smaller than 257 nm will also have sufficient energy to break the O–H bond.

Exercises

1. Calculate the maximum photon wavelength that would break the N≡N triple bond.

Solution: See Section A.1 on page 234.

Review Example 1.11: Gas Solubility

The solubility of molecular oxygen in water is vital to the survival of aerobic aquatic life. As a result, it is one of the measures of water quality that you will learn about later in Chapter 5 (dissolved oxygen or DO). The solubility of other gases, such as carbon dioxide, will also play an important role in the environment. Thus, it is important for you to know some general rules about gas solubility.

General Rules

- As the temperature of the liquid solvent increases, the solubility of gases decrease. This is easy to remember if you think about cold soft drinks that are carbonated – cold drinks retain the fizz longer and warm drinks get flat quicker.
- As the pressure on the system decreases (pressure of the atmosphere, e.g., or within the container), the solubility of gases decreases. This is easy to remember if you think about mountain climbing – it is hard to breathe on the top of mountains because the partial pressure of oxygen is much lower than at sea level.

Exercises

1. How can you keep an open soft drink container carbonated? Would you put it in a refrigerator or put it on a counter at room temperature?
2. If the atmospheric pressure decreases with altitude, explain why a hiker might have trouble breathing (dissolved oxygen in the blood) on the top of a mountain.
3. Which of the following conditions would result in a high amount of dissolved oxygen – warm water or cold water?

Solution: See Section A.1 on page 235.

Review Example 1.12: Balancing Chemical Reactions

Chemical reactions, just like nuclear reactions, must be balanced. The principle used to balance reactions is often referred to as the Law of Conservation of Mass, which states that the total mass of the reactants must equal the total mass of the products. In other words, the total number of each element on the reactants' side must be the same as that of each element on the products' side. Our modern understanding of the relationship between mass and energy (via the famous $E = mc^2$ equation) tells us that conservation of mass is not strictly true. In nuclear reactions, mass is often lost or gained since large quantities of energy are produced or absorbed. In chemical reactions, the mass lost or gained is so small that conservation of mass can be considered as practically inviolable. To balance a chemical reaction, the best place to start is to look for an element that occurs only once in the reactants and once in the products. Once this element's coefficient is set, then working through the other elements usually becomes inevitable. For example, the combustion of octane (C_8H_{18}), the main component in gasoline, is achieved by the following process (combustion reactions *always* include molecular oxygen as a reactant and ideally carbon dioxide and water).

- Since carbon is in only one reactant and one product, it is a good place to start (H is also a good place to start, but O is not). Start by just assuming that the first coefficient is 1.

$$1C_8H_{18} + \_O_2 \rightarrow \_H_2O + \_CO_2 \qquad (R1.33)$$

- Eight carbons on the left must result in eight carbons in the products.

$$1C_8H_{18} + \_O_2 \rightarrow \_H_2O + 8CO_2 \qquad (R1.34)$$

- Now look at the H. Eighteen H's on the left means 18 on the right, but water has H_2, so a coefficient of nine balances it.

$$1C_8H_{18} + \_O_2 \rightarrow 9H_2O + 8CO_2 \qquad (R1.35)$$

- Since the products are done, add up all of the O's. Twenty five O's on the right means 25 on the left, but since the reactant is O_2, a fraction is required. The math should work out: $25 \div 2 \times 2 = 25$

$$1C_8H_{18} + 25/2O_2 \rightarrow 9H_2O + 8CO_2 \qquad (R1.36)$$

- To get rid of the fractions, just multiply all coefficients by 2.

$$2C_8H_{18} + 25O_2 \rightarrow 18H_2O + 16CO_2 \qquad (R1.37)$$

Exercises

1. Balance the following reactions.
   (a) $\_C_4H_{10} + \_O_2 \rightarrow \_H_2O + \_CO_2$
   (b) $\_Na_2CO_3 + \_HCl \rightarrow \_CO_2 + \_NaCl + \_H_2O$
   (c) $\_Al + \_HCl \rightarrow \_AlCl_3 + \_H_2$

Solution: See Section A.1 on page 235.

Review Example 1.13: Redox Reactions

The chemical analysis that allowed geologists to determine that the Cr-53 in certain BIFs was a by-product of terrestrial oxidation and not microbial or hydrothermal vent contributions is the result of a special type of chemical reaction called a redox reaction. Redox reactions derive their name because the reaction is a result of an oxidation and a reduction of certain reactants. This type of reaction is very important to all sorts of scientists because redox reactions are usually associated with large energy transfers and are responsible for biochemical metabolism as well as reactions within batteries. These are tremendously important reactions and deserve a closer examination.

Reduction and oxidation involve a transfer of electrons from one reactant to another. Reduction is the gain of electrons, and oxidation is the loss of electrons. It is easy to remember this if you can remember one of the two mnemonics given as follows.

Loss of	Oxidation
Electrons is	Is
Oxidation	Loss of electrons
goes	
Gain of	Reduction
Electrons is	Is
Reduction	Gain of electrons

With either LEO GER or OIL RIG, you can easily remember how electrons are transferred for each type of reaction. Be careful since reduction intuitively means loss, but in this case, reduction is a gain of electrons and a reduction in positive charge.

Recognizing oxidation and reduction can also be a challenge. It is easy in reactions such as the following.

$$Cu^{2+}(aq) + Zn(s) \rightarrow Cu(s) + Zn^{2+}(aq) \quad \text{(R1.38)}$$

Clearly, copper goes from an oxidation state of +2 to an oxidation state of 0, while zinc goes from an oxidation state of 0 to one of +2. Thus, copper(II) has been reduced and zinc has been oxidized. Another way to describe the reactants is that copper(II) plays the role of an oxidizing agent since it caused zinc to be oxidized. Zinc plays the role of a reducing agent since it caused copper(II) to be reduced. When compounds are participants in a reaction instead of elements, the identification of oxidation and reduction is more difficult. Examine the following reaction.

$$SiCl_4(l) + 2Mg(s) \rightarrow Si(s) + 2MgCl_2(s) \quad \text{(R1.39)}$$

It is easy to identify the oxidation state of pure, neutral elements – they are always zero. So, both the oxidation states of the reactant Mg and the product Si are zero. The oxidation state of the elements in $SiCl_4$ require some assumptions, as the most electronegative element will get its most stable oxidation state if possible. Chlorine is the most electronegative element, and since it is a halogen, its most stable oxidation state is −1 (it gains one electron to obtain a noble gas configuration). Since there are four Cl's in the compound, there is a −4 charge contribution from Cl's. Since the molecule is neutral overall, each Si must have an overall charge of +4, so the oxidation state of Si in $SiCl_4$ is +4. This means that Mg and Cl in $MgCl_2$ have oxidation states of +2 and −1, respectively. Now we can see that Si in $SiCl_4$ was reduced and Mg was oxidized. This means that $SiCl_4$ was the oxidizing agent and Mg was the reducing agent.

With inorganic compounds, redox reactions are recognized by the loss or gain of electrons. This can be done as well in organic reactions, but oxidation in organic reactions most often results in an increase in the number of oxygen atoms on a compound. Reduction results in an increase in the number of hydrogens on a compound. This is often easier to identify than looking for a change in the oxidation state of each element. Here is an example.

$$CH_3CHO + Ag_2O \rightarrow CH_3COOH + 2Ag \quad \text{(R1.40)}$$

Notice that the second carbon in the CH_3CHO loses an H and gains and O in the reaction, which is a sign of oxidation. If oxidation numbers were to be assigned, then CH_3CHO would have oxidation states of C:−3, H:+1, C:+1, H:+1, and O:−2; and in CH_2COOH, the oxidation numbers would be C:−3, H:+1, C:+3,O:−2, O:−2, and H:+1.

Here is a summary of the rules for assigning oxidation states.

Rules

- The oxidation state of all pure, neutral elements is zero.
- Hydrogen can only have oxidation states of −1, 0, +1.
- Assign the first oxidation state to the most electronegative element, giving it its most stable charge according to the octet rules you learned; if this rule causes the violation of the previous rule concerning hydrogen, then give hydrogen a +1 oxidation state and reassign the most electronegative element the lowest possible charge.
- The sum of the oxidation states of all of the elements in a species must equal its overall charge.

Exercises

1. Determine the oxidation state of S in each of the following species.

   (a) S = _____ in $S_2O_3^{2-}$
   (b) S = _____ in SO_3
   (c) S = _____ in H_2S
   (d) S = _____ in S_8

2. Which of the following reactions is a redox reaction?
   (a) $3Na_2SO_3 + 2H_3PO_4 \rightarrow 3H_2SO_3 + 2Na_3PO_4$
   (b) $C_3H_8 + 5O_2 \rightarrow 4H_2O + 3CO_2$
   (c) $Na_2CO_3 + 2HCl \rightarrow CO_2 + 2NaCl + H_2O$
   (d) $2Al + 6HCl \rightarrow 2AlCl_3 + 3H_2$

Solution: See Section A.1 on page 235.

Review Example 1.14: Thermodynamics

The second law of thermodynamics governs the energetics of all processes. It states that any process that increases the entropy of the universe will be spontaneous. This means that in order for a reaction to proceed, to be spontaneous, the sum of the entropy from the system and surroundings must be positive. Gibbs free energy is a more convenient statement of this law.

$$\Delta G_{sys} = \Delta H_{sys} - T\Delta S_{sys} \tag{1.16}$$
$$\Delta S_{universe} > 0 \text{ (for any spontaneous process)}$$

This new measure, ΔG_{sys} , replaces $\Delta S_{universe}$ as an easier way to determine if something is spontaneous since only the system needs to be measured and not the surroundings. Thus, a spontaneous reaction can be determined from knowing about the enthalpy and entropy of the system, and nothing else. Note the change in the sign – for a spontaneous reaction, the $\Delta G_{sys} < 0$ while $\Delta S_{universe} > 0$.

The Gibbs free energy of a chemical reaction can be easily determined at room temperature (25 °C) by using the tabular values of ΔG_f° for each of the chemicals in a reaction.

$$N_2(g) + 3H_2(g) \rightarrow 2NH_3(g) \tag{R1.41}$$

$$\begin{aligned}\Delta G^\circ_{rxn} &= \Sigma\Delta G^\circ_{f\ \text{products}} - \Sigma\Delta G^\circ_{f\ \text{reactants}}\\ &= ((2\ \text{mol})(-16.4\ \text{kJ/mol}))\\ &- ((1\ \text{mol})(0\ \text{kJ/mol}) + (3\ \text{mol})(0\ \text{kJ/mol}))\\ &= -32.8\ \text{kJ}\end{aligned}$$

This reaction is therefore spontaneous at room temperature. The free energy values for some common chemicals can be found in Table I.2 on page 321, and also note that the state of the chemical is important. If water had been a liquid in Reaction R1.41, then the value of ΔG_f° would have been different.

Free energy values for reactions are very temperature dependent, so this method of calculating the free energy of a reaction is only valid at 25 °C. At any other temperature, the free energy must be calculated by using the final equation in Eq. (1.16) where the entropy and enthalpy must be determined. Thus, if the free energy of the aforementioned reaction were measured at 0.0 °C, then the calculation would require the determination of ΔH°_{rxn}, ΔS°_{rxn}, and then ΔG°_{rxn}.

$$\begin{aligned}\Delta H^\circ_{rxn} &= \Sigma\Delta H^\circ_{f\ \text{products}} - \Sigma\Delta H^\circ_{f\ \text{reactants}}\\ &= ((2\ \text{mol})(-45.9\ \text{kJ/mol}))\\ &- ((1\ \text{mol})(0\ \text{kJ/mol}) + (3\ \text{mol})(0\ \text{kJ/mol}))\\ &= -91.8\ \text{kJ}\end{aligned}$$

$$\begin{aligned}\Delta S^\circ_{rxn} &= \Sigma\Delta S^\circ_{f\ \text{products}} - \Sigma\Delta S^\circ_{f\ \text{reactants}}\\ &= ((2\ \text{mol})(192.8\ \text{J/(K}\cdot\text{mol)}))\\ &- ((1\ \text{mol})(191.6\ \text{J/(K}\cdot\text{mol)}) + (3\ \text{mol})(130.7\ \text{J/(K}\cdot\text{mol)}))\\ &= -198.1\ \text{J/K}\end{aligned}$$

$$\Delta G_{sys} = \Delta H_{sys} - T\Delta S_{sys}$$
$$= -91.8\text{ kJ} - (273.15 + 0.0\text{ K})\left(-198.1\text{ J/K} \times \tfrac{1\text{ kJ}}{1000\text{ J}}\right)$$
$$= -37.7\text{ kJ}$$

As you can see, the value for the free energy at 0.0 °C is different than at 25.0 °C. While ΔG is temperature-dependent, ΔH and ΔS are temperature-independent (for most reasonable temperatures).

In summary, think about what the second law of thermodynamics and ΔG imply – by performing a few simple calculations you can determine whether or not a reaction will occur given a set of environmental conditions and, more importantly, what the long-term fate of a reaction or chemicals might be. For example, examine the chemicals in Table 1.12. If you wanted to determine the most stable form of nitrogen, you would look for the species with the largest negative ΔG_f° value. The nitrate ion is the most stable form of nitrogen when oxygen is available (as it is in our atmosphere).

Species	ΔG_f° (kJ/mol)
NH_3(g)	−16.4
NO_2^-(aq)	−32.2
NO_3^-(aq)	−108.7

Table 1.12 Free energy of some nitrogen compounds.

Exercises

1. Calculate the free energy change for the following reactions (assume 25 °C). For each reaction, also state whether it is spontaneous or nonspontaneous.
   (a) $N_2(g) + O_2(g) \rightarrow 2NO(g)$
   (b) $2C_4H_{10}(g) + 13O_2(g) \rightarrow 8CO_2(g) + 10H_2O(g)$
   (c) $2HCl(aq) + Fe(s) \rightarrow H_2(g) + Fe^{2+}(aq) + 2Cl^-(aq)$
2. Calculate the free energy change for the following reaction at 415 °C.

$$N_2(g) + 3H_2(g) \rightarrow 2NH_3(g) \qquad \text{(R1.42)}$$

3. Determine the temperature at which the following reaction becomes spontaneous. Hint: Determine the ΔH and ΔS, then assume ΔG is zero (the point at which a reaction starts to become spontaneous, and then solve for T).

$$N_2(g) + 3H_2(g) \rightarrow 2NH_3(g) \qquad \text{(R1.43)}$$

4. Which is the most stable form of sulfur in the hydrosphere? SO_3^{2-} or SO_4^{2-}

Solution: See Section A.1 on page 235.

Review Example 1.15: Molar Mass

In the laboratory, chemicals are measured out using a mass balance. While this results in a measure of the mass of the compound, it does not directly provide information about the number of molecules in the sample. This is inconvenient because chemical reactions are balanced by assuming that elements and compounds react on the basis of individual molecules interacting, not their masses interacting. Conveniently, however, the periodic table provides a conversion from mass to the number of molecules. The atomic mass listed in each element box represents the mass of the element per mole of atoms. The mole is similar to a term such as dozen or gross, representing a countable number of things. Of course, a dozen is 12 things and a gross is 144 things, but a mole is 6.022×10^{23} things.

If you examine the periodic table, you will see that 1 mole of carbon atoms has a mass of 12.011 g. One mole of carbon dioxide (CO_2) has a mass of

$$12.011\text{ g/mol} + 2 \times 15.9994\text{ g/mol} = 44.0098\text{ g/mol}$$

Armed with a periodic table and the chemical formula of any compound, you should be able to determine the *molar mass* or *formula weight* of the compound.

☞ The *molar mass* of a compound is the mass of a mole of molecules. The masses on the periodic table are equivalent to g/mol or daltons/atom. For example, a mole of carbon atoms has a mass of 12.0107 g and one atom of carbon has a mass of 12.0107 Da.

Exercises

1. Determine the molar mass of the following compounds.
   (a) Co_2O_3
   (b) $CaSO_4 \cdot 4\,H_2O$
   (c) sodium dichromate
   (d) C_8H_{18}
   (e) How many moles are in 2.19 g of C_8H_{18}?

Solution: See Section A.1 on page 237.

Review Example 1.16: Stoichiometry

Stoichiometry is an incredibly important concept for chemistry, and it is analogous to cooking. If a cook wants to recreate a favorite dish, he or she would need to add exactly the right proportions of all of the ingredients to get the same result. Imagine making cookies but adding twice as much flour as the recipe requires! If an industrial chemist wants to make hardened steel, he or she needs to add precisely the correct ratio of ingredients or the steel will not perform up to the exacting standards that are required to build a bridge. If a pharmaceutical chemist wants to make a drug and one of the ingredients is particularly expensive, he or she will want to know the minimum amount to add to make the number of pills that are required. Adding more will mean higher costs and more waste. That is stoichiometry.

Here are some rules and terms that you need to know.

- All reactions must be balanced.
- The coefficients in the reaction represent molecules or moles, not mass.
- One of the reactants is likely to run out first leaving all other reactants in excess and stopping the production of the products. This reactant is called the limiting reactant.

Here is an example.

$$2C_2H_4 + 2HCl + O_2 \rightarrow 2C_2H_3Cl + 2H_2O \tag{R1.44}$$

Assuming that you start this reaction with 25.0 g C_2H_4, 60.0 g HCl, and 20.0 g O_2, answer the following questions.

1. Which reactant is the limiting reactant? Start by finding out how much product can be obtained from each reactant.

$$25.0 \text{ g } C_2H_4 \times \frac{1 \text{ mol } C_2H_4}{28.053 \text{ g } C_2H_4} \times \frac{2 \text{ mol } C_2H_3Cl}{2 \text{ mol } C_2H_4} = 0.89117 \text{ mol } C_2H_3Cl$$

$$60.0 \text{ g HCl} \times \frac{1 \text{ mol HCl}}{36.461 \text{ g HCl}} \times \frac{2 \text{ mol } C_2H_3Cl}{2 \text{ mol HCl}} = 1.6456 \text{ mol } C_2H_3Cl$$

$$20.0 \text{ g } O_2 \times \frac{1 \text{ mol } O_2}{31.9988 \text{ g } O_2} \times \frac{2 \text{ mol } C_2H_3Cl}{1 \text{ mol } O_2} = 1.2500 \text{ mol } C_2H_3Cl$$

The limiting reactant (LR) always produces the least amount of product, so the LR is C_2H_4. Notice that the LR is not necessarily the reactant with the least starting mass. Now that we know the LR, the other two calculations are useless. From now on, all calculations will start with the limiting reactant.

2. How many grams of vinyl chloride (C_2H_3Cl) can be produced assuming a 100% efficient reaction? Since the aforementioned calculation shows that 0.89117 mol of C_2H_3Cl can be obtained, we just need to convert this mole amount to a mass.

$$0.89115 \text{ mol } C_2H_3Cl \times \frac{62.4979 \text{ g } C_2H_3Cl}{1 \text{ mol } C_2H_3Cl} = 55.6950 = 55.7 \text{ g } C_2H_3Cl$$

Exercises

1. Balance the combustion reaction for octane (C_8H_{18}).
   (a) If the reaction is started with 10 mol of octane and 10 mol of oxygen, which chemical would be the LR?
   (b) If the reaction is started with 10.0 g of octane and 10.0 g of oxygen, which chemical would be the LR?
   (c) What mass of carbon dioxide would be produced from 10.0 g of octane and 10.0 g of oxygen with 100% efficiency?

Solution: See Section A.1 on page 237.

Review Example 1.17: Rate Laws

How do we know what the activation energy really is for the combustion of methane and what the actual mechanism is? Determining mechanisms for chemical reactions is part of the study of reaction kinetics. Many mechanisms for a particular reaction can be proposed based on logical steps, but ultimately, the mechanism that fits with experimental data is the correct mechanism.

For example, consider the proposed mechanism for the following overall reaction.

$$NO_2 + CO \rightarrow NO + CO_2 \qquad (R1.45)$$

Based on this overall reaction, the general Rate Law would be

$$Rate = k[NO_2]^x[CO]^y \qquad (1.17)$$

Mechanism I

$$\text{step1}: NO_2 \rightarrow NO + O \text{ (slow)} \qquad (R1.46)$$

$$\text{step2}: O + CO \rightarrow CO_2 \text{ (fast)} \qquad (R1.47)$$

$$Rate = k[NO_2]^1[CO]^0$$

The Rate Law of a mechanism is determined by the slow step. The coefficients on the reactants in the slow step become exponents in the specific Rate Law. In the first (slow) step, NO_2 had a coefficient of 1, so it has an exponent of 1 in the Rate Law, and since CO was not present in the slow step, its exponent in the Rate Law is 0. Also note that the steps in the mechanism must add together to yield the overall reaction. All of the *reaction intermediates* (O, for example) must cancel out. Mechanisms cannot start with anything but one or more reactants (except in the case of a catalyst). For example, the first step must start with either NO_2 or CO or both but nothing else since there are only reactants present at the beginning of a reaction.

☞ *Reaction intermediates* are species that are generated in early steps and then consumed in later steps and are not present in the overall reaction.

Mechanism II

$$\text{step 1}: NO_2 + CO \rightarrow NO + CO_2 \text{ (slow)} \qquad (R1.48)$$

$$Rate = k[NO_2]^1[CO]^1$$

Mechanism III

$$\text{step 1}: 2NO_2 \rightarrow NO + NO_3 \text{ (slow)} \qquad (R1.49)$$

$$\text{step 2}: NO_3 + CO \rightarrow NO_2 + CO_2 \text{ (fast)} \qquad (R1.50)$$

$$Rate = k[NO_2]^2[CO]^0$$

Clearly, the three mechanisms are quite different and result in different specific Rate Laws – all of which are a subset of the general Rate Law. Since all mechanisms are reasonable and logical, the only way to determine which mechanism is correct is to measure the Rate Law experimentally. If it turns out that the experimental Rate Law supports a second-order reaction with respect to NO_2 and no rate dependence on CO, then Mechanism III is the correct mechanism.

Exercises

1. Write the Rate Law for the following steps.
   (a) (an overall reaction) $H_2 + Cl_2 \rightarrow 2HCl$
   (b) (a slow step in a mechanism) $SO_2 + O_2 \rightarrow SO_3 + O$
   (c) (a slow step in a mechanism) $2NO \rightarrow N + NO_2$
2. Propose two mechanisms for the production of NO from molecular oxygen and nitrogen.

$$N_2(g) + O_2(g) \rightleftharpoons 2NO(g) \qquad \text{(R1.51)}$$

Write out the steps of each mechanism, and derive the correct Rate Law for each mechanism.

Solution: See Section A.1 on page 238.

Review Example 1.18: Lewis Structures

How do we know that carbon dioxide has two C=O double bonds and molecular nitrogen has a single N≡N triple bond? The answer is twofold: spectroscopic evidence shows that the bonds in each molecule are broken or excited by photons that correspond to the electronic and vibrational energies of the double and triple bonds. Single C–O bonds and single N–N bonds, in other compounds, show much weaker bonds. This is the empirical evidence.

The theoretical construct that allows chemists to predict the types of bonds a particular molecule might have comes from following Lewis Dot Structure rules. These rules establish an electron accounting system that, in most cases, correctly predicts bond orders that are observed in simple molecules, such as carbon dioxide. Here are the rules:

☞ The *octet rule* refers to the stability gained when elements fill their *s* and *p* orbitals, which require eight electrons. A similar stability is gained by hydrogen when it fills its 1*s* orbital with two electrons. Elements in the *p*-block beyond the third row of the periodic table still follow the octet rule despite having filled *d* electrons.

Lewis Structure Rules

1. Treat ions separately (NH_4NO_3 becomes $NH_4^+ + NO_3^-$).
2. Count the valence e^-; only *p* & *s* orbitals (represented as "h" for electrons the structure "has").
3. Determine # of e^- needed (represented as "n" for "need": n = # atoms × 8 or × 2 for H).
4. Determine # of shared e^- (represented by "s" for "shared": s = n − h).
5. Determine the minimum # of bonds (represented by "b" for "bonds": b = s ÷ 2).
6. Put the first atom at the center (except when H) and the other atoms on the exterior and add bonds as calculated.
7. Add the remaining bonds to the structure.
8. Add the rest of the e^- as pairs to all atoms ($Z \geq 5$) to achieve an octet (exceptions can be Be, B, N, Al).
9. Calculate and minimize formal charges (FCs) by expanding the octet of the central atom, if possible (hypervalency).
10. Identify equivalent or near-equivalent Lewis structures (resonance structures).

Remember that the number of valence electrons an atom has only involves the electrons in the last row of the element's position in the periodic table. For example, fluorine has a total of nine electrons, but it only has seven valence electrons.

Also, remember that elements in the p-orbital block of the periodic table gain stability by adopting a full set of electrons, usually referred to as the *octet rule*. Oxygen, phosphorus, and other elements in the p-block are stable in compounds when they possess eight valence electrons – thus the term "octet". Exceptions to this include H and He, which gain stability by possessing two valence electrons (a duet), and elements in the third row and below of the p-block (such as sulfur and bromine), which can violate the octet rule if necessary and carry up to 18 electrons.

Here is an example that demonstrates this technique up to step 6.

The Lewis Structure of O_2

- The number of valence electrons the structure has (h): h = 6 + 6 = 12
- The number of valance electrons the structure needs (n): n = 8 + 8 = 16
- The number of electrons that must be shared (s): s = n − h = 16 − 12 = 4
- The minimum number of bonds in the structure (b): b = s ÷ 2 = 4 ÷ 2 = 2

Thus, the structure is

$$\ddot{\underset{..}{O}}=\ddot{\underset{..}{O}}$$

Exercises

1. Determine the Lewis Dot Structure for the following species. Make sure you specify the h, n, s, b numbers for each structure.
   (a) F_2
   (b) CO_2
   (c) H_2O

Solution: See Section A.1 on page 238.

Review Example 1.19: Electronegativity

Many of the compounds that are important for environmental issues are constructed with covalent bonds. The degree to which these bonds polarize is reflected in an element's electronegativity, which is defined as the ability of an atom to attract electron density toward itself while participating in a covalent bond. Polar bonds often lead to polar molecules, and a molecule's polarity is directly related to its solubility in water or other solvents. Determining bond polarity is a simple subtraction of the electronegativities of the two elements involved in the bond. A couple of examples will help to explain this definition. Refer to the electronegativity table of the p-block elements that is on the first page of this book.

In the diatomic compounds H_2 and Cl_2, the bond holding the two atoms together is covalent – the two electrons are shared. Are they shared evenly and equally? In these two cases, the answer is yes, because both H atoms have the same electronegativity and each "pull" on the covalent electrons with the same "force." The same is true in Cl_2. In HCl, however, while the bond holding H and Cl together is covalent, Cl has an electronegativity of 3.16 and H has an electronegativity of 2.20. This means that the attractive force on the covalently bonded electrons is about 50% stronger for Cl than for H; thus, the electron density is higher near the Cl end of the bond and lower near the H end of the bond. This polarity has a large effect on the physical and chemical properties of HCl. It is polar, with the Cl end being slightly negative compared to the more positive H end of the molecule.

Exercises

1. Identify the element in each pair that is more electronegative.
   (a) C or O
   (b) H or C
   (c) Br or Be
2. Which is the most electronegative element?
3. Identify the molecular compounds that have polar covalent bonds.
   (a) HF
   (b) Br_2
   (c) CS
   (d) CO
   (e) N_2

(f) O_2

(g) NO

(h) F_2

Solution: See Section A.1 on page 239.

Review Example 1.20: Resonance, Hypervalency, and Formal Charges

A slight complication in Lewis Structures develops when considering the issue of resonance and hypervalency.

Resonance occurs when there are one or more structures, such as in ozone (O_3). The Lewis Structure rules for ozone produce a number of bonds (b) of three for three atoms. Since there are two ways to distribute three bonds between two atoms, there must be two different structures possible. The actual structure of ozone is an average of the two resonance structures, resulting in each ozone bond having a bond order of 1.5.

Hypervalency occurs for elements in row three and below on the p-block of the periodic table. These elements have a set of empty d-orbitals that can be filled with electrons beyond the octet. An example of this can be seen in the structure of the phosphate ion as follows.

Finally, drawing the Lewis Structures of some molecules results in several structures that seem to be equivalent. In these instances, the use of formal charges (FC) helps to determine which structure is the most stable. A *FC* is the effective charge an atom has in a Lewis Structure given the number of electrons it shares and the number it has in lone pairs. Stable structures tend to have electron density spread out evenly and have the highest electron density on the most electronegative atoms. Operationally, these rules can be stated in the following way:

1. The *FC* of an atom is calculated in the following manner:

$$FC = (\#\ \text{valence}\ e^-\text{'s}) - (\#\ \text{non-bonding}\ e^-\text{'s}) - (\#\ \text{bonds})$$

2. The sum of the *FC*'s of all of the atoms equals the charge on the molecule or species.
3. The most stable structure has
   (a) the smallest ΔFC (largest *FC* minus the smallest *FC*)
   (b) the lowest *FC* on the most electronegative elements, when possible

Here are some examples.

The Lewis Structure of NO_3^-

- The number of valence electrons the structure has: $h = 5 + 6 \times 3 + 1 = 24$
- The number of valance electrons the structure needs: $n = 8 + 8 \times 3 = 32$
- The number of electrons that must be shared: $s = 32 - 24 = 8$
- The minimum number of bonds in the structure: $b = 8 \div 2 = 4$

Thus, the initial structure is given as follows. Since none of the atoms exhibits hypervalency, this initial structure is also the final structure. There are three places to put the double bond, so there must be three equivalent resonance structures.

Formal Charges

- $FC_N = 5 - 0 - 4 = +1$
- $FC_{O_{\text{single bond}}} = 6 - 6 - 1 = -1$
- $FC_{O_{\text{double bond}}} = 6 - 4 - 2 = 0$
- $\Delta FC = (+1) - (-1) = 2$

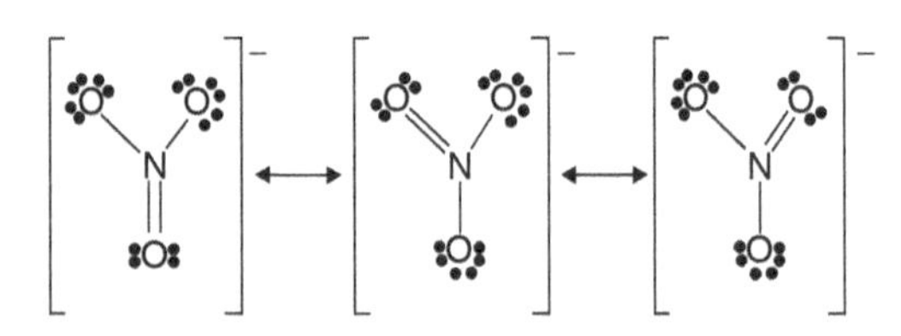

The Lewis Structure of PO_4^{3-}

- The number of valence electrons the structure has: $h = 5 + 6 \times 4 + 3 = 32$

- The number of valence electrons the structure needs: n = 8 + 8 × 4 = 40
- The number of electrons that must be shared: s = 40 − 32 = 8
- The minimum number of bonds in the structure: b = 8 ÷ 2 = 4

Thus, the initial structure is

Formal Charges

- $FC_P = 5 - 0 - 4 = +1$
- $FC_{O_{\text{single bond}}} = 6 - 6 - 1 = -1$
- $\Delta FC = (+1) - (-1) = 2$

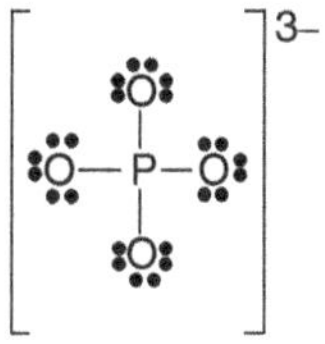

Since P can exhibit hypervalency, we can increase the number of bonds to the central atom in order to spread out the highly negative charge on the oxygen atoms. So, let us add a double bond to see what happens to the *FCs*.

Formal Charges

- $FC_P = 5 - 0 - 5 = 0$
- $FC_{O_{\text{single bond}}} = 6 - 6 - 1 = -1$
- $FC_{O_{\text{double bond}}} = 6 - 4 - 2 = 0$
- $\Delta FC = (0) - (-1) = 1$

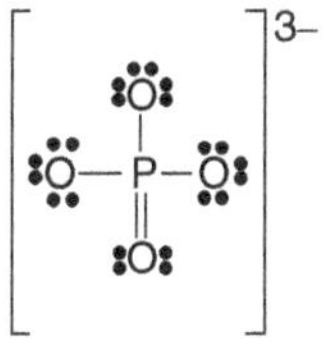

The ΔFC has decreased by 1 (the best ΔFC is 0, so it is getting better). Let us see if adding another double bond is better.

Formal Charges

- $FC_P = 5 - 0 - 6 = -1$
- $FC_{O_{\text{single bond}}} = 6 - 6 - 1 = -1$
- $FC_{O_{\text{double bond}}} = 6 - 4 - 2 = 0$
- $\Delta FC = (0) - (-1) = 1$

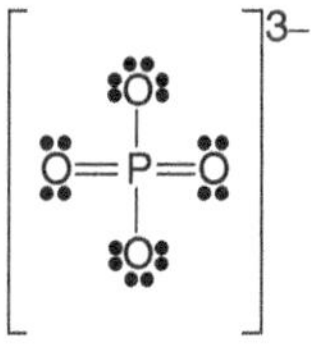

The ΔFC did not change, so nothing was gained. Adding a third double bond would make things even worse. In fact, the structure is less stable because it violates electronegativity rules – the least EN element (P) has a more negative charge compared to the most EN element (O). This is not the most stable structure; the previous structure with one double bond is the most stable (note that there would be a total of four resonance structures here).

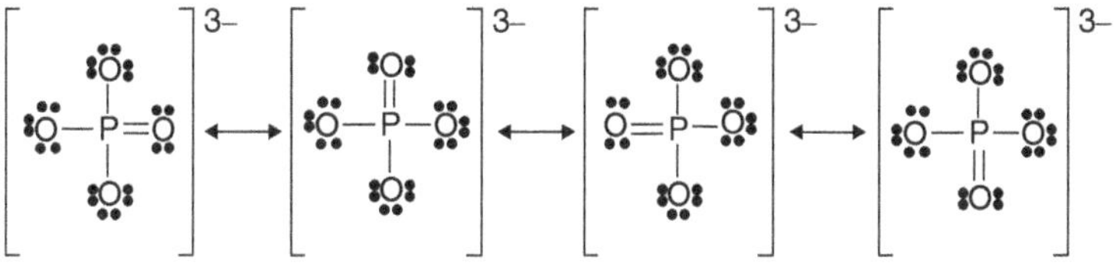

Exercises

1. Draw the Lewis Dot Structure for the following species. Make sure you specify the h, n, s, b numbers for each structure. Calculate the *FCs* for the elements, and find the most stable structure when considering *FCs* and hypervalency.
   (a) SO_4^{2-}
   (b) CNO^- (atoms in that order: C–N–O)
   (c) CO_3^{2-}

Solution: See Section A.1 on page 239.

Review Example 1.21: Acids and Bases

The chemistry of acids and bases is vitally important to the environment and to physiology. The proteins in any organism fold in very specific ways when they are produced. Small changes in the pH of the solution in which the proteins are folding can cause misshapen proteins. Proteins without the correct shape lose their functionality, which can be detrimental to the organism. In this text, the discussion of the role that acidic rain had in the chemical weathering of terrestrial chromium is yet another example of the central role that acid/base chemistry plays.

Water is of primary importance to environmental chemistry since much of the globe is covered by water and most organisms are composed predominantly of water. Among the interesting properties of water is its ability to autohydrolyze.

$$2H_2O(l) \rightleftharpoons H_3O^+(aq) + OH^-(aq) \tag{R1.52}$$

In the aforementioned reaction, one water molecule is donating an H^+ to the other water molecule, making the first water molecule an Arrhenius acid and the second water molecule is an Arrhenius base. Under these definitions, anything that produces H_3O^+ (the hydronium ion) in solution is an acid and anything that produces OH^- (the hydroxide ion) in solution is a base. At 25 °C the equilibrium constant for Reaction 1.18 has a value of 1.0×10^{-14} and is given a special designation as K_w. Since the expression for $K_w = [H_3O^+][OH^-]$, in a pure solution of water, the $[H_3O^+]$ and $[OH^-]$ must be equal since the formation of each hydronium ion leads to the formation of a hydroxide ion. Thus, taking the square root of the K_w value will provide the equilibrium concentration of the hydronium and hydroxide ions.

$$\begin{aligned} K_w &= [H_3O^+][OH^-] = 1.0 \times 10^{-14} \\ &= (x)(x) = x^2 = 1.0 \times 10^{-14} \\ x &= 1.0 \times 10^{-7} \end{aligned} \tag{1.18}$$

This value of x, representing the molar concentration of the hydronium and hydroxide ions at equilibrium, establishes a quantitative scale for expressing the amount of acid or base that is in solution. The pH scale, loosely defined as the power of hydrogen, is calculated in the following manner.

$$pH = -\log([H_3O^+]) \tag{1.19}$$

At equilibrium, in a pure water solution, the pH is 7.0 from $-\log(1.0\times10^{-7})$. This is considered a neutral solution, where the hydronium and hydroxide ions have the same concentration. An acidic solution is one where pH < 7, and a basic (or alkaline) solution is one where the pH > 7. It is less common to calculate the pOH of a solution, but it is still a useful measure of the state of a solution.

$$pOH = -\log([OH^-]) \tag{1.20}$$

The pH and pOH of a solution are directly related to each other. If $-\log$ is applied to both sides of Eq. (1.18), then the relationship becomes apparent.

$$\begin{aligned} -\log(K_w) &= -\log([H_3O^+][OH^-]) \\ -\log(1.0 \times 10^{-14}) &= -\log([H_3O^+]) - \log([OH^-]) \\ 14 &= pH + pOH \end{aligned}$$

Chemists categorize acids and bases with three sets of definitions. The Arrhenius definition is most appropriate in environmental systems since they are mostly aqueous in nature. Under this definition, an acid is any compound that donates an H^+ to water to produce H_3O^+. Bases are those compounds that donate a hydroxide ion. Below is an example of an Arrhenius acid and base.

$$HCl(aq) + H_2O(l) \rightleftharpoons H_3O^+(aq) + Cl^-(aq) \tag{R1.53}$$

$$NaOH(aq) \rightleftharpoons Na^+(aq) + OH^-(aq) \tag{R1.54}$$

Under the Brønsted–Lowry definitions, acids are produced the same way, but bases are produced by species that accept a proton from water, which results in the production of the hydroxide ion.

$$CN^-(aq) + H_2O(l) \rightleftharpoons HCN(aq) + OH^-(aq) \quad (R1.55)$$

Since the hydronium and hydroxide ions are the two principal actors in aqueous acid/base chemistry, the Arrhenius and Brønsted–Lowry definitions will be the most important. All Arrhenius acids and bases are a specific case of Brønsted–Lowry acids and bases in aqueous solutions. When dealing with organic solvents, the Brønsted–Lowry definition becomes more important.

Not all acids and bases have the same effect when dissolved in water. Some acids, such as HCl, completely ionize when dissolved in water and produce the maximum number of hydronium ions. Other acids, such as acetic acid (CH_3COOH), ionize to a small extent and produce very little hydronium ion in comparison. Tables 1.13 and 1.14 list a few examples of acids and bases along with their equilibrium constants in the case of the weak acids and bases. The magnitude of the equilibrium constant is a measure of the acid or base strength. A more complete table can be found in Appendix I. With so many acids and bases available, it becomes difficult to memorize which are strong and which are weak; thus, the best way to manage this skill is to know the common strong acids and bases and to assume that all others are weak. In Table 1.14, the equilibrium constants of strong acids and bases are listed because they are effectively infinite, which means that the equilibrium arrows used in Reaction R1.53 really should be a right arrow (→) because the reaction goes to 100% completion, whereas the equilibrium arrow in Reaction R1.55 is appropriate because the reaction favors the reactants but does produce some hydroxide ions. After this brief review, you should be able to list the strong acids and bases and write the equilibrium acid/base reaction for any acid or base. They all follow an easy schema.

Common strong acids	Common strong bases
H_2SO_4	NaOH
HNO_3	KOH
HCl	LiOH
HBr	
HI	
$HClO_4$	

Table 1.13 Strong acids and bases commonly used in the laboratory.

$$\text{Acids: Acid(aq)} + H_2O(l) \rightleftharpoons H_3O^+(aq) + \text{Conjugate Base(aq)}$$
$$\text{Bases: Base(aq)} + H_2O(l) \rightleftharpoons OH^-(aq) + \text{Conjugate Acid(aq)}$$

Acid	Formula	Conjugate base	Short-hand reaction	K_a
Acetic	CH_3COOH	CH_3COO^-	$CH_3COOH(aq) \rightleftharpoons H^+(aq) + CH_3COO^-(aq)$	1.75×10^{-5}
Ammonium ion	NH_4^+	NH_3	$NH_4^+(aq) \rightleftharpoons H^+(aq) + NH_3(aq)$	5.6×10^{-10}
Carbonic	H_2CO_3	HCO_3^-	$H_2CO_3(aq) \rightleftharpoons H^+(aq) + HCO_3^-(aq)$	4.5×10^{-7}
Hypochlorous	HClO	ClO^-	$HClO(aq) \rightleftharpoons H^+(aq) + ClO^-(aq)$	4.0×10^{-8}
Hydrocyanic	HCN	CN^-	$HCN(aq) \rightleftharpoons H^+(aq) + CN^-(aq)$	6.2×10^{-10}
Hydrofluoric	HF	F^-	$HF(aq) \rightleftharpoons H^+(aq) + F^-(aq)$	6.3×10^{-4}
Hydrogen carbonate ion	HCO_3^-	CO_3^{2-}	$HCO_3^-(aq) \rightleftharpoons H^+(aq) + CO_3^{2-}(aq)$	4.7×10^{-11}
Nitrous	HNO_2	NO_2^-	$HNO_2(aq) \rightleftharpoons H^+(aq) + NO_2^-(aq)$	5.6×10^{-4}
Phenol	C_6H_5OH	$C_6H_5O^-$	$C_6H_5OH(aq) \rightleftharpoons H^+(aq) + C_6H_5O^-(aq)$	1.0×10^{-10}

For more acids, see Table I.5 on page 323.

Table 1.14 Some common weak acids, their formulas, weak acid reactions, and equilibrium constants.

Acids and bases always come in conjugate pairs. For example, CN^- is acting as a base in Reaction R1.55, and it produces OH^- and HCN. Notice how similar CN^- and HCN are – they are only different by a proton (H^+). The CN^- ion is the base and HCN is the conjugate acid. It is appropriate to state this in a different way - HCN is an acid and CN^- is its conjugate base. Another item to notice is that CN^- could come from several different ionic compounds, such as NaCN or KCN. When the focus of the chemistry is on acids and bases, it is irrelevant whether NaCN or KCN is used because once dissolved in solution, the compounds will ionize and produce K^+ and CN^-, for example. Only the CN^- matters for acid/base chemistry. The sodium and potassium ions are spectator ions and do not play an important role in the chemistry – in fact, they can be ignored for the most part. Thus, it is important for you to be able to recognize when a salt of an acid is a base, as is the case for KCN and NaCN.

Finally, some acids can contribute more than one proton, as is the case for phosphoric acid.

$$H_3PO_4(aq) \rightleftharpoons H^+(aq) + H_2PO_4^-(aq) \quad K_{a1} \quad (R1.56)$$

$$H_2PO_4^-(aq) \rightleftharpoons H^+(aq) + HPO_4^{2-}(aq) \quad K_{a2} \tag{R1.57}$$

$$HPO_4^{2-}(aq) \rightleftharpoons H^+(aq) + PO_4^{3-}(aq) \quad K_{a3} \tag{R1.58}$$

Phosphoric acid can ionize three times and contribute three protons to water, making it a polyprotic acid. Carbonic acid (H_2CO_3) is a diprotic acid. Another item to note when comparing Reaction R1.52 and R1.58 is the presence or absence of water. The hydronium ion is most correctly represented by H_3O^+, but often chemists will write H^+ as a shorthand way of writing the hydronium ion. Bare protons (H^+) are not stable in an aqueous solution and immediately attach to a water molecule to produce a hydronium ion. In this textbook, you will see both notations but realize that the hydronium ion is a more accurate representation of the solution. The following two reactions are equivalent.

$$HNO_2(aq) \rightleftharpoons H^+(aq) + NO_2^-(aq) \tag{R1.59}$$

$$HNO_2(aq) + H_2O(l) \rightleftharpoons H_3O^+(aq) + NO_2^-(aq) \tag{R1.60}$$

Exercises

1. Determine the pH of the following situations.
   (a) $[H_3O^+] = 1.0 \times 10^{-4}$ M
   (b) $[H_3O^+] = 4.52 \times 10^{-5}$ M
   (c) $[OH^-] = 6.13 \times 10^{-6}$ M
2. Determine the $[H_3O^+]$ for each of the following conditions.
   (a) pH = 9.84
   (b) pOH = 5.39
   (c) $[OH^-] = 4.37 \times 10^{-3}$ M
3. Label each of the following as a strong acid (SA), a weak acid (WA), a strong base (SB), a weak base (WB), or an electrolyte (E).
   (a) ____ HNO_2
   (b) ____ HNO_3
   (c) ____ KCl
   (d) ____ NH_3
   (e) ____ NaOH
   (f) ____ NaCN
4. In each of the following comparisons, circle the acid or base that is the strongest of the pair.
   (a) CH_3COOH or HNO_2
   (b) NH_3 or CN^-
   (c) H_3PO_4 or H_2SO_4
   (d) NaOH or NH_3
5. Write the weak acid equilibrium reaction for $HClO_2$.
6. Write the weak base equilibrium reaction for ClO_2^-.

Solution: See Section A.1 on page 241.

Review Example 1.22: Chemical Equilibrium

Even when chemical reactions seem to stop producing products, they are in a state of dynamic equilibrium, where reactants are forming products at the same rate that products are reforming reactants. The mathematical representation of this state is called the equilibrium constant, generally symbolized by K_{eq}, but it can have other specific forms as seen in Table 1.15. While the equilibrium constant represents a reaction at a singularly unique state, the reactant quotient (Q) represents that state of a reaction in any nonequilibrium state. If the state of equilibrium is the goal of the reaction, the reaction quotient indicates how far and in which direction the reaction will proceed to equilibrium.

Solubility product		
General form:	$MX(s) \rightleftharpoons M^+(aq) + X^-(aq)$	$K_{sp} = [M^+][X^-]$
Specific example:	$PbCl_2(s) \rightleftharpoons Pb^{2+}(aq) + 2Cl^-(aq)$	$K_{sp} = [Pb^{2+}][Cl^-]^2$
Weak acid		
General form:	$HA(aq) + H_2O(l) \rightleftharpoons H_3O^+(aq) + A^-(aq)$	$K_a = \frac{[H_3O^+][A^-]}{[HA]}$
Specific example:	$HCN(aq) + H_2O(l) \rightleftharpoons H_3O^+(aq) + CN^-(aq)$	$K_a = \frac{[H_3O^+][CN^-]}{[HCN]}$
Weak base		
General form:	$B(aq) + H_2O(l) \rightleftharpoons BH^+(aq) + OH^-(aq)$	$K_b = \frac{[BH^+][OH^-]}{[B]}$
Specific example:	$CN^-(aq) + H_2O(l) \rightleftharpoons HCN(aq) + OH^-(aq)$	$K_b = \frac{[HCN][OH^-]}{[CN^-]}$
Water auto-dissociation		
General form:	$2H_2O(l) \rightleftharpoons H_3O^+(aq) + OH^-(aq)$	$K_w = [H_3O^+][OH^-]$
Formation constant		
General form:	$M(aq) + nL(aq) \rightleftharpoons ML_n(aq)$	$K_f = \frac{[ML_n]}{[M][L]^n}$
Specific example:	$Ni^{3+}(aq) + 6CN^-(aq) \rightleftharpoons Ni(CN)_6^{3+}(aq)$	$K_f = \frac{[Ni(CN)_6^{3+}]}{[Ni^{3+}][CN^-]^6}$
Henry's law constant		
General form:	$X(g) \rightleftharpoons X(aq)$	$K_H = \frac{[X]}{P_X}$
Specific example:	$CO_2(g) \rightleftharpoons CO_2(aq)$	$K_H = \frac{[CO_2]}{P_{CO_2}}$

Table 1.15 Common equilibrium constant expressions.

For example, the reaction for the formation of nitrogen monoxide from molecular nitrogen and oxygen

$$N_2(g) + O_2(g) \rightleftharpoons 2NO(g) \tag{R1.61}$$

has an equilibrium expression of

$$K_p = \frac{(P_{NO})^2}{(P_{N_2})(P_{O_2})} \tag{1.21}$$

where K_p is a specific equilibrium constant for a gas reaction (P = pressure), and the pressure of each gas in the reaction is represented by P. The expression is always formulated with products in the numerator and reactants in the denominator. Also, notice that the coefficients in the reaction become exponents in the equilibrium expression and that solids and pure liquids are excluded from the expression. The reaction quotient has the same general form.

Similarly, K_c is the equilibrium expression for a general reaction where the amounts of each chemical are specified by [ ] and usually in molar concentrations.

$$2Fe^{2+}(aq) + Cu^{2+}(aq) \rightleftharpoons 2Fe^{3+}(aq) + Cu(s) \qquad K_c = \frac{[Fe^{3+}]^2}{[Fe^{2+}]^2[Cu^{2+}]} \tag{R1.62}$$

Suppose that the equilibrium constant for Reaction R1.61 has a value of 7.2×10^{-5} and the reaction began with 0.78 atm of N_2, 0.21 atm of O_2, and 0.00 atm of NO. The reaction quotient would have the following value.

$$Q = \frac{(0.00)^2}{(0.78)(0.21)} = 0 \tag{1.22}$$

Thus, $Q_p < K_p$, and in order for the reaction to reach equilibrium, the concentration of the products must increase while the concentration of the reactants must decrease – in other words, the reaction must shift to the products. If $Q_p > K_p$, then the reaction must shift toward the reactants. Finally, if $Q_p = K_p$, then the reaction is at equilibrium.

Being able to write equilibrium reactions and understand equilibrium expressions will be important for later chapters. Also, knowing the meaning of each of the specific equilibrium expressions will help you to identify the types of chemistry involved.

Exercises

1. Write the equilibrium expression for the following reactions.
   (a) $H_2SO_4(aq) + H_2O(l) \rightleftharpoons H_3O^+(aq) + HSO_4^-(aq)$
   (b) $CaSO_4(s) \rightleftharpoons Ca^{2+}(aq) + CO_3^{2-}(aq)$
   (c) $Hg^{2+}(aq) + 4Cl^-(aq) \rightleftharpoons HgCl_4^{2-}(aq)$
2. Write the reverse of Reaction R1.61. How is the K_p expression of the reverse reaction related to the expression for Reaction R1.61? If the forward reaction has a $K_p = 7.2 \times 10^{-5}$, then what would be the value for the reverse reaction?
3. Complete Table 1.16 and predict which way the reaction will shift in each case.

$$2NO_2(g) \rightleftharpoons N_2O_4(g) \quad \text{(R1.63)}$$

Write the equilibrium expression for the above reaction:

	Initial Conc.(M)			Equilibrium Conc.(M)			The reaction
Exp. #	$[NO_2]$	$[N_2O_4]$	Q	$[NO_2]$	$[N_2O_4]$	K_{eq}	will shift ...
1	0.0200	0.0000		0.0172	0.00139		
2	0.0300	0.0300		0.0244	0.00280		
3	0.0400	0.0000		0.0310	0.00452		
4	0.0000	0.0200		0.0310	0.00452		

Table 1.16 Simulated equilibrium experiments.

(a) Calculate Q and K_{eq}, and enter these values in the table.
(b) Which way will the reaction shift? Enter ← or → in the last column.

Solution: See Section A.1 on page 242.

Review Example 1.23: Equilibrium Calculations

Now that you have reviewed acid/base definitions and had a refresher on writing equilibrium constants, it is time to put these to use in solving an environmental problem. Imagine there is a soft drink factory (such as Coca Cola or Pepsi) that has two retaining ponds to catch excess water in the event of rain. One day, two tanker trucks (each with a capacity of 1.0 kL) were driving on the factory campus en route to deliver phosphoric acid (H_3PO_4) and carbonic acid (H_2CO_3), each at a concentration of 10.0 M, to make a batch of soda (two of the major ingredients in carbonated soft drinks). As each truck approached the ponds, there was an accident and the contents of the truck hauling phosphoric acid spilled into one pond (call it pond "P") and the carbonic acid spilled into the other pond (call it pond "C"). If each pond originally contained 100.0 kL of pure water at pH 7, what would be the pH of the ponds after the acids were thoroughly mixed (assuming that each pond had no buffer capacity)?

The first step to solve this problem is to recognize that a dilution of the acids has occurred. Dilutions are calculated using the $M_1V_1 = M_2V_2$ equation. Note that the total volume of the two ponds has changed with the *addition* of the acids.

Pond "P"

$$M_2 = \frac{M_1 V_1}{V_2}$$

$$M_2 = \frac{(10\text{ M})(1.0\text{ kL})}{100.0\text{ kL} + 1.0\text{ kL}}$$

$M_2 = 0.09901$ M phosphoric acid

Pond "C"

$$M_2 = \frac{M_1 V_1}{V_2}$$

$$M_2 = \frac{(10\text{ M})(1.0\text{ kL})}{100.0\text{ kL} + 1.0\text{ kL}}$$

$M_2 = 0.09901$ M carbonic acid

Since both of these acids are weak acids, an equilibrium will be established when the acids have thoroughly mixed, and in order to solve this problem, we need to employ ICE tables (which stands for Initial, Change, Equilibrium). A chemical reaction is required to set up an ICE table. Given the chemical formulas of the aforementioned acids, you should be able to write the weak acid reactions for each (then find the K_{a1} for each in Table I.5 on page 323 in Appendix I). The shorthand version of the reactions is as follows. The initial concentration for each acid was calculated earlier and, before any reaction starts, the products are zero.

	$H_3PO_4(aq)$	$\rightleftharpoons$ $H^+(aq)$ +	$H_2PO_4^-(aq)$
I	0.09901	0	0
C	$-x$	$+x$	$+x$
E	$0.09901-x$	x	x

	$H_2CO_3(aq)$	$\rightleftharpoons$ $H^+(aq)$ +	$HCO_3^-(aq)$
I	0.09901	0	0
C	$-x$	$+x$	$+x$
E	$0.09901-x$	x	x

We knew that the reaction was going to shift to the right because the products are zero and can only increase (no negative concentrations are possible!). Now we need to write the K_a expressions for each ICE table.

☞ The *5% Rule* is a rule of thumb used to simplify an equation. As long as the value of the unknown (x) is less than 5% of the value it is subtracted from (or added to), we will accept a 5% error (or less) in the accuracy of the solution in exchange for simplifying the mathematics.

$$K_{a1} = \frac{[H^+][H_2PO_4^-]}{[H_3PO_4]}$$

$$6.9 \times 10^{-3} = \frac{(x)(x)}{(0.09901 - x)}$$

$$K_{a1} = \frac{[H^+][HCO_3^-]}{[H_2CO_3]}$$

$$4.5 \times 10^{-7} = \frac{(x)(x)}{(0.09901 - x)}$$

We need to solve for x, but this will result in a quadratic formula, which is solvable in this case but would have resulted in a higher order equation if the coefficients in the reactions were greater than 1. To simplify the math, we can use the *5% Rule*: assume $x \ll 0.09901$ so that $0.09901 - x \approx 0.09901$. Thus, our assumptions become the following.

$$K_{a1} = \frac{(x)(x)}{(0.09901 - \cancel{x}_0)} = 6.9 \times 10^{-3}$$

$$x^2 = (0.09901)(6.9 \times 10^{-3})$$

$$x = \sqrt{(0.09901)(6.9 \times 10^{-3})} \quad (1.23)$$

$$x = 0.02614$$

$$K_{a1} = \frac{(x)(x)}{(0.09901 - \cancel{x}_0)} = 4.5 \times 10^{-7}$$

$$x^2 = (0.09901)(4.5 \times 10^{-7})$$

$$x = \sqrt{(0.09901)(4.5 \times 10^{-7})} \quad (1.24)$$

$$x = 2.111 \times 10^{-4}$$

Now we need to check the 5% Rule assumption to see if it is valid.

☞ *Successive Approximations* is an iterative process that uses a previous approximation in order to determine a more accurate approximation.

$$\%\text{err} = \frac{x}{0.09901} \times 100\% = \frac{0.02614}{0.09901} \times 100\% = 26\%$$

$$\%\text{err} = \frac{x}{0.09901} \times 100\% = \frac{2.111 \times 10^{-4}}{0.09901} \times 100\% = 0.21\%$$

Clearly, the assumption failed for Pond "P" but was good for Pond "C." Thus, there is more work to be done in the case of Pond "P." When the 5% Rule fails, we need to find a better approximation for x by using *Successive Approximations*. We just need to rewrite Eq. (1.23) and reinsert the "$-x$" term that was dropped when we assumed that x was effectively zero (5% rule), then plug in our first approximation for x ($x_1 = 0.02614$ for Pond "P").

$$x_2 = \sqrt{(0.09901 - x)(6.9 \times 10^{-3})}$$

$$= \sqrt{(0.09901 - 0.02614)(6.9 \times 10^{-3})}$$

$$= 0.02242$$

The second approximation is fed right back in for the "$-x$" term repeatedly until the approximations stop changing to the number of significant figures we will carry in the end (which is 2).

$$x_3 = \sqrt{(0.09901 - 0.02242)(6.9 \times 10^{-3})} = 0.02299$$

$$x_4 = \sqrt{(0.09901 - 0.02299)(6.9 \times 10^{-3})} = 0.02290$$

$$x_5 = \sqrt{(0.09901 - 0.02290)(6.9 \times 10^{-3})} = 0.02292$$

$$x_6 = \sqrt{(0.09901 - 0.02292)(6.9 \times 10^{-3})} = 0.02291$$

The best approximation of x is 0.02291 M (we could have stopped the approximations for rounding purposes at x_3 but going to x_5 demonstrates the principle more effectively). This value of x is still going to fail the 5% rule, but it is a more accurate value and the 5% rule no longer applies.

Now we need to calculate the pH, and examining the ICE tables reveals that $[H^+] = x$, so all we need is the x to determine the pH.

$$\text{pH} = \log[H^+] = \log[0.02291\ \text{M}] = 1.64$$

$$\text{pH} = \log[H^+] = \log[2.111 \times 10^{-4}\ \text{M}] = 3.68$$

Note that an extra significant figure (sig fig) is added to the pH because it is a log scale and the digits to the right of the period match the sig figs of the x.

We have determined that the pH of one pond is about 2 units lower than the other pond. The reason Pond "P" is more acidic than Pond "C" is because of the difference in the K_a's of the acids – phosphoric acid is stronger than carbonic acid. Remember that the pH scale is a logarithmic scale, which means that a difference of 2 pH units means Pond "P" is 100 times more acidic.

The aforementioned ICE calculation is not strictly correct since phosphoric and carbonic acids are polyprotic. Thus, we have ignored the contribution of the second H (and third in the case of phosphoric acid) to the acidity. Let us be more rigorous here and see if it makes a difference. We have to consider the second ionization of each acid and write a new ICE table. We already have a good estimate for the $[H^+]$ from the first ionization.

	$H_2PO_4^-$(aq)	⇌	H^+(aq) +	HPO_4^{2-}(aq)
I	0.02291		0.02291	0
C	$-x$		$+x$	$+x$
E	$0.02291-x$		$0.02291+x$	x

assume $x \ll 0.02291$

$$K_{a2} = \frac{(0.02291 + x)(x)}{0.02291 - x} = 6.2 \times 10^{-8}$$

$$6.2 \times 10^{-8} = \frac{(0.02291 + \cancel{x}_{0})(x)}{0.02291 - \cancel{x}_{0}} = \frac{(0.02291)(x)}{0.02291} = \frac{\cancel{(0.02291)}(x)}{\cancel{0.02291}}$$

$$6.2 \times 10^{-8} = x$$

check the assumption:

$$\%\text{err} = \frac{x}{0.02291} \times 100\% = \frac{6.2 \times 10^{-8}}{0.02291} \times 100\% = 0.00027\%$$

	HCO_3^-(aq)	⇌	H^+(aq)	+	CO_3^{2-}(aq)
I	0.0002111		0.0002111		0
C	$-x$		$+x$		$+x$
E	$0.0002111-x$		$0.0002111+x$		x

assume $x \ll 0.0002111$

$$K_{a2} = \frac{(2.111 \times 10^{-4} + x)(x)}{2.111 \times 10^{-4} - x} = 4.7 \times 10^{-11}$$

$$4.7 \times 10^{-11} = \frac{(2.111 \times 10^{-4} + \cancel{x}_{0})(x)}{2.111 \times 10^{-4} - \cancel{x}_{0}} = \frac{(2.111 \times 10^{-4})(x)}{2.111 \times 10^{-4}} = \frac{\cancel{(2.111 \times 10^{-4})}(x)}{\cancel{2.111 \times 10^{-4}}}$$

$$4.7 \times 10^{-11} = x$$

check the assumption:

$$\%\text{err} = \frac{x}{2.111 \times 10^{-4}} \times 100\% = \frac{4.7 \times 10^{-11}}{2.111 \times 10^{-4}} \times 100\% = 2.2 \times 10^{-5}\%$$

Clearly, both assumptions were good, and the resulting x adds insignificantly to the previous x determined from the first calculation. The conclusion is that the second ionization of these polyprotic acids is insignificant compared to the first. The pH has not changed significantly. This means that we can certainly ignore the third ionization of phosphoric acid.

Exercises

1. Determine the pH of a 0.0100 M solution of HCN.
2. Determine the pH of a 0.0100 M solution of NH_3.
3. (Challenge) Determine the pH of a 0.0100 M solution of HNO_2 (nitrous acid, not nitric acid!!).
4. (Challenge) Determine the pH of a 0.100 M solution of H_2SO_4 (the first H comes off as a strong acid, but the second H is part of a weak acid).

Solution: See Section A.1 on page 243.

Review Example 1.24: Balancing Redox Reactions

Balancing redox reactions also presents a challenge because they often cannot be simply balanced by examining the elements on each side of the reaction. Redox reactions must also have balanced electron transfers since the oxidation half-reaction produces electrons that the reduction half-reaction must absorb. There cannot be extra electrons in either case. Following is a list of steps that prescribe one way to balance redox reactions without having to assign oxidation states to every single element. In this book, most of the chemical reactions discussed will be aqueous reactions since water is such a large part of the environment; therefore, the balancing of redox reactions will be assumed to be in aqueous solutions.

Steps for balancing redox reactions

1. Separate the reaction into half-reactions.
2. Balance each half-reaction for each element except H, O.
3. Balance for O by adding H_2O.
4. Balance for H by adding H^+.
5. If the reaction is under basic conditions, add enough OH^- to ***both sides*** to neutralize all H^+; $H^+ + OH^- \rightarrow H_2O$
6. Balance each half-reaction for charge by adding electrons.
7. Find the least common multiple (LCM) of the electrons in the two half-reactions and equate them.
8. Add the half-reactions together and cancel any species that is found on both sides.

Here are two examples with stepwise changes marked by **bold font**.

Example 1: $\mathbf{MnO_4^- + Br^- \rightarrow Mn^{2+} + Br_2}$ (acidic reaction)

(step 1 – half-reactions)
$MnO_4^- \rightarrow Mn^{2+}$ $\quad$ $Br^- \rightarrow Br_2$

(step 2 – balance for all elements except H, O)
$MnO_4^- \rightarrow Mn^{2+}$ $\quad$ $\mathbf{2}Br^- \rightarrow Br_2$

(step 3 – balance for O by adding H_2O)
$MnO_4^- \rightarrow Mn^{2+} + \mathbf{4H_2O}$ $\quad$ $2Br^- \rightarrow Br_2$

(step 4 – balance for H by adding H^+)
$\mathbf{8H^+} + MnO_4^- \rightarrow Mn^{2+} + 4H_2O$ $\quad$ $2Br^- \rightarrow Br_2$

(step 5 – under basic conditions... the reaction is acidic – nothing to do!)
(step 6 – balance each half-reaction for charge by adding electrons)
$\mathbf{5e^-} + 8H^+ + MnO_4^- \rightarrow Mn^{2+} + 4H_2O$ $\quad$ $2Br^- \rightarrow Br_2 + \mathbf{2e^-}$

(step 7 – find the LCM of the electrons in the two half-reactions and equate them)
$\mathbf{2\times}(5e^- + 8H^+ + MnO_4^- \rightarrow Mn^{2+} + 4H_2O)$ $\quad$ $\mathbf{5\times}(2Br^- \rightarrow Br_2 + 2e^-)$

(step 8 – add the half-reactions and cancel like terms)
$10e^- + 16H^+ + 2MnO_4^- + 10Br^- \rightarrow 2Mn^{2+} + 8H_2O + 5Br_2 + 10e^-$
$\cancel{10e^-} + 16H^+ + 2MnO_4^- + 10Br^- \rightarrow 2Mn^{2+} + 8H_2O + 5Br_2 + \cancel{10e^-}$
$16H^+ + 2MnO_4^- + 10Br^- \rightarrow 2Mn^{2+} + 8H_2O + 5Br_2$

Example 2: $\mathbf{MnO_4^- + SO_3^{2-} \rightarrow MnO_4^{2-} + SO_4^{2-}}$ (basic reaction)

(step 1 – half-reactions)

$MnO_4^- \rightarrow MnO_4^{2-}$ $\qquad$ $SO_3^{2-} \rightarrow SO_4^{2-}$

(step 2 – balance for all elements except H, O) - nothing to do

(step 3 – balance for O by adding H_2O)

$MnO_4^- \rightarrow MnO_4^{2-}$ $\qquad$ $\mathbf{H_2O} + SO_3^{2-} \rightarrow SO_4^{2-}$

(step 4 – balance for H by adding H^+)

$MnO_4^- \rightarrow MnO_4^{2-}$ $\qquad$ $H_2O + SO_3^{2-} \rightarrow SO_4^{2-} + \mathbf{2H^+}$

(step 5 – under basic conditions, add enough OH^- to ***both sides*** to neutralize all H^+)

$MnO_4^- \rightarrow MnO_4^{2-}$ $\qquad$ $\mathbf{2OH^-} + H_2O + SO_3^{2-} \rightarrow \mathbf{2OH^-} + SO_4^{2-} + 2H^+$

$MnO_4^- \rightarrow MnO_4^{2-}$ $\qquad$ $2OH^- + H_2O + SO_3^{2-} \rightarrow \mathbf{2H_2O} + SO_4^{2-}$

(step 6 – balance each half-reaction for charge by adding electrons)

$\mathbf{1e^-} + MnO_4^- \rightarrow MnO_4^{2-}$ $\qquad$ $2OH^- + H_2O + SO_3^{2-} \rightarrow 2H_2O + SO_4^{2-} + \mathbf{2e^-}$

(step 7 – find the LCM of the electrons in the two half-reactions and equate them)

$\mathbf{2}\times(1e^- + MnO_4^- \rightarrow MnO_4^{2-})$ $\qquad$ $\mathbf{1}\times(2OH^- + H_2O + SO_3^{2-} \rightarrow 2H_2O + SO_4^{2-} + 2e^-)$

(step 8 – add the half-reactions and cancel like terms)

$2e^- + 2MnO_4^- + 2OH^- + H_2O + SO_3^{2-} \rightarrow 2MnO_4^{2-} + 2H_2O + SO_4^{2-} + 2e^-$

$\cancel{2e^-} + 2MnO_4^- + 2OH^- + \cancel{H_2O} + SO_3^{2-} \rightarrow 2MnO_4^{2-} + \cancel{2}_1 H_2O + SO_4^{2-} + \cancel{2e^-}$

$2MnO_4^- + 2OH^- + SO_3^{2-} \rightarrow 2MnO_4^{2-} + H_2O + SO_4^{2-}$

Exercises

1. Balance the following redox reactions
   (a) $Cr_2O_7^{2-}$ (aq) + NO_2^- (aq) → Cr^{3+} (aq) + NO_3^- (aq) (acidic)
   (b) Cu(s) + HNO_3(aq) → Cu^{2+}(aq) + NO(g) (basic)
   (c) $Cr(OH)_3$(s) + ClO_3^-(aq) → CrO_4^{2-}(aq) + Cl^-(aq) (basic)
   (d) Pb(s) + PbO_2(s) + HSO_4^-(aq) → $PbSO_4$(s) (acidic)

Solution: See Section A.1 on page 246.

1.6 IMPORTANT TERMS

★ metric prefixes
★ biotic, abiotic
★ anthropogenic
★ blackbody radiation
★ electromagnetic radiation
  · amplitude
  · wavelength
  · frequency
  · energy
★ microwave background
★ spectroscopy
  · transmission
  · absorption
  · emission
  · absorbance
  · Beer's Law
★ atomic structure
  · protons
  · neutrons
  · electrons
  · atomic number
  · mass number
  · isotopes
★ electronegativity
★ nucleosynthesis
★ Solar Nebular Model
★ Hadean Eon
  · accretion
  · differentiation
  · condensation of the atmosphere
  · young Sun
★ Archean Eon
  · GOE
  · BIFs
  · emergence of life

★ Proterozoic eon
  · transition to an oxic atm
★ Phanerozoic eon
  · Cambrian explosion
★ Periodic Table
  · alkali metals
  · alkaline earth metals
  · transition metals
  · halogens
  · noble gases
★ oxidation states
★ chemical kinetics
  · radioisotope half-life
  · isotopic dating
  · Rate Law
  · first-order reactions
  · second-order reactions
  · integrated rate laws
  · activation energy
★ solubility rules
★ chemical equilibrium
  · equilibrium constant (K_{eq})
  · reaction quotient (Q)
  · specific equilibria: K_{sp}, K_a, K_b, K_H, K_p, K_c, K_f, K_w
  · ICE tables and calculations
★ stoichiometry
  · balancing reactions
  · formula weight
  · limiting reactant
  · percent yield
★ Lewis structures
  · octet & duet rule
  · valence electrons
  · resonance
  · hypervalency
  · formal charge
★ redox chemistry
  · reduction
  · oxidation
  · reducing agent
  · oxidizing agent
★ acid/base chemistry
  · the pH scale
  · hydronium ion
  · hydroxide ion
  · acidic
  · basic or alkaline
  · Arrhenius acids & bases
  · Brønsted–Lowry acids & bases
  · strong acids & bases
  · weak acids & bases
  · conjugate acid/base pairs
  · spectator ions
★ nomenclature
  · polyatomic ions
  · ionic compounds
  · hydrates
  · covalent/molecular compounds
★ life
  · the Miller–Urey experiment
  · biomolecules
  · macromolecules
  · replication
  · molecular evolution
★ thermodynamics
  · Gibbs free energy
  · entropy
  · enthalpy
  · endothermic
  · exothermic
  · spontaneous
  · nonspontaneous
  · second law of thermodynamics
★ amphiphilic

EXERCISES

Questions about The Big Bang, Solar Nebular Model, and the Formation of the Earth

1. How old is the universe?
2. How old is our solar system?
3. What are the two major pieces of evidence for the Big Bang theory mentioned in this chapter?
4. What elements are predominantly the result of the Big Bang?
5. What elements are predominantly the result of the nuclear fusion in massive stars?
6. What elements are predominantly the result of supernova explosions of massive stars?
7. What are the two most abundant elements in the Sun?
8. Which stars live longer lives – low-mass or high-mass stars?
9. Which element has the most stable nucleus?

10. What element is the main product of hydrogen fusion?
11. What element is the main product of helium fusion?
12. List a few of the elements produced from carbon fusion.
13. What element is formed as a result of oxygen fusion and is the primary reactant in the final fusion process in a massive stars?
14. Why can't a massive star use iron and heavier elements as a source of nuclear fusion?
15. What happens to a massive star once all of the nuclear fuel has been used up?
16. What two general categories of nuclear reactions are predominantly responsible for the synthesis of the elements heavier than iron?
17. Our solar system was formed (select the correct one):
    (a) from the remains of one or more small to medium star(s)
    (b) from the remains of one or more massive star(s)
    (c) at the beginning or shortly after the beginning of the universe.
18. Describe the main evidence for your choice in the preceding problem.

The Solar Nebular Model

1. What is the solar wind, and what did it do to the early solar system?
2. List two pieces of evidence stemming from the orbits and axial spins of the planets that corroborate the solar nebular model.
3. Why is the solar system mostly a two-dimensional structure?
4. What is the region of the solar system called that separates the terrestrial planets from the gas giants?
5. If you were to classify elements as volatile and nonvolatile, how were these two categories of elements distributed between the two regions of the solar system separated by the asteroid belt?
6. What two planets are exceptions to the normal planetary axis of rotation?

The Ages of the Earth

1. Very briefly, list the major changes to the Earth during the Hadean eon.
2. Specify the ages of the Earth covered by the Hadean eon.
3. Very briefly, list the major changes to the Earth during the Archean eon.
4. Specify the ages of the Earth covered by the Archean eon.
5. Very briefly, list the major changes to the Earth during the Proterozoic eon.
6. Specify the ages of the Earth covered by the Proterozoic eon.
7. Very briefly, list the major changes to the Earth during the Phanerozoic eon.
8. Specify the ages of the Earth covered by the Phanerozoic eon.
9. How did the Moon form?
10. The first atmosphere of the Earth was made of what gases? How did the Earth lose this first atmosphere?
11. Where did the second atmosphere come from?
12. What gases were the major components of the second atmosphere?
13. Why does the Earth have a core? Why isn't the Earth homogeneous in composition? What is the core composed of?
14. What elements compose most of the Earth's crust? Why are these elements so abundant?
15. How is the age of rocks measured?
16. Where did most of the CO_2 from the second atmosphere go?
17. The early atmosphere of the Earth was believed to be lacking what gas that is tremendously important to the current biosphere? The current atmosphere contains what percentage of this gas?

18. Describe the geological evidence that is the result of the growing concentration of this gas.

For the answers, see Section A.1 on page 249.

BIBLIOGRAPHY

Bada, J. L. *Earth and Planetary Science Letters* **2004**, *226*, 1–15.

Ball, W. W. R. *A Short Account of the History of Mathematics*, 1st ed.; Dover Publications: New York, 1960.

Battistuzzi, F. U.; Feijao, A.; Hedges, S. B. *BMC Evolutionary Biology* **2004**, *4(1)*, 44.

Belcher, C. M.; McElwain, J. C. *Science* **2008**, *321*, 1197–1200.

Berner, R. A. *Proceedings of the National Academy of Sciences of the United States of America* **1999**, *96*, 10955–10957.

Birkey, A. GOP's Beard Wants More Coal Plants Because God Will Fix Global Warming. Electronic, 2011; http://www.tcdailyplanet.net/gops-beard-wants-more-coal-plants-because-god-will-fix-global-warming/.

Butterfield, N. J. *Geobiology* **2009**, *7*, 1–7.

Canfield, D. *Annual Review of Earth and Planetary Sciences* **2005**, *33*, 1–36.

Chiosi, C. *AIP Conference Proceedings* **2010**, *1213*, 42–63.

Cleaves, H. J.; Aubrey, A. D.; Bada, J. L. *Origins of Life and Evolution of Biospheres* **2009**, *39(2)*, 109–126.

Dennett, D. *Darwin's Dangerous Idea: Evolution and the Meaning of Life*; Simon & Schuster: New York, 1996.

Fitz, D.; Reiner, H.; Rode, B. M. *Pure and Applied Chemistry* **2007**, *79(12)*, 2101–2117.

Frei, R.; Gaucher, C.; Poulton, S. W.; Canfield, D. E. *Nature* **2009**, *461*, 250–254.

Frei, R.; Polat, A. *Earth and Planetary Science Letters* **2007**, *253*, 266–281.

Frischknecht, U. S. Nucleosynthesis in massive rotating stars. PhD thesis, Philosophisch-Naturwissenschaftlichen Fakult at der Universit at Basel, 2012.

Grogan, P. *Free Inquiry* **2013**, *33(2)*, 16–19.

Hanczyc, M. M.; Fujikawa, S. M.; Szostak, J. W. *Science* **2003**, *302*, 618–622.

Harada, K. *Naturwissenschaften* **1970**, *57*, 114–119.

Hazen, R. M. *American Mineralogist* **2006**, *91*, 1715–1729.

Hecht, J. *Sky and Telescope* **2010**, *120(2)*, 21–26.

Herzberg, C.; Condie, K.; Korenaga, J. *Earth and Planetary Science Letters* **2010**, *292*, 79–88.

Horneck, G. *Acta Astronautica* **2008**, *63*, 1015–1024.

Horowitz, A. Paul Broun: Evolution, Big Bang 'Lies Straight From The Pit Of Hell'. Electronic, 2012; http://www.huffingtonpost.com/2012/10/06/paul-broun-evolution-big-bang_n_1944808.html.

Hoyle, F. *Astrophysical Journal* **1960**, *132*, 565–590.

Imai, E.; Honda, H.; Hatori, K.; Brack, A.; Matsuno, K. *Science* **1999**, *283*, 831–833.

Isley, A. E.; Abbott, D. H. *Journal of Geophysical Research* **1999**, *104*, 15461–15477.

Joyce, G. F. *Nature* **2002**, *418*, 214–221.

Kapur, S. Santorum Takes Climate Change Denial to a Biblical Level. Blog, 2012; http://talkingpointsmemo.com/election2012/santorum-takes-climate-change-denial-to-a-biblical-level.

Kass-Simon, G., Farnes, P., Eds. *Women of Science: Righting the Record*; Indiana University Press: Bloomington, IL, 1990.

Kasting, J. F. *Nature* **2006**, *443*, 643–644.

Lal, A. K. *Astrophysics and Space Science* **2008**, *317*, 267–278.

Lazcano, A.; Miller, S. L. *Cell* **1996**, *85*, 793–798.

Leman, L.; Orgel, L.; Ghadiri, M. R. *Science* **2004**, *306*, 283–286.

Limongi, M.; Chieffi, A. *The 10th International Symposium on Origin of Matter and Evolution of Galaxies*, *1269*, **2010**, 110–119.

Lide, D. R., Ed. *CRC Handbook of Chemistry and Physics*, 87th ed.; CRC Press, 2007.

McFadden, K. A.; Huang, J.; Chu, X.; Jiang, G.; Kaufman, A. J.; Zhou, C.; Yuan, X.; Xiao, S. *Proceedings of the National Academy of Sciences of the United States of America* **2008**, *105*, 3197–3202.

Makalkin, A. B.; Dorofeeva, V. A. *Solar System Research* **2009**, *43(6)*, 508–532.

Martin, R. B. *Biopolymers* **1998**, *45*, 351–353.

Marty, B. *Earth and Planetary Science Letters* **2012**, *313-314*, 56–66.

Matzke, N. J. Kitzmiller v. Dover: Intelligent Design on Trial. Electronic, 2013; http://ncse.com/book/export/html/11798.

Nisbet, E.; Fowler, M. R. *Chinese Science Bulletin* **2011**, *56*, 4–13.

Nisbet, E. G.; Sleep, N. H. *Nature* **2001**, *409*, 1083—1091.

Ohara, S.; Kakegawa, T.; Nakazawa, H. *Origin of Life and Evolution of the Biosphere* **2007**, *37(3)*, 215–223.

Orgel, L. E. *Prebiotic Chemistry and the Origin of the RNA World* **2004**, *39*, 99–123.

Overway, K. *Journal of Chemical Education* **2007**, *84(4)*, 606–608.

Plankensteiner, K.; Reiner, H.; Rode, B. M. *Peptides* **2005**, *26*, 535–541.

Romano, D.; Matteucci, F. *Monthly Notices of the Royal Astronomical Society* **2003**, *342*, 185–198.

Romano, D.; Matteucci, F. *Nuclear Physics A* **2005**, *758*, 328c–331c.

Rosenberg, A.; Shea, D. W. *Philosophy of Biology*; Routledge: New York, 2008.

Rosing, M. T.; Bird, D. K.; Sleep, N. H.; Bjerrum, C. J. *Nature* **2010**, *464*, 744–747.

Sagan, C. *Cosmos*; Random House: New York, 1980.

Scott, C. T.; Bekker, A.; Reinhard, C. T.; Schnetger, B.; Krapeþ, B.; Rumble, D.; Lyons, T. W. *Geology* **2011**, *39(2)*, 119–122.

Shibuya, T.; Komiyac, T.; Nakamuraa, K.; Takaia, K.; Maruyamad, S. *Precambrian Research* **2010**, *182*, 230–238.

Stojanoski, K.; Zdravkovski, Z. *Journal of Chemical Education* **1993**, *70*, 134–135.

The Pale Blue Dot. photograph, The Voyager Spacecraft As Operated by NASA, 1990.

Tomaschitz1, R. *International Journal of Theoretical Physics* **2005**, *44(2)*, 195–218.

Trigo-Rodriquez, J. M.; Martin-Torres, F. J. *Planetary and Space Science* **2012**, *60*, 3–9.

Westheimer, F. H. *Science* **1987**, *235*, 1173–1178.

Wiescher, M.; Gorres, J.; Uberseder, E.; Imbriana, G.; Pignatari, M. *The Annual Review of Nuclear and Particle Science* **2010**, *60(1)*, 381–404.

Wilde, S. A.; Valley, J. W.; Peck, W. H.; Graham, C. M. *Nature* **2001**, *409*, 175–178.

Wood, B. J.; Halliday, A. N. *Nature* **2005**, *437*, 1345–1348.

Wood, B. J.; Walter, M. J.; Wade, J. *Nature* **2006**, *441*, 825–833.

Woosley, S. E.; Weaver, T. A. *The Astrophysical Journal of Supplement Series* **1995**, *101*, 181–235.

2

MEASUREMENTS AND STATISTICS

> "I often say that when you can measure what you are speaking about, and express it in numbers, you know something about it; but when you cannot measure it, when you cannot express it in numbers, your knowledge is of a meager and unsatisfactory kind; it may be the beginning of knowledge, but you have scarcely in your thoughts advanced to the state of Science, whatever the matter may be."
>
> —Lord Kelvin, 1883

> "There are three kinds of lies: lies, damned lies, and statistics."
>
> —Mark Twain, 1904 (quoting others in his autobiography)

> "If you screw something up in the first few steps then the rest of the experiment is just a waste of time!"
>
> —Lord Overway, 2013 (at least he wishes he was a lord)

2.1 INTRODUCTION

Have you ever experienced the following situation? I was on a hike in the Blue Ridge Mountains, and I was getting tired. Each step I took seemed to require more and more energy. I finally reached the end and was ready to congratulate myself for finishing. It seemed to be a long hike – at least 9 miles. As I left the trailhead, I noticed the sign announcing the trail's length – it was only 5 miles. My perception of the hike was way off!

Philosophers and neuroscientists tell us that we live in a world that is fabricated from our senses ("representative realism"). We do not experience reality but our perception of reality. While this is a great topic for discussion in a philosophy class, my point in raising this issue is that our perceptions and intuitions are often inaccurate. We experience time, distance, and temperature differently depending on the situation and our mood at the moment.

When examining the state of the environment, it is important to remove as much of our personal bias as possible from observations, such that several people observing and measuring the same thing will come to the same conclusions. This is only possible when measurements are calibrated to some externally validated source. You would never have a group of people pick up random sticks, mark off their estimates of an inch, and then use these sticks to start measuring objects. You would expect everyone to use rulers instead, rulers that have been manufactured under controlled conditions and calibrated so that the rulings on each are accurate and all rulers are precisely similar. The same holds true when making environmental measurements.

Lord Kelvin's quote establishes the basis for analytical measurements of two kinds: *qualitative* and *quantitative*. Examples of qualitative observations are distinguishing color and comparisons such as "Kendra is taller than Wanda." These observations are kind of fuzzy. It is what Lord Kelvin meant when he said "it may be the beginning of knowledge." What is the wavelength of light that is generating the red color in that laser pointer? How

☞ *Qualitative* measurements are those that describe the quality of an object such as color, general size, or binary information (present or not present).

Environmental Chemistry: An Analytical Approach, First Edition. Kenneth S. Overway.

Companion website: www.wiley.com/go/overway/environmental_chemistry

much taller is Kendra compared to Wanda? These are quantitative questions that demand measurements, which can be expressed as numbers. Lord Kelvin, if he were alive today, might say, "Now we're talkin' Science!"

☞ *Quantitative* measurements are those that describe the quantity of an object such as concentration, mass, and pressure.

The purpose of this chapter is to prepare you for making quantitative measurements in the laboratory. In order to do this, you must understand the nature of making measurements, the types of errors you may encounter, how to make measurements that are as accurate and precise as possible, and how to interpret the results.

2.2 MEASUREMENTS

Every measurement contains some noise and error, unless you are counting marbles. Even in this case, given enough marbles, there could arise an error – you could count them incorrectly. When making measurements of environmentally important chemicals, it is impossible to measure anything exactly. Atoms and molecules are so small that it would be impossible to count them similarly to marbles. They usually need to be measured in bulk, which means trillions times trillions of molecules – plus or minus a few billion. Further, molecules in bulk are often invisible to the naked eye, such as salt dissolved in water. If you are presented with two glasses of water, one containing pure water and the other containing salty water, the casual observer would not be able to tell the difference. Careful tests would show that one was less dense than the other and the optical density (refractive index) of one would be greater than that of the other, but the salt does not show any obvious color. You could take a sip from each glass and taste the difference. This would result in a qualitative measurement (it is salty or not), but you would not be able to measure the concentration of salt by taste.

Chemists employ a large array of instruments to make measurements that are impossible or dangerous for humans to perform with our five senses. These instruments often involve light, flames, extractions, and a vacuum chamber. Appendix F contains an explanation of these instruments. Since the purpose of this chapter is to prepare you to perform careful measurements in the lab, you need to understand the challenges you will face. You will find noise and error in every nook and cranny in the lab, so understanding the nature of these parameters and how to minimize them will improve the quality of your measurements. It is also likely that you will generate a lot of numerical data during your time in the lab, so making effective use of the tools available to you will make the entire enterprise less daunting.

2.2.1 Random Noise

☞ The *analyte* is the element or compound that is of interest.

☞ The *signal* is the portion of the measurement that is linked to the presence of the *analyte*. The *noise* is the portion of the measurement that is linked to parameters unrelated to the *analyte*, such as temperature, interference from external sources, and quantum events.

Most of the measurements you will make in the study of environmental chemistry will be with modern chemical instrumentation. Instruments such as an absorbance spectrophotometer, a fluorometer, or a gas chromatograph are all fairly sophisticated tools that measure various properties of chemicals. All of them have one thing in common – they all report to the analyst electrical signals, not the actual physical property being measured. For example, if you grab a beaker and place it on a common analytical balance, you might see the readout display 253 g. That number is not the actual mass of the beaker but an estimate of the mass that results from some complicated electronics. The philosophers have struck again! The result of any measurement is not really the property of the *analyte* but a calculated representation of the property. This should not surprise you. How can you write down the mass of a beaker in your notebook, since mass is a property of matter that expresses inertia, weight, fills space, and so forth? You can only write down a numerical representation of the mass.

Before you lose yourself in a philosophical rabbit hole, there is a practical reason for starting this chapter in this way. Every measurement you will perform in the lab depends on an instrument's ability to measure an electrical signal precisely and accurately. Unfortunately, every electrical signal that is not measured at 0 K has thermal noise in it. Electrons, just as the gas molecules you learned about in general chemistry, express their thermal energy by jumping around chaotically. This means that any measurement will be noisy. If we call the thing we want to measure the *signal* and the other stuff that makes it hard to measure the signal the *noise*, then the following can be stated unambiguously: the signal is good and the noise is bad!

When the signal is so big compared to the noise, we often do not even notice the noise. Analytical balances you have used in lab before, for example, usually have such low noise that when you place the beaker on the pan, the digital display eventually settles on a number that does not change no matter how long you stare at it. If, however, you could move the decimal place over a few notches, you would see the noise. Sometimes, you see this noise when you are measuring the mass of something that is small – the number bounces around in the last digit. If the number bounces between 253.4567 and 253.4568 g, then it might not really matter which number you write down because this is less than a millionth of a percent difference. In the language of an analytical chemist, the signal-to-noise ratio (SNR) is very high. If the measurement bounced between 253 and 278 g (and all of the numbers in between), what would you write down for the mass of the beaker? An analytical chemist would say that in this situation, the SNR is very low. If your lab partner asked you, "What is the mass of the beaker?", under which situation (high SNR or low SNR) would you be the most *certain* about reporting a mass to your partner? Clearly, in the high SNR situation, your certainty would be higher, but at what point would you be forced to declare that you cannot report any mass because the noise is too high?

☞ *Certainty* refers to precise and accurate knowledge. In science, all knowledge is provisional and certainty is unobtainable. Expressing *uncertainty* numerically is typically a measure of the precision of a measurement. If the measurement was repeated, it would most likely (68% of the time) fall within the ± boundaries. The number placed after the ±, such as 12.14±0.27, is often referred to as the uncertainty. The official definition, from the 2008 Joint Committee for Guides in Metrology (JCGM) Guide to the Expression of Uncertainty in Measurement (GUM), of uncertainty is a "parameter, associated with the result of a measurement, that characterizes the dispersion of the values that could reasonably be attributed to the measurand (the quantity intended to be measured)."

Underlying all measurements is a background level of random noise. The obvious question now arises: how do we know when something is noise and when something is a signal? Examine the graph in Figure 2.1. The measured data points jump around randomly, but every once in a while, there is something that appears to rise above the noise, as labeled on the graph. Did sample A provide a measurable signal, or is it just noise? How about sample B? Clearly, sample C rises above both A and B, so if either of those is considered signal, then sample C is definitely signal.

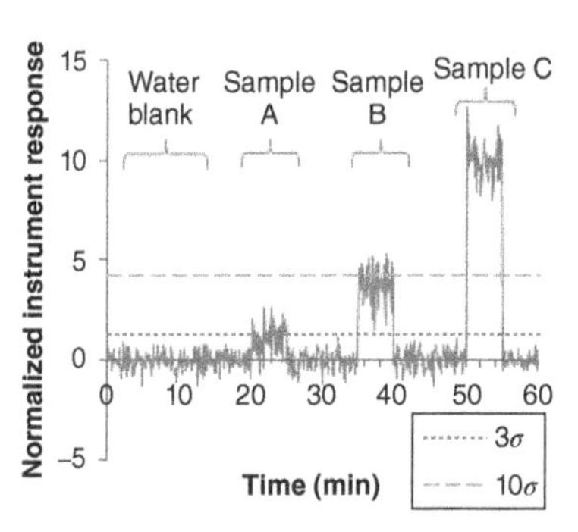

Figure 2.1 Signal and noise can sometimes be indistinguishable. The 3σ line above the baseline is the threshold between signal and noise.

Think about this in a common application. You have a sample of pond water that you think contains a blue algicide – at least it looks blue to you. You turn on a spectrophotometer, an instrument that measures the intensity of colors, and you put in a sample that contains pure deionized water from the chemistry lab (a water *blank*). You watch the readout that resembles the instrument response from time 0 to 15 min in Figure 2.1. Since this sample is "pure" water with no visible color, it should give no signal, so the random variations in the measurement can be attributed to noise. You replace the water blank with a sample of the pond water (sample A) and look at the readout (from time = 20–25 min). Did it change enough for you to exclaim, "Eureka! There is an algicide dye in my sample!?" When sample A is replaced by sample B from another pond, the signal definitely got stronger. Sample C from a third pond gave an even stronger signal. Which of the sample signals is different enough from the water blank to be considered detectable? What constitutes "different enough?"

☞ The *blank* is a solution that establishes the baseline. It usually contains everything except the analyte.

Before "different enough" can be quantified, the limits of the noise must be understood. Where can a line be drawn such that all of the noise will stay below it? Well, unfortunately nothing is for sure, but analytical chemists have agreed on where this line is placed. The line placement depends on being able to predict where the noise will be and where it will not be. At this point, you might exclaim, "It is *random* noise, for goodness sakes! How can anyone predict random noise?!?" It is actually quite easy.

The measurements in Figure 2.2 represent measurements of a water blank from a few different instruments. Fluorometers measure light emitted from an excited sample, while an atomic absorption spectrophotometer measures the amount of light a sample absorbs. The x axis represents the response the instrument gave for the samples. It is not important that the instrument response from each instrument was different because all measurements of this sort are relative – they depend on specific parameters of the instrument (temperature, dust levels, age of the lamp and detectors, and so forth), which change from day to day. Despite

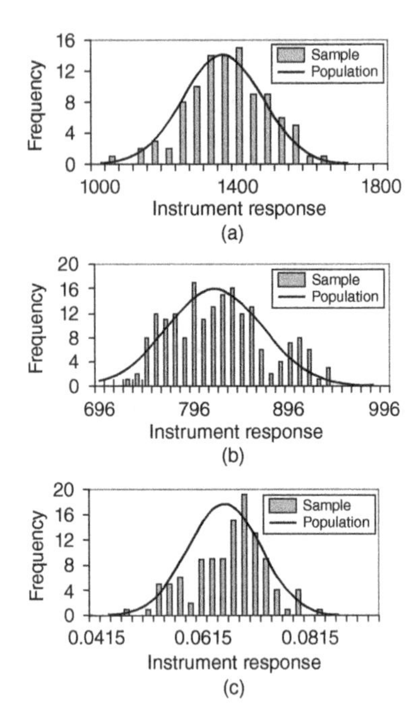

Figure 2.2 Measurements from a blank in (a) a cuvette fluorometer, (b) a 96-well plate fluorometer, (c) an atomic absorption spectrophotometer (N>100 in each case). The bars represent the distribution of the actual measurement, and the line is a Gaussian curve fitted to the average and standard deviation of the data.

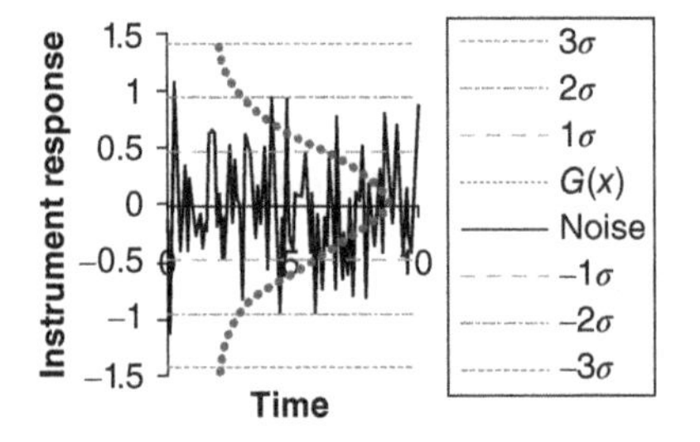

Figure 2.3 The noise in most measurements has a Gaussian distribution based on the calculated average and standard deviation of the noise. This is due to the Central Limit Theorem.

the differences between these instruments, the distribution of noise in each case fits, in general, within the confines of a Gaussian curve (the population) or Normal Distribution. This sort of "bell curve" is a common occurrence in nature, and it turns out that it is a good model for the thermal noise in electrical signals. This fact of nature was first established by Pierre Laplace (who developed the Solar Nebular Theory) as the Central Limit Theorem, which demonstrates that the average of several randomly drawn samples from any population distribution will be normally distributed. Thus, what is commonly observed in the lab has theoretical underpinnings.

If the average and standard deviation of the baseline noise from Figure 2.1 are calculated, they produce a Gaussian curve that fits the data (see Figure 2.3). In this case, the average of the data points is close to zero and the noise bounces around zero. Most of the data points never exceed the upper and lower tails of the Gaussian curve. Figure 2.3 represents the fundamental assumption of noise analysis for the measurements you will make in this class – that the noise is contained within a known range of values represented by a Gaussian curve. While each measured value of the instrument response cannot be predicted precisely, the probability that any measurement of noise will fall outside of the Gaussian curve (based on the calculated average and standard deviation of the noise) is extremely small. There is a 68.3% probability that any single measurement will fall within $\pm 1\sigma$ from the center, a 95.4% probability that any single measurement will fall within $\pm 2\sigma$ from the center, and so on (see the horizontal lines in Figure 2.3). Table 2.1 contains a partial list of these probabilities and their normalized distance from the center (column 1). This normalized distance from the center is defined as z. The z term just represents the number of steps with a size of σ away from the center. It is a term more commonly used by statisticians but will become important later when you learn about Hypothesis Testing (Section 2.5).

The Gaussian function, seen in Eq. (2.1), contains two parameters that were referred to earlier. The distribution average (μ) represents the center of the distribution, and the standard deviation (σ) is a measure of the width of the distribution. Figure 2.3 demonstrates that μ simultaneously represents the average of the data points and the *baseline* of the measurement when no signal is present (the blank).

Look back to Figure 2.1. Most measurements you will perform will be similar to those in this figure in that the sample signals will only add to the baseline (some signals will subtract from the baseline, but it is rare for a signal to do both). As a result, we only have to consider probabilities on one side of the Gaussian distribution (called one-tailed instead of two-tailed), so our $+n\sigma$ lines have slightly larger probabilities compared to those given in column 2 of Table 2.1 because the area contains all of the negative side of the baseline (from $-\infty$) to the positive σ value. Thus, at the 3σ line above the baseline, the area contains 99.9% of the noise (99.7% + $\frac{1}{2}$ × 0.3%). It appears that sample A pokes up above the 3σ line only sometimes. The average of the data points from sample A is just below the 3σ line. Sample B pokes up above the 3σ line entirely, and sample C is above the 10σ line. We can confidently state that the likelihood that sample A is noise is slightly more than 0.1%. For sample B, the probability that it is noise is less than 0.1%. Analytical chemists have decided that the threshold between signal and noise is 3σ[1] (see featurette on Warren Crummett). This means that what is universally considered a signal is only likely to be noise 0.1% of the time. That is a quite low chance of being wrong (where wrong means that the signal is actually noise). This puts sample A below the minimum level of signal, so it is indistinguishable from noise. The instrument response from samples B and C can definitely be called signal. This value of 3σ represents an SNR of 3 (alternatively, a z of 3) and is considered to be the minimum SNR for qualitative measurements (the Limit Of Detection (LOD)) – where the analyst can say, "I detected something that was not noise." Analytical chemists have also drawn another line at 10σ and have agreed that SNR = 10 (a z of 10) is the Limit Of Quantitation (LOQ) – the limit where the analyst can say, "I detected something that was not noise *and* I can assign a number to its concentration." Only sample

[1] MacDougall, D.; Crummett, W. B. *Analytical Chemistry* **1980**, *52*, 2242–2249.

C is above the quantitative detection limit. This implies that while a concentration can be calculated for sample B, it is really not very reliable and can be used only as an estimate.

To summarize, you will be making measurements as part of your study of environmental chemistry. You will need to report your measurements to your lab partners and to your instructor. Your report will need to convey some level of confidence in the measurement. If noise in the measurement is assumed to have Gaussian distribution, any measurement that falls below an SNR of 3 is considered not detected (ND). Any measurement between SNR of 3 and 10 is considered detectable (above the LOD), but the numerical value given to that measurement needs to be viewed with a high degree of uncertainty. Any measurement that is above an SNR of 10 is considered quantifiable (above the LOQ), and its numerical value is trustworthy. The random noise, which is related to the precision of the measurement, is characterized by the standard deviation of the noise and represented by σ in the Gaussian formula (Eq. (2.1)). The signal usually rises out of the baseline, which is calculated by taking the average of the noise and represented by μ in the Gaussian formula. Finally, you should reflect on the distinction between signal and noise and how that distinction is related to the statistical confidence that is built into the Gaussian distribution (as in "what is the probability you would be wrong in declaring a signal, that rises above a SNR of 3 or a SNR of 10, to be qualitatively or quantitatively detectable?"). While the qualitative and quantitative detection limits are applicable to experiments where many measurements are performed (Mocak *et al.*, 1997), they will serve as a good starting point for the more limited number of measurements you will perform in this course.

$\pm n\sigma$ (or z)	Two-sided % prob.	One-sided % prob.
0.67	50	75
1.00	68.3	84.1
1.64	89.9	95.0
2.00	95.4	97.7
3.00	99.7	99.9
10.0	100	100

You can be confident that 99.7% of all noise will be contained within the bounds of $\pm 3\sigma$ (two-sided), or 99.9% of all noise will fall below the $+3\sigma$ line (one-sided). A more complete table can be found in Appendix H.

Table 2.1 These probabilities represent the *confidence levels* for a Gaussian distribution.

The Gaussian equation is

$$G(x) = \frac{1}{\sigma\sqrt{2\pi}} e^{-\frac{1}{2}\left(\frac{x-\mu}{\sigma}\right)^2} \quad (2.1)$$

☞ The *baseline* is calculated by taking the average of the instrument response given by the blank and is the level from which any signal is measured.

Warren B. Crummett (1922–2014) was the subcommittee chair on environmental analytical chemistry and coauthor of the seminal paper that defined the qualitative and quantitative detection limits for analytical chemists in 1980. He is especially important to me because he was also a 1943 graduate of Bridgewater College of Virginia, my home institution. Warren also wrote *Decades of Dioxin*, in which he shares his perspective of heightened interest in dioxin by the government and industry as analytical advancements allowed the detection of pollutants at extremely low levels.

Example 2.1: General Statistics

Use a spreadsheet to determine the average ($\bar{x}$), standard deviation (s), relative standard deviation (RSD), maximum value, minimum value, the range, and the number of data points of the numbers in Table 2.2.

7.05	6.00	4.77	6.69	7.10	9.56	4.66	5.12
8.16	7.25	9.46	12.09	8.95	3.91	7.03	10.29
6.62	7.61	6.27	8.77	4.16	8.66	8.20	11.04

Table 2.2 A set of numbers from a Gaussian distribution.

Solution: See Section B.1 on page 253.

Example 2.2: Signal-To-Noise Ratio (SNR)

Calculate the baseline, noise, signal level, and SNR for two different samples (Sample A and Sample B) using the data given in Table 2.3. When plotted in a graph, the data will look like Figure 2.1. The questions below will step you through the process.

Blank	Sample A	Sample B
1.1844	1.9831	9.4616
0.5324	1.2627	9.1638
0.5516	1.9993	8.5835
1.1929	1.1270	9.5922
0.6368	1.4045	10.2634
0.8920	1.1045	9.1483
0.7263	1.8622	9.1141
0.3387	1.5284	8.8063
0.7667	1.3332	9.2327

Table 2.3 A set of simulated data for the measurement of a blank and two samples.

1. What is the baseline value of the blank?
2. What is the noise level calculated from the measurement of the blank?

☞ The *mean* or *average* ($\bar{x}$) of a data set represents the center of a distribution and is calculated by summing all of the data points and dividing by the total number of points. The *standard deviation* (s) represents a measure of the precision of a sample distribution and is best calculated using a spreadsheet (STDEV function) or a calculator (σ_{n-1} or s_x). The *relative standard deviation* is calculated as $\frac{s}{\bar{x}} \times 100\%$). The *range* of a data set is simply the difference between the largest and smallest numbers in the set.

3. What is the signal level for Sample A? Remember that signal height for Sample A is measured from the baseline.
4. What is the noise level in Sample A?
5. Calculate the SNR for Sample A.
6. Calculate the noise, signal, and SNR for Sample B.
7. Given the SNR for Sample A, is Sample A ND, only above the Limit of Detection (>LOD), or above the Limit of Quantitation (>LOQ)?
8. Given the SNR for Sample B, is Sample B ND, only above the Limit of Detection (>LOD), or above the Limit of Quantitation (>LOQ)?

Solution: See Section B.1 on page 253.

Example 2.3: Pooled Standard Deviation

Some measurements can be performed on different days, by different analysts, yet are really part of the same distribution. On occasion, it is necessary to combine these smaller sets into one larger pool of data. When this occurs, it is important to incorporate the size of each data set in the descriptive statistics that make the larger set. If one set contains 10 measurements and another contains 5, then any average or standard deviation must be weighted by the size of each data set. Of special use is the pooled standard deviation, which is represented by

$$s_{pooled} = \sqrt{\frac{s_1^2(N_1-1)+s_2^2(N_2-1)+s_3^2(N_3-1)+\cdots+s_n^2(N_n-1)}{N_1+N_2+N_3+\cdots+N_n-n}} \tag{2.2}$$

where s_n is the standard deviation of data set n and N_n is the number of measurements in data set n. Basically, the standard deviation from each set is squared and then multiplied by the degrees of freedom (DOF) for each set (which is usually $N-1$), then the sum of all of these terms is divided by the DOF of the combined data set (represented by the entire divisor). This formula can be used for any number of combinations of data sets. For pooling the standard deviations of three data sets, the formula would be

$$s_{pooled} = \sqrt{\frac{s_1^2(N_1-1)+s_2^2(N_2-1)+s_3^2(N_3-1)}{N_1+N_2+N_3-3}} \tag{2.3}$$

2.2.2 Significant Figures (Sig Figs)

Recall the example described earlier where a beaker was placed on a balance in order to determine its mass. Watching the digital readout either settle down to a single value or bounce around randomly would give you a feel for the certainty of the measurement. What if, however, the balance was designed to freeze the display after 5 s? With the single number, you would not have any feel for how "good" or certain that number represented the mass of the beaker. Could you really trust that single measurement? If you took the beaker off the balance and then put it back on again and got the exact same number, how would your certainty change? If you repeatedly measured the beaker over and over and got the same number, your certainty should be very high. If the number highly fluctuated then your certainty would be very low. In order for any result to communicate certainty to the reader, replicate measures are essential.

Figure 2.4 contains 20 simulated measurements of a penny made on a noisy mass balance. A spreadsheet or calculator can easily calculate the average and standard deviation. The average conveys the mass of the penny, and the standard deviation conveys the uncertainty. If the measurement is, instead, shooting arrows at a target then the average would stand in relation to the bullseye and be a measure of the accuracy of the measurement. The standard deviation would be related to the scatter of the arrows and a measure of the precision of the measurement.

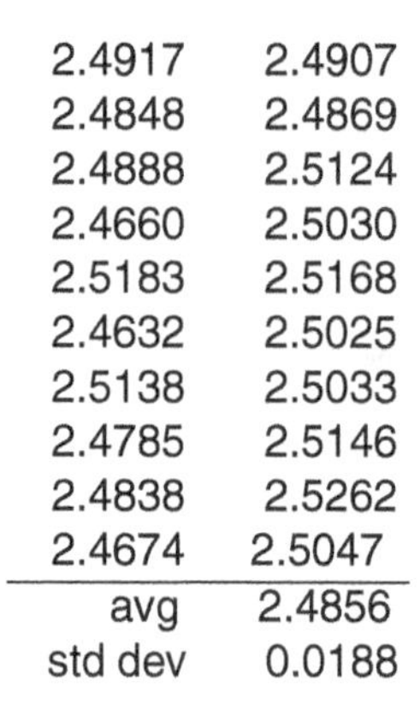

2.4917	2.4907
2.4848	2.4869
2.4888	2.5124
2.4660	2.5030
2.5183	2.5168
2.4632	2.5025
2.5138	2.5033
2.4785	2.5146
2.4838	2.5262
2.4674	2.5047
avg	2.4856
std dev	0.0188

Figure 2.4 Replicate measurements allow an analyst to convey a level of uncertainty in the reported results. In this case, one can conclude that a penny weighs, on average, 2.486±0.019 g.

If the standard deviation were very small, then the uncertainty in the measurement would be very low. Conversely, if the standard deviation were very large, then the uncertainty would be very large. Look at the digits of the standard deviation presented in Figure 2.4. If this number is an expression of the uncertainty, the first significant figure (sig fig) ("1") really conveys most of the uncertainty – the penny's mass cannot be trusted after the hundredths place. The next sig fig of the standard deviation is barely useful because if the uncertainty is in the hundredths place, then what does the thousandths place even communicate? It is the noise of the noise or an even more precise measure of the imprecision of the measurement – another philosophical rabbit hole!

There seems to be an unofficial convention to format statistical results to one or two sig figs in the uncertainty, with the same precision given to the measurement. This book will always use two sig figs in the uncertainty.

Going back to the penny example in Figure 2.4, the resulting standard deviation of the measurements is 0.0188 g, so the proper format for this number is 0.019 g. This means that this expression of the uncertainty has its least precise digit in the thousandths place. The penny mass must have the same precision, so its last digit must also be in the thousandths place. The steps to follow for proper formatting are as follows.

- Round the uncertainty (standard deviation in this case) to two sig figs.
- Round the measurement so that it has the same precision as the uncertainty.

The rules for sig figs you might have learned in general chemistry applied to numbers that were always measured once and may have been multiplied by or subtracted from other numbers. For example, a number with 4 sig figs multiplied by a number with 3 sig figs will result in a number with 3 sig figs – you always go with the sig figs from the least precise number. These rules assume that the precision comes from the measuring instrument with clearly stated precision limits. Most of the measurements performed in an environmental chemistry lab are complicated enough that there is no standard number of sig figs to use – we will let the statistical noise in the replicate measures determine the number of sig figs. It makes no sense, for example, to use all four decimal places from an analytical balance if the last two decimal places are constantly changing because of noise. It is clear to the observer that the nominal precision of the measurement supersedes the assumed precision of the instrument.

Example 2.4: Significant Figures

For each of the following examples, the result of a data analysis is given as *result* $\pm$ *uncertainty*. In each case, format this pair of numbers with the proper sig figs. Assume the two sig fig rule on the uncertainty.

1. 624.694 ± 228.357
2. 3840.2 ± 4739.17
3. 470.254 ± 42.5641

Examine the correctly formatted results, and label each as ND, *detectable*, or *quantitatively detectable*. Assume that the number of trials was well over 20 in each case.

1. 0.1422 ± 0.0017
2. 640 ± 170
3. 4.7 ± 1.8
4. 8.3 ± 1.3

Solution: See Section B.1 on page 254.

2.2.3 Systematic Errors

The percent error equation is

$$\%err = \frac{(obs. - true)}{true} \times 100\% \quad (2.4)$$

Error and uncertainty are not the same. Error is the difference between the measurement and the true value, as in a percent error calculation (see Eq. (2.4)), and thus, it is a measure

Instrumental Errors: Temperature changes, baseline drift, impure primary standard, low voltage in the power source, calibration offset
Method Errors: Losses of the analyte due to inefficient derivatization, filtration or extraction, evaporation, precipitation, contaminants, interferences, matrix mismatch
Personal Errors: Mistakes in judgment due to poor training, color blindness, number bias, conscious/unconscious prejudice

The above list provides a few examples of each.

Table 2.4 Systematic errors generally depend on the training of the analyst, the instrumentation, and the method used to process the samples and standards.

☞ A *standard* is a pure substance to which unknowns are compared when measured.

☞ *Standardization* is the analytical process by which a measured parameter of a sample, such as concentration, is fixed in reference to a standard.

☞ A *calibration standard* is a prepared solution or mixture that is used in a quantitation method to calibrate an instrument.

Vendor	Descrip.	Purity (%)	Price ($/g)
NIST	SRM 915b	99.907 ± 0.021	19.80
FS[a]	ACS[b]	≥ 99.95% to 100.05%	71 ¢
FS	ACS	≥ 99.0%	66 ¢
FS	Chalk powder	98.0% to 100.5%	9 ¢
FS	$CaCO_3$	97%	3 ¢
FS	Marble chips	Unknown	2 ¢

[a] FS = Fisher Scientific.
[b] ACS = American Chemical Society certified.
It should be obvious that the price of the sample scales up with its purity and the level of certification. Prices were obtained on December 26, 2012.

Table 2.5 Some prices for various samples of calcium carbonate.

of accuracy. Uncertainty is a measure of the precision and associated with random noise. Unlike random error, systematic errors introduce a consistent bias in the measurement such as a positive offset or a negative offset (but never both for the same source). These errors are the result of problems with instrumentation (such as a calibration error), the method (such as using the wrong indicator), or the analyst (such as incorrect usage of an instrument). The good news is that systematic errors can be eliminated from the measurement. The bad news is that their source is often difficult to discover. Several common sources of systematic error are listed in Table 2.4. One of the most common sources is poor calibration as a result of using an impure stock chemical as a standard. Standardization is one of the most important steps you will complete during a measurement of an experimental unknown, and it is vital that the proper chemical be used. Fortunately, the government is here to help.

2.3 PRIMARY AND SECONDARY STANDARDS

Every measurement requires an analyst to use tools, such as rulers and spectrometers. Take a ruler, for example – how do you know that the markings on the ruler are accurate? If you bought five rulers from different companies and none of them had identical markings, which one is correct? These are questions you might not have ever asked yourself, but what if your job was on the line or your company would lose a client or a lawsuit if the measurement you were making was not accurate? Accuracy is very important in science, and due to the interdisciplinary nature of science, it is important that every collaborator on a project use the same "ruler" when making measurements and sharing the results. A disastrous example of two labs using different "rulers" was the Mars Climate Orbiter, which burned up in the Martian atmosphere on September 23, 1999, because two National Aeronautics and Space Administration (NASA) teams working in different locations used different units when communicating with each other and the satellite. It was a $193 million mistake! (Isbell, Hardin, and Underwood, 1999).For analytical measurements performed in the United States, the National Institute of Standards and Technology (NIST) is the government agency that produces and sells *standards* that have certified levels of purity – known as standard reference materials (SRMs). These standards can be used to calibrate instruments, such as rulers and spectrometers, so that there is no systematic error present due to a miscalibration. These SRMs can be quite expensive because they have been thoroughly analyzed, and the purity of the material is known to a high degree. Other agencies, such as the American Chemical Society, also certify the purity of some compounds. Table 2.5 gives a relative comparison of the purities and cost.

Besides the purity differences among standards, there are differences in use. A *primary standard* is a stable chemical with a very precisely known purity. As you can see from Table 2.5, they can also be very expensive. A primary standard can be used to *standardize* a *secondary standard*, which is less pure or whose purity is less precisely known but presumably less expensive and perfectly useful for calibrating a chemical method when cost is an important factor. For example, imagine you have been assigned to analyze a few hundred samples of soil that have been shipped to you from a SuperFund site contaminated with cadmium. This will take you several weeks, and you will need to calibrate your instrument at least two or three times per day. The NIST SRM for cadmium is a solution that costs $380 per 50 mL ($7.60 per mL) – way too expensive for you to use every day for several days. You decide to make a 2 L solution of cadmium using $Cd(NO_3)_2 \cdot 4\,H_2O$ (56¢ per gram) and deionized water (with a little nitric acid). This solution, made at the same concentration as the NIST SRM, would cost 1.6¢ per milliliter – 475 × less expensive than the NIST SRM. Once it is standardized with the NIST SRM, it could be used for many days, if properly stored, and save a considerable amount of money. Both solutions are essentially the same – cadmium nitrate in acidified water – but one has a certified concentration (NIST SRM) and the other has a concentration that can be determined and certified "in-house." Secondary standards, while not the preferred standards to use, are sometimes useful or necessary.

Procedurally, your first quantitative experiment will require a primary or secondary standard. Typically, the standard, if it is a solid such as calcium carbonate, is placed in a drying oven overnight to remove any moisture that may have been adsorbed from room humidity. After cooling in a desiccator, the solid is weighed to make a stock solution containing the analyte of interest (such as calcium, given a calcium carbonate primary standard). This stock solution is made using a high-precision analytical balance and volumetric glassware. In later sections, you will learn about how the primary or secondary standard is used in quantitation methods, such as the method of external standards (see Section 2.6). For now, it is just important that you know all quantitative measurements begin with a standard.

☞ A *primary standard* is a pure and stable compound that should be your first choice to calibrate an instrument or method.

☞ A *secondary standard* is a stable compound that can be used for calibration if it is standardized using a primary standard.

Example 2.5: Solution Preparation

A calcium carbonate primary standard has a certified purity of 99.97%. A stock standard was obtained by adding 0.1764 g of the primary standard to a 250-mL volumetric flask. After dissolving the calcium carbonate with a few milliliters of HCl, the solution was diluted to volume with deionized water. This standard was used to obtain a calibration curve according to the volumes given in Table 2.6.

Std #	Vol. of stock (mL)	Total vol. (mL)
0	0	25.00
1	0.100	25.00
2	0.250	25.00
3	0.500	25.00
4	1.000	25.00
5	1.500	25.00

Table 2.6 A table of volumes used to make a set of calibration standards.

1. What is the concentration, in units of mg Ca/L, of the stock solution?
2. Calculate the concentration of each of the calibration standards.

Solution: See Section B.1 on page 255.

☞ For a review of molecular mass calculations, see Review Example 1.15 on page 33.

2.3.1 Other Reagents

In many methods, reagents other than the primary/secondary standard (the analyte) and the solvent are added to the sample or standards. It should be obvious that whenever the standard and solvent are added, accurate, and precise equipment should be used since the amounts of these items are crucial to the concentration of the solution. Other reagents, however, can generally be added with much less care as long as they are added in excess. If you recall your knowledge of stoichiometry, you should be familiar with limiting reagents and reagents in excess. All calculations concerning the amount of products formed or reactants consumed start from the limiting reagent. It usually does not matter whether other reagents are in excess by 1 g or 1 kg; if they are in excess, then they do not affect the outcome. It is the same with other reagents in quantitative measurements. It is generally appropriate to use volumetric glassware (see Section 2.7.2) to deliver *aliquots* of the standard and to set the total volume of the solution and then to use a low-precision graduated cylinder to add a reagent that is part of the method. Analytical chemists have much to worry about when making quantitative measurements and making careful measurements takes much care and time. A good analytical chemist knows when to worry about accuracy and precision and when not to worry.

☞ For a review of stoichiometric conversions, see Review Example 1.16 on page 34.

☞ An *aliquot* is a fixed portion of a certain solution. It is synonymous with "a specific amount."

Often, you will read instructions in the laboratory that describe how to make solutions or mix reagents. If you think about what each statement is implying, and which compounds are analytes and which are reagents, you should be able to determine where to apply your most careful skills and where to be less careful. Here are some examples.

- **"add approximately 3 mL of liquid X":** If liquid X is not the analyte, then approximately really means approximately – you can use less care, and the volume added only needs to be close to 3 mL. Do not waste valuable time trying to obtain exactly 3.00 mL. If, however, liquid X is the analyte, then the writer means that it does not matter whether you add 2.84 or 3.21 mL – what matters is that you know exactly how much you added and that it is approximately close to 3 mL.
- **"dissolve 1.4482 g of compound X":** Since the author included so many decimal places in the step, he or she must really want you to be careful to obtain exactly that amount. It could be that later calculations are simplified to assume this exact mass, so you should be careful to get it right or be prepared to recalculate if necessary. If compound X is not the analyte, then there may be an obscure reason for the precision, or the author does not understand that it does not really matter what mass is dissolved. Everyone makes mistakes, so you should not assume that just because it is written down this way, that it is really important.
- **"quantitatively mass about 3.0 g of compound Y":** This statement conveys the importance of a careful measurement by using the word *quantitatively* while suggesting that the amount be somewhere near 3 g. You should use care in this case but not worry about obtaining 3.0000 g on the analytical balance.

You might recall your experiences in general chemistry lab, or you might want to drop by a lab (and say hi to your former professor!) and just watch the general chemistry students working. Your general chemistry lab proctor probably found it entertaining (and maybe frustrating) to watch general chemistry students spend 10 min at the analytical balance moving single grains of compound Y back and forth between the weighing dish and the stock in order to obtain ***exactly*** 3.0000 g when another student spent less than a minute to obtain a random 2.8735 g, which worked equally well in that case. Experienced analytical chemists know when to worry and when not to worry! If you share this knowledge with your former professor, she or he will revel in the part she or he played in helping to produce such a bright student (you!) who knows when to worry!

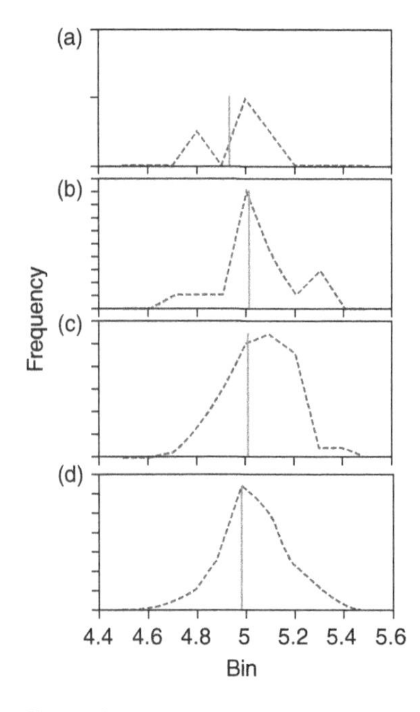

Figure 2.5 Histograms of (a) 4, (b) 20, (c) 100, and (d) 200 randomly generated numbers. The vertical line represents the calculated average (the true average is 5).

2.4 SAMPLE AND POPULATION DISTRIBUTIONS

Trusting in the accuracy of single measurements is always a risky endeavor. It is much better to perform many measurements, but in the real world, each measurement consumes the analyst's time, material resources, and instrument time. There is a trade-off between the need for many measurements and the cost of these measurements. In some experiments, it is easy to perform many measurements, and in others, each measurement may consume an hour or more of time. Clearly, there cannot be any standard number of measurements. The problem is that we are relying on Gaussian statistics in order to distinguish between signal and noise and establish a basis for the level of certainty. Three or four measurements do not even come close to representing the underlying Gaussian distribution of the noise, which means that the certainty by which we can make claims for distinguishing signal and noise is much less than if we could make a thousand measurements.

Look at the histograms in Figure 2.5. The first histogram represents four measurements and is obviously not Gaussian in shape. The other histograms represent increasing numbers of measurements, and each histogram increasingly is more similar to a Gaussian distribution. William Gosset, a Master Brewer at Guinness (see featurette on William Gosset), encountered this same problem when dealing with various measures of quality in beer and barley production. He was responsible for improving all aspects of the business of brewing by applying science and statistics to the entire process. He developed a relationship between the *sample distribution*, the small number of measurements that are feasible for any given parameter, and the *population distribution*, which represents all possible measurements of that parameter. The population distribution is well-defined and represented by μ and σ, the population mean and standard deviation, respectively, from the Gaussian equation. The sample distribution is not well-defined and is represented by $\bar{x}$ and s, the sample mean and standard deviation, respectively, of the sample distribution. As the number of measurements (N) increases, the sample distribution becomes more similar to that of the population and $\bar{x}$ becomes μ while s becomes σ.

☞ The *population* is the set containing all possible measurements.

What happens with the area under the Gaussian curve when the distribution is ill-defined? We know that $\pm$ 1.64σ (z = 1.64) on the Gaussian curve covers 90% of the area (for a two-tailed consideration), but how does this translate to the sample distribution? Gosset developed the Student's t as a "fudge factor" to stand in for the z term in the population. If you look at Table 2.7, you will see that for a given confidence level, the Student's t value is initially much larger than the z term for small sample sizes. As the sample size increases, the t value gets closer to the z term and at $N = \infty$, t and z are the same. The Student's t is larger at small N because it expresses an increased uncertainty. In order for us to confidently label something as a signal that lies outside of the 3σ boundary, we need to have a well-defined boundary, which is what we do not have with a sample distribution. Thus, to be just as confident in our assertion, we need to extend the 3σ boundary outward a bit (to 4.143σ at N = 10).

☞ The *sample* is a subset of the population.

	Confidence levels		
α (one tail)	90%	95%	99.9%
α (two tail)	80%	90%	99.8%
DOF	t	t	t
1	3.078	6.314	318.313
2	1.886	2.920	22.327
3	1.638	2.353	10.215
4	1.533	2.132	7.173
5	1.476	2.015	5.893
10	1.372	1.812	4.143
20	1.325	1.725	3.552
50	1.299	1.676	3.261
100	1.290	1.660	3.174
(z) ∞	1.282	1.645	3.090

The DOF represents degrees of freedom, which is $N-1$, where N is the number of samples. The z values occur at DOF = ∞ (NIS, 2013).

Table 2.7 This table contains Student's t values for different sample sizes and confidence levels.

In summary, any confidence we can express in labeling something "signal" depends on how many measurements have been performed and how certain we are that the noise in the measurement represents a Gaussian distribution. As seen in Table 2.7, if the number of measured samples is less than 20 (as a rough estimate), then we are dealing with a sample distribution and must concede a little extra uncertainty than if we had measured 100 samples. This is the difference between the t and z values that will be used in the next section.

2.5 HYPOTHESIS TESTING

Have you ever taken the Pepsi Challenge? It is that blind taste test where the contestant is given two unmarked cups – one containing Coca Cola and the other containing Pepsi – and then he or she is asked to identify which soft drink he or she prefers. Some of you might say that they both taste the same, but soft drink aficionados would scoff at this and insist that there is a *significant difference* between the two sodas. To a statistician, the words *significant difference* have a deeper, nonarbitrary meaning, and mathematical proof is required to use them. Is there really a difference in taste between the two sodas, and can anyone really detect the difference with taste? Could someone actually prove a difference?

☞ Two measurements are *statistically different* if the difference between the measurements is larger than the threshold value set by a chosen confidence level. The comparison is made in a nonarbitrary way, by which anyone following the same rules would come to the same conclusion. The rules are set in an *hypothesis test*, which uses the measurement means, standard deviations, the DOF of a measurement, and a statistical threshold value, such as a Student's t, a z-term, or F-test value.

These sorts of comparisons are made all of the time in science, and, as you might have guessed, scientists have developed a nonarbitrary way to make these decisions. By "nonarbitrary," I mean that two people examining the same comparison would come to the same conclusion every time, unlike what might happen in a taste test. This process is called *hypothesis testing*, and it is a miniature cycle of the scientific method where a hypothesis is stated, numbers are crunched, and the hypothesis is either accepted or rejected. This process can definitively render an answer to the question "are they different?".

Let us go back to the Pepsi Challenge to demonstrate a hypothesis test. The question of soft drink preference is ill-defined and not practically testable. What if the test was something more precise like whether one soda has a higher sugar content compared to the other? Specifically, let us test for D-fructose since these sodas (and seemingly, every other processed food item in America) contain high-fructose corn syrup. Is the human tongue sensitive enough to detect the difference? Maybe. Then there is the issue of human bias – if the two cans were displayed in the open with their corporate labels, would a person be able to remove all bias and actually pick the drink with the most fructose? Even if that person

☞ The DOF is the number of data points minus the number of constraints placed on the data set. In our context, the constraint is the sample mean since if the entire data set were to be randomly resampled, the last data point would have to have a known value in order to maintain the sample mean.

☞ The *null hypothesis* (H_0) is the default hypothesis that starts any hypothesis test.

William Gosset (1876–1937) received a chemistry and mathematics degree at New College, Oxford and went on to invent the t test to express the extra uncertainty required when considering a small number of measurements. He was employed by the Arthur Guinness Sons and Company during a time when the company wanted to apply scientific methodology to the business of brewing beer. The name "Student's t" came from a decision by one of the Guinness executives to let Gosset publish his statistical findings without identifying the brewing company. Thus, Gosset adopted the pseudonym of Student when he published his work on the t distribution. So it was the quest for the best, most reproducible, and most cost-effective beer that produced the statistics of small samples. Let's raise a toast to Gosset and his gift to environmental chemistry. Cheers! Courtesy Wujaszek, 2006.

was an employee of one company or was raised in a family that always preferred one brand over another? Humans are influenced by all kinds of conscious and unconscious factors that get in the way of unbiased determinations. That is the advantage of the hypothesis test – it should produce an unambiguous result that cannot be denied by anyone, regardless of bias.

To start the hypothesis test, a hypothesis must be stated. By convention, this hypothesis is called the *null hypothesis*, which always assumes that the two items being compared are the same. This is so that anyone who might run this test would start with the same hypothesis and would come to the same conclusion because the comparison threshold is now defined by a Student's t value, given the DOF for the test. I will show a specific example and then show the general cases later.

Example 2.6: Hypothesis Testing: Comparing Two Experimental Means

Does one brand of soda contain more D-fructose than another? The null hypothesis for this test needs to be defined.

H_o: The D-fructose concentrations in Brand A and Brand B are the same. ($\bar{x}_{\text{Brand A}} = \bar{x}_{\text{Brand B}}$)

Now that we have a starting place, we need to make the measurements in order to draw a conclusion. One of the best methods for testing for simple sugars is to use an enzyme assay because it can specifically target the D-fructose molecule and ignore all of the other ingredients that might interfere with other measurements. So, let us say that we send the samples off to the lab to be tested for D-fructose, with the results in Table 2.8.

Brand A $\left(\frac{\text{g fructose}}{100\text{ mL}}\right)$	Brand B $\left(\frac{\text{g fructose}}{100\text{ mL}}\right)$
7.35	7.83
7.37	7.36
7.01	7.41
7.27	7.36

Table 2.8 Simulated fructose measurements for two brands of soft drinks.

Solution:

We must find the average of each set of measurements.

$$\bar{x}_{\text{Brand A}} = 7.25 \quad \bar{x}_{\text{Brand B}} = 7.49 \text{ (g/100 mL)}$$

We must find the standard deviation of each set of measurements.

$$s_{\text{Brand A}} = 0.1657 \quad s_{\text{Brand B}} = 0.2279 \text{ (g/100 mL)}$$

In order to compare the two sets, we need to find the pooled standard deviation (see Eq. (2.2 Appendix 2.3 on page 58).

$$s_{pooled} = 0.1992 \text{ g/100 mL}$$

We are comparing two experimental means here, so the following equation must be applied.

$$t_{calc} = \frac{\bar{x}_{\text{Brand A}} - \bar{x}_{\text{Brand B}}}{s_{pooled}} \sqrt{\frac{N_{\text{Brand A}} \cdot N_{\text{Brand B}}}{N_{\text{Brand A}} + N_{\text{Brand B}}}}$$

$$t_{calc} = \frac{7.25 - 7.49}{0.1992} \sqrt{\frac{4 \cdot 4}{4+4}}$$

$$t_{calc} = 1.70 \text{ (take the absolute value)}$$

To complete the hypothesis test, we must now compare t_{calc} to the Student's t value for a specific confidence level, known as the critical t or t_{crit}. It comes from the Student's t table, and it is a threshold value that, if exceeded, means the two values being compared are statistically different. First, however, we must determine the DOF of the experiment. The DOF for a single set is $N - 1$, or for two sets, is $N_1 + N_2 - 2$. Basically, one DOF is lost per set. For our test, the DOF is 6. Finally, we need to choose a confidence level. Let us use 95% for this test. If you examine Table H.1 (page 315), you will see that for our test, $t_{crit} = 2.447$ (DOF = 6, 95% confidence level for a two-tail α value).

Now we make the comparison between t_{calc} and t_{crit}. Remember that t_{crit} is the threshold value for the comparison. Since $t_{calc} < t_{crit}$, the null hypothesis is accepted, and the D-fructose levels of the two samples are statistically indistinguishable. The t_{calc} falls somewhere between t_{crit} at 90% confidence (1.943) and t_{crit} at 95% confidence. This infers that there is around a 7% chance that the two fructose levels are different.

☞ In situations where you have no reason to suspect that one $\bar{x}$ is larger than the other, it is appropriate to choose a t_{crit} from a *two-tailed* confidence interval. This t_{crit} will be larger than one from a *one-tailed* confidence interval. Practically, this means that in order for the null hypothesis to fail, the difference between the two means must be larger in the two-tailed case than in the one-tailed case. Use a one-tailed t_{crit} when you know one mean will be always be smaller or always larger than the other (or one mean compared to the true value μ).

From this example, you can see that a hypothesis test involves a t test followed by the interpretation of the results. The soft drink example compared two experimental means to see if they were statistically different from each other. Because the threshold value of t_{crit} was not exceeded by the t_{calc}, the two sample means were close enough, given the precision and confidence level, that they can be considered statistically indistinguishable.

Other comparisons that you might make as part of your environmental chemistry class could involve comparisons of other sorts.

- Comparing an experimental mean to a true value, such as a calibration standard
- Comparing the standard deviations from two experimental sets

In each of these comparisons, the equation to use is different. The equations are listed in the following box and examples follow.

comparing two experimental means:

$$t_{calc} = \frac{\bar{x}_1 - \bar{x}_2}{s_{pooled}} \sqrt{\frac{N_1 \cdot N_2}{N_1 + N_2}} \qquad (2.5)$$

Comparing an experimental mean with a true value:

$$t_{calc} = \frac{\bar{x} - \mu}{s} \sqrt{N} \qquad (2.6)$$

Comparing the standard deviations from two experimental samples:

$$F_{calc} = \frac{s_1^2}{s_2^2} \qquad (2.7)$$

Here are some more examples to demonstrate the comparisons.

Example 2.7: Hypothesis Testing: Comparing a True Value to an Experimental Mean

A calcium carbonate primary standard purchased from NIST was certified to have a purity of 99.95%. Using an atomic absorption spectrophotometer, an analyst measured the purity with five replicate analyses given in Table 2.9. Using a hypothesis test, determine whether there is a systematic error in the measurement.

Trial	Percent purity
1	99.780
2	99.858
3	99.766
4	99.734
5	99.622

Table 2.9 Simulated replicate analyses of a $CaCO_3$ primary standard.

Solution: This is a comparison involving a true value (the primary standard from NIST) and an experimental mean. The solution is determined by using Eq. (2.6). Our stepwise process for solving this problem is to state the null hypothesis, identify and calculate all of the variables in Eq. (2.6), and then compare our t_{calc} to a t_{crit} from the Student's t table.

State the null hypothesis.

> H_0: *The experimental percent purity is the same as the certified value given by NIST for the primary standard; therefore, there is no systematic error, and the instrument is calibrated properly.* $(\bar{x} = \mu)$

We need to determine the experimental mean and standard deviation and identify the number of trials and the true value (μ). Since these are intermediate calculations, extra sig figs will be used.

$\bar{x} = 99.752\%$ $\quad s = 0.0858\%$ $\quad N = 5$ $\quad \mu = 99.95\%$

Let us plug in the values into Eq. (2.6) to determine t_{calc}.

$$t_{calc} = \frac{99.752-99.95}{0.0858}\sqrt{5} = -5.1602$$

Since our concern is not whether the experimental mean is higher or lower than the true value, we only care if it is different, we can drop the negative sign. Let us compare this t_{calc} to a t_{crit} at the 99% confidence level. Since there were five measurements made, the DOF is 4, yielding a t_{crit} of 4.604 (two tail). Thus, $t_{calc}>t_{crit}$, and we must reject the null hypothesis – the experimental mean is statistically different than the certified true value, and thus, there is a systematic error in the measurement.

When a comparison between an experimental mean and a primary standard shows the presence of a systematic error, the analyst has gained some valuable information. This result implies that unless the systematic error is found, the measurements for calcium carbonate using this instrument and method will usually result in a value that is lower than the true value (measurements of concentration will be inaccurate). The challenge to the analyst is to find the error – it could be in the instrument calibration, it could be in the method used to prepare the sample, or it could be an error made by the analyst.

Example 2.8: Hypothesis Testing: Comparing Two Experimental Standard Deviations

A small company specializing in trace measurements of mercury in sediment has been given a large contract by the government to measure mercury levels in the sediment of a local watershed as part of a criminal investigation. The director of the project wants to purchase a new instrument to dedicate to this project. There are two instruments on the market to make this measurement; one manufacturer claims that their instrument (Instrument A) has the lowest noise levels and thus provides the most precise measurements, but their instrument costs $5000 more than the other. An analyst is given both instruments on a trial basis and a comparison is made to test the claims of the manufacturers. Given the measurements in Table 2.10, is one instrument more precise than the other (is there a significant difference between the noise of the two instruments)?

Trial	Instrument A (ppb Hg)	Instrument B (ppb Hg)
1	12.654	12.531
2	12.790	12.743
3	12.690	12.508
4	12.689	12.756
5	12.635	12.877
6	12.742	12.824
7	12.742	
8	12.655	

Table 2.10 Simulated replicate mercury analyses of sediment.

Solution: This is a comparison involving two experimental standard deviations, and the solution involves an F test (see Eq. (2.7)), which is similar to a hypothesis test t's. Our stepwise process for solving this problem is to state the null hypothesis, identify and calculate all of the variables in Eq. (2.7), and then compare our F_{calc} to a F_{crit} from the F test table (see Appendix H on page 316). Let us specify that we will require a confidence level of 99% for this comparison.

State the null hypothesis.

> H_0: *The standard deviations of the two instruments (A & B) are the same; therefore, there is no difference in the precision of the two instruments.*

We need to determine the standard deviations of each set of trials and identify the DOF for each set. Since these are intermediate calculations, extra sig figs will be used in the standard deviations.

$s_A = 0.0537$ ppb Hg $\quad s_B = 0.1529$ ppb Hg $\quad \text{DOF}_A = 7 \quad \text{DOF}_B = 5$

The F test formula (Eq. (2.7)) requires that the largest standard deviation be placed in the numerator and the smallest placed in the denominator. This yields the following ratio.

$$F_{calc} = \frac{0.1529^2}{0.0537^2} = 8.11$$

Using the DOF values from the numerator and the denominator, the F_{crit} value at the 99% confidence level is 7.406 (numerator DOF = 5, denominator DOF = 7). Thus, $F_{calc} > F_{crit}$, and the null hypothesis is rejected. There is a statistically significant difference between the precision of the two instruments at the 99% confidence level. It may be worth paying the extra money for Instrument A since its precision is nearly three times smaller than that of Instrument B.

2.6 METHODS OF QUANTITATION

In some situations, being able to confirm the presence or absence of an environmentally important chemical is useful. The actual concentration levels of the analyte may not be as important as knowing that contamination has occurred. For example, performing a fecal coliform test on a stream sample will tell you whether livestock have gotten access to the stream or whether there are leaky sewer pipes somewhere along the watershed. While this is valuable information, knowing the level of contamination will help you to determine

The fundamental relationship that analysts rely on is

$$S = kC \qquad (2.8)$$

where S is the analytical signal that the analyte causes in the measurement, k is some constant of proportionality that usually is factored into other items and is the measure of sensitivity, and C is the concentration of the analyte, which is generally the focus of any quantitation. In absorbance measurements, for example, this translates to Beer's Law.

$$Abs = \varepsilon bC$$

whether the contamination is a health concern. This is the difference between qualitative and quantitative measurements.

Analytical chemists have devised several different methods of quantitation, each with its specific set of advantages and disadvantages. The general purpose of quantitative methods is to determine the concentration of an analyte. Because the advantages and disadvantages of each method vary, the analyst will have to evaluate whether or not the circumstances of the proposed measurement require one method over the others. In an ideal world, they would all yield the same result, but none of us live in an ideal world. The problems embedded in each section and at the end of the chapter will present simulated data for each method so that you can test your skills at analysis. The spreadsheet exercise that is presented at the end of the chapter is ready-made to assist you in this endeavor.

☞ A *stock solution* is made directly from the primary standard and is usually very concentrated. Aliquots of the stock are diluted to make the *calibration standards*. The calibration standards are the solutions that are measured in the process of a quantitation. When the method of measurement is very sensitive, it is often necessary to dilute the stock solution and then make the calibration standards from that diluted stock solution known as a *working stock solution.*

☞ A *matrix mismatch* occurs when an interference that can be found in the sample is not also found at the same level in the calibration standards, leading to an accuracy error.

☞ The *matrix* is everything in the solution (solvent and contaminants) except the analyte.

2.6.1 The Method of External Standards

The method of external standards refers to a quantitative method where the calibration standards are made from a primary standard, which is "external" to the sample. The sample, or unknown, is distinct from the calibration standards, and they are never mixed. A *stock solution* is made from a primary or secondary standard (see Section 2.3), and from this stock solution, all of the *calibration standards* are made. These solutions contain increasing amounts of the analyte, the concentration of which is known to a high degree of precision and accuracy because an analytical balance and volumetric glassware (see section 2.7.2) are used to make all of these solutions. The calibration standards are measured using an instrument that responds to some physical or chemical property of the analyte, such as absorption of light or redox potential. The resulting relationship between the instrument signal and the known concentration of the analyte forms a mathematical relationship (seen in Eq. (2.8)), which is hopefully linear. This equation allows measured signals from unknown samples to be converted into concentrations, thus allowing the analyst to quantitate the amount of the analyte in samples that yield a signal. Figure 2.6 summarizes this method pictorially, and Section 2.8 presents the calculations necessary for a complete analysis with the method of external standards.

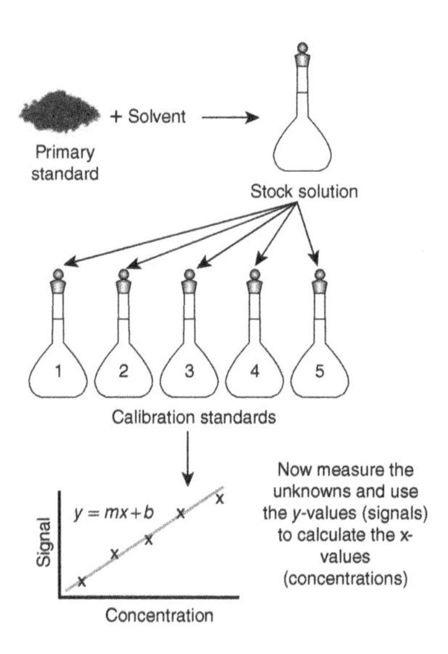

Figure 2.6 The method of external standards requires a stock solution and calibration standards produced separately from any unknown samples.

The advantage of this method is one of efficiency. One set of calibration standards can be used to measure a limitless number of unknown samples, assuming that the instrument maintains its calibration. The signal from most instruments will drift with time, so recalibration with the same calibration standards would suffice to extend the time frame of the measurements. The disadvantage comes in the form of a potential systematic error (see Section 2.2.3) as a result of something called a *matrix mismatch*. Since the calibration standards are usually made with a pure solvent and a very pure primary standard, the matrix of the standards will not match the matrix of the unknown sample, which is likely to be impure if it is an environmental sample. Some chemical species in the matrix of the sample may either enhance or attenuate the signal of the analyte. Signal enhancement would lead to an overestimate of the analyte concentration while a signal attenuation would lead to an underestimate. The magnitude of this systematic error is impossible to quantify and thus would lead to an inaccurate quantitation of the analyte.

Example 2.9: The Method of External Standards

The nitrite ion is commonly determined by a colorimetric procedure using a reaction called the Griess reaction. In this reaction, the sample containing the nitrite ion is reacted with sulfanilamide and *N*-(1-naphthyl)ethylenediamine to form a colored species that absorbs light at 550 nm. A set of calibration standards containing the nitrite ion were made using volumetric glassware. Each standard was derivatized according to the Griess method, and a portion was transferred to a special 10-cm cuvette to be measured using a UV-Vis spectrophotometer. The results in Table 2.11 were obtained for the standards.

Conc. (μM)	Absorbance
2.00	0.065
6.00	0.205
10.00	0.338
14.00	0.474
18.00	0.598
25.00	0.821

Table 2.11 Simulated calibration curve results using the method of external standards.

A single unknown sample was prepared in the same way and also measured with the spectrophotometer, resulting in an absorbance of 0.658.

1. Determine the slope of the best-fit line using a linear regression.
2. Determine the y intercept of the best-fit line using a linear regression.
3. What is the concentration of nitrite ion in the unknown?
4. What is the uncertainty in the unknown concentration?

Solution: See Section B.1 on page 256.

F F F F (a)

OH H_3C OH H_3C (b)

Figure 2.7 Internal standards (IS) are usually related to the analyte. In example (a), the analyte is naphthalene and the IS is 2,3,6,7 tetrafluoronaphthalene. In example (b), the analyte is ethanol and the IS is 1-propanol. In each example, the IS is structurally similar to the analyte but exotic enough not to be found in the matrix. Fluorinated hydrocarbons are unusual in nature and would not be found with naphthalene in a coal sample, for example. The natural fermentation in wine production would produce ethanol but not 1-propanol.

2.6.2 Internal Standards

An *internal standard* (IS) is a chemical that is added uniformly to all calibration standards and/or unknown samples in order to create a reference point internal to the solution. The IS is often similar to the analyte, but it is sufficiently different as to be distinguishable from the analyte and exotic enough that it would not be normally found in the standard or sample matrix. See Figure 2.7 for some examples of analyte/IS pairs. The IS is mixed with the analyte but must be measured separately from the analyte so that two distinct signals result. The analyte signal is always measured in reference to the IS signal, usually in the form of a ratio (analyte signal ÷ IS signal). The result is that variations in the analyte signal due to random or systematic errors can be factored out while intentional variations in the analyte concentration, inherent in calibration standards and different unknown concentrations, can be seen. The use of an IS to remove random and systematic errors from calibration standards is the basis of the method of multipoint IS, which is basically the method of external standards with an IS added to it. The use of an IS to measure the intentional changes of analyte concentrations in unknown samples is the basis of the method of single-point IS. Each of these methods will be explained in the following sections.

☞ An *internal standard* is a chemical that is added uniformly to all calibration standards and/or unknown samples. Since its concentration is known, all measurements are made in reference to it.

2.6.2.1 The Method of Multipoint Internal Standard While the method of external standards is a very efficient method for measuring the concentration of an analyte in many unknown samples, it can suffer from systematic errors and from random variations that may occur in producing and measuring each calibration standard. These random variations would increase the scatter of the points on the calibration curve and increase the uncertainty in each measurement. These random variations can appear as a result of inconsistent volumes of each standard analyzed or reproducibility problems in derivatizing the unknown. Here is an example that will help to explain this problem and the use of an IS to remove it.

Example 2.10: Using an Internal Standard to Improve Reproducibility

Chromatography is an analytical method that separates analytes in a mixture and can then quantitate each. A more thorough explanation of chromatography can be found in Appendix F. One of the features of chromatography is that the sample is injected into the instrument by means of a small syringe, usually doling out a few microliters of sample. When manually injecting samples, it is difficult to measure out a consistent volume each time because the graduated scale on the syringe body is not very precise. Before the advent of autosamplers,

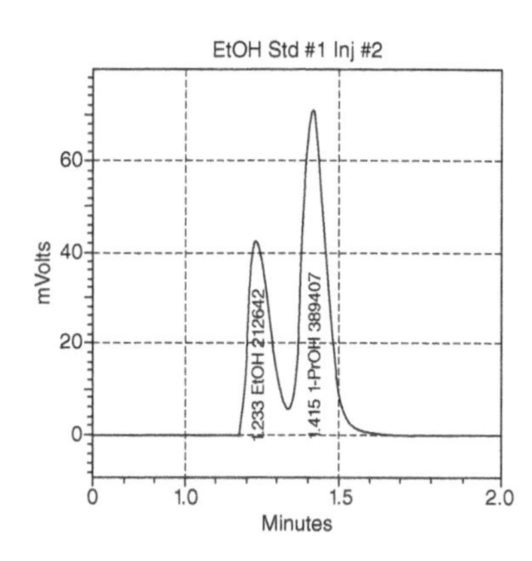

Figure 2.8 A chromatogram for ethanol (EtOH) analysis of a fermentation process might resemble this with 1-propanol (1-PrOH) as the IS.

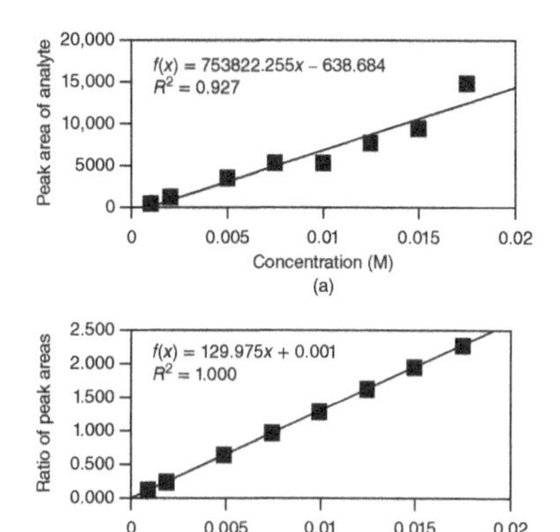

Figure 2.9 The Actual Analyte Peak Areas of the calibration standards are plotted versus their concentration in (a). Using the ratio between the Actual Analyte Peak Area and the Actual IS Peak Area, a corrected calibration curve is plotted in (b).

which robotically measure out the sample volume, it was likely that a significant amount of imprecision would be introduced to the analysis as a result of the variations in injection despite the use of very precise volumetric glassware in the earlier steps of the method.

Conc. (M)	Ideal peak area	Ideal sample inj. vol. (μL)	Actual sample inj. vol. (μL)	Actual analyte peak area	Actual IS peak area	Ratio of peak areas (analyte/IS)
0.001	650	1	0.74	481	3,698	0.130
0.002	1,300	1	0.92	1,202	4,625	0.260
0.005	3,250	1	1.10	3,566	5,487	0.650
0.0075	4,875	1	1.10	5,352	5,489	0.975
0.01	6,500	1	0.83	5,375	4,135	1.300
0.0125	8,125	1	0.94	7,675	4,723	1.625
0.015	9,750	1	0.97	9,455	4,848	1.950
0.0175	11,375	1	1.31	14,929	6,562	2.275

Table 2.12 Simulated calibration curve results using the method of multipoint internal standards.

Consider the data given in Table 2.12. It demonstrates the random error problem that the method of external standards is susceptible to and how using the method of multipoint IS, a variant of the method of external standards, solves this problem. The first two columns of the table represent the ideal calibration curve – external standards with known concentrations and the measured signal (in this case, Peak Area, see Figure 2.8). If the analyst could perfectly and reproducibly inject 1 μL of each of the calibration standards, represented by the third column, into the chromatograph, then the ideal peak area would be measured. This is sometimes a difficult proposition, so column 4 represents the actual injection volume delivered to the chromatograph. These are all close to 1 μL but different enough to affect the amount of analyte measured and the resulting signal (given as the Actual Analyte Peak Area). For example, the injection volume of the second standard is 8% smaller than the ideal volume, while the third standard injection volume is 10% greater than the ideal. This 18% variation will cause these points to deviate from the ideal linear trendline, as seen in Figure 2.9a. In fact, you will see the random variations in column 4 reflected in the scatter of Figure 2.9a. The beauty of including an IS in this experiment is that when standard 2 varies from the ideal by 8%, the IS will experience the same variation. In this case, 8% less analyte **and** 8% less IS are delivered to the chromatograph. The signal for the analyte and IS will decrease proportionally. If you examine the Actual Analyte and Actual IS Peak Area columns, you will see this proportional change because the IS concentration is kept constant throughout the standards, so the IS peak area should also be constant (ideally equal to 5000 in this example). If the ratio of the two peak areas is calculated and plotted, as in Figure 2.9b, then the random variations are removed and the ideal calibration curve emerges. The use of an IS can cut down the uncertainty in an unknown concentration by a factor of 10 in some cases.[2]

The random variations that can creep into a calibration curve could come from other sources besides injection variations: detector drift over the time of the measurements (later measurements have enhanced signal compared to earlier points, for example) and sample mass variations for solids. These errors would lead to an erosion of the precision of the measurements. Systematic errors that can be introduced into the calibration curve could come from analyte loss due to extraction or derivatization. If uncorrected, this could lead to an inaccurate measurement of the analyte. An IS is affected by these errors in the same way as the analyte. Thus, measuring the analyte relative to the IS removes the errors.

[2]Based on simulated data with an 8% RSD noise factor.

2.6.2.2 The Method of Single-Point Internal Standard A single-point use of an IS is similar to using a proportion to determine an unknown amount. In high school math classes, you probably solved something like the following with a proportion.

Example 2.11: A Single-Point Internal Standard

If five cucumbers at a farmers market cost \$2.50, how many cucumbers can be purchased for \$4.00?

Solution: A proportionality can be used to solve this problem assuming that the relationship between cost and cucumbers is linear. Since the price and amount change, the problem has two different states.

$$\frac{\text{Amount}_1}{\text{Price}_1} = \frac{\text{Amount}_2}{\text{Price}_2}$$

Substituting the values into the states yields the following.

$$\frac{5 \text{ cucumbers}}{\$2.50} = \frac{x \text{ cucumbers}}{\$4.00}$$

$$x = 8 \text{ cucumbers}$$

The same process is used in the method of single-point IS, except that a ratio of the IS to an analyte standard is used to determine the concentration of an unknown. This method requires the measurement of two different samples – one standard and IS and the other, unknown and IS – so the name of the method is a little deceptive.

☞ To *spike* a sample is to add some amount of a standard to the solution. Usually, the spike is very small, such as in the μL range, such that the overall volume changes insignificantly. When you think of spiking a sample, think of using the very sharp tips of a volumetric pipette. They somewhat resemble spikes.

Example 2.12: Another Single-Point Internal Standard

A calibration standard containing 0.0275 mg/L of Mg^{2+} (the analyte) and 4.21 mg/L of Ni^{2+} (the IS) was analyzed using atomic absorption spectroscopy and yielded absorbance signals of 0.4532 and 0.1723, respectively. A sample containing an unknown concentration of Mg^{2+} was *spiked* with the Ni^{2+} standard such that the final IS concentration was 4.21 mg/L Ni^{2+} and analyzed as before, yielding absorbance signals of 0.3245 for Mg^{2+} and 0.1727 for Ni^{2+}. What was the concentration of Mg^{2+} in the unknown?

Solution: A proportional relationship can be set up using the ratio of the analyte (A) and internal standard (IS) signals and concentrations. This IS method involves using ratios for the two different states, resulting in the following.

$$\frac{\text{Ratio of Amounts}_1}{\text{Ratio of Signals}_1} = \frac{\text{Ratio of Amounts}_2}{\text{Ratio of Signals}_2} \tag{2.9}$$

If the signals are replaced with S and the amounts (concentrations in this case) are replaced with C, then the following results are obtained.

$$\frac{\left(\frac{C_A}{C_{IS}}\right)_1}{\left(\frac{S_A}{S_{IS}}\right)_1} = \frac{\left(\frac{C_A}{C_{IS}}\right)_2}{\left(\frac{S_A}{S_{IS}}\right)_2} \tag{2.10}$$

If the standard solution is assigned to the first state and the unknown is assigned to the second state, solving for the unknown concentration yields the following.

$$(C_A)_{unk} = \frac{\left(\frac{C_A}{C_{IS}}\right)_{std}}{\left(\frac{S_A}{S_{IS}}\right)_{std}} \times \left(\frac{S_A}{S_{IS}}\right)_{unk} \times (C_{IS})_{unk} \tag{2.11}$$

$$(C_A)_{unk} = \frac{\left(\frac{0.0275 \frac{mg}{L} Mg^{2+}}{4.21 \frac{mg}{L} Ni^{2+}}\right)_{std}}{\left(\frac{0.4532}{0.1723}\right)_{std}} \times \left(\frac{0.3245}{0.1727}\right)_{unk} \times \left(4.21 \frac{mg}{L} Ni^{2+}\right)_{unk} \quad (2.12)$$

This results in a Mg^{2+} concentration of 0.0196 mg/L. This should make sense since the analyte signal in the standard (0.4532) was larger than the signal for the unknown (0.3245), so the concentration of the analyte should be smaller in the unknown. Notice that if the IS had given the same signal in the standard and in the unknown, then the IS terms could be completely eliminated from the equation. This demonstrates that in the ideal situation where there is no systematic or random error, the IS is not needed. A proportion involving just the analyte in a standard and an unknown would suffice to solve this problem. The IS is necessary, however, to remove systematic errors and minimize random variations. The concentration of the IS does not have to be the same from sample to sample, but it is usually convenient to do so since spiking both the standard and the unknown sample in the same way with the IS would result in the same concentration. Rarely do the signals from two measurements of the same concentration yield the same result due to random error.

2.6.3 The Method of Standard Additions

Let us pick on the method of external standards a little more. It is susceptible to random and systematic errors caused by factors external to the actual samples, such as variation in injection volumes. The previous section showed how an IS can take care of these issues. Another common problem that is especially likely to occur with environmental samples is a difference between the matrix of the standards and the matrix of the unknown sample – mentioned previously as a matrix mismatch. This problem cannot be fixed with an IS since the existence of an *interfering* chemical in the unknown sample and not the standard will still cause an error with or without an IS.

☞ An *interference* is a chemical that is in the matrix of the sample that either enhances or attenuates the signal of the analyte.

The method of standard additions is an alternative method that offers immunity to a certain type of matrix interference. In order to describe this immunity, it is important that you understand the method first. There are at least nine different variants of the method of standard additions as reported by one author (Bader, 1980). Presented here are two methods that I feel are the most useful for the typical measurements that are made in an undergraduate laboratory. The basic procedure for these two variant methods is to add increasing amounts of a primary standard to an unknown sample. The result is a series of data points that increase from a baseline measurement of just the unknown sample to measurements of increasing amounts of the standard. A couple of examples will be helpful in setting up each variant.

2.6.3.1 The Equal-Volume Version of the Method of **Multiple** *Standard Additions* The procedure for measuring one unknown sample using the equal-volume variant of the method of multiple standard additions starts with making a stock solution of the primary standard just the same as in the method of external standards. Next, several volumetric flasks are filled with equal amounts of the unknown sample, and then the stock solution of the standard is added to these flasks in increasing amounts (using volumetric pipettes), starting with zero. This step gives the method its namesake – multiple additions are made to the set of flasks using the standard solution. The amount of the addition is not crucial as long as the signal of each solution increases perceptively and the instrument response remains linear across all additions. The final steps are to fill all the flasks to the mark with the solvent. Figure 2.10 provides a graphical summary of the method.

Assuming that the unknown sample contains a measurable amount of the analyte, the first solution will yield a nonzero signal, and each solution after that will yield an increasing signal. In this method, it is the x intercept that yields the important information about the concentration of the unknown. The x intercept will be a negative volume, but if the absolute value is put into the dilution equation (Eq. (2.14) on page 74) as V_2 along with the stock concentration as C_2 and the sample volume as V_1, then the C_1 variable yields the unknown concentration. To state this succinctly (but possibly incomprehensibly),

> The x intercept is the volume of the stock solution of the standard that yields the unknown concentration if diluted to the original volume of the unknown sample that was added to all of the solutions.

This definition does not always make sense to the nascent analyst, so the following analogy might make more sense.

> Imagine it were possible to add antimatter analyte such that for each molecule of antimatter analyte added to the unknown sample, one molecule of analyte would be destroyed. Since the first solution in the multiple standard addition series contains only the unknown sample and solvent, it gives a nonzero signal because it contains analyte. In order to bring this signal down to zero (the x intercept), all of the analyte must be removed. If the stock solution of the standard could be magically turned into a stock solution of the antimatter standard, it could be added to the first solution of the series to destroy the analyte. The x intercept specifies the volume of this antimatter solution that would do the trick.

While being fanciful, this description conveys the meaning of the negative sign of the x intercept. It is mathematically subtracting the analyte from the first solution in order to get to zero analyte and zero signal.

There is an alternative method for calculating the unknown concentration in this version of the method of multiple standard additions. It involves plotting Signal versus Concentration of Added Standard (instead of Volume of Added Standard). In this graph, the x intercept gives the concentration of the unknown in solution #0 – the solution that did not receive any added standard (it was just diluted to some total volume). In this method, the original concentration is determined by mathematically "undiluting" the solution. In the following example problem, the answer will use both calculation methods to demonstrate that they are equivalent.

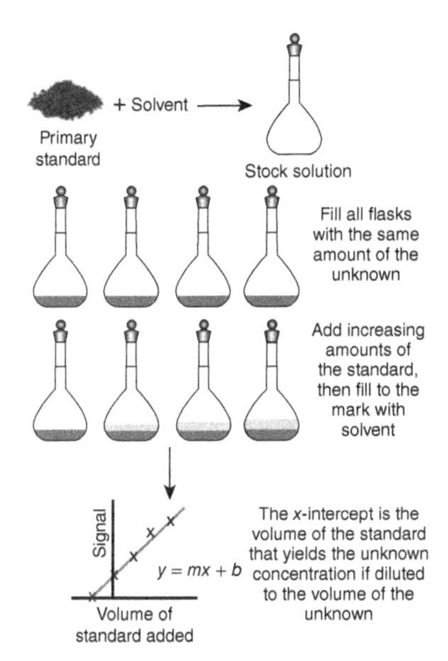

Figure 2.10 The method of multiple standard additions requires a stock solution of the primary standard to be added, in increasing amounts, to several flasks of equal aliquots of the unknown sample. The x intercept of the Signal versus Volume of Standard Added graph yields an indirect measurement of the unknown.

☞ To see the derivation for the method of multiple standard additions, see Appendix G.

Example 2.13: The Method of Multiple Standard Additions

Quinine, an antimalarial compound that was first used by the indigenous peoples of Peru and Bolivia, is a flavoring agent in tonic water. Several 20.00-mL aliquots of a tonic water sample were added to four 100.00-mL volumetric flasks. Each of these flasks was spiked with a 200.0 mg/L standard solution of quinine according to the following table. Each solution was filled to the mark with deionized water, and then each solution was poured into its own four-window quartz cuvette and measured in a fluorescence spectrometer ($\lambda_{excitation}$ = 350 nm and $\lambda_{emission}$ = 450 nm), resulting in the signal levels (in cps or counts per second) mentioned in Table 2.13. What was the concentration of quinine in the undiluted tonic water?

Solution #	Added standard (mL)	(Optional column) conc. of added standard (mg/L)	Fluorescence signal (cps)
0	0	0.00	7,499.1
1	5	10.00	13,145.8
2	10	20.00	17,794.7
3	15	30.00	23,172.8

Table 2.13 Results from a simulated method of multiple standard additions.

Solution:

Examine the table, and you will hopefully see the design of the method of multiple standard additions. The second column shows how much of the standard, the 200.0 mg/L quinine solution made with a primary standard, was added to 20.00 mL aliquots of the tonic water (which is the undiluted unknown sample). Each of these solutions was diluted to a total volume of 100.0 mL using a volumetric flask, and the fluorescence signals from these solutions were recorded. Reexamine Figure 2.10 to make sure you see the connection. Plotting signal versus volume added yields graph in Figure 2.11a.

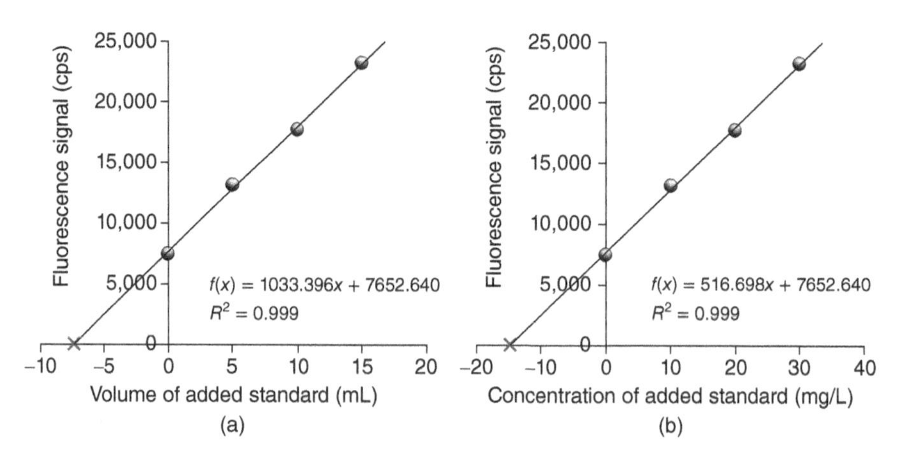

Figure 2.11 Graphs that result from the method of multiple standard additions. In (a), the x axis is the Volume of Added Standard, which yields an x intercept that is the volume. In (b), the x axis is the Concentration of Added Standard (after dilution) yielding an x intercept that is the concentration.

☞ The dilution equation is

$$C_1V_1 = C_2V_2 \tag{2.14}$$

where C_1 is the solution with the high concentration (stock), C_2 is the diluted solution that is made from the stock, and V_1 is the volume of the stock solution that was diluted to a total volume (V_2) to make the diluted solution.

Notice that the first solution (solution # 0), which contains zero added quinine standard, has a nonzero signal because it contains quinine from the tonic water. The other solutions contain increasing amounts of quinine and thus increasing signals. The x intercept is extrapolated from the linear fit – just set $y = 0$ and solve for x in the $y = mx+b$ linear equation.

$$x\text{ intercept} = \frac{-b}{m} = \frac{-7652.640\text{ cps}}{1033.396\frac{\text{cps}}{\text{mL}}} = -7.40533\text{ mL} \tag{2.13}$$

Remember that this represents the volume of the standard that would give the concentration of quinine in the undiluted tonic water if diluted to the volume of the unknown aliquot. Using the dilution equation, we solve for C_1.

$$C_1 = \frac{C_2V_2}{V_1} = \frac{(200.0\text{ mg/L})(7.40533\text{ mL})}{20\text{ mL}} = 74.053\text{ mg/L} \tag{2.15}$$

Linear regression analysis (see Section 2.8.3) would provide an uncertainty value in the unknown concentration of 4.0, yielding a final answer of 74.1 ± 4.0 mg/L quinine for the unknown tonic water. Some random error was added to the data to simulate real results. The true concentration of quinine is 80 mg/L.

Now let us consider the alternative method for determining the unknown concentration. The third column of the aforementioned table contains the calculated concentration of the quinine standard after each solution is diluted to the 100.0 mL mark. It is a simple matter of using the dilution equation. Plotting signal versus concentration of the added standard (instead of the volume of the added standard) yields graph in Figure 2.11b. The x intercept calculated from this graph is −14.81067 mg/L quinine. Note that this is not a volume as graph in Figure 2.11a since the x axis is in units of concentration (mg/L). The x intercept is the *diluted* concentration of the unknown quinine (in the 100.0 mL solution). To determine the *undiluted* concentration, just use the dilution equation solving for C_1 as before, except the values for C_2 and V_2 are different.

$$C_1 = \frac{C_2V_2}{V_1} = \frac{(14.81067 \text{ mg/L})(100.0 \text{ mL})}{20 \text{ mL}} = 74.053 \text{ mg/L} \tag{2.16}$$

Notice that this gives us the same answer, as it should. The difference here is that the x intercept is the concentration of "antimatter" quinine that would completely remove all of the quinine (from the unknown) that exists in the first solution made with zero added standard quinine. The absolute value of this concentration just needs be "undiluted" from the total volume (100 mL) to the aliquot volume (20 mL).

Your lab instructor may prefer that you use one method or the other, or you may prefer one method over the other because it makes more sense to you. Either way, the two calculation methods yield the same results.

2.6.3.2 The Variable-Volume Version of the Method of Standard Additions In this variant of the method of standard additions, an aliquot of the unknown solution is measured, then the aliquot is spiked with a small amount of a primary standard, and then this spiked solution is measured. This process can be repeated or just done once. Repeating the process requires graphing the results and determining the line of best fit, but it also provides a measurement of uncertainty, which the single spike variant does not.

This method does not require the use of several volumetric flasks – only one container (such as a beaker). The aliquot of the sample is delivered with a volumetric pipette (or equivalent), and the spikes are added using volumetric pipettes, so the container does not need to be calibrated given that the analyst knows the exact volumes added and the total volume can be calculated. This method does require, however, that the method of measurement does not consume the sample. Spectrophotometers, where the sample is placed in a cuvette or delivered to a flow cell, or electrochemical measurements where an electrode is placed in the solution, work very well with this variant of the method of standard additions (or the method of multiple standard additions if multiple additions are added). This method does not work well for chromatographic or atomic absorption measurements where some of the sample is consumed during analysis since the total volume of the sample is uncertain after each measurement. It also requires that the addition be sufficiently small in volume such that the concentration of the sample matrix remains practically unchanged (a large addition of volume from a standard might dilute the matrix and thus dilute the effects of the interferences). The following example will demonstrate the process involved in the variable-volume version of the method of standard additions.

☞ To see the derivation for the two-point variable-volume version of the method of standard additions, see Appendix G.

Example 2.14:

Let us redo the quinine assay in the previous example on the equal-volume method of multiple standard additions and see how this method gives the same result. In this example, no random error was added to the data. I will explain the reason later on in the solution section.

One 20.00-mL aliquot of the tonic water sample was placed into a 50-mL beaker. A portion of this sample was poured into a four-window quartz cuvette and measured in a fluorometer (λ_{ex} = 350 nm and λ_{em} = 450 nm). The contents of the cuvette was poured back into the beaker, and 0.100 mL of a 1000.0 mg/L quinine standard solution was added to the beaker. After the beaker was thoroughly mixed, a portion of the contents was poured into the cuvette and remeasured. This process was repeated two more times with additional 0.100 mL spikes of the quinine standard. Table 2.14 contains the resulting data. What was the concentration of quinine in the undiluted tonic water?

Solution #	Added standard (mL)	Fluorescence signal (cps)	$(V_{unk} + N \cdot V_s) \times S$ (mL · cps)
0	0	40,000.0	800,000
1	0.100	42,288.6	850,000
2	0.200	44,554.5	900,000
3	0.300	46,798.0	950,000

Table 2.14 Results from a simulated variable-volume method of multiple standard additions.

Solution: Let me first solve this problem by assuming that only one addition was made, so only solutions #0 and #1 will be considered. We need to use Eq. (G.4) in Appendix G on page 312.

$$C_{unk} = \frac{\left[C_{std}\left(\frac{V_{spike}}{V_{unk}+V_{spike}}\right)\right]}{\left[\left(\frac{S_{spike}}{S_{unk}}\right) - \left(\frac{V_{unk}}{V_{unk}+V_{spike}}\right)\right]}$$

$$C_{unk} = \frac{\left[(1000.0\ \mathrm{g/mL})\left(\frac{0.100\ \mathrm{mL}}{20.00\ \mathrm{mL}+0.100\ \mathrm{mL}}\right)\right]}{\left[\left(\frac{42{,}288.6\ \mathrm{cps}}{40{,}000.0\ \mathrm{cps}}\right) - \left(\frac{20.00\ \mathrm{mL}}{20.00\ \mathrm{mL}+0.100\ \mathrm{mL}}\right)\right]}$$

$$C_{unk} = 80.0\ \mathrm{mg/L} \tag{2.17}$$

Without any random or systematic error added, the result is a perfect 80.0 mg/L quinine, which is correct. Now let us consider the other additions – solutions #0 through #3. Using Eq. (G.6) from Appendix G on page 313, a graph of $(V_{unk} + NV_{spike})\,S$ versus N will give a linear response. Figure 2.12 is a result.

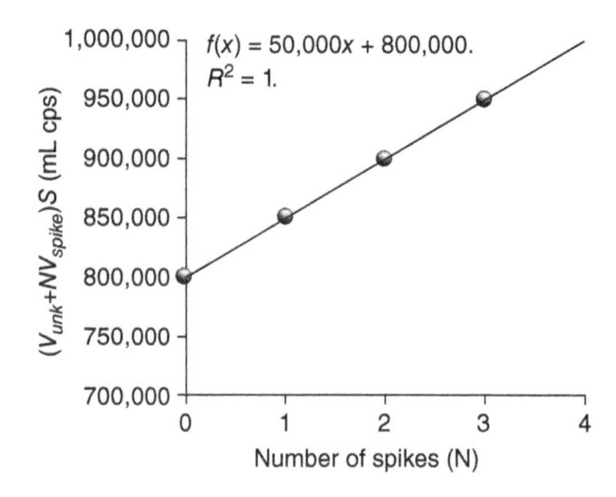

Figure 2.12 The result of a variable-volume multiple standard additions experiment.

☞ To see the derivation for the variable-volume version of the method of multiple standard additions, see Section G.3 in Appendix G on page 313.

Using the slope, intercept, and Eq. (G.8), the C_{unk} can be found.

$$C_{unk} = \frac{bC_{std}V_{spike}}{mV_{unk}} = \frac{(800,000\ \mathrm{mL \cdot cps})(1000.0\ \mathrm{mg/L})(0.100\ \mathrm{mL})}{(50,000\ \mathrm{mL \cdot cps})(20.00\ \mathrm{mL})} = \left(80.0\ \tfrac{\mathrm{mg}}{\mathrm{L}}\right) \tag{2.18}$$

Both methods result in the same answer, so why would one ever add four or five spikes if only one is sufficient? There are two reasons why more spikes are better.

First, both methods are equivalent when *there is no random error present*, but when random error is present (which is **always**!), the method using multiple spikes gains the advantage of being able to average out the random error. This leads to an unexpected increase in accuracy. Usually, random error is associated with precision, but remember that the two-point method uses two points to determine the ratio between the points. If there is random variation in one or both of these points, then the ratio can be drastically off. If one or two points are randomly high in the multipoint method, the odds are that the other one or two points will be low (since random error expresses 50% of its deviations high and 50% low). Thus, the multipoint method can average these highs and lows out to get closer to the true ratio (or slope, in the case of the multipoint). Table 2.15 contains the results of a comparison between the two methods when different levels of noise are added to the measured signal. You can see that the multipoint variant out performs the two-point variant in every case and the two-point method performs even worse as the noise level increases. Thus, if random noise is above the level of a few percent, it will be well worth the effort to use the multipoint variant.

The inputs		The results	
Item of comparison	Level of random error (%)	Multi-point method	Two-point method
% of trials less accurate	1.3	40	60
Average % error	1.3	3.13	4.68
% of trials less accurate	4	25	75
Average % error	4	8.60	22.1
% of trials less accurate	10	10	90
Average % error	10	18.8	73.7

When random error, at the level specified in column 2, was added to the measured signal and 20 trials were conducted, the multipoint method was less likely to be inaccurate and the average % error from the true value was smaller. The differences between the multipoint and two-point methods increased as the level of noise increased.

Table 2.15 Two important comparisons were made between the two-point and the multipoint variants of the method of standard additions.

Second, the two-point variant assumes that there is a linear response between the unspiked solution and the spiked solution (similar to the two-point IS method). With only two points, it is impossible to see a nonlinear response, whereas nonlinearity in the data will be obvious when the graph from the multipoint method is produced. If nonlinearity occurs, it is often because analyte concentrations are too high and all solutions need to be diluted. While it is impossible to see the nonlinearity with the two-point variant, nonlinearity is likely to occur in spectroscopic measurements where the absorbance signal is above 1.5 or 2.0. A nonlinear measurement means that calculating the unknown concentration is more difficult, and the error analysis becomes impossible without sophisticated software.

Finally, the multipoint method provides enough data to allow an estimate of uncertainty to be calculated. This is vitally important if a comparison of methods, samples, or instruments is needed or simply to communicate to someone else how reliable the result is. If you need to just make a quick measurement and get a rough idea of the concentration of an unknown, then the two-point method is quite appropriate. If you are going to report your results to some regulatory agency, archive the results as part of a long-term study, or if the results will be used in a legal dispute, then multipoint methods are the most appropriate.

2.6.3.3 How the Method of Standard Additions Eliminates Proportional Errors Remember that one of the advantages of the method of standard additions is that it is not susceptible to one of the systematic errors caused by a matrix mismatch. The first kind of error is a proportional error, where a certain chemical is in the matrix of the sample that does not contribute to the signal directly (so the sample blank would give a zero signal) but increases the signal of the analyte in proportion to the concentration of the analyte. Imagine two samples from two different lakes – one which is known to contain copper(II) species and the other that is copper free. If the sample blank from the copper-free lake was measured spectroscopically for copper, it should result in a zero signal (because there is not any copper in it). If increasing amounts of a copper standard were added to this sample blank, then increasing signals should result, giving rise to the "no error" set in Figure 2.13. A proportional error is one that mathematically multiplies the signal from each measurement. So, for the copper-free lake sample blank, there would still be no signal because there is no copper. Once some copper is added from a standard solution, then the proportional interference would cause a proportional error, as seen in the "proportional error" set in Figure 2.13. Notice that this line shares the same point at zero concentration but has a different slope. A second error is seen as the "constant-error" set in Figure 2.13, where the data set has the same slope but different y intercepts. This type of interference contributes to the signal independent of the analyte. In the case of a spectrophotometric analysis of copper, which has a light blue color, a constant error would be present if blue food dye were present in the lake (many of the blue algicides that are sometimes visible in ponds contain copper). The blue food dye would provide an absorbance signal in the absence of copper and an additional signal for all samples that contain copper.

☞ It is not a good idea to assume linearity (or not) with spectrophotometers. Many atomic absorption instruments are nonlinear above absorbance levels of 1.2, whereas a good UV-vis spectrophotometer is usually linear above an absorbance of 2.0.

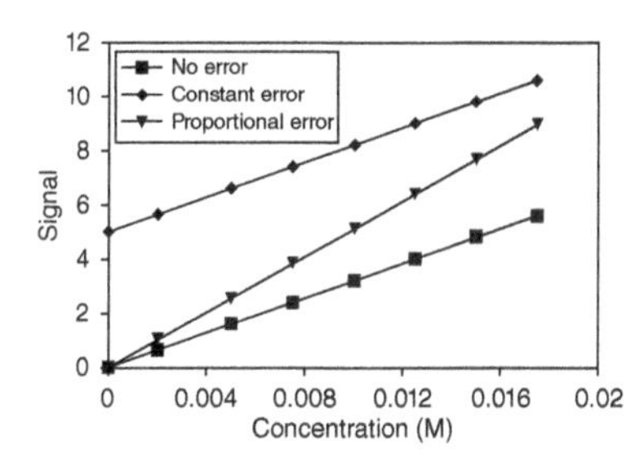

Figure 2.13 The two errors caused by chemical interferences in the matrix of the sample are constant errors and proportional errors – here compared to the error-free sample blank spiked with known amounts of a primary standard in the method of standard additions.

From Figure 2.13, it is obvious that there is a problem with interferences that cause both proportional and constant errors – the measurements are not the same as the "no-error" data set. Since the equal-volume version of the method of standard additions depends on the x intercept for the concentration of the unknown, if a proportional error were present in the matrix, the intercept would remain unchanged. This is not the same for a constant error in the matrix. Since the variable-volume version of the method of standard additions depends on the y intercept and slope for the concentration of the unknown, a proportional error would result in a multiplicative effect on both variables and would cancel in the final analysis. Again, the same is not true for the constant-error interference, which would contribute an additive effect to each variable and would not cancel out.

In conclusion, the only time the method of standard additions can eliminate a systematic error caused by a component in the matrix is when the sample blank (with zero concentration of the analyte) gives a zero signal (this is the test for a proportional error). Unfortunately, sample blanks are not usually available for environmental measurements. One further complication this presents is that a sample blank is often hard to obtain. If you are testing a lake for copper, for example, you could use a sample from a lake that you know is copper-free, but you must assume that this sample blank has the same matrix as the lake that you are testing – rarely will this be strictly true, but it may be practically true. Thus, the method of standard additions is a useful method of quantitation that can be quicker to use on a single sample if the sample is not consumed by the analysis because of the ease with which spiking a sample and remeasuring can be accomplished.

2.7 QUANTITATIVE EQUIPMENT

Quantitative methods of measurement are lengthy endeavors that you should not plan on completing in some spare moment of time you have before going onto something else. Most quantitative lab experiments you will complete as part of this class will take a minimum of 3 h and sometimes can stretch across multiple sessions. Imagine you have been working for 4 h and have finally finished the measurements, and while you are cleaning up, you remember that you used low-precision glassware and a noisy electronic balance. Those early steps should not have that much effect on the final results, since you were so careful in calibrating the instrument and making measurements, right? ***Wrong!*** You have heard the saying that a chain is no stronger than the weakest link, right? The same holds true for quantitative methods – the errors from every step accumulate and rarely do they ever cancel each other out. It is important to minimize the errors every step along the way (reread the "Lord Overway" quote at the beginning of the chapter!). Thus, it is important for you to use the proper chemicals and equipment so that your hard work is not wasted. You have already learned about primary and secondary standards – chemicals that have a high purity worthy of quantitative work. In order to make effective use of these standards, you need to use the proper equipment.

2.7.1 Analytical Balances

The most indispensable piece of equipment in a preparatory lab is an analytical balance. You should be using one that has 0.1 mg precision (four decimal places in the gram mode). This will imply that the balance has a glass enclosure around the pan to prevent air currents from interfering with the measurement. While it is okay to leave one of the doors open while you are moving the weigh boat in and out of the balance to scoop out the sample and then place it back on the pan to register the mass, you should always close all of the doors and let the balance settle before you record the final mass of the substance.

You should never place warm or cold objects on the balance pan and expect to get an accurate mass. The temperature difference between the container and the air inside the balance will produce air currents that may affect the buoyancy of the container. You should also make sure that the balance is leveled. If you notice the digital display drifting either down or up without settling, it may be the result of the balance level. There is usually a leveling device at the back of the balance, which indicates the level while you twist two of the feet on the balance.

Finally, you should never place anything that is wet (on the outside) on the balance since some liquid remaining on the pan will affect the mass.

2.7.2 Glassware

Common glassware that you have seen in a general chemistry laboratory is a variety of fairly accurate and precise glassware and very coarsely precise glassware. It is usually easy to tell the difference if you think about how liquids are typically measured. You are likely aware that water and many other liquids form a meniscus when in contact with glassware, thus when a container is filled with water, you must get on eye-level with the meniscus in order to place the level of water within some graduated set of lines. If the surface area of the meniscus is large, then the uncertainty in the volume is going to be large. Beakers, for example, have wide mouths and very coarse rulings on the side. This is because it is a waste of manufacturing time and expense to make the rulings any more precise – it just does not matter with so much surface area. Beakers and Erlenmeyer flasks are great for mixing and containing solutions, but they are not designed to impart any useful precision or accuracy to a measurement.

A graduated cylinder moves much further up the evolutionary scale of precise glassware. Have you ever wondered why they are so tall and narrow? If they were short and wide, they would be beakers with low precision! Graduated cylinders are narrow because they reduce the area of the meniscus and allow for a more precise measurement to be made. They often have finer rulings than beakers because they can truly deliver that level of precision. It is appropriate to use graduated cylinders for reagents and solvent volumes when precision does not matter (see Section 2.3.1 for more detail on this), but for the analyte and quantitatively important solution volumes, graduated cylinders are still not precise enough. If they are certified as Class A or Class B, then they are manufactured with some precision. Typically, a 10-mL graduated cylinder with $\pm$ 0.5 mL markings that is Class A certified has a $\pm$ 0.1 mL precision tolerance (Class B is $\pm$ 0.2 mL). There will be a capital "A" or "B" on the cylinder if it is certified. If it does not have a letter, then you should assume that it is even less precise.

If "narrower is better" worked for graduated cylinders, then it will not be a surprise when you look at a volumetric pipette, burette, or volumetric flask. Pipettes are usually very narrow because narrower means better precision. Most volumetric pipettes are designed to deliver (designated by TD on the pipette) and flasks to contain (designated by TC on the flask) a nominal amount of liquid. It is, therefore, inappropriate to use a 10-mL volumetric flask to deliver 10.00 mL of solution to another container, because the flask was designed to contain and not deliver volumes. Too much liquid remains in the flask, because of intermolecular adhesion forces, for the delivery to be accurate and precise. Pipettes and flasks also come as Class A or B certified. A 10-mL Class A volumetric pipette will have a $\pm$ 0.02 mL tolerance. If it is a Class A graduated pipette it will have a $\pm$ 0.05 mL tolerance. Each piece of glassware should have its class designation and tolerance printed on the neck of the pipette or flask near the top. It is important for you to know this since you will need to record these volumes in your lab notebook. If a 10-mL pipette has $\pm$0.02 mL tolerance, you should record four sig figs in your notebook (10.00 mL). If you just write "10 mL" and do not indicate the precision, you will be limiting the number of sig figs you can express in the concentration of the solutions you make. If an analytical balance provides you five sig figs and the volumetric flask provides you four sig figs, then you can confidently write the final concentration of your solution with four sig figs. Your instructor may be looking for items such as this when your lab notebook is graded. Even if your notebook will not be graded, it is a good lab practice to provide as much information as possible in your lab notebook.

Since you have completed a general chemistry lab course, you are likely familiar with most of this glassware. In a quantitative environmental chemistry lab, you will be expected to be more careful in making measurements than you have in the past. You should really try hard to get the bottom of the meniscus right on the mark instead of "sort of close." You should make sure that all of your glassware is thoroughly cleaned before you use it. It is especially important to use deionized or reverse-osmosis water in all of your aqueous solution preparations since city water can vary in its hardness, and the electrolytes in the water can affect your measurements.

You also need to remember that when draining volumetric pipettes you always let the pipette drain by gravity (not by force), and after you touch the tip of the pipette to the inside of the receiving container, you leave the tip of the pipette filled with liquid. It is calibrated to take into account this small amount of liquid. These glass pipettes and flasks are also calibrated at 20 °C, so their accuracy decreases as the room temperature deviates significantly from 20 °C. They are usually designed for volumes larger than 1 mL. Graduated pipettes that can deliver tenths of a milliliter or larger are available, and flasks as large as 2 L are common in labs.

Each measurement you make is only as good as the equipment you use and the care with which you use it. You are now in an upper-level science course, so you should expect more from yourself than you might have during your earlier experiences in the laboratory.

2.7.3 Pipettors

No quantitative laboratory would be complete without a set of micropipettors. These devices are often not seen by students in general chemistry and biology courses for a very good reason – they are expensive and hard to clean! If your instructor brings them out for use in your lab, then you know you have crossed a figurative line in your professor's mind from "Let the heathens only use cheap glassware!" to "These are trusted students, so give them the good stuff!". Micropipettors make delivering small volumes, from a few μL to few milliliters, very easy, precise, and accurate. They use disposable tips to avoid contamination. Each pipettor has a certain range of volumes that it can deliver, and a knob at the top is used to dial the desired volume. You need to be careful not to dial any volume outside of the specified range or you will risk jamming some of the gears inside the device.

Your instructor will hopefully give you some specific training on the use of these pipettors. When you push down on the knob, you will feel two spring stages. The first is a light spring, which goes almost to the bottom range of the knob, then you will feel the second (stiffer) spring. Its range is smaller and stops at the very bottom of the knob range. The first spring is for delivering or loading the desired volume, and the second spring is to eject the very last drop of liquid that remains in the tip. Here is a general procedure to follow.

For single-volume deliveries

1. Dial the pipettor to the desired volume.
2. Grab the pipettor so that the barrel is firmly in your grip and your thumb is on the spring-loaded knob.
3. Find the tray of pipette tips that fit your pipettor, and firmly insert the bottom of the pipettor into a tip. When you lift up the pipettor, the tip should be attached – if you do it correctly, you will not have to touch the tip (which may be sterile).
4. Hold the pipettor vertically with the tip on the bottom, and push the knob down to the bottom of the first spring.
5. Insert the tip below the surface of the liquid, and ***slowly*** release the knob. The liquid will be drawn into the tip. If you release the knob too quickly, the liquid can squirt up into the body of the pipettor and foul it (and really tick off your lab instructor!).
   (a) Once liquid is in the tip ***never***, hold the pipettor horizontally or set it down on the benchtop. Liquid in the tip could migrate into the pipettor.

6. Examine the liquid in the tip to ensure that there are no bubbles. If there are, it is likely that you did not keep the tip submerged during the acquisition of the liquid. If there are bubbles, push the liquid out and try again.
7. With your thumb completely off of the knob, lift the pipettor out of the liquid and move it to the receiving container.
8. Slowly press the knob down to the bottom of the first spring. You will notice that there is still some liquid remaining in the tip.
9. Press the knob until you reach the bottom of the second spring. The remaining drop of liquid should leave the tip.
10. Examine the tip, and make sure that there is no residual liquid in the tip. If so, you probably pushed the knob down too fast. This can happen with viscous liquids such as glycerol.
11. If you are going to deliver another aliquot of the same liquid, then you can reuse the tip. If you are going to pipette a different liquid, eject the tip into the trash and reload a new tip.

Most pipettor manuals describe a slightly different procedure for delivering multiple aliquots of the same liquid. You might need to put 500 μL of the same solution into five different containers. If this is the case, you can use the following procedure, which tends to be a little faster than the previous method.

For multivolume deliveries

1. Dial to the desired aliquot and load the tip as described before.
2. Push the knob down to the bottom of the first spring and then down to the bottom of the second spring.
3. Insert the tip into the liquid. Call this the stock solution.
4. Release the knob slowly all the way up to the top to draw in the liquid.
5. With your thumb off of the knob, remove the tip from the liquid and move the pipettor to the receiving container.
6. Push the knob down to the bottom of the first spring. *Do not go down to the bottom of the second spring.* You should see some liquid still in the tip.
7. While still holding the knob at the bottom of the first spring, move the pipettor back over to the stock solution and acquire another aliquot.
8. With your thumb off of the knob, repeat the delivery procedure going down to the bottom of the first spring.
9. Repeat until all aliquots have been delivered.
10. Move the pipettor to the container holding the stock solution and push the knob down to the bottom of the second spring to eject the residual liquid back into the stock container.

Pipettors, just as any other lab equipment, should be cleaned on occasion since there will always be someone who lets up on the spring too quickly and fouls the inside of the pipettor body. If you want to clean the pipette, ask your instructor for the pipettor manual, which contains all of the details for disassembly. There is often a tool that is necessary to pop open the pipettor case.

Pipettors can be calibrated, unlike glass pipettes, and should be calibrated occasionally. The pipettor manual usually describes this procedure, but it can be simply done by using an electronic balance, some deionized water, a weigh boat, and a calculator. Look up the density of water based on the ambient temperature, and you should be able to verify the calibration of the pipettor.

2.7.4 Cleaning

If you have ever cooked a meal for a large group of people, you know that the cleanup is a time-consuming task. Quantitative analyses are twice as worse – along with the after-experiment cleanup is usually a pre-experiment cleaning. The correct glassware for an analysis must also be free of any contaminants that could degrade the measurement. Since most quantitative measurements are made on samples with low concentrations, glassware that looks clean may be harboring contaminants at significant levels. If glassware has been scrubbed with soap and rinsed with deionized water, it is likely to be clean enough for most of your work. Some soaps contain phosphates, so soap and water may not be sufficient for phosphate measurements.

Your instructor will be the arbiter of the definition of clean. Since your measurements will likely not be sent to a government agency as part of a longitudinal study or used in a courtroom to settle a charge of pollution, it may not be an effective use of time to thoroughly clean glassware (you will be busy enough with preparing and measuring standards and samples). Professional environmental scientists do not have this luxury. Since their measurements are often part of an official record that may be used to substantiate criminal charges or grant construction permits, they must have scrupulously clean containers that are certified by an analytical lab. The Environmental Protection Agency has guidelines available if you are interested in reading or following them for your experiment (EPA, 1992).

If soap and water are not sufficient, there are a few common detergents that are specially designed for analytical glassware. If you are working with biological samples that could accumulate proteins, then a soak in a peptidase cleaner will work. If the glassware has some tough organic or metal stains, then a soak in a chromic acid/sulfuric acid bath will work. A less toxic alternative is to use NoChromix, which is an inorganic oxidizer that functions the same way as the chromic acid bath without the hazardous chromium waste. Organic contamination can also be cleaned with a strong base bath – a saturated solution of NaOH or KOH dissolved in ethanol or methanol. Each of these baths presents personal safety issues that will require you to wear special clothing, special rubber gloves, and eye protection (sometimes a splash mask). Acid baths are very corrosive, base baths are flammable and caustic, and both can produce fumes that are noxious. Be especially careful about soaking volumetric glassware in base baths – strong bases can etch glass, destroying the effectiveness of the ground glass stopcock and changing the volume of the glassware. If you need to use a base bath with volumetric glassware, soak the pieces for only a few minutes and then rinse immediately. Make sure that you get sufficient training before you use base or acid baths, and wear clothes that you do not care much about since acid baths can easily burn holes in the fabric.

2.7.5 Sample Cells and Optical Windows

Many of the environmentally important chemicals that you will investigate in the lab during the course of this semester will involve spectroscopic methods. The interaction of light and matter, introduced in Chapter 1, is one of the most powerful and convenient ways to investigate the structure and amount of matter in a given mixture. Some instruments that you might use, such as the atomic absorption spectrophotometer or one of the forms of chromatography, require that the sample be injected into the instrument in some fashion. In these cases, the containers you use to store and deliver the sample do not have to have any special optical properties – they just need to be clean and inert. Other methods of spectroscopic analysis require that the sample be placed into a cuvette or well plate and then placed in the optical path of the instrument. These containers must be made out of optically transparent material that is inert to the solvent and contributes little to no background signal to the measurement.

The most common materials for these containers are various plastics, glass, and quartz. Each material has its advantages and disadvantages, so let us examine where and when each material is appropriate.

2.7.5.1 ***Plastic*** Plastic cuvettes are the least expensive type of cuvettes. Polystyrene cuvettes, poly(methyl methacrylate) (PMMA), and related polymers cost about 15¢ per cuvette. These cuvettes are mostly limited to aqueous solutions since some organic solvents will dissolve the plastic. Both varieties come in two-window or four-window versions – absorbance measurements require only two-window version, whereas fluorescence measurements require four-window version. The polystyrene cuvettes are limited to visible spectrum measurements only since the plastic begins to absorb light around 320 nm (remember that visible radiation is from about 400 to 750 nm and UV is <400 nm). The PMMA cuvettes allow measurements in the >280 nm region of the UV, and the proprietary polymers sold as UV-Cuvette® and UVette® allow measurements in the >240 nm region of the UV spectrum. See Figure 2.14 for a visual display of cuvette absorbance ranges.

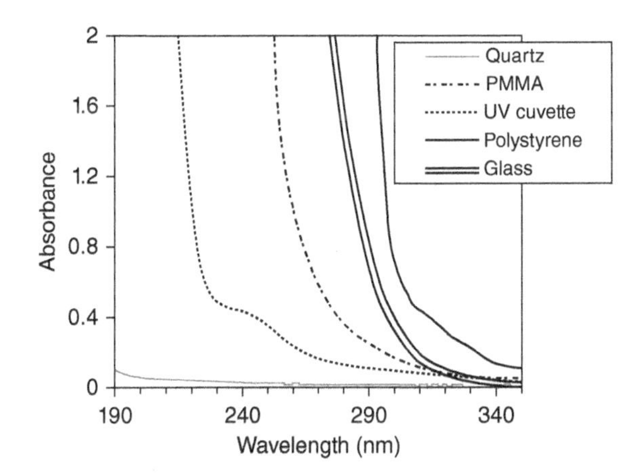

Figure 2.14 The absorbance spectra of the most common cuvettes are shown here. All measurements in the visible region can be accomplished with the inexpensive, disposable polystyrene cuvettes. Glass is useful in the visible region when organic solvents are needed. For aqueous solutions measured in part of the UV spectrum, the polymethylmethacrylate (PMMA) and other UV cuvette brands are appropriate and inexpensive. Quartz cuvettes provide the best spectral transmission and solvent compatibilities but are very expensive.

2.7.5.2 ***Glass and Quartz*** While restricting optical measurements to the visible spectrum and just the low edge of the UV spectrum, glass offers a wider tolerance for solvents than plastic. Glass cuvettes are more expensive than plastic cuvettes, priced around $10 per cuvette. Quartz cuvettes offer the best of both worlds – a wide range of solvent compatibility and optical transmission below 200 nm. Quartz cuvettes, however, are around $30 per cuvette.

2.7.5.3 ***Well Plates*** Some spectroscopic instruments require that samples be placed in a well-plate format. These plates come with different amounts of wells, but 96 wells (8×12), 384 wells (16×24), and 1536 wells (32×48) are common. They come in different formats: various clear plastics for visible and UV-vis measurements, opaque white and black plastic for luminescence measurements (the opaque material prevents "cross talk" between cells – the luminescence in one cell does not show up in the measurement of another cell), and removable rows for reducing waste and expense when only a few samples need to be measured. Costs of these plates range from a couple of dollars per plate for 96-well formats and more other formats.

These plates are extremely useful when replicate measurements need to be made with very expensive analytes. Well volumes range from 300 to 12 μL, so very small amounts of precious samples can be measured. Multichannel micropipettes with 8, 12, and 16 tips can be used to inject replicate solutions into the plates, reducing the time for loading the plates and increasing the number of replicate samples (for why measuring replicate samples is important, see Section 2.4). Well plates have a small path length compared to 1-cm cuvettes, so they result in a less sensitive measurement.

2.8 LINEAR REGRESSION LITE

Most spreadsheets provide all of the necessary functions to perform linear regression on the data that you will collect as part of your laboratory experience. Since the purpose of this textbook is not to teach you the derivation of the linear regression method but to teach you how to use linear regression, the following description will make use of the functions that are built into MS Excel and many other spreadsheets. You will see these functions being employed in the following exercises. The data you will collect for each lab will start out with a calibration curve of some sort (using either the method of external standards or the method of multiple standard additions) followed by measurements of the unknowns (such as water samples from a local pond). Keep this in mind as you figure out how the following spreadsheet formulas are used in an actual data set. A more thorough derivation of the equations involved in linear regression can be found in an analytical textbook (Moore and McCabe, 1999; Skoog, Holler, and Crouch, 2007).

2.8.1 The Method of External Standard Regression Template

Example 2.15:

Perform a full regression analysis on the calibration standards and samples given in Table 2.16 using the method of external standards.

Calibration standards	
Concentration (mg Ca/L)	Signal (absorbance)
2	0.065
6	0.205
10	0.338
14	0.474
18	0.598
25	0.821

Nonreplicate samples	
Label	Signal (absorbance)
River 1	0.658
River 2	0.350
River 3	0.550
River 4	0.800

Replicate measurements	
Label	Signal (absorbance)
River 5	0.421, 0.422, 0.419, 0.420

Table 2.16 Comprehensive simulated results from a method of external standards experiment, including the calibration standards, nonreplicate sample measurements, and replicate sample measurements.

Solution: To solve this problem, go to the textbook homepage and download the Regression-Template.xls spreadsheet by right-clicking on the link and saving the spreadsheet to a network or flash drive. Open up the file with Microsoft Excel, find the "External Stds blank" sheet, and follow the steps as follows. Follow the use of these equations by referring to Figures 2.15–2.17.

Regression Analysis		
Slope	***	0.032813
Std. Dev. in Slope (s_m)	***	0.000345
Y-intercept	***	0.006671
Std. Dev. in Y-int (s_b)	***	0.005045
R^2	***	0.999559
Standard error in y (s_r)	***	0.006426
N (count of stds)	***	6
S_{xx}	***	347.5
y-bar (avg signal of stds)	***	0.416833

Figure 2.15 In this section of the regression spreadsheet for the method of external standards, you will use various built-in functions to establish a minimal regression analysis.

Procedures (We will start with the easy formulas first.)

1. Enter the concentrations and signals from the standards into Conc. and Signal columns (A and B). The yellow cells are designated for raw data.
2. Use the *SLOPE* and *INTERCEPT* functions to calculate the slope (m) and intercept (b) of the equation of best fit for the Signal versus Conc. data. Enter these formulas in the regression analysis section of the spreadsheet template.
   (a) In the cell to the right of the Slope label, enter the slope formula (delete the *** first or just type over it). The *SLOPE(y values,x values)* function takes two arguments: a range of y values and a range of x values. Type `=slope(`, then use your mouse to highlight the signal values,[3] then push the comma key, and then use your mouse to highlight the concentration values, then type `)`, and then press ENTER.
   (b) Repeat the same procedure for the *INTERCEPT* formula, which takes the same arguments. Use the cell just to the right of the "Y-intercept" label. Use rows 5–36.
3. The *RSQ* function takes the same arguments as *SLOPE* and *INTERCEPT*. Enter the *RSQ* formula in the cell next to the R^2 label. Use rows 5–36.
   (a) The next useful function, *RSQ*, calculates the R^2 value for the fit, which estimates the degree to which the calibration data fits to the line of best fit. If the calibration standards fall perfectly on the line, then the $R^2 = 1$, whereas a very bad fit would be a fractional number approaching zero. Your instructor can decide what is an unacceptable fit.
4. The *STEYX* function, which finds the standard deviation in the measured y values, takes arguments similarly to *SLOPE*. Enter this formula in the cell to the right of the "Standard error in y" (often referred to as s_r) cell. Use rows 5–36.
5. The *COUNT* function counts the number of nonempty cells, so it is convenient to use it to count the number of standards used, since this can change from experiment to experiment. Enter this formula into the cell to the right of the "*N* (count of standards)" cell. *COUNT* only takes one range of values, so count either the Conc. cells or the Signal cells (not both). Use rows 5–36.
6. The *AVERAGE* function can be used to calculate the $\bar{y}$ value at the bottom of the regression analysis section. This is the average of the signals from the standards. Use rows 5–36.

Now, let us move on to the hard formulas in the regression analysis section.

[3]The yellow cells for the Conc. and Signal columns go down to row 36. This leaves excess room for experiments where you might have 4 standards or 12 standards. I would suggest that you use all of the yellow cells in your formulas (rows 5–36) so that you will not have to change the formulas in the sheet when you change the number of standards from experiment to experiment. If the cells are empty, then the spreadsheet will ignore them, so with a little extra effort, you can make the spreadsheet flexible for nearly every situation.

7. The S_{xx} term will become important in other formulas. It is defined as

$$S_{xx} = \sum (x_i - \bar{x})^2 = \sum x_i^2 - \frac{(\sum x_i)^2}{N} \tag{2.19}$$

Microsoft Excel has a related function called *VARP* (variance of the population of x values), defined as

$$VARP = \frac{\sum (x_i - \bar{x})^2}{N} \tag{2.20}$$

Notice that $S_{xx} = N \times VARP$, which is the formula you should enter into the cell to the right of the "S_{xx}" label. You have already calculated the N in the cell above. The arguments going to *VARP* are the range of concentration values for the standards found in column A. Use rows 5–36.

8. The standard deviation in the slope (s_m) has the following formula (Skoog, Holler, and Crouch, 2007).

$$s_m = \sqrt{\frac{s_r^2}{S_{xx}}} = \sqrt{\frac{(STEYX)^2}{S_{xx}}} \tag{2.21}$$

You can easily enter this formula into the cell to the right of the "Std. Dev. in Slope" label by referring to the *STEYX* cell and the S_{xx} cell.

9. The standard deviation in the y intercept (s_b) has the following formula.

$$s_b = s_r \sqrt{\frac{\sum x_i^2}{N \sum x_i^2 - (\sum x_i)^2}} \tag{2.22}$$

Notice that the denominator of this radical is NS_{xx}. This formula can be implemented in a spreadsheet, using the function *SUMSQ* as the following.

$$s_b = s_r \sqrt{\frac{SUMSQ(x \text{ values})}{N \cdot S_{xx}}} \tag{2.23}$$

You can enter this formula into the cell to the right of the "Std. Dev. in Y-int" label by referring to the *STEYX* cell, using the *SUMSQ* function and inputting the concentration of the standards in column A, and referring to the S_{xx} cell. Use rows 5–36 in the *SUMSQ* function.

This completes the formulas in the regression analysis section, which are interesting to look at but are functionally more useful in determining the unknown sample concentration and uncertainty. If you did everything correctly, your formulas should yield the red numbers to the right of the cells, assuming that you used the standard Conc. and Signal values given in the given table. You can delete the red numbers once you have verified this.

Let us move on now to the non-replicates section of the spreadsheet, which is the place where unknown samples, which were only measured once, will be analyzed.

10. Enter the labels and signal values for the unknown into the Label and Signal columns in the spreadsheet (columns G and H).

11. Now for the concentration of the unknown samples. Remember that the concentration of the calibration standards, which are given in this case but are normally were calculated, represents the x values and the Signal represents the y values of the linear regression. The goal is to fit this data to a line represented by

$$y = mx + b \tag{2.24}$$

Once this relationship has been established, the signals from the unknowns (y values) can be used to determine the concentration of the unknowns (x values). The formula comes from solving for x in the equation of the line.

$$x = \frac{y - b}{m} \tag{2.25}$$

Non-replicates Section						
Label	Signal	Conc.		Unc. (s_c)		
---	0.658	***	±	***	19.84974 ±	0.22518
---	0.35	***	±	***	10.46320 ±	0.21261
---	0.55	***	±	***	16.55836 ±	0.21578
---	0.8	***	±	***	24.17730 ±	0.24453
---			±			
---			±			
---			±			
---			±			
---			±			
---			±			

Figure 2.16 The non-replicates section includes all of the singly measured unknown samples. Replace the - - - in the Label column with a description of the sample, and then add the signals to the Signal column. The other shaded cells contain formulas for the concentration and uncertainty in place of the *** markers.

Thus in the Conc. column (column I), you need to use the slope and y intercept calculated in the regression analysis section along with the signal to find the x value in Eq. (2.25). If you copy this formula and paste it into the other cells in the column, you will get incorrect results because the spreadsheet is not keeping the slope and y intercept fixed in the formula. You can accomplish this by using $ in the cell reference, which locks the row or column. The formula in cell I5, for example, would be `= (H5-$E$6)/$E$4`. This will allow you to copy the formula down as far as the Signal values go.

12. Once the unknown concentration is calculated, we must calculate some estimate of the uncertainty for this value. For the method of external standards, this equation takes the following form (Skoog, Holler, and Crouch, 2007).

$$s_c = \left(\frac{STEYX}{SLOPE}\right)\sqrt{\frac{1}{M}+\frac{1}{N}+\frac{(\bar{y}_c-\bar{y})^2}{SLOPE^2\cdot S_{xx}}} \tag{2.26}$$

where M is the number of replicate measurements of the unknown, N represents the number of calibration standards in the calibration curve, $\bar{y}_c$ represents the average signal value of the unknown, and $\bar{y}$ represents the average signal value of all of the standards. The other formulas, $STEYX$, S_{xx}, $SLOPE$, you have already calculated. Since this formula will be placed in the non-replicates section of the spreadsheet, it should be obvious that $M = 1$. Related to this, if the unknown sample is only measured once, then the $\bar{y}_c$ is just the signal value since the average of a single value is just that value. You can use the cells in the regression analysis section for the values of N, $STEYX$, S_{xx}, $SLOPE$, and $\bar{y}$. Make sure that you use $'s in the formula when you refer to the cells in the regression analysis section since you want these references to remain fixed when copying the formula down in column K. If everything is done correctly, then you should get the red numbers that are in columns L and N.

You should now take a breather, pat yourself on the back, and marvel at the intricacy of the spreadsheet you are preparing... then get ready for the next section – the replicates section.

In the replicates section, the formulas will be very similar, except that you need to realize that all of the signals and concentrations calculated are for a single unknown that was measured in the replicates.

13. Complete the Conc. formula just as you did for the non-replicates section. Copy these formulas down as far as the replicate signals. If there are only four signals, then copy the formulas down so that there are only four calculations in each column. If you complete an experiment with more than four replicate measures, then just copy the formula Conc. down as far as needed.

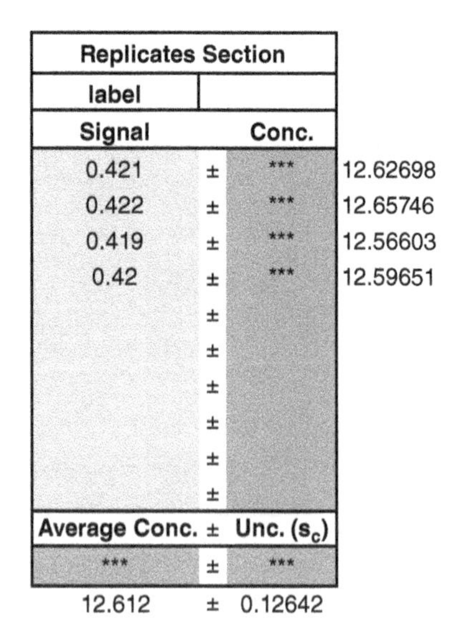

Figure 2.17 The replicates section includes all of the replicate measurements for a single unknown sample. The signals go in the signal column, and formulas replace the *** markers in the other columns.

14. Locate the Average Conc. label and use the *AVERAGE* function to take the average of all of the calculated concentration values in column K. You should extend the range for the *AVERAGE* function to the bottom of the section (rows 19–28) so that this spreadsheet template can accommodate up to 10 replicate measurements for a single sample.
15. Now for the final formula – the uncertainty calculation. You will use Eq. (2.26) again, except this time, the value of the M term is not 1 since this section is for replicate measures. The M term in the formula should be replaced by the *COUNT* function, which will count the number of replicate measures in this section. You can count either the Signal column or the Conc. column – it will not matter. Another difference is that the $\bar{y}_c$ should be replaced by the average of all of the signal values. Use rows 19–28 just as you did for the Average Conc. and *COUNT* functions.

    If you have completed all of the formulas correctly, you should get the red numbers to the side and bottom of the replicates section. Delete the red numbers when you have verified this.

    If you examine the spreadsheet in its entirety, you should notice a few things:

    - All of the bright yellow cells receive raw data from the experiment.
    - The bright green cells contain the final results of the experiment, which are the concentration and uncertainty calculations for nonreplicate and replicate measures of unknowns. These values are to be copied into any lab report (and formatted to the proper sig figs) when the final results are presented.
    - The graph that lies in the middle of the sheet can be used for a report. You will have to modify the x axis and y axis labels to fit the measurement (sometimes, it will be Absorbance versus Concentration or Peak Area versus Partial Pressure). You may want to remove the title if you plan to place the graph in a report document and give it a figure caption. Graphs with captions usually do not have titles since this would provide redundant information.

You should save this spreadsheet and put it in a convenient and safe place. You will likely use it repeatedly for homework and lab experiments. When you have completed the other sections of the spreadsheet (method of multipoint IS & method of multiple standard additions), you should consider changing the file property to *READ-ONLY*. Each time you use the spreadsheet, you may need to modify it slightly, and you do not want to lose the results of the previous experiments or the original format. Each new lab experiment should start by making a copy of this regression spreadsheet, saving it with a filename that reflects the nature of the experiment. You can feel confident in modifying each of these files without destroying the hard work you put in to completing the template.

2.8.2 The Method of Multipoint Internal Standard Regression Template

Example 2.16:

Perform a full regression analysis on the calibration standards and samples given in Table 2.17 using the method of multipoint IS.

Calibration standards		
Concentration (mg Mg/L)	Analyte signal	IS signal
0.001	1,570	4,384
0.002	3,114	4,596
0.005	6,585	6,570
0.0075	6,367	4,568
0.018	2,185	4,790
0.0125	9,405	4,605
0.015	7,497	3,022
0.0175	12,883	4,801

Nonreplicate samples		
Label	Analyte signal	IS signal
River 1	4124	4765
River 2	5033	4908
River 3	7301	5058
River 4	1784	5125

Replicate measurements		
Label	Analyte signal	IS signal
River 5	5123	4865
River 5	5265	5043
River 5	4997	5234
River 5	5023	4987

Table 2.17 Comprehensive simulated results from a method of multipoint internal standard experiment, including the calibration standards, nonreplicate sample measurements, and replicate sample measurements.

Solution: To solve this problem, go to the RegressionTemplate.xls and find the "Multi-point IS blank" sheet. Follow the use of the equations by referring to Figures 2.18–2.20.

Regression Analysis		
Slope	***	139.25739
Std. Dev. in Slope (s_m)	***	4.3271868
Y-intercept	***	0.3161812
Std. Dev. in Y-int (s_b)	***	0.0452225
R^2	***	0.9942401
Standard error in y (s_r)	***	0.0687558
N (count of stds)	***	8
S_{xx}	***	0.0002525
y(avg signal of stds)	***	1.543387

Figure 2.18 This is the regression analysis section of the method of multipoint internal standard.

Non-replicates Section									
	Analyte	IS	Ratio						
Label	Signal (A)	Signal (IS)	(A÷IS)	Conc.		Unc. (s_c)			
---	4124	4765	***	***	±	***	0.00394	±	0.00055
---	5033	4908	***	***	±	***	0.00509	±	0.00054
---	7301	5058	***	***	±	***	0.00809	±	0.00052
---	1784	5125	***	***	±	***	0.00023	±	0.00059
---					±				
---					±				
---					±				
---					±				
---					±				
---					±				

Figure 2.19 This is the non-Replicates section of the method of multipoint internal standard.

Replicates Section				
sample label		---		
Analyte Signal (A)	IS Signal (IS)	Ratio (A÷IS)	Conc.	
5123	4865	***	***	0.0052913
5265	5043	***	***	0.0052266
4997	5234	***	***	0.0045853
5023	4987	***	***	0.0049623
Average Conc.		±	Unc. (s_c)	
***		±	***	
0.0050164			0.0003245	

Figure 2.20 This is the replicates section of the method of multi-point internal standard.

Procedures

1. Enter the concentrations and signals from the standards and IS into the Conc., Analyte Signal, and IS Signal columns.
2. Since this is an IS method, we need to calculate the ratio of the analyte signal to the IS signal. In column C, you should enter a formula that will divide the analyte signal by the IS signal.
3. The regression analysis is the same for this method as it is for the method of external standards, except that instead of using the analyte signal for the y values, you will use the ratio calculation. Follow the steps listed in Example 2.15 to complete the *SLOPE*, *INTERCEPT*, *RSQ*, *STEYX*, *COUNT*, $\bar{y}$, S_{xx}, "Std. Dev. in Slope", and "Std. Dev. in Y-int."
4. This completes the formulas in the regression analysis section, now we will complete the non-replicates section. This is also going to follow the same steps as outlined in the method of external standards. Calculate the concentration and uncertainty here by using the ratio as the y value and the concentration as the x value.
5. Repeat this process in the replicates section by calculating the ratio and then using it to find the concentration and the uncertainty.

If you have completed all of the formulas correctly, you should get the red numbers to the side and bottom of the replicates section. Delete the red numbers when you have verified this. Make sure that you save the spreadsheet!

2.8.3 The Equal-Volume Variant of the Method of Multiple Standard Addition Regression Template

Example 2.17:

Perform a full regression analysis on the following solutions produced by the method of multiple standard additions. All solutions were made with 20.00 mL of the unknown. Additions of a 200.0 mg Ca/L were made according to Table 2.18. All solutions were diluted to a total volume of 100.00 mL.

Volume of Std. Added (mL)	Signal (arbitrary)
0	7,499.1
5.00	13,145.8
10.00	17,794.7
15.00	23,172.8

Table 2.18 Simulated results from an experiment using the equal-volume variant of the method of multiple standard additions.

aliquot of unknown	20	mL
std conc	200	mg/L
total vol	100	mL
	(optional column)	
Volume of std	Conc of std	Signal
0	***	7499.1
5	***	13145.8
10	***	17794.7
15	***	23172.8

Figure 2.21 This is the data section of the regression spreadsheet for the method of multiple standard additions. The "Conc of Std" column is useful if you are going to use the alternative method of plotting the result of this method (Signal versus Conc. of Added Standard). You can disregard it if you are going to plot Signal versus Volume Standard Added.

Solution: To solve this problem, find the "Multiple Std Addns blank" sheet in the Regression-Template.xls. Follow the use of the equations by referring to Figures 2.21-2.22.

Procedures

1. Enter the raw data into the bright yellow cells as before.
2. If you want to calculate the concentration of the diluted standard, use the dilution formula by solving for C_2 and using the stock standard concentration as C_1, the volume added as V_1, and the total volume as V_2.
3. Fill in the formulas for the regression analysis as before using *SLOPE, INTERCEPT, STEYX, N*, S_{xx}, and $\bar{y}$. See the directions in Example 2.15.
4. The x intercept formula is found by taking the equation of a line and replacing the y value with a zero and then solving for x. This should give a negative value.
5. The uncertainty equation for the method of multiple standard additions is related to Eq. (2.26). Since all of the solutions made in this method make the measurement of only one sample possible, there is, by definition, only one unknown measured. Thus, the M term in Eq. (2.26) is not necessary and is not included in the equation.

$$s_c = \left(\frac{STEYX}{SLOPE}\right)\sqrt{\frac{1}{N} + \frac{(\bar{y}_c - \bar{y})^2}{SLOPE^2 \cdot S_{xx}}} \tag{2.27}$$

   The other difference is in the value of $\bar{y}_c$. Since the uncertainty is calculated at the x intercept, the y value is, by definition, zero. Thus, the $\bar{y}_c$ drops out. Enter this formula into the cell to the right of the "uncertainty in x-int" label. Use *STEYX*, the slope, the count of the standards (N), $\bar{y}$, and the S_{xx} values to calculate the uncertainty.
6. The Results section contains the final analysis, which is the concentration and uncertainty of the unknown sample. To find the formula for the concentration, we need to go back to Eq. (2.15) on page 74. Multiply the negative of the x intercept by the stock standard concentration and divide by the volume of the unknown aliquot.
7. To calculate the uncertainty, we just need to convert the "uncertainty in x-int" in the same way we converted the x intercept calculation in the previous step. It is a matter of scaling the uncertainty in the x intercept volume to the uncertainty in the unknown concentration.

$$\text{conc. unc.} = (\text{x} - \text{int unc.})\left(\frac{\text{conc. stock std}}{\text{vol. unknown aliquot}}\right)$$

$$= (0.39833)\left(\frac{200}{20}\right) = 3.9833$$

Results		
slope	***	1033.4
Intercept	***	7652.6
x-int	***	-7.4052642

Error Analysis		
STEYX	***	289.1046
N (count of additions)	***	4
Sxx	***	125
ybar	***	15403.1
uncertainty in x-int	***	0.39833579

Results		
***	±	***
74.05264	±	3.98336

Figure 2.22 This is the regression analysis section of the spreadsheet. It does not contain any calculation of errors in the slope or intercept.

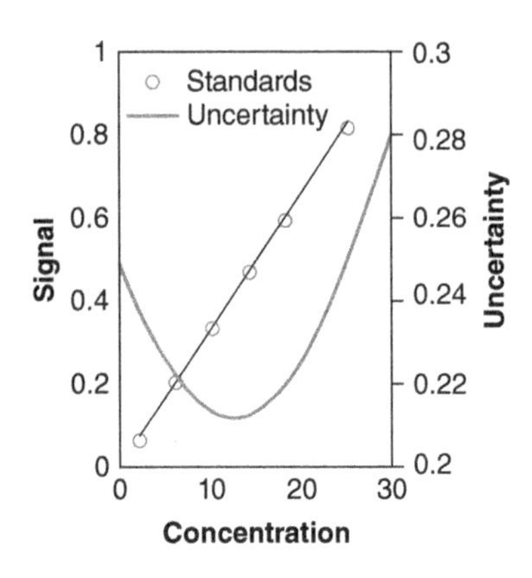

Figure 2.23 Unknown samples whose concentration is close to the middle of the calibration curve will have the smallest uncertainty. The uncertainty curve shows that the uncertainty is at a minimum in the middle of the standards and increases as the unknown concentration gets larger or smaller than the center.

2.8.4 Where Unknowns Should Fall on the Calibration Curve

One consequence of Eq. (2.26) is that the uncertainty is affected by the distance the unknown signal falls away from the center of the calibration curve (represented by $\bar{y}$). Figure 2.23 shows this trend, which is driven by the $\left(\bar{y}_c - \bar{y}\right)^2$ term. As you can see, the uncertainty reaches a minimum as the resulting measurement falls in the middle of the calibration standards and increases as the measurement moves away from the center going in either direction. The "take-home message" is that, as much as possible, the calibration curve standards should be made to fit the concentration of the unknown. The problem is that often the unknown is, well, unknown. This could mean that the unknown may turn out to be too concentrated or too dilute for the standards. If the unknown is too concentrated, then it can easily be diluted and remeasured with the purpose of inputting the signal near the middle of the calibration standards. If the unknown is too dilute, more standards can be made to extend the low end of the calibration curve – assuming that the LOD has not been reached.

2.9 IMPORTANT TERMS

- ⋆ blank
- ⋆ signal
- ⋆ noise
- ⋆ random error
- ⋆ Gaussian or normal distribution
- ⋆ baseline
- ⋆ Central Limit Theorem
- ⋆ signal-to-noise ratio
- ⋆ systematic error
  - instrumental errors
  - method errors
  - personal errors
- ⋆ accuracy
- ⋆ precision
- ⋆ NIST
- ⋆ standard
- ⋆ stock solution
- ⋆ working stock solution
- ⋆ calibration standard
- ⋆ primary standard
- ⋆ secondary standard
- ⋆ standardize
- ⋆ Student's *t*
- ⋆ *z* term
- ⋆ hypothesis testing
  - null hypothesis
  - degrees of freedom (DOF)
  - comparing two experimental means
  - comparing an experimental mean to a certified standard
  - comparing two experimental standard deviations
- ⋆ quantitative
- ⋆ qualitative
- ⋆ analyte
- ⋆ matrix
- ⋆ matrix mismatch
- ⋆ internal standard
- ⋆ interference
- ⋆ aliquot
- ⋆ spiking samples
- ⋆ the dilution equation
- ⋆ methods of quantitation
  - the method of external standards
  - the method of multipoint IS
  - the method of single-point IS
  - The equal-volume version of the method of multiple standard additions
  - The variable-volume version of the method of standard additions
    (i) proportional errors
    (i) constant errors
- ⋆ analytical balance
- ⋆ volumetric pipette
- ⋆ volumetric flask
- ⋆ burette
- ⋆ micropipettors
- ⋆ optical materials
  - transmission range of quartz
  - transmission range of glass
  - transmission range of polystyrene
  - transmission range of PMMA
  - transmission range of quartz
- ⋆ linear regression

Note: The readers are requested to refer to the online version of this chapter for color indication.

EXERCISES

1. Describe the difference between qualitative and quantitative measurements?
2. Describe random noise, how it can be reduced or eliminated from measurements, whether it is associated with accuracy or precision, and which parameter of a Gaussian distribution is it associated with.
3. Describe systematic errors, how they can be reduced or eliminated from measurements, whether they are associated with accuracy or precision, and which parameter of a Gaussian distribution are they associated with.
4. What mathematical distribution describes most measurements? Why?
5. Under what conditions can a signal be first identified from noise?
6. How is a SNR calculated?
7. Describe the LOD.
8. Describe the LOQ.
9. In two steps, summarize the rule for specifying sig figs in the results of a measurement.
10. What are the differences between a primary and a secondary standard?
11. Why is a standard necessary in most measurements?
12. Describe the differences between a population and a sample distribution. What statistical variables are used in each?
13. What is the purpose of a pooled standard deviation?
14. What is the purpose of performing a hypothesis test?
15. List three different scenarios where a hypothesis test would be useful in an environmental measurement.
16. List the advantages and disadvantages of each of the following methods of quantitation:
    (a) the method of external standards
    (b) the method of multiple standard additions
    (c) the method of single-point IS
    (d) the method of multipoint IS.
17. What method is being employed if a sample or standard is spiked with a stock standard solution of the analyte?
18. What method is being employed if a sample or standard is spiked with a stock standard solution of a chemical similar to, but different from, the analyte?
19. Why is the variable-volume method of multiple standard additions convenient to use with a micropipette?
20. How does the method of multiple standard additions achieve a set of standards with the same matrix?
21. Why must objects be at room temperature when placed on an electronic balance?
22. What physical difference between a 100 mL graduated cylinder and a 100 mL volumetric flask results in the difference in the respective accuracy of the glassware?
23. List the advantages and disadvantages of glass pipettes compared to micropipettes?
24. If an analyte had a strong absorbance at 240 nm, which cuvette materials would be most appropriate?
25. If glass cuvettes are useful only in the visible radiation range and are more expensive than the various plastic cuvettes, why would anyone ever use glass?
26. Why is it useful to go through all of the trouble of a full regression analysis of an unknown and calibration standards?

BIBLIOGRAPHY

Bader, M. *Journal of Chemical Education* **1980**, *57(10)*, 703–706.

Box, J. F. *Statistical Science* **1987**, *2(1)*, 45–52.

Caballero, J. F.; Harris, D. F. *Journal of Chemical Education* **1998**, *75(8)*, 996.

Cardone, M. J. *Analytical Chemistry* **1986**, *58(2)*, 438–445.

Ellison, S. L. R.; Thompson, M. *The Analyst* **2008**, *133(8)*, 992–997.

EPA, *Specifications and Guidance for Contaminant-Free Sample Containers*; 1992.

Ingle, J. D. *Journal of Chemical Education* **1974**, *51(2)*, 100–105.

Isbell, D.; Hardin, M.; Underwood, J. Mars Climate Orbiter Team Finds Likely Cause of Loss. Electronic, 1999; http://mars.jpl.nasa.gov/msp98/news/mco990930.html.

Long, G. L.; Winefordner, J. D. *Analytical Chemistry* **1983**, *55(7)*, 712A–724A.

Mocak, J.; Bond, A. M.; Mitchell, S.; Scollary, G. *Pure & Applied Chemistry* **1997**, *69(2)*, 297–328.

Moore, D. S.; McCabe, G. P. *Introduction to the Practice of Statistics*; W. H. Freeman and Company: New York, 1999.

1.3.6.7.2. Critical Values of the Student's t Distribution. Electronic, 2013; http://www.itl.nist.gov/div898/handbook/eda/section3/eda3672.htm.

Skoog, D. A.; Holler, F. J.; Crouch, S. R. *Principles of Instrumental Analysis*, 6th ed.; Thomson Brooks/Cole: Belmont, CA, 2007.

Thompson, M. *The Royal Society of Chemistry* **2009**, *AMCTB No 37*, 1–2.

Thomsen, V.; Schatzlein, D.; Mercuro, D. *Spectroscopy* **2003**, *18(12)*, 112–114.

3

THE ATMOSPHERE

I have of late, (but wherefore I know not) lost all my mirth, forgone all custom of exercises; and indeed, it goes so heavily with my disposition; that this goodly frame the earth, seems to me a sterile promontory; this most excellent canopy the air, look you, this brave o'erhanging firmament, this Majestical roof, fretted with golden fire: why, it appears no other thing to me, then a foul and pestilent congregation of vapours. What a piece of work is a man! How noble in reason, how infinite in faculty! in form and moving how express and admirable! In action how like an Angel! In apprehension how like a god! The beauty of the world! The paragon of animals! And yet to me, what is this quintessence of dust?

—William Shakespeare, from *The Tragedy of Hamlet, Prince of Denmark*

Behave so the aroma of your actions may enhance the general sweetness of the atmosphere.

—Henry David Thoreau, 1817–1862

For the first time in my life I saw the horizon as a curved line. It was accentuated by a thin seam of dark blue light - our atmosphere. Obviously this was not the ocean of air I had been told it was so many times in my life. I was terrified by its fragile appearance.

—Ulf Merbold, German astronaut

3.1 INTRODUCTION

If you consider the size of the Earth (about 6×10^{24} kg in mass and 12,700 km in diameter), the thin layer of vapor clinging to the surface of the globe that we call our atmosphere is negligible by mass and size. If the Earth were the size of a soccer ball (9 in. in diameter), the atmosphere would extend less than 1 cm from the surface (for comparison, your pinky finger is about 1 cm wide). If you consider just the layer of the atmosphere in which we live, the troposphere, nearly all of human experience took place in the volume that would extend about 0.2 mm from the surface of the ball! Yet, this thin collection of gases is tremendously important to us and is the home to some very complicated chemical, physical, and biological activities.

Gas	Mixing ratio (%)
N_2	78.08
O_2	20.95
Ar	0.93
H_2O	< 0.5–3.5
CO_2	≥0.04

Other important components, which are much less than 1%, will be central to this chapter.

Table 3.1 The general composition of the atmosphere is summarized.

As you have already learned in Chapter 1, the atmosphere of the Earth has gone through drastic transformations during its 4.6 Gyr history. After losing the first atmosphere, composed of mostly hydrogen and helium, to the relentless solar wind, the gases dissolved in the molten Earth erupted out of volcanoes to form the second atmosphere. The gases were predominantly water, carbon dioxide, and molecular nitrogen with trace amounts of hydrochloric acid, sulfur dioxide,, and other gases. Once the atmosphere had sufficiently cooled, the water condensed and washed out most of the carbon dioxide. It is estimated that the current concentration of CO_2 represents 0.001% of the original amount in the early atmosphere, with 99.999% having been lost to the lithosphere as carbonate rocks and the hydrosphere as carbonic acid (Seinfeld and Pandis, 2006). Remember that this former state

Environmental Chemistry: An Analytical Approach, First Edition. Kenneth S. Overway.

Companion website: www.wiley.com/go/overway/environmental_chemistry

is similar to the state of the atmosphere of Venus, having an oppressive atmospheric pressure 90 times greater than that on the Earth and an atmospheric composition of 96.5% CO_2 and 3.5% N_2. Venus never cooled enough to form liquid water and has thus maintained its original concentration of CO_2. Since molecular nitrogen is so inert and relatively insoluble in water (about 50 times less soluble in water compared to CO_2), it persisted in our atmosphere and has become the dominant component. Molecular oxygen, as mentioned in Chapter 1, has accumulated over the eons as a result of photosynthetic organisms. The current atmospheric composition of the Earth is listed in Table 3.1.

The purpose of this chapter is to introduce the complex chemistry of the atmosphere so that you have an understanding of the dynamic processes that occur naturally and anthropogenically. The chapter starts with an overview of each layer of the atmosphere, progressing from the top of the atmosphere down to the Earth's surface. Finally, this chapter focuses on the three most important atmospheric issues, namely stratospheric ozone, tropospheric smog, and the tropospheric greenhouse effect.

3.2 AN OVERVIEW OF THE ATMOSPHERE

Since the Earth's atmosphere forms the most exterior layer of the planet's composition, it is the first portion of the Earth to receive radiation from the Sun. Remember that the

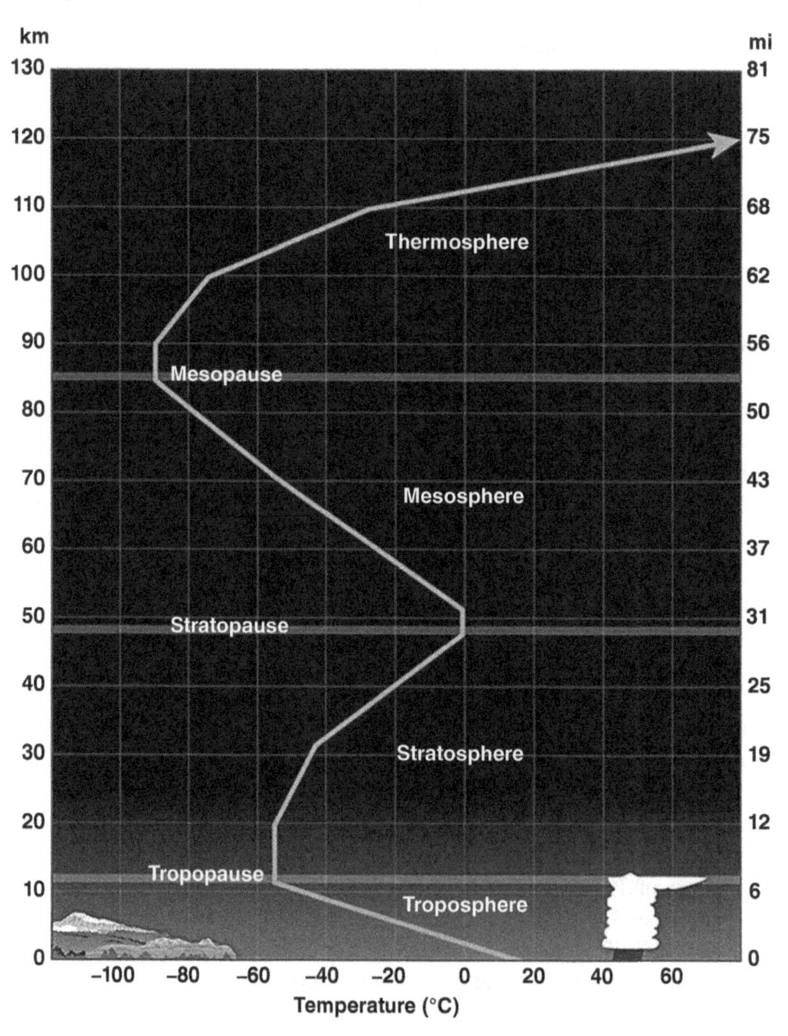

Figure 3.1 Gases in the atmosphere of the Earth absorb solar radiation in different regions of the electromagnetic spectrum and form distinct layers as a result. Source: http://www.srh.noaa.gov/jetstream/atmos/atmprofile.htm.

Sun's photosphere behaves similarly to a blackbody radiator and emits electromagnetic (EM) radiation according to its temperature. The interaction between the solar radiation and the gases in the atmosphere results in a zigzagged vertical profile of temperatures that forms layers within the atmosphere. Figure 3.1 shows the vertical profile of these layers. When the solar radiation gets absorbed by the molecules in the atmosphere, much of the energy turns into heat. The absorption of radiation occurs unevenly in the layers of the atmosphere, so a temperature profile results. Heat changes the density of the gases, so certain layers will not mix with others to a large extent. These layers tend to contain different amounts of trace components and possess different chemical and physical properties.

To understand the formation of the layers, it is important to consider (1) the type of radiation being absorbed and (2) the gases found in that layer. Each layer of the atmosphere is separated by a *pause* (such as the tropopause, stratopause, and mesopause). These regions of the atmosphere form where the temperature of gases reverses a previous trend. The altitude of each pause and layer is dynamic due to variations in the incident solar radiation and latitude, so all of the stated altitudes must be viewed as rough approximations.

3.3 THE EXOSPHERE AND THERMOSPHERE

The *exosphere* extends beyond the *thermosphere* from about 500 km to interplanetary space. Combined with the exosphere, the thermosphere contains about 0.001% of the mass of the atmosphere and has gas pressures less than 0.3 mtorr (0.4 μatm). While temperatures can reach 1200 °C as measured by the kinetic energy of the gases (thus the prefix "thermo"), the very low density of this layer would make it feel cold because very few collisions of gas particles occur in order to transfer heat. The thermosphere extends from the mesopause (about 90 km) upward to about 500 km. With an altitude of about 400 km, the International Space Station orbits the Earth within the thermosphere.

☞ *Photolytic* reactions, or *photolysis*, are chemical reactions involving a photon of energy as a reactant.

The thermosphere receives the full spectrum and intensity of solar radiation that reaches the Earth. Figure 1.1 shows that the Sun emits a significant amount of ultraviolet and some X-ray radiation. While the radiation with the highest intensity has a wavelength greater than 250 nm, the radiation that is less than 250 nm is very high in energy per photon and dangerous to any living organism on the Earth's surface. Fortunately, most of this radiation never reaches the surface because it is absorbed by the two most abundant neutral gases in the thermosphere, molecular nitrogen (N_2) and molecular oxygen (O_2), and several minor species listed in Table 3.2. The important *photolytic* reactions that result from the interaction between these species and solar radiation can be seen in Table 3.3, and the following example problem describes how these reactions correspond to the peaks in Figure 3.2.

Rxn enthalpy (kJ/mol)	Photolytic reaction	Rxn #
945	$N_2 + h\nu \longrightarrow 2N$	(R3.1)
1503	$N_2 + h\nu \longrightarrow N_2^+ + e^-$	(R3.2)
1402	$N + h\nu \longrightarrow N^+ + e^-$	(R3.3)
1165	$O_2 + h\nu \longrightarrow O_2^+ + e^-$	(R3.4)
1314	$O + h\nu \longrightarrow O^+ + e^-$	(R3.5)
1209	$O_3 + h\nu \longrightarrow O_3^+ + e^-$	(R3.6)
1218	$H_2O + h\nu \longrightarrow H_2O^+ + e^-$	(R3.7)
1329	$CO_2 + h\nu \longrightarrow CO_2^+ + e^-$	(R3.8)
1352	$CO + h\nu \longrightarrow CO^+ + e^-$	(R3.9)
894	$NO + h\nu \longrightarrow NO^+ + e^-$	(R3.10)

Source: Torr and Torr (1979)

Table 3.3 Some of the photolytic reactions that predominate in the thermosphere where photon energies are very high.

Species	Thermospheric max. density (mcs/cm^3)	Mesosphere max. density (mcs/cm^3)	Literature source
O_3	1.1×10^8	8.7×10^9	Batista *et al.* (1996)
NO	1.0×10^9	3.0×10^7	Hedin *et al.* (2012), Baker *et al.* (1977)
CO	5.5×10^8	- - -	Emmert *et al.* (2012)
CO_2	2.2×10^9	6.9×10^{11}	Kostsov and Timofeyev (2003)
H_2O	1.1×10^8	8.7×10^9	Shapiro *et al.* (2012)

For comparison, O_2 and N_2 have number densities in the lower thermosphere of 10^{11} and 10^{12} mcs/cm^3, respectively.

Table 3.2 Estimated maximum number density (concentration) values for several important trace gases in the upper atmosphere.

☞ For a review of photolysis calculations see Review Example 1.10 on page 28.

☞ For a review of naming covalent compounds, see Review Example 1.9 on page 27.

☞ In photolytic reactions, a photon is denoted as $h\nu$, representing Planck's constant (h) and the photon frequency (ν), which combine to give the energy of the photon, as in the equation $E = h\nu = \frac{hc}{\lambda}$.

Example 3.1: Photolysis

Reaction R3.1 (in Table 3.3) shows how a high-energy photon can break the triple bond in molecular nitrogen, corresponding to an energy of 945 kJ/mol. This can be calculated using the thermodynamic values found in Appendix I (Table I.2), or it can be looked up in Table I.7 found on page 325.

$$
\begin{aligned}
& N_2 + h\nu \longrightarrow 2N \\
\Delta H_{rxn} &= \Sigma\Delta H^{\circ}_{f\,products} - \Sigma\Delta H^{\circ}_{f\,reactants} \qquad (3.1) \\
&= \left[(2\text{ mol})(472.68\text{ kJ/mol})\right] - \left[(1\text{ mol})(0\text{ kJ/mol})\right] \\
&= 945.36\text{ kJ}
\end{aligned}
$$

Find the maximum wavelength of a photon that can supply this bond-breaking energy. See Exercise 1.10 for help on this problem.

Solution:

The aforementioned problem describes a photolysis of N_2. The bond energy (B.E.) is given in units of kJ per 1 mol of N_2, which describes the total energy needed to break all of the triple bonds of a mole of molecular nitrogen. We want to focus on the photolysis of a single molecule of nitrogen, which absorbs a single photon. The first step is to convert the B.E. (as calculated by ΔH_{rxn}) per mole to per molecule (mcs).

$$\Delta H_{rxn}(\text{per 1 mol of } N_2)\left(\frac{1\text{ mol}}{6.02214129\times10^{23}\text{mcs}}\right)$$

$$\left(945\frac{\text{kJ}}{\text{mol}}\right)\left(\frac{1\text{ mol}}{6.02214129\times10^{23}\text{mcs}}\right) = 1.5698\times10^{-21}\text{kJ/mcs}$$

Now we want to know what sort of photon could deliver this much energy, so we need the equation that relates photon energy and wavelength. We just need to plug in the energy and solve for wavelength. We have to be careful because the units need to be canceled out. When calculating the wavelength, it is assumed that the formula is per photon. Since one molecule absorbs one photon, the per mcs unit cancels. Look up the values for h and c (Table I.1 on page 317), and make sure that you specify the units for these constants.

$$
\begin{aligned}
E &= \frac{hc}{\lambda} \\
\lambda_{max} &= \frac{hc}{E} = \frac{(6.62606957\times10^{-34}\text{ J}\cdot\text{s})(2.99792458\times10^{8}\text{ m/s})}{1.5698\times10^{-21}\text{ kJ/mcs}} \\
&= \frac{(6.62606957\times10^{-34}\text{ J}\cdot\text{s})(2.99792458\times10^{8}\text{ m/s})}{(1.5698\times10^{-21}\text{ kJ/mcs})\left(\frac{1000\text{ J}}{1\text{ kJ}}\right)\left(\frac{1\text{ mcs}}{1\text{ photon}}\right)}\left(\frac{10^{9}\text{ nm}}{1\text{ m}}\right) \\
&= 126.54 = 127\text{ nm}
\end{aligned}
$$

This photon, with a wavelength of 127 nm, represents the minimum energy and the maximum wavelength since energy and wavelength are inversely proportional.

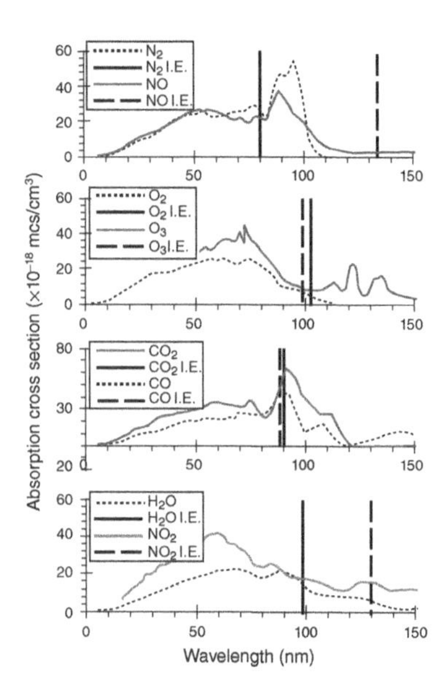

Figure 3.2 Solar radiation up to about 100 nm strikes the thermosphere and is absorbed by several different gases, the most concentrated of which are molecular nitrogen and oxygen. The vertical axis, labeled Absorption Cross Section, represents the absorption efficiency. The traces labeled I.E. represent the low energy end of the ionization energy continuum. Source: O_2 data from Hegelund *et al.* (2005); O_3 data from Ackerman (1971); N_2 data from Chan *et al.* (1993c).

If you look for an absorption at 127 nm for molecular nitrogen in Figure 3.2, you will notice that it is not there. This is because the B.E. calculation in the previous example uses a thermodynamic calculation to determine a kinetic parameter. Examine Figure 3.3. In order for the reaction to proceed, the photon must usually exceed the enthalpy of the reaction – it is the activation energy that is really required. Further, this method of estimating photolysis energies only works for simple endothermic reactions. From the experimental measurements that are seen in Figure 3.2, we can assume that the activation energy requires at least a 100 nm photon (equivalent to about 1190 kJ/mol of photons). While this calculation is always an underestimate of the activation energy, it yields an estimate that is useful. The actual activation energy must be determined experimentally and is not always available.

Reaction R3.2 represents the ionization of N_2, which also requires a photon to initiate. Ionization is the process of ejecting an electron from an atom or molecule, and in this case, it is the energy to eject the highest energy valence electron. The same calculation is made to convert the ionization energy into the wavelength of a photon.

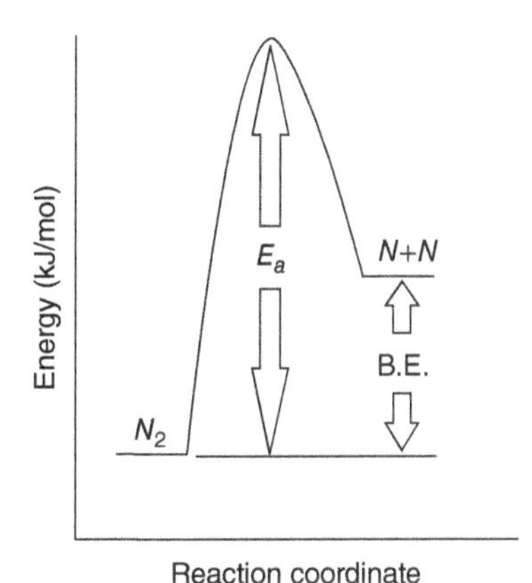

Figure 3.3 Thermodynamic calculations only reveal the difference between the initial and final states (B.E. calculated by ΔH_{rxn}), not the activation energy. Photolysis requires activation energy. The difference between the E_a and the B.E. goes to the kinetic energy of the products and manifests itself as heat.

Example 3.2: Ionizing Radiation

The first ionization energy of molecular nitrogen is 1503 kJ/mol, represented by Reaction R3.2. Calculate the maximum wavelength of the photon that could carry the energy required to ionize N_2. What happens to the additional energy if a photon with a shorter wavelength gets absorbed?

Solution: We will follow the same process as in the previous example.

$$\lambda_{max} = \frac{(6.62606957 \times 10^{-34}\ \text{J} \cdot \text{s})(2.99792458 \times 10^8\ \text{m/s})}{(1503\ \text{kJ/mol})\left(\frac{1\ \text{mol}}{6.02214129\times10^{23}\ \text{mcs}}\right)\left(\frac{1\ \text{mcs}}{1\ \text{photon}}\right)\left(\frac{1000\ \text{J}}{1\ \text{kJ}}\right)}\left(\frac{10^9\ \text{nm}}{1\ \text{m}}\right) = 79.59\ \text{nm}$$

The aforementioned calculation shows that the minimal energy needed to ionize N_2 corresponds to a photon with a wavelength of about 80 nm. What would happen if N_2 absorbed a photon with a wavelength of 75.0 nm? The extra energy could go into the kinetic energy (*K.E.*) of the electron.

$$K.E._{\text{electron}} = E_{\text{photon}} - E_{B.E.} \tag{3.2}$$

If the photon has the minimal energy, then the electron has a kinetic energy of zero. If the photon has more than enough energy, then the electron has a nonzero kinetic energy. Given a photon of 75.0 nm and a binding energy (or ionization energy) of 1503 kJ/mol, the electron could have the following kinetic energy.

$$E_{\text{photon}} = \frac{(6.626 \times 10^{-34}\ \text{J} \cdot \text{s})(2.9979 \times 10^8\ \text{m/s})}{(75.0\ \text{nm})\left(\frac{1\ \text{m}}{10^9\ \text{nm}}\right)} = 2.6485 \times 10^{-18}\ \text{J}$$

$$E_{B.E.} = (1503\ \text{kJ/mol})\left(\frac{1\ \text{mol}}{6.022 \times 10^{23}\ \text{mcs}}\right)\left(\frac{1\ \text{mcs}}{1\ \text{photon}}\right)\left(\frac{1000\ \text{J}}{1\ \text{kJ}}\right) = 2.4959 \times 10^{-18}\ \text{J}$$

$$K.E._{\text{electron}} = 2.6485 \times 10^{-18} - 2.4959 \times 10^{-18}\ \text{J} = 1.53 \times 10^{-19}\ \text{J}$$

Thus, photolysis with photon energies beyond the ionization continuum, as seen in Figure 3.2, can be explained.

The implication of the aforementioned calculation is that the λ_{max} that can ionize N_2 is the beginning of a continuum of wavelengths that can be absorbed. If you examine Figure 3.2, you can see the vertical line that represents the λ_{max} for the ionization of N_2, and you will also see that the absorption of nitrogen forms a continuous band from 80 nm and lower – all of those very high-energy photons are capable of being absorbed by molecular nitrogen.

A similar situation exists for molecular oxygen. Reaction R3.4 shows the first ionization of molecular oxygen and the other trace gases, which are marked on Figure 3.2 by vertical lines. Each line starts a continuum of absorbed photons with wavelengths shorter than the position of the line. Thus, all photons with wavelengths smaller than about 100 nm can be absorbed by the gases in the atmosphere.

The upper region of the thermosphere contains some very exotic species as a result of the high-energy photolysis and low pressures. The concentration of these ions peaks above 120 km and implies that the ionizing radiation is less abundant at lower altitudes and thus produces the thermospheric temperature trend of decreasing temperatures until the mesopause (since less photolysis means less heat). This layer of ions within the thermosphere is called the ionosphere, and it is useful for land-based radio communications with the other side of the globe. The moving ions generate a magnetic field which reflects radio waves and helps them to travel around the curvature of the planet.

3.4 THE MESOSPHERE

Compared to other layers of the atmosphere, very little is known about the mesosphere. The Greek word mesos means "middle," which describes its position between the thermosphere and the stratosphere. The mesopause at the top of this layer is home to the coldest recorded temperatures on the planet. The mesosphere holds about 0.1% of the atmosphere's total mass and contains unusually high concentrations of sodium, iron, and other ionized metals as a result of the vaporization of frequent meteors, contributing to the yellow portion of the night glow seen in some of the stunning photographs taken from the International Space Station. The mesosphere ranges from about 80–90 km in altitude to about 50–60 km where it meets the stratopause. The temperature trend in the mesosphere is the opposite to that in the thermosphere. Coldest at the top, the mesosphere increases in temperature as altitude decreases because of the increasing concentration of gases (mostly N_2 and O_2) and the absorption of solar radiation in the 100–200 nm region (see Figure 3.4). As the atmospheric pressure increases by a factor of 100 from the top to the bottom of the mesosphere, this radiation begins to be absorbed by molecular oxygen and is converted to heat. Much of the heat is generated by the absorption band of molecular oxygen from about 130 to 175 nm as well as by the photolysis of nitric oxide (see Table 3.4). Oxygen is different from nitrogen in this case because it can exist in different energetic states (Table 3.5).

Rxn enthalpy (kJ/mol)	Photolysis reaction	Rxn #
688	$O_2 + h\nu \longrightarrow O + O^*$	(R3.11)
632	$NO + h\nu \longrightarrow N + O$	(R3.12)
821	$NO + h\nu \longrightarrow N + O^*$	(R3.13)
552	$CO_2 + h\nu \longrightarrow CO + O$	(R3.14)
741	$CO_2 + h\nu \longrightarrow CO + O^*$	(R3.15)
1076	$CO + h\nu \longrightarrow C + O$	(R3.16)
1265	$CO + h\nu \longrightarrow C + O^*$	(R3.17)

Table 3.4 Some of the photolytic reactions that predominate in the mesosphere.

Excited states of oxygen

Ground-state atomic oxygen: $O(^3P)$ will be referred to as O

Excited-state atomic oxygen: $O(^1D)$ will be referred to as O^*

Excited-state molecular oxygen, level 1: $O_2(^1\Delta)$ will be referred to as O_2^*

Excited-state molecular oxygen, level 2: $O_2(^1\Sigma)$ will be referred to as O_2^{**}

Table 3.5 The various forms of oxygen and the way they will be referred to in this textbook.

You may recall learning about the incorrect, yet instructive, model of the atom developed by Niels Bohr showing how electrons absorb EM radiation and jump to higher energy levels that resemble orbits around the nucleus. Molecular oxygen and atomic oxygen are found in the ground and excited states in the mesosphere. The various states of oxygen play an important role in ozone chemistry, which will be described later in the chapter. These states are labeled as $O(^3P)$ and $O(^1D)$ for atomic oxygen by atmospheric chemists, representing the ground state and an excited state, respectively. For molecular oxygen, the ground state is given as simply O_2 while the two common excited states are labeled as $O_2(^1\Delta)$ and $O_2(^1\Sigma)$. The labels ^3P, ^1D, $^1\Delta$, and $^1\Sigma$ represent atomic and molecular orbitals with special symmetry and electron spin designations. While these forms do add a slight complication to atmospheric reactions, they do play an important role in explaining reactions that are important to the environment. One visually stunning feature of the Earth that is explained, in part, by excited states of oxygen is something called air glow or night glow. After solar radiation excites and breaks up molecular oxygen and nitrogen, these species recombine and relax down to their ground states at night and in the process emit radiation that makes the atmosphere appear to glow.

The reaction that generates these different states of atomic oxygen is represented by Reaction R3.11, resulting in a ΔH_{rxn} of 688 kJ/mol and a corresponding energy of a 174 nm photon. If you look at Figure 3.4 for this transition, you will see it is at the low energy end of the major absorption peak for molecular oxygen (130–175 nm).

A result of the increasing pressure of the major and minor gases, the reactions listed in Table 3.4 and the associated spectral absorptions screen radiation below 200 nm. Absorption reaches its maximum at the bottom of the mesosphere and the stratopause.

Figure 3.5 shows the radiation that makes it through the thermosphere and mesosphere. The 50 km altitude represents the stratopause. Nearly all of the radiation with wavelengths less than 180 nm have been eliminated from the spectrum. The remaining light from 180 to 400 nm still contains some dangerous UV radiation. Figure 3.5 foreshadows the next stage of the story, where ozone finishes the job of clearing the spectrum of the remaining dangerous UV radiation.

There is a small amount of ozone in the lower mesosphere, so some of the photons between 200 and 300 nm are absorbed, but most of these make it through the mesosphere and into the stratosphere – the subject of the next section.

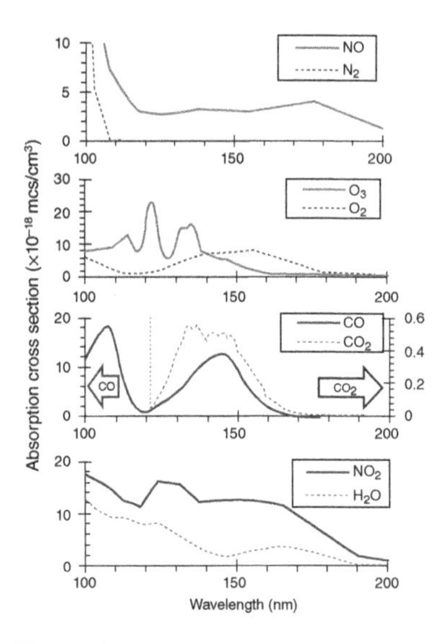

Figure 3.4 Major and minor components of the mesosphere begin to absorb solar radiation that penetrates the thermosphere – mostly with wavelengths greater than 100 nm. Source: Data from Keller-Rudek *et al.* (2013).

3.5 THE STRATOSPHERE

The story of the stratosphere is inextricably tied to the story of ozone (O_3). The name derives from the Latin word stratus meaning "to spread out." The stratopause, at the top of the stratosphere, sees a turnabout in the temperature profile. The increase in temperature as the altitude decreases in the mesosphere reaches a maximum at the stratopause of around 0 °C, and the temperature continues to decrease until the tropopause at about 10–15 km of altitude. Because of this decreasing temperature profile and the fact that gas density is very dependent on temperature, there is very little mixing in the stratosphere. The most dense gases are at the bottom and the least dense gases are at the top of the stratosphere, causing this layer to have very little turbulence compared to the troposphere. Most commercial aircraft fly in the upper troposphere or tropopause in order to avoid the tropospheric turbulence. Supersonic jets achieve their record speeds in the stratosphere because of the low pressures and smooth air currents. The stratosphere contains about 19.9% of the mass of the atmosphere by mass.

The reason for the temperature profile of the stratosphere is the presence of a trace gas, ozone, produced by photolytic reactions involving molecular oxygen. Ozone is only produced significantly below the stratopause. To find out why, we need to understand ozone production, first resolved by the geophysicist Sidney Chapman (1888–1970) in the 1930s. The cycle of ozone production and destruction is called the Chapman Cycle, and it is so important to the stratosphere and the protection of life on the surface from harmful radiation that it deserves its own subsection.

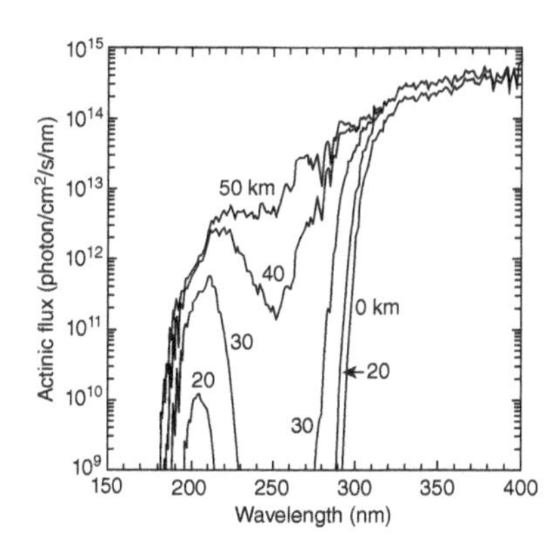

Figure 3.5 The radiation available to cause atmospheric reactions is referred to as the actinic flux. The thermosphere and mesosphere remove most of the <180 nm radiation (see the 50 km line). Much of the rest is absorbed by the stratosphere at various altitudes. Source: NASA.

3.5.1 The Chapman Cycle

The genesis of ozone comes from the presence of molecular oxygen and radiation energetic enough to break the oxygen double bond.

$$O_2 + h\nu \xrightarrow{k_1} O + O \tag{R3.18}$$

In the aforementioned photolysis reaction, the two oxygen atoms produced are in the ground state (with the rate constant above the arrow indicating the first step in the Chapman Cycle).

☞ The *Chapman Cycle* describes the steps that occur for natural ozone production and destruction in the stratosphere.

Example 3.3: Enthalpy of Photolysis

Calculate the standard enthalpy of Reaction R3.18, then determine the wavelength of the photon with the minimal energy to break this bond.

Solution: The answer is 240 nm. To see the process of the calculation, go to Section C.1 on page 261.

☞ For a review of the kinetics rate laws and mechanisms, see Review Example 1.17 on page 35.

We know from Figure 3.5 that the thermosphere effectively blocks most of the radiation up to 180 nm, so we can assume that radiation between 180 nm and 240 nm is responsible for causing this reaction since higher energy photons are unavailable below 50 km.

The ground-state oxygen atom is a reactive species and will react with nearly any molecule it collides with. Given that molecular oxygen is the second most abundant species in the atmosphere, the following reactions are very likely.

$$O + O_2 \rightleftharpoons O_3^* \tag{R3.19}$$

$$O_3^* + M \longrightarrow O_3 + M \tag{R3.20}$$

☞ *Chaperone* molecules collide with excited species and remove excess vibrational energy upon collision. The result is a ground-state product, which is more stable.

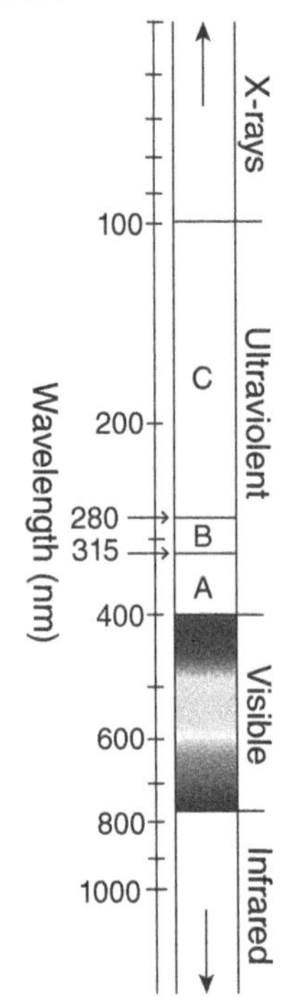

Figure 3.6 The UV spectrum is divided into ranges. The UV-C range will be the focus of discussions from the top of the atmosphere until the troposphere. Source: NASA.

Reaction R3.19 forms an excited form of ozone, which can fall apart if it cannot lose the extra energy. It will often collide with molecular oxygen or nitrogen, since they are the most abundant species. Reaction R3.20 contains a *chaperone* molecule, denoted by M, which represents either O_2 or N_2. Upon collision, O_3^* transfers some of its excess energy (in the form of vibrational energy) to M, resulting in a ground state O_3, which is much more stable, and an M species with extra vibrational energy. Since O_2 and N_2 have much stronger bonds than O_3 (double and triple bonds compared to ozone's 1.5 bond), they are able to dissipate this energy with subsequent collisions without the risk of falling apart (as O_3^* would in Reaction R3.19). Reactions R3.19 and R3.20 are often combined for brevity.

$$O + O_2 + M \xrightarrow{k_2} O_3 + M \qquad \text{(R3.21)}$$

Example 3.4: Rate Laws for the Chapman Cycle

What are the rate laws for Reactions R3.18 and R3.21?

Solution: See Section C.1 on page 261.

Thus far, the Chapman Cycle has produced ozone by absorbing radiation that is between 180 and 240 nm, which is in the UV-C section of the UV spectrum (see Figure 3.6). Once ozone is available, it can absorb UV radiation.

$$O_3 + h\nu \xrightarrow{k_3} O + O_2^* \qquad \text{(R3.22)}$$

This reaction can produce a combination of excited-state and ground-state forms for atomic and molecular oxygen. These variants will be discussed in more detail in Section 3.7.3.

Example 3.5: Ozone Photolysis

Estimate the wavelength of the photons necessary in order to photolyze ozone according to Reaction R3.22 at the standard temperature assuming all combinations of ground- and excited-state atomic oxygen and molecular oxygen.

Solution: The various wavelengths are 307, 264, 1123, 454, 596, 405 nm. To see the work for these calculations, go to Section C.1 on page 262.

The excited-state atomic oxygen and molecular oxygen predominantly react with a chaperone species and lose their excited-state energy. These reactions are usually excluded from the Chapman Cycles because they do not generate ozone.

$$O^* + M \longrightarrow O + M \qquad \text{(R3.23)}$$
$$O_2^* + M \longrightarrow O_2 + M \qquad \text{(R3.24)}$$

Since M (O_2 and N_2) represents over 99% of all of the gases in the atmosphere, the excited-state forms of molecular and atomic oxygen will predominately follow Reactions R3.23 and R3.24 to regain their ground state because of the high probability that the next collision will be with a chaperone. A small portion of these excited-state species do participate in other reactions, as we will see in the section on the troposphere.

The final reaction in the Chapman Cycle involves the collision of ozone with a ground-state oxygen atom to reform molecular oxygen.

$$O + O_3 \xrightarrow{k_4} 2O_2 \qquad \text{(R3.25)}$$

Thus, the four steps in the Chapman Cycle describe the generation of ozone from molecular oxygen, and the destruction of ozone by photolytic and nonphotolytic means to produce molecular oxygen, absorbing photons predominately in the range of 180–300 nm in the process. This cyclic process is driven by solar radiation and is responsible for the relatively large quantities of ozone that are found in the stratosphere. As a result, decreasing altitudes in the stratosphere have less of the 180–300 nm radiation available, leading to a decrease in temperature. This trend continues until the tropopause, where most of the radiation with wavelengths less than 300 nm has been absorbed (see Figure 3.5 on page 101).

But why does this only occur in the stratosphere? There is molecular oxygen in the thermosphere and mesosphere along with plenty of UV-C radiation powerful enough to break the O_2 bond. Shouldn't this lead to significant amounts of ozone in all of the layers above the stratosphere? Satellite and weather balloon measurements show that significant ozone production only occurs below 50 km. This is because the formation of ozone requires a sufficient concentration of molecular oxygen and nitrogen, and Eq. (C.5) (on page 262) giving the rate law for Reaction R3.21 tells the story. This equation gives the rate of the formation of ozone. Another way to look at this is to determine the *lifetime* of the oxygen atom in the ozone-forming step of the Chapman Cycle (Reaction R3.21). If the oxygen atom has a long lifetime, then the rate of ozone formation will be slow. Conversely, if the lifetime of the oxygen atom is very short, then this means that it is forming ozone at a high rate.

Using Eq. (C.5), we can find the lifetime of atomic oxygen in Reaction R3.21 using the model set by Eq. (3.3) (see the box on the margin).

$$\tau_O = \frac{[O]}{Rate} = \frac{1}{k_2[O_2][M]} \tag{3.4}$$

With this equation for calculating the lifetime of atomic oxygen, we just need the value of k_2 and the values of $[O_2]$ and [M] as a function of altitude. Table 3.6 contains the experimental data needed to determine the lifetime.

The *lifetime* (τ) of a chemical species in a particular reaction is a kinetic property which describes how long a species persists. It is derived from the rate law. The lifetime of A in a reaction with B is

$$A + B \xrightarrow{k} C$$

$$Rate = k[A][B]$$

$$\tau_A = \frac{[A]}{Rate} = \frac{1}{k[B]} \tag{3.3}$$

The lifetime is determined by rearranging the rate law to find the ratio of the concentration of the species of interest to the rate.

Example 3.6: The Lifetime of Atomic Oxygen

Calculate the lifetime of atomic oxygen in Reaction R3.21 at altitudes 100, 70, and 50 km. Refer to the values in Table 3.6. What do these lifetimes say about the rate of the formation of ozone at each altitude?

Solution: The lifetime of atomic oxygen in Reaction R3.21 is given in Eq. (3.4). What remains is to substitute the values in Table 3.6 into the equation. Here is a demonstration of the first lifetime at 100 km.

$$\begin{aligned} \tau_O &= \frac{1}{k_2[O_2][M]} \\ &= \frac{1}{\left(1.62 \times 10^{-33} \frac{\text{cm}^6}{\text{mcs}^2\text{s}}\right)\left(2.56 \times 10^{11} \frac{\text{mcs}}{\text{cm}^3}\right)\left(1.22 \times 10^{12} \frac{\text{mcs}}{\text{cm}^3}\right)} \\ &= 1.97 \times 10^9 \text{ s} \end{aligned}$$

Here are the calculated values for all altitudes.

Altitude (km)	τ_O (s)
100	1.97×10^9
70	1.07×10^4
50	2.35

Let us examine the results to ensure that they make sense. First, look at Table 3.6. It should make sense that as the altitude decreases, the concentrations of molecular oxygen and M increase since lower altitudes have higher pressures and thus more molecules of gas. Second,

Altitude (km)	k_2 $\left(\frac{\text{cm}^6}{\text{mcs}^2\text{s}}\right)$	$[O_2]$ (mcs/cm^3)	[M] (mcs/cm^3)
100	1.62×10^{-33}	2.56×10^{11}	1.22×10^{12}
90	1.84×10^{-33}	6.63×10^{11}	3.16×10^{12}
80	1.73×10^{-33}	5.77×10^{12}	2.75×10^{13}
70	1.29×10^{-33}	1.23×10^{14}	5.89×10^{14}
60	9.02×10^{-34}	1.86×10^{15}	8.89×10^{15}
50	7.51×10^{-34}	1.09×10^{16}	5.20×10^{16}
40	9.02×10^{-34}	2.76×10^{16}	1.32×10^{17}

Concentration values are given as molecules per cubic centimeter since pressures in the upper atmosphere are very low.

Table 3.6 Kinetic data for Reaction R3.21 as a function of altitude.

since O_2 is 20.95% of the atmosphere, each of the $[O_2]$ values is $[M]\times0.2095$. Finally, let us examine the lifetime values. At an altitude of 100 km, atomic oxygen persists about 2 billion seconds or 62 years. It is likely that the real lifetime of atomic oxygen is not that long, but the point is that the formation of ozone from atomic oxygen is very, very slow at this altitude. This is because the pressure and gas density are so low that it takes a long time for an atom of oxygen to collide with a molecule of oxygen. At an altitude of 50 km, an atom of oxygen persists for about 2.35 s. This should make sense since at 50 km, the atmospheric pressure is much higher than at 100 km (by more than a factor of 10^4), making a collision between atomic oxygen and molecular oxygen (and eventually, another collision with M) likely to occur in a very short period of time.

The aforementioned example calculation demonstrates why ozone does not form in significant quantities above the stratopause (at about 50 km).

Example 3.7: The Lifetime of O_2^*

Derive the equation for the lifetime of O_2^* in Reaction R3.24. How does the concentration of M affect the lifetime of O_2^* ?

Solution: To find the lifetime, we need to know the rate law for the reaction. Let us restate the reaction for clarity.

$$O_2^* + M \xrightarrow{k} O_2 + M$$

The rate law is given as follows.

$$Rate = k[O_2^*][M]$$

We just need to find the ratio of the $[O_2^*]$ to the rate.

$$\tau_{O_2^*} = \frac{[O_2^*]}{Rate} = \frac{1}{k[M]}$$

Since the lifetime of O_2^* is inversely proportional to [M], high concentrations of M would lead to very short lifetimes of O_2^*. This should make intuitive sense since high concentrations of M would mean that a collision between O_2^* and M is very likely, leading to a short lifetime for O_2^*.

In summary, much chemistry goes on in the stratosphere. If you harken back to Figure 3.5 on page 101, you will see the spectral window from 230 to 300 nm being absorbed by stratospheric ozone starting at 40 km. A significant spectral window from 180 to 230 nm is only slowly attenuated and makes some penetration into the troposphere. The spectral window is absorbed by a minor absorption band in oxygen that can be seen in Figure 3.7. This band, while several orders of magnitude less efficient at absorbing UV radiation compared to the 130–175 nm band, is nonetheless able to knock out this segment of the UV spectrum because of the sheer distance the light must travel through increasingly higher pressures of molecular oxygen. As a result, the troposphere sees only a small slice of the low-energy UV-B photons (280–315 nm) and most of the UV-A photons (315–400 nm). This has not always been the case since the formation of the Earth. Appreciable atmospheric oxygen levels began rising during the Proterozoic eon (2.5–0.5 Ga), beginning with about 1% O_2 and ending around 10%. We know that the formation of ozone is sensitive to the concentration of molecular oxygen (given the calculation in Example 3.6), so the ozone layer would not have been at its current level until sometime after that.

3.6 THE TROPOSPHERE

The final layer of the atmosphere is most familiar to us all – the troposphere. The Greek word tropos, meaning turning or mixing, is an apt description of this layer. Its temperature profile promotes large-scale vertical mixing and contains most of the weather patterns. The

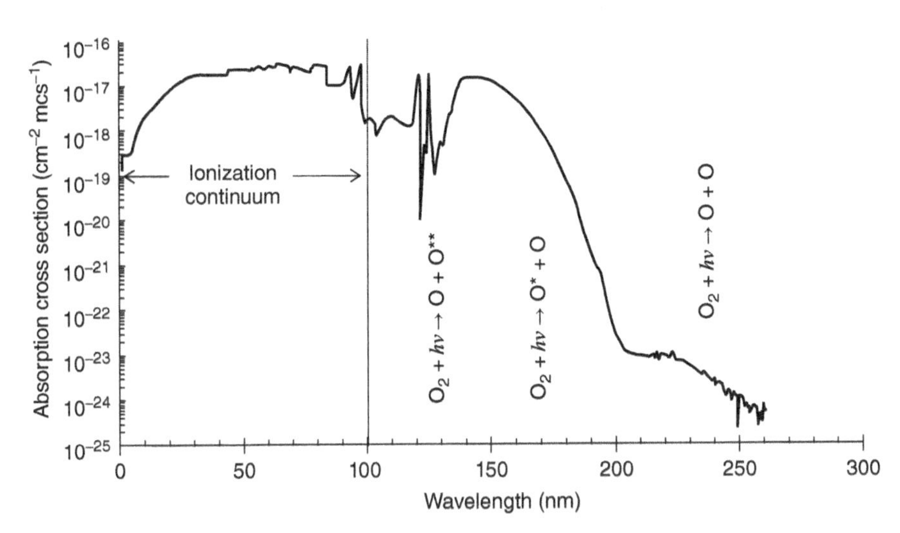

Figure 3.7 This absorbance spectrum of molecular oxygen contains two of the features mentioned in Section 3.3 – the ionization continuum beginning at 103 nm and the absorbance from 130–175 nm. Much less intense but still significant is the absorption from 200 nm to 250 nm. Each region is labeled with a representational reaction. Source: Keller-Rudek http://www.earth-syst-sci-data.net/5/365/2013/essd-5-365-2013.html. Used under CC BY 3.0 https://creativecommons.org/licenses/by/3.0/.

troposphere contains 80% of the mass of the atmosphere and about 99% of water vapor. The height of the troposphere varies from about 7 km above the poles to about 17 km above the equator.

From the previous sections, you already know that most of the radiation below 300 nm is absorbed by the previous layers. The rest of the solar radiations, consisting of a sliver of the UV-B radiation, all of the UV-A and visible radiation, and much of the IR and microwave frequencies reach the top of the troposphere. The UV and visible manage to penetrate down to the surface of the planet without being absorbed. The IR and microwave regions of the spectrum do get absorbed and affect the temperature of the troposphere, but the overall temperature of the troposphere is predominately affected by the temperature of the Earth's surface. About 70 % of the radiation that the Earth receives from the Sun is absorbed, but about 30 % is reflected back into space by the clouds and the surface features (water, snow, land, and so forth). This reflected fraction is referred to as the *albedo* and assigned a fractional number of 0.29 (a planetary average).

The radiation that does not get reflected is absorbed by the surface, and the energy gets converted into heat. Because of this absorption, the surface of the Earth is warmer than the surrounding gases and sets the temperature profile of troposphere, with the warmest gases generally on the surface and the coldest gases at the tropopause. The solar radiation that reaches the surface is intermittent, however, due to the rotation of the planet, and sometimes this temperature profile can become inverted. More details on these thermal inversions will be presented in Section 3.8.

If the troposphere were composed entirely of molecular oxygen and nitrogen, the temperature description of the atmosphere and the surface would be complete. The average surface temperature of the planet, which can be calculated based on the solar irradiance and the average albedo, would be about −17 °C (about 1 °F) – well below freezing. The planet would be an ice ball with that temperature, so we know that there is some other factor that makes our planet more balmy.

☞ The *albedo* refers to the global average planetary reflectance. It is derived from the Latin word for whiteness. The albedo of the Earth has changed in the past, reaching its highest value during ice ages. Table 3.7 gives the albedo averages for some planets and albedo estimates of surface features on the Earth.

Planet or feature	Albedo
Mercury	0.119
Venus	0.75
Earth	0.306
Moon	0.123
Mars	0.25
Fresh snow	0.85
Desert sand	0.40
Grassland	0.25
Deciduous trees	0.15
Ocean	0.06

Table 3.7 Planetary average albedos and albedo estimates of surface features. Source: Lissauer, 2013.

3.6.1 The Planetary Energy Budget

Looking back to Figure 1.1 on page 2, you can see the ideal blackbody radiation curve overlaid on the measured solar irradiance. Planck's Law states that the irradiance of a blackbody is related to the temperature (T) and is a function of the emitted radiation wavelength (λ).

$$I(\lambda) = \frac{2\pi c^2 h}{\lambda^5 \left(e^{\left(\frac{ch}{k_B \lambda T}\right)} - 1 \right)} \tag{3.5}$$

where c is the speed of light, h is Planck's constant, and k_B is Boltzmann's constant. In order to determine the total energy emitted from the Sun across all wavelengths, the area under the curve must be determined through integration of Eq. (3.5) over all wavelengths.

$$E_{\text{Sun}} = \int_0^\infty \frac{2\pi c^2 h}{\lambda^5 \left(e^{\left(\frac{ch}{k_B \lambda T}\right)} - 1 \right)} d\lambda = \frac{2\pi^5 k_B^4}{15 h^3 c^2} T^4 = \sigma_{SB} T^4 \tag{3.6}$$

This is the energy, given in units of W/m^2, at the surface of the Sun. The complicated term that includes h, c, and k_B is replaced by the Stefan–Boltzmann constant (σ_{SB}).

Example 3.8: The Stefan–Boltzmann Constant

Determine the value of the Stefan–Boltzmann constant (σ_{SB}) and the value of the solar irradiance (E_{Sun}) given that the Sun's surface temperature is 5778 K.

Solution: To calculate the value of the Stefan–Boltzmann constant (σ_{SB}), we just need to use the terms in Eq. (3.6) and then substitute in their numerical values.

$$\begin{aligned} \sigma_{SB} &= \frac{2\pi^5 k_B^4}{15 h^3 c^2} \\ &= \frac{(2)(3.14159)^5 (1.38066 \times 10^{-23}\ \text{J/K})^4}{(15)(6.62606957 \times 10^{-34}\ \text{J} \cdot \text{s})^3 (2.99792458 \times 10^8\ \text{m/s})^2} \\ &= 5.67053 \times 10^{-8}\ \text{J/s/m}^2/\text{K}^4 = 5.67054 \times 10^{-8}\ \text{W/(m}^2 \cdot \text{K}^4) \end{aligned}$$

Note that 1 W = 1 J/s. For the solar irradiance, we just need to plug in the Stefan–Boltzmann constant and the Sun's temperature into Eq. (3.6).

$$E_{\text{Sun}} = \sigma_{SB} T^4 = \left(5.67053 \times 10^{-8}\ \text{W/(m}^2 \cdot \text{K}^4)\right)(5778\ \text{K})^4 = 6.320 \times 10^7\ \text{W/m}^2$$

This energy is radiated in all directions by the Sun and must travel a great distance to reach the Earth (see Figure 3.8). By the time the energy reaches the Earth, it is much more diffuse, scaled down by the ratio between the surface area of the Sun ($4\pi r_{\text{Sun}}^2$) and the surface of a sphere at the distance of the Earth's orbit ($4\pi r_{\text{orbit}}^2$).

$$E_{\text{sc}} = E_{\text{Sun}} \frac{4\pi r_{\text{Sun}}^2}{4\pi r_{\text{orbit}}^2} = E_{\text{Sun}} \frac{r_{\text{Sun}}^2}{r_{\text{orbit}}^2} \tag{3.7}$$

where the Sun's radius is 6.96342×10^5 km and the Earth's average orbit is 150×10^6 km. The Earth's orbit is slightly elliptical, with the Earth closest to the Sun in January at 147×10^6 km and farthest from the Sun in July at 152×10^6 km. The value of E_{sc} is known as the solar constant and is about 1360 W/m^2. The Earth, having a certain size, receives a portion of the solar constant according to the cross-sectional area the globe represents (see the bottom of Figure 3.8) spread over the spherical surface area of the planet, leading to the input energy the Earth receives from the Sun.

$$E_{\text{input}} = E_{\text{sc}} \frac{\pi r_{\text{Earth}}^2}{4\pi r_{\text{Earth}}^2} = \frac{E_{\text{sc}}}{4} \tag{3.8}$$

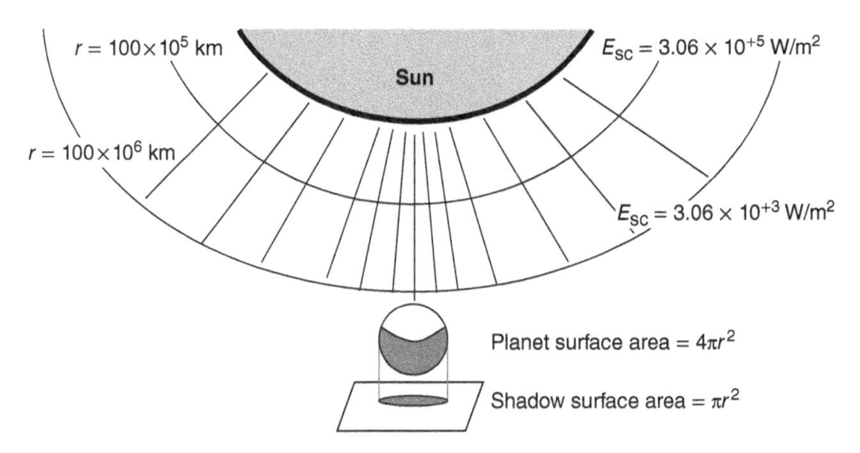

Figure 3.8 The Sun's photosphere produces energy according to Planck's Law, but it diffuses over a great distance when it arrives at the Earth. This diffuse energy strikes a cross-sectional area the size of Earth's shadow, but since the Earth is a globe, the area is spread out over an area four times larger than the shadow. Courtesy K. Overway, 2013.

This input energy, about 341 W/m^2, drives nearly all systems on the planet (weather, climate, most life, almost everything). The Earth also acts as a blackbody radiator. If the input energy never left the planet, then it would continue to heat up. There must be an output energy to balance the input energy in order for the Earth to maintain a relatively stable temperature. When the output energy is equal to the input energy, the temperature will stabilize. We can estimate the temperature of any planet based on Planck's Law by assuming that all of the energy the planet has comes from its star[1] and knowing how much of its input radiation is absorbed. Using the (1-albedo) as an estimate of the absorbed energy, we get the following.

$$\begin{aligned} E_{\text{output}} &= E_{\text{input}} \times (1\text{-albedo}) \\ \sigma_{SB} T^4 &= \frac{E_{\text{sc}}}{4}(1\text{-albedo}) \end{aligned} \quad (3.9)$$

Solving for T results in the following.

$$T = \left[\left(\frac{E_{\text{sc}}}{4}\right)(1\text{-albedo})\left(\frac{1}{\sigma_{SB}}\right)\right]^{\frac{1}{4}} \quad (3.10)$$

Assuming an albedo of 0.306 and using the solar constant of 1360 W/m^2, the calculated temperature of the Earth should be 254 K (about −19 °C or −2.5 °F). The actual average temperature of the surface is 288 K (about 15 °C or 60 °F), so something else must be interfering with the Earth's ability to shed heat into the outer space. This "something" is called the greenhouse effect.

Example 3.9: Predicting Planetary Temperatures

Extrasolar planets are being discovered all of the time. Much is usually known about their host stars, which are easier to detect and study. One such star, HD 4308, is in the Tucana constellation and has the following properties.

- Distance from the Earth: 72 light years
- Solar mass: 0.833 × the mass of our Sun
- Radius: 0.916× the radius of our Sun

[1] This is not quite correct for the Earth since it obtains some heat from the nuclear reactions that occur in the core and mantle.

- Surface temperature: 5597 K
- Age: 7.072 Gyr
- Our Sun's radius is 6.96342×10^5 km
- Our Sun's mass = 1.9891×10^{30} kg.

If an extrasolar planet at an orbit of 234×10^6 km from HD 4308 has an albedo of 0.44 and a planetary radius of 5931 km, what is the expected planetary temperature (ignoring any greenhouse effect)?

Solution: See Section C.1 on page 263.

☞ The *greenhouse effect* is a natural phenomenon whereby certain gases in the atmosphere absorb outgoing IR radiation and trap the heat near the surface.

☞ *Greenhouse gases* (GHGs) are those that possess a vibrational mode that generates an oscillating electric field that has the same frequency as IR radiation, allowing the radiation to be absorbed by the molecule.

3.6.2 The Greenhouse Effect

At the beginning of Chapter 1, you were introduced to Wien's Displacement Law (Eq. (1.1)), which predicts the most intense wavelength of radiation that a blackbody would emit. If Earth's actual temperature were used to calculate the wavelength of radiation that it emits as a blackbody, the following results.

$$
\begin{aligned}
\lambda_{max} &= \frac{2.897768 \times 10^{-3}\ \text{m/K}}{T} \\
&= \frac{2.897768 \times 10^{-3}\ \text{m/K}}{288\ \text{K}} \\
&= 1.006 \times 10^{-5}\ \text{m} = 10.06\ \mu\text{m}
\end{aligned}
$$

This calculation implies that most of the radiation that the Earth is emitting from its surface is in the IR region of the EM spectrum (750 nm–10,000 μm). In the previous sections, the focus was on UV radiation that photolyzed molecular bonds and ionized atoms and molecules. This high-energy radiation comes from a blackbody radiator with a very high temperature (such as the Sun at 5778 K). The Earth's surface is much cooler, so the radiation it emits has much less energy per photon, yet these photons still carry a significant amount of energy away from the Earth's surface. Since the Earth's actual surface temperature is about 288 K and higher than the expected temperature (256 K), our atmosphere is trapping enough heat to raise the surface temperature by over 30 K. This phenomenon, called the *greenhouse effect* and first described by French mathematician Joseph Fourier (1768–1830) in 1824, elevates the surface temperature of the Earth as a result of certain trace gases in the troposphere and to a smaller extent in the stratosphere. These trace gases, called *greenhouse gases* (GHGs), absorb this outgoing radiation and trap the resulting heat near the surface. The result is a surface temperature that is elevated compared to what is expected, much the same as an agricultural greenhouse is much warmer on the inside compared to the ambient temperature surrounding it. Venus has a runaway greenhouse effect, with its surface temperature at 730 K (about 480 K hotter than expected). Mars has a small greenhouse effect, with an average surface temperature of 218 K (about 1 K hotter than expected).

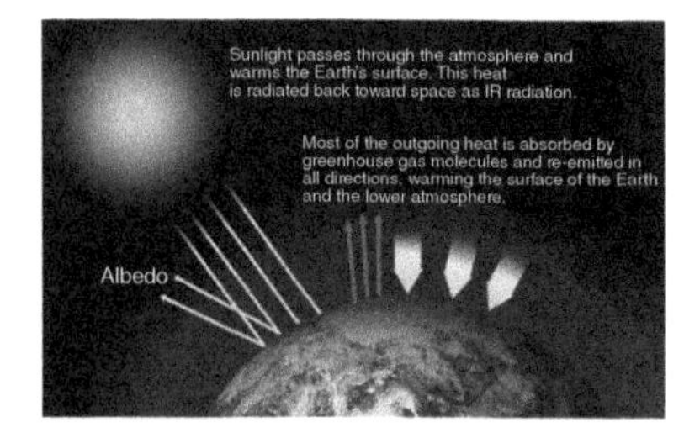

Figure 3.9 Solar radiation that is absorbed by the surface causes the Earth to emit IR radiation as a blackbody radiator. Greenhouse gases in the troposphere absorb a significant portion of the IR radiation coming from the surface. Source: NASA.

To be absolutely clear, the greenhouse effect involves the absorption of IR radiation that is emitted by the Earth's surface because of the heat it contains. It is not the absorption of radiation from the Sun that is reflected off the Earth's surface (see Figure 3.9). Most of this is visible radiation. The reflected radiation, a function of the Earth's albedo, can be absorbed by the atmosphere, but it often just radiates back into space. The Earth (and any blackbody radiator) produces radiation as a function of its temperature, according to Wien's Law, which is an indirect effect of the radiation coming from the Sun.

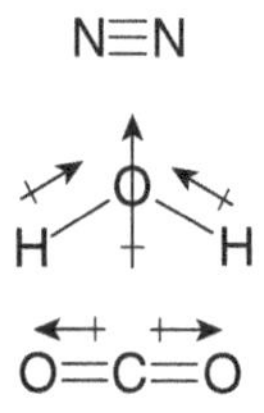

Figure 3.10 The Lewis structures of N_2, H_2O, and CO_2 show that N_2 and CO_2 are nonpolar while H_2O is polar if viewed as static molecules.

GHGs are able to absorb IR radiation because both the gases and the radiation produce an oscillating electric field. Examine Figure 3.10 and remember the principle of electronegativity as the ability of an atom to pull electron density from a bond toward itself. In the Lewis structure for molecular nitrogen, there is no electronegativity difference between two atoms of the same element, so N_2 is a nonpolar molecule. Water has two bond dipoles pointing toward the oxygen atoms and a net dipole forming a vertical axis through the center of the molecule. Carbon dioxide has two bond dipoles pointing in opposite directions toward the two oxygen atoms, so the net dipole is zero since the bond dipoles are equal in magnitude and opposite in direction. The Lewis structures in Figure 3.10 do not tell the whole story, however, since molecules with temperatures above zero Kelvin express their heat energy by vibrating, as seen in Figure 3.11. Even though carbon dioxide is nonpolar when drawn under the Lewis model, in reality its atoms are vibrating. One such vibration is a flapping motion where the two oxygen atoms flap up and down while the central carbon atom moves in the opposite direction. When the two oxygen atoms flap upward, the bond dipoles no longer completely cancel and result in an upward-pointing overall dipole arrow for the structure. This results in an electric field as demonstrated by the sine wave below the structure. When the two oxygen atoms flap back and form a linear structure, the bond dipoles cancel and the electric field goes to zero (as indicated by the sine wave crossing the x axis). The downward flap produces a net dipole and an electric field pointing in the opposite direction. You can imagine the molecule flapping back and forth, producing an electric field that oscillates in synchrony. This vibrational mode of carbon dioxide has an *oscillating dipole moment* and is the key characteristic of a GHG.

EM radiation also possesses an oscillating electric field. If the electric field of a photon passes through the oscillating electric field of a GHG, the photon can be absorbed if the frequencies of the two oscillations are the same. The oscillating electric field of the molecule acts as a handle through which the photon can interact and add to the energy of the vibrating molecule. The molecule absorbs a photon of IR radiation and jumps up to a higher vibrational mode, causing bonds to stretch or bend more vigorously. Examples of the vibrational modes of carbon dioxide and water can be seen in Figures 3.12 and 3.13.

Each molecule can have multiple vibrational modes, and each mode has a characteristic IR absorption (if IR active). Figure 3.14 shows the IR absorption spectrum for carbon dioxide with its absorption peaks labeled. Several GHGs in the troposphere absorb the IR radiation emitted (not reflected) by the surface of the Earth and trap the radiation as heat. As a result, the Earth is 33 K warmer than expected. Carbon dioxide concentrations before humans came on the scene ranged from about 200 to 300 ppmv (at least in the past 800 kyrs). GHGs have been a part of the Earth's atmosphere since the second atmosphere formed from volcanism during the Hadean eon.

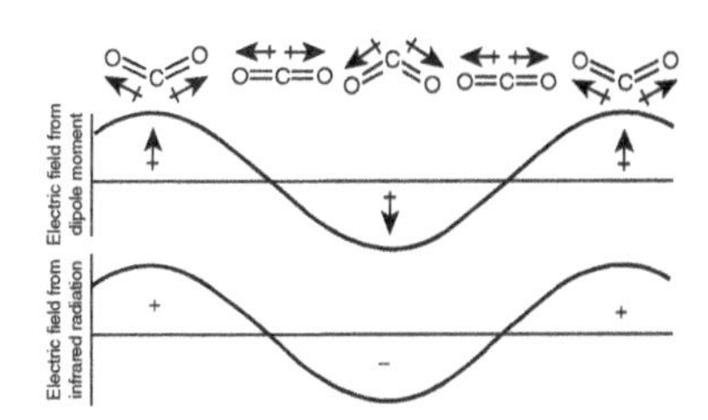

Figure 3.11 During one of its vibrational modes, the normally nonpolar CO_2 becomes temporarily polar and generates an electric field, which matches the frequency of IR radiation.

☞ An *oscillating dipole moment* is a characteristic of a vibrational mode of a polar or nonpolar molecule that generates an oscillating electric field.

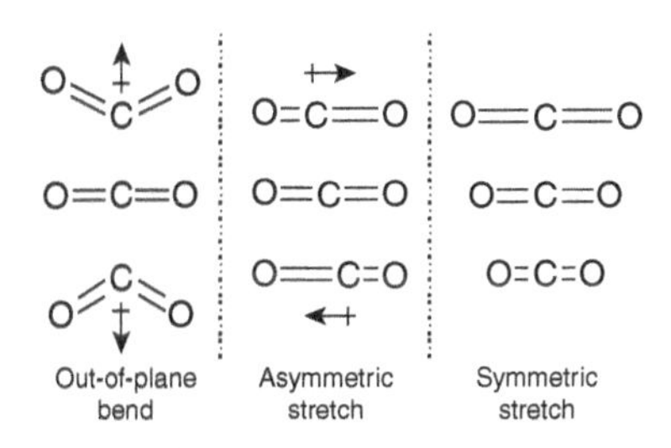

Figure 3.12 Carbon dioxide has three different vibrational modes. Two modes are "IR active" and the other (symmetric stretch) is "IR inactive."

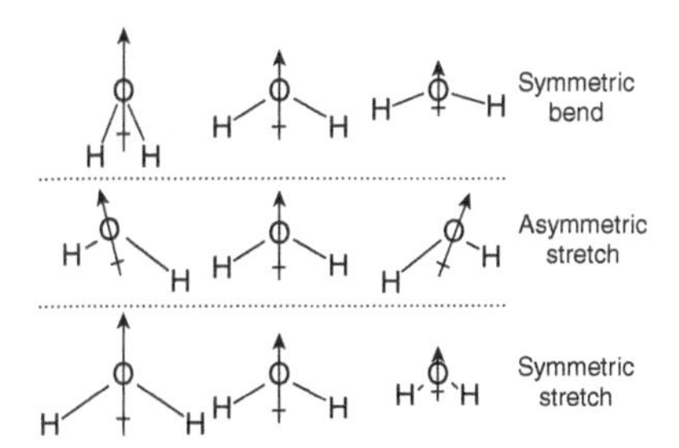

Figure 3.13 Water has a permanent dipole moment, but its vibrational modes show that the dipole oscillates during the vibrations. Examine the length or angle of the dipole arrow carefully and you will see it changes as the molecule vibrates. All modes are IR active.

Example 3.10: Lewis Structures of Greenhouse Gases

Which of the following molecules would you expect to have IR-active vibrational modes and thus be categorized as a greenhouse gas? Draw each molecule using the Lewis structure methodology, and then imagine the molecules bending and vibrating.

$$O_2, SO_2, CCl_2F_2, SO_3, H_2, Ar, CH_4$$

Solution: A greenhouse gas is a molecule that produces an oscillating electric field when it vibrates. This means that it could be a molecule that is polar or nonpolar. Let us examine the gases one by one.

(a) O_2: Molecular oxygen is similar to molecular nitrogen in that it cannot have a fixed or oscillating dipole because there is no electronegativity difference between two atoms of the same element. It does still vibrate, but that vibration is not IR active and therefore O_2 is not a greenhouse gas.

☞ For a review of drawing Lewis structures, see Review Example 1.18 on page 36.

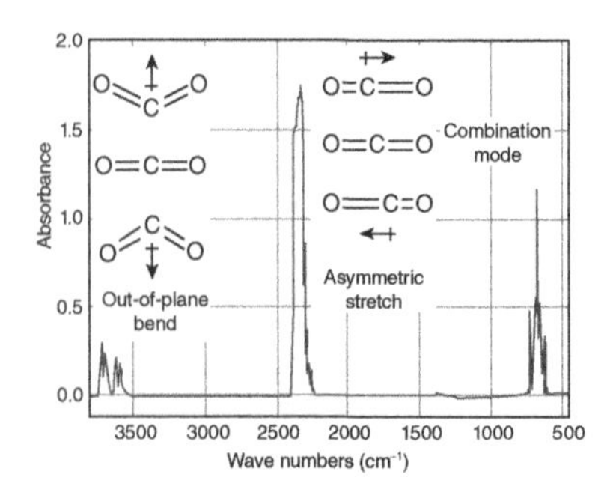

Figure 3.14 The IR absorption spectrum for carbon dioxide. The *x* axis is in wave numbers. The two IR-active vibrational modes from Figure 3.12 (plus a combination of the two modes) show absorption peaks in the spectrum. Source: Original spectrum by NIST, modified by K. Overway.

Have (h): Sum of the *valence* electrons from each atom
Need (n): Sum of the electrons needed to fill the octet (or duet) of each atom
Share (s): Subtract the electrons you have from the electrons you need
Bonds (b): Divide the shared electrons by 2

Formal charge (*FC*)

$$FC = \text{(v.e.)} - \text{(n.b.e.)} - \text{(b)}$$

$$\Delta FC = \text{(largest } FC) - \text{(smallest } FC)$$

where v.e. is the total number of valence electrons, n.b.e is the total number of nonbonding electrons, and b is the total number of bonds.

Table 3.8 The electron counting system for drawing Lewis structures.

(b) SO_2: The electron accounting for SO_2 is given as follows along with the Lewis structures (Table 3.8) in Figure 3.15.

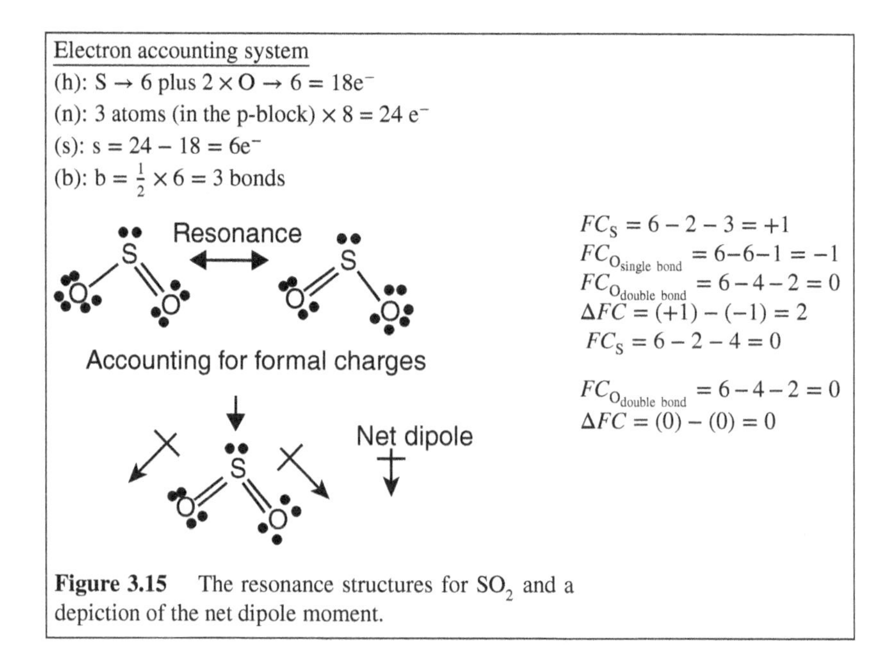

Figure 3.15 The resonance structures for SO_2 and a depiction of the net dipole moment.

The structure with two double bonds is the most stable; sulfur can break the octet rule and exhibit hypervalency. Sulfur dioxide appears to have a bent structure just as water (see Figure 3.13) with the net dipole in the opposite direction. We know water has three IR-active vibrational modes (symmetric, asymmetric, and bending modes). Sulfur dioxide will also have three IR-active modes, as seen in Figure 3.16.

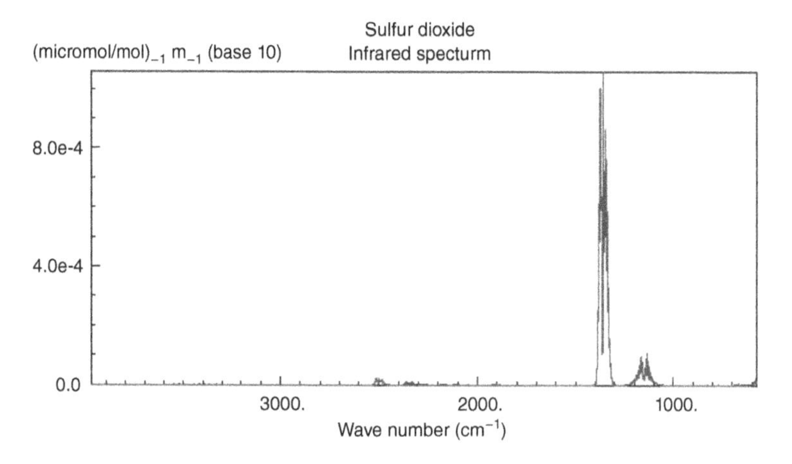

Figure 3.16 The IR spectrum of SO_2. (Courtesy NIST)

(c) CCl_2F_2: The electron accounting for CCl_2F_2 is given as follows along with the Lewis structure in Figure 3.17.

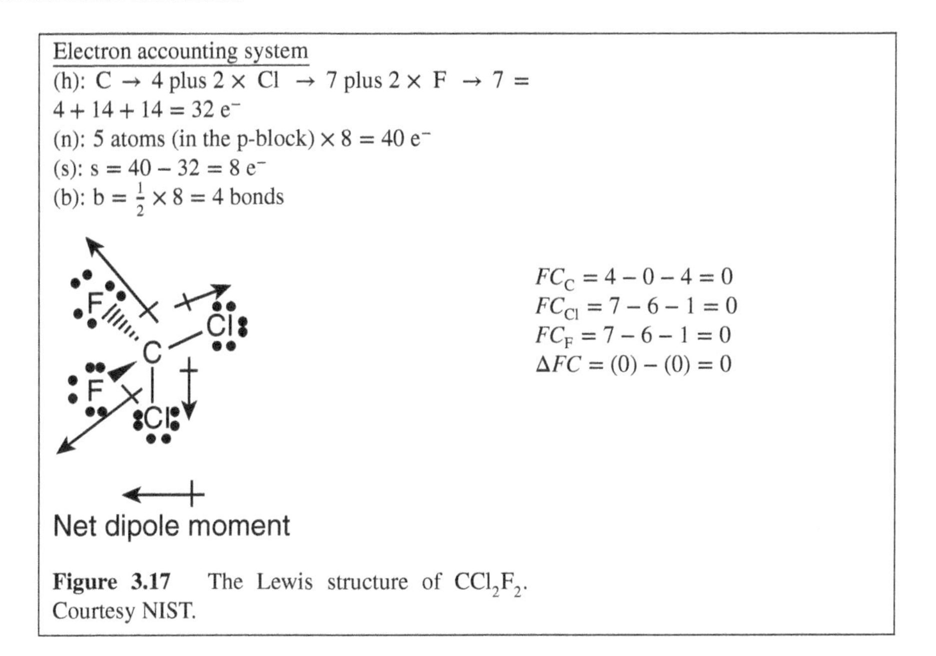

Figure 3.17 The Lewis structure of CCl_2F_2. Courtesy NIST.

There is no possibility or need to reduce *FC* states by adding double bonds since carbon cannot break the octet rule. CCl_2F_2 is a chlorofluorocarbon and appears to have a net dipole in its tetrahedral shape. Hopefully, you can imagine those bonds bending and stretching in various ways to change the angle or length of the dipole arrow. CCl_2F_2 is definitely a greenhouse gas as seen in Figure 3.18.

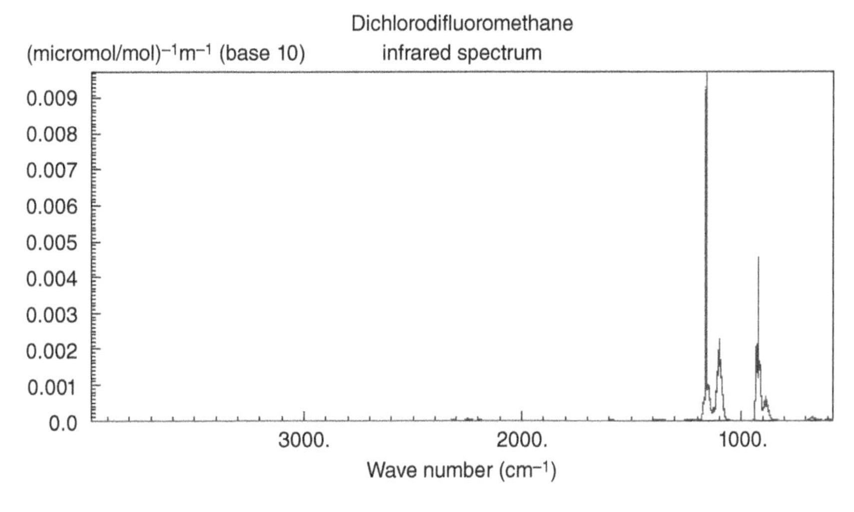

Figure 3.18 The IR spectrum of CCl_2F_2.

For the remaining solutions, see Section C.1 on page 264.

GHGs are all trace components of the atmosphere and yet have a tremendous impact on the surface temperature of the planet and on other factors that will be discussed in more detail in Section 3.13.

3.7 TROPOSPHERIC CHEMISTRY

David Foster Wallace told an insightful joke during his 2005 commencement speech at Kenyon College.

> There are these two young fish swimming along, and they happen to meet an older fish swimming the other way, who nods at them and says, "Morning, boys, how's the water?" And the two young fish swim on for a bit, and then eventually one of them looks over at the other and goes, "What the hell is water?" (Wallace, 2008)

This quote is insightful because most people on most days go about their lives without thinking of the "container" of air that we all share. Before space travel, humans had never left the troposphere, except a few test pilots who flew supersonic flights in the stratosphere. Humans rely heavily on the troposphere to provide us with life-giving oxygen, and yet we discharge about 70% of all of our pollution into the air. We seem to rarely think about fouling it because it is so large, we are so small, and it is economically expedient to discharge wastes into the air instead of properly disposing of them. Ulf Merbold's quote at the beginning of the chapter encapsulates a realization that we (collectively) have not had yet. To be fair, vegetation emits a tremendous amount of hydrocarbons into the atmosphere as well, giving places such as the Great Smoky Mountains their name. What are these pollutants, where do they come from, and what is their fate? Let us start with the car.

3.7.1 The Internal Combustion Engine

The modern internal combustion engine is widely employed in three different variations: the four-stroke engine, the two-stroke engine, and the diesel engine. Each has its advantages, disadvantages, and characteristic emission profile. The purpose of the next few sections is to focus on the pollutants while leaving the mechanics and thermodynamics for another textbook.

Most cars use octane (and related alkanes) as fuel and, in the presence of compressed air, cause the following reactions.

$$2C_8H_{18}(g) + 25O_2(g) \rightarrow 16CO_2(g) + 18H_2O(g) \quad \text{(R3.26)}$$
$$2C_8H_{18}(g) + 17O_2(g) \rightarrow 16CO(g) + 18H_2O(g) \quad \text{(R3.27)}$$
$$N_2(g) + O_2(g) \rightleftharpoons 2NO(g) \quad \text{(R3.28)}$$

Reaction R3.26 shows the ideal combustion reaction, producing only carbon dioxide and water. Under nonideal conditions, some carbon monoxide is produced (Reaction R3.27). Octane is supplied from the gas tank and oxygen is supplied from the ambient air, which contains mostly nitrogen gas. Some of the nitrogen is oxidized as a result of high combustion temperatures.

☞ For a review of Gibbs Free Energy and other thermodynamic terms, see Review Example 1.14 on page 32.

Example 3.11: Standard Free Energy

Calculate the standard free energy (assume 298 K) for Reactions R3.26 and R3.28.

Solution: A standard free energy calculation requires a balanced reaction and a table of thermodynamic values (see Table I.2 in Appendix I on page 318).

<u>For Reaction R3.26</u>

$$\begin{aligned}\Delta G^\circ_{rxn} &= \Sigma\Delta G^\circ_{f\,products} - \Sigma\Delta G^\circ_{f\,reactants}\\ &= \left[(16 \text{ mol } CO_2)(-394.4 \text{ kJ/mol}) + (18 \text{ mol } H_2O)(-228.6 \text{ kJ/mol})\right]\\ &\quad - \left[(2 \text{ mol } C_8H_{18})(+16.40 \text{ kJ/mol}) + (25 \text{ mol } O_2)(0 \text{ kJ/mol})\right]\\ &= -10{,}458 \text{ kJ}\end{aligned}$$

<u>For Reaction R3.28</u>

$$\begin{aligned}\Delta G^\circ_{rxn} &= \Sigma\Delta G^\circ_{f\,products} - \Sigma\Delta G^\circ_{f\,reactants}\\ &= ([(2)(87.6)] - [(1)(0) + (1)(0)]) \text{ kJ}\\ &= 175.2 \text{ kJ}\end{aligned}$$

This example demonstrates why octane is a fuel – it spontaneously combusts and provides a lot of energy for mechanical work. Reaction R3.28 is nonspontaneous and absorbs energy to produce nitric oxide (NO). Since NO is a toxic gas, it is convenient that it is not a favorable reaction. So why does it happen at all? Nitric oxide is thermodynamically unstable compared to N_2 and O_2, so why does it form?

These calculations tell us a story that would imply that automotive NO_x pollution is not a problem and would not accumulate to levels that are sometimes seen in cities such as Los Angeles and Mexico City. We know that these calculations, which rely solely on thermodynamic principles, tell us that the formation of NO is unfavorable, so NO will immediately revert back to N_2 and O_2, and give an overestimate of the loss of nitric oxide in the re-equilibration that occurs after exhaust is expelled from a car. There must be other factors involved since NO_x (representing NO and NO_2) pollution persists in certain localities.

A closer examination of Reaction R3.28 shows that this reaction cannot possibly occur as written and must be an overall reaction. Think about the Lewis structures of each molecule, and try to imagine N_2 and O_2 colliding to yield NO.

Example 3.12: Lewis Structures, Reaction Mechanisms, and Plausibility

Draw the Lewis structures for each molecule in Reaction R3.28, and then use arrows to propose a collision that would result in the formation of NO.

Solution: Once the structures are drawn, we can look at collisions.

Electrons for N_2
(h): $5 + 5 = 10\ e^-$
(n): $2 \times 8 = 16\ e^-$
(s): $16 - 10 = 6\ e^-$
(b): $b = \frac{1}{2} \times 6 = 3$ bonds

:N≡N:

$FC_N = 5-2-3=0$

Electrons for O_2
(h): $6 + 6 = 12\ e^-$
(n): $2 \times 8 = 16\ e^-$
(s): $16 - 12 = 4\ e^-$
(b): $b = \frac{1}{2} \times 4 = 2$ bonds

O=O

$FC_O = 6 - 4 - 2 = 0$

Electrons for NO
(h): $5 + 6 = 11\ e^-$
(n): $2 \times 8 = 16\ e^-$
(s): $16 - 11 = 5\ e^-$
(b): $b = \frac{1}{2} \times 5 = 2.5$ bonds

Since only whole bonds can be shown, there will be a double bond. Since O is more electronegative than N, O will get a full octet and N will get an unfilled octet.

N=O

$FC_N = 5 - 3 - 2 = 0$
$FC_O = 6 - 4 - 2 = 0$

There are only a few ways N_2 and O_2 can collide.

If the two molecules line up in parallel, bonds could form between the N and O while the N_2 and O_2 bonds break. This would form the two NO as expected, but it would require the breaking of a double and triple bond simultaneously, requiring 945+498=1443 kJ/mol of energy. This is an enormous amount of energy, and it is very unlikely to happen. Another option is for the N_2 and O_2 to collide head-on or one of them to collide head-on. It would require one of the bonds to break (either double or triple) to form an intermediate species (either N_2O or NOO). This implies that there is at least another step to form the two NO that are in the overall reaction.

This exercise reveals how unrealistic Reaction R3.28 is and how it is not useful for predicting NO persistence and concentrations. Studies of the formation of nitric oxide show that the following sequence is really responsible (Bowman, 1973; Iverach, Basden, and Kirov, 1973). First, molecular oxygen is formed as a result of the radical decomposition of a hydrocarbon, such as octane.

☞ The reaction arrow $\xrightarrow{\Delta}$ means that a chemical reaction driven by heat.

$$RCH_3 + M \overset{\Delta}{\rightleftharpoons} R\dot{C}H_2 + H\cdot + M \qquad \text{(R3.29)}$$
$$H\cdot + O_2 \rightarrow \cdot OH + O \qquad \text{(R3.30)}$$

Once the reactive atomic oxygen is formed, nitric oxide production follows.

$$O + N_2 \overset{k_f}{\rightleftharpoons} NO + N \qquad \text{(R3.31)}$$
$$N + O_2 \xrightarrow{k_5} NO + O \qquad \text{(R3.32)}$$

The sum of Reactions R3.31 and R3.32 gives the overall reaction in Reaction R3.28. Notice also that Reactions R3.31 and R3.32 form a cycle. Given that collisions between atomic oxygen and nitrogen are more likely to occur with molecular oxygen and nitrogen, this cycle is driven by the entropy of dilution. Once the first atom of oxygen is released in Reaction R3.30, it is statistically likely that NO will form. There is only one possibility that NO will not form and that is, the two atomic oxygen atoms finding each other and reforming the O=O bond. Most of the other molecules the atomic oxygen will find are molecular nitrogen since it represents a majority of all the gases present. Once mixing occurs, which happens thoroughly at elevated temperatures, NO formation is inevitable and nearly irreversible. This is a clear example of why a deeper understanding of underlying kinetics is necessary for explaining an important conclusion.

3.7.1.1 The Four-Stroke Gasoline Engine If you have ever driven a car, it is most likely the engine was a four-stroke engine. Its cycle is divided into four segments in which the tightly fitted piston slides up and down the walls of the engine cylinder as it loads fresh fuel and air, compresses the mixture, moves in a power stroke resulting from the explosion of the fuel and air, and flushes out the combustion exhaust.[2] In the first segment, called the intake stroke, the piston moves down, generates a negative pressure, and pulls in the fuel-air mixture that flows into the expanding cylinder through an open intake valve. Next, the intake valve closes and the piston moves upward in the compression stroke. Just after the piston reaches its maximum height the spark plug fires and ignites the fuel-air mixture; the explosion drives the piston down and converts a portion of the released energy into mechanical energy in the power stroke. Finally, the piston moves upward and pushes the exhaust out of an open exhaust valve in the last segment of the cycle. All throughout the engine cycle, motor oil is sprayed onto the engine components from a pool of oil located below the engine.

The four-stroke engine, which has an overall energy efficiency of about 25%, can be characterized as having less hydrocarbon (HC) emissions, but more NO_x emissions compared to a two-stroke engine. The features that help it achieve this are explained below.

Separated intake and exhaust stages: Since the four-stroke engine fires every other cycle of the crankshaft, the intake of fresh fuel and air is completely separated from the expulsion of the exhaust – there is no mixing of the two. Thus, the unburned HCs in the exhaust are kept to a minimum.

Lubrication is kept separate from the fuel: The four-stroke engine is designed to run upright and therefore places the lubricating oil in the crankshaft

[2]Very fine diagrams of the four-stroke, two-stroke, and diesel engines can be found at www.animatedengines.com.

area of the engine. Only a small film of oil that lubricates the piston ring is burned during combustion, keeping HC emissions to a minimum.

High operational temperatures: Four-stroke engines operate at relatively high temperatures and cylinder compression, compared to two stroke engines, and so they produce more NO_x emissions.

3.7.1.2 The Two-Stroke Gasoline Engine Two-stroke engines are commonly deployed in power tools, such as weed eaters and chain saws. These engines are preferred for small tools because they offer more power per weight compared to four-stroke engines (because the power-stroke occurs twice as often) and because two-stroke engines can operate at any angle. The engine cycle is divided into two segments, as the name implies. In the charging/power stroke, the downward moving piston compresses a fresh fuel-air mixture that is located below the piston and pushes it into the top of the cylinder head space by way of an open passageway in the chamber. This action flushes the combustion exhaust from the previous cycle out of an exhaust port. It is inevitable that the fresh fuel and exhaust will mix, so some fuels leaks out with the exhaust. In the compression stroke, the piston moves up, blocks the exhaust port, and compresses the fresh fuel-air mixture. The movement of the piston upward generates a reduced pressure in the space below the piston and pulls in fresh fuel. At maximum compression above the piston, there is the compressed fuel-air mixture used in the current cycle and a decompressed fuel-air mixture below the cylinder for the next cycle. Just after maximum compression, the spark plug fires and the explosion drives down the piston, converting the chemical energy to mechanical energy and heat in this power-stroke. The engine repeats the process by flushing out the exhaust with the next batch of fresh fuel as the piston completes its downward charging/power stroke.

In two-stroke engines, the space above and below the moving piston is actively used in the process. Thus, lubricant cannot be sprayed on the mechanical components because they are sealed inside of the piston chamber. A high-molecular-weight two-cycle motor oil is added to the fuel so that the components can be lubricated internally, as opposed to externally as in the four-stroke engine. Two-stroke engines generally have high HC emissions and low NO_x emissions, as explained below.

Intake and exhaust mix: Fresh fuel is used to flush out the exhaust gases and there is an inevitable emission of unburned fuel. This significantly increases the HC emissions.

Lubrication is mixed with the fuel: Because the design of the engine, which allows it to run at any angle, a lubricant must be added to the gasoline. When the fuel mix cycles between the lower and upper chambers it provides lubrication to the piston and crankshaft. The lubricant is not as flammable as the gasoline, so much of it remains unburned after the combustion of the fuel and contributes to high HC emissions.

Cooler running temperatures: The two-stroke engine operates at a cooler temperature than a four-stroke engine, which decreases the amount of NO_x produced during combustion.

3.7.1.3 The Four-Stroke Diesel Engine Diesel engines are often associated with applications that require a lot of power, and traditionally these engines delivered more power than gasoline engines, although engine differences are decreasing. Diesel engines produce their maximum power at lower RPMs and, because they use a higher compression ratio than gasoline engines, they are more fuel efficient.

A diesel engine functions much like a four-stroke gasoline engine. When the piston moves down in the intake stroke, it pulls in fresh air (not mixed with fuel). In the compression stroke the piston moves upward and compresses the air. At the maximum compression, the fuel injector sprays in the diesel fuel. The temperature and pressure of the compressed air is enough to ignite the fuel without the aid of a spark plug. The power-stroke is followed by an upward exhaust-stroke that expels the exhaust gases. The space above and below the piston are isolated, so the lubricating oil is not mixed with the fuel and not burned in significant quantities with the fuel. The exhaust and fuel-air mixture are also isolated, so there is no fresh fuel expelled with the exhaust as with the two-stroke engine. Thus, the HC emissions are much lower than a two-stroke gasoline engine and even lower than a four-stroke gasoline engine. Diesel engines run very hot and therefore have the highest emissions of NO_x. The final, and unique, characteristic of diesel engines is their high emission of particulate matter (PM). Petroleum-based diesel fuel is about 50% cetane ($C_{16}H_{34}$), with the rest of the components being shorter and longer chain hydrocarbons. Cetane is much less volatile and flammable than octane (the principle component of gasoline), so when cetane is injected into the cylinder at the last moment it remains in larger droplets than in a gasoline/air mixture, and therefore it does not burn completely. The incomplete combustion on the surface of the diesel droplets forms soot, technically called PM.

Separated intake and exhaust stages: Just like the four stroke gasoline engine, this reduces HC emissions.

Lubrication is kept separate from the fuel: Just like the four stroke gasoline engine, lubricant oil is not present in the combustion chamber contributing to significantly lower HC emissions.

High running temperatures: Diesel engines run hot and therefore produce significant NO_x emissions.

Diesel fuel is not as flammable; fuel is less dispersed when it burns: When the fuel is injected into the cylinder at the last moment is does not disperse into a gas but remains an aerosol. This results in an incomplete combustion and the formation of soot.

3.7.1.4 Engine Emission Comparison I have alluded to some comparative differences between the engine types. Now let us look at some empirical data. Remember that the purpose of describing these engines is so that you can understand the impact that each has on air quality. The first comparison between a two-stroke and four-stroke engine can be seen in Table 3.9.

The * next to the word diesel in Table 3.10 represents just one of many grains of salt that need to be taken when considering the diesel emissions from Volkswagen cars. In 2015, a scientist from University of West Virginia noticed that the emissions from the VW cars he tested on the road did not match the results of emission tests in the lab. VW engineers had installed software to detect testing conditions and retune the engine performance in order to satisfy emission standards. Normal running conditions produced better fuel economy and more power but much higher emissions of NO_x. The illegal NO_x emissions were reduced under testing conditions. When something seems too good to be true... Source: Glinton (2015)

Pollutant	Two-stroke (g/kWhr)	Four-stroke (g/kWhr)
CO	495	380
HC	267	22
NO_x	0.9	2.2

Table 3.9 Emissions from a Yamaha 232-cc two-cylinder, four-stroke outboard motor and a Suzuki 211-cc two-cylinder, two-stroke outboard motor without catalytic converters.

Pollutant (mg/km)	VW Golf gasoline	VW Golf diesel*
HC	520	27
NO_x	33	550
PM	5.3	32.4

The measurements of hydrocarbons (HC), nitrogen oxides (NO_x), and particulate matter (PM) imply mass per km traveled.

Table 3.10 Emissions from two Volkswagen Golf cars with different four-stroke engines.

The two-stroke engine emits more carbon monoxide and many more hydrocarbons compared to the four-stroke engine, as a result of the separation of the lubricant, fuel mixture, and exhaust in the four-stroke engine. The NO_x emissions from the four-stroke engine are higher because of its higher operating temperatures. Table 3.10 compares two models of the Volkswagen Golf. The gasoline engine emitted much less NO_x and PM compared to the diesel engine because of the cooler operating temperatures and more flammable/dispersed fuel. The diesel has very high PM emissions due to the nature of the fuel and design of the engine.

3.7.1.5 Fuel Alternatives and Additives The gasoline used to power four-stroke engines is a mixture of many different hydrocarbons. The major component of gasoline is isooctane (2,2,4-tetramethylpentane), and the octane number you see on the pump at the gas station means that the gasoline behaves as if it were made of a specific percent of isooctane (hereafter referred to as octane). Nowadays, "regular" gasoline is usually 87 octane or close to 87% octane (although prior to the 1980s, "regular" gasoline meant leaded gasoline). Engines are finely timed machines, and if the fuel mixture is not tightly controlled, then the explosion that drives the motor could be irregular and cause damage. Preignition and knocking are two such problems that occur when the fuel ignites before the spark plug fires or when the flame front after ignition is not uniform. Several additives have been formulated to prevent engine knock, one of which is tetraethyl lead (TEL).

Tetraethyl lead was introduced to the consumer market in the 1920s as an inexpensive way to prevent knocking. Its toxic effects were known since 1904, and a comprehensive study in 1943 showed that lead exposure caused long-term learning difficulties. Despite this, TEL was used in the United States until the 1970s when catalytic converters were coming into use (TEL poisons the catalyst and eventually inactivates it). Many other countries used it until the 1990s and 2000s. Despite efforts by the United Nations, a handful of countries used TEL as of 2015. NASCAR had a federal exemption and used TEL until 2008. As a result, millions of tons of lead have been dispersed over the entire United States (remember the quote from David Wallace?). As late as 2005, the use of leaded gasoline was being defended in the media by the Cato institute (Milloy, 2005) while exposed NASCAR pit crews sported blood levels above the pediatric intoxication level (O'Neil *et al.* (2006)). Lead is a neurotoxin because it binds strongly to thiol groups on proteins (where cysteine and methionine are present) and distorts the shape. This denaturing effect degrades or inactivates the function of the protein, leading to toxic effects from inhibiting calcium absorption to behavioral disorders and learning disabilities.

Fortunately, less toxic additives have replaced TEL as an antiknock agent and have concomitantly resulted in the reduction of carbon monoxide and lead emissions since the TEL phaseout. One of the more infamous additives is methyl tert-butyl ether (MTBE), which has been used since the late 1970s and has been fouling drinking water as a result of leaking underground gasoline tanks. A more popular additive has been ethanol, an agricultural product from the fermentation of corn. In 2011, more corn was grown in the United States to make ethanol for the fuel supply than to make feed for livestock. These additives effectively increase the octane number, preventing knocking, and introduce more oxygen into the combustion reaction (since each additive contains oxygen in its structure). Table 3.11 shows the result of two studies comparing four-stroke engines using gasoline and ethanol/gasoline blends (E10 means 10% ethanol). There is some advantage to gasoline/ethanol blends for reductions in CO and NO_x, but other emissions increase and fuel economy decreases because ethanol is less energy dense than octane.

Table 3.12 shows the improvements to emission that are made for diesel fuel when blended with biodiesel. Biodiesel, such as ethanol and MTBE, contains oxygen atoms in its molecular structure (usually in the form of an ester). These oxygen atoms assist the fuel to burn cleaner. As a result, incomplete combustion, which is responsible for the formation of CO and PM, is reduced.

Pollutant	E10[a] (% change)	E10[b] (% change)	E85[b] (% change)
CO	−17	−15±28	−5±51
HC	−36	+11±18	−45±24
NO_x	−7	+4±26	−42±15
acetaldehyde	- - -	130±110	+2900±1200
benzene	- - -	12±31	−72±15
fuel economy[c]	- - -	−3 to 4%	−25 to 30%

[a] From Jia *et al.* (2005) using a 124-cc monocylinder, four-stroke, air-cooled motorcycle engine using regular gasoline and E10 (10% ethanol) gasoline while running at 50 km/hr without a catalytic converter.
[b] From Graham, Belisle, and Baas (2008) using flex fuel vehicles with catalytic converters. Results reflect average values across 43 vehicles for E10 and 11 vehicles for E85.
[c] Fuel economy reflects a change compared to 100% petroleum gasoline (U.S. Department of Energy (2013)).

Table 3.11 Emission differences for various blends of gasoline and ethanol compared to 100% gasoline.

Pollutant	20% SME (% change)	20% YGME (% change)	SME (% change)	YGME (% change)
CO	−7.5	−7.0	−18.2	−17.8
HC	−3.1	−2.3	−42.5	−46.3
NO_x	+1.5	+1.1	+13.1	+11.6
SN	−15.8	−16.8	−61.1	−64.2

Source: Canakci and Gerpen (2003).
The biodiesel fuels tested include blends with diesel and 100% soybean oil methyl ester (SME) and yellow grease methyl ester (YGME). The SN parameter is the Bosch Smoke Number, a measure of the particulate matter.

Table 3.12 Emissions of a John Deere 4276T, four-cylinder, four-stroke, turbocharged diesel engine running at 1400 RPM compared to using No. 2 diesel fuel.

☞ Atmospheric organic compounds, both anthropogenic and biogenic, tend to be referred to as volatile organic compunds (VOCs) or simply hydrocarbons (HCs). While not strictly the same because HC implies only hydrogen and carbon, sometimes these two acronyms are synonyms.

☞ *Photochemical smog* is a type of air pollution that is characterized by the presence of O_3, NO_x, CO, and VOCs. It is typically associated with the combustion of petroleum fuels.

Another automotive advancement mentioned before is the catalytic converter. These devices, introduced in the 1970s, use precious metals (Pd, Pt) to assist the conversion of CO to CO_2, NO_x to N_2 and O_2, and unburned HCs to CO_2 and H_2O.

3.7.2 Ground-Level Ozone and Photochemical Smog

The emissions from cars and other anthropogenic sources mix with VOC emissions from the biosphere in the troposphere to form a soup of thousands of trace components. Only 0.04% of the atmosphere is anything other than N_2, O_2, and Ar, but these trace components are significant enough to affect the air quality. The troposphere is home to most of the water vapor in the atmosphere, and because the tropopause is much colder than the freezing point of water,[3] most of the water condenses into liquid or icy clouds. When it does this, its density increases, its buoyancy decreases, and, therefore, it tends to stay in the troposphere.

The study of tropospheric oxidation reactions began in the late 1940s when cities, such as Los Angeles, began to experience *photochemical smog*. Scientists soon discovered that ozone was one of the main ingredients. We learned earlier that ozone is formed in the lower stratosphere by the photolysis of molecular oxygen with UV-C radiation. How is it also formed in significant quantities in the troposphere when UV radiation is very limited?

[3] At the tropopause, the pressure is about 50 torr or 6.6 kPa and the temperature is about −50 °C or 220 K, but the freezing point of water is still about 0 °C since the solid–liquid line on the phase diagram is so vertical.

The reddish color of NO_2 in the troposphere gives cities such as Los Angeles a colorful (and toxic) skyline. Since NO_2 has a visible color, it must absorb some light in the visible portion of the spectrum (otherwise, it would appear colorless). Nitrogen dioxide is formed as a secondary pollutant from the primary combustion pollutant nitric oxide, which is colorless.

$$2NO + O_2 \rightarrow 2NO_2 \qquad \text{(R3.33)}$$

This reaction is likely an overall reaction with an underlying mechanism involving the formation of an intermediate NO_3 or N_2O_2 (or ONNO).

Example 3.13: Reaction Mechanisms

Propose a mechanism for the oxidation of nitric oxide (Reaction R3.33) that involves the formation of the intermediate N_2O_2. Sketch the transition state structure for each step.

Solution: Remember that reaction intermediates are produced in an early step and consumed in a later step. It is also required that the summation of the mechanism must reduce to the overall reaction. The molecule N_2O_2 looks like two NO molecules smashed together, so let us propose it.

$$\textit{step 1}: \ NO + NO \underset{k_{-1}}{\overset{k_1}{\rightleftharpoons}} ONNO$$

This step is given the equilibrium arrows because it seems to be a very reversible reaction with the N–N bond forming and breaking. Once this intermediate forms, we have used up 2 NO from the overall reaction, and now we need to consume an O_2.

$$\textit{step 2}: \ ONNO + O_2 \overset{k_2}{\longrightarrow} 2NO_2$$

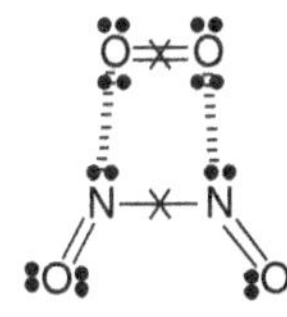

The transition state for the second step would look something as the figure to the left, where a successful collision would involve the two molecules approaching each other with parallel O=O and N–N bonds. The N–N bond would break *homolytically* and the oxygen double bond would break.

☞ *Homolytic* bond cleavage occurs when a bond splits in such a way that one electron stays with one atom and the other electron in the bond goes with the other atom, leaving each neutral in charge. This is different from *heterolytic* cleavage where the electron pair of the chemical bond goes with one atom, leaving one atom negatively charged and the other positively charged.

Once formed, nitrogen dioxide absorbs a photon at 400 nm (thus its reddish color) in the following cycle.

$$NO_2 + h\nu \overset{k_1}{\longrightarrow} NO + O \qquad \text{(R3.34)}$$

$$O + O_2 + M \overset{k_2}{\longrightarrow} O_3 + M \qquad \text{(R3.35)}$$

$$O_3 + NO \overset{k_3}{\longrightarrow} NO_2 + O_2 \qquad \text{(R3.36)}$$

The three reactions represent the NO_x cycle that contributes to tropospheric or ground-level O_3. We know that the rate laws for these reactions are given by the following equations.

$$Rate_1 = k_1[NO_2]$$
$$Rate_2 = k_2[O][O_2][M]$$
$$Rate_3 = k_3[O_3][NO]$$

$Rate_1$ depends on radiation and is the rate-limiting step. Atomic oxygen is very reactive, so once it is produced from the first reaction, it will react quickly. The ozone produced collides with NO. Since these reactions form a cycle, the reactions will proceed until $Rate_1$ balances $Rate_3$, when the cycle will come to equilibrium and maintain steady concentrations of NO_2, NO, and O_3. So we can set $Rate_1$ equal to $Rate_3$ and solve for $[O_3]$.

$$k_1[NO_2] = k_3[O_3][NO]$$

$$[O_3] = \frac{k_1[NO_2]}{k_3[NO]} \qquad (3.11)$$

If we assume that $[O_3]$ and [NO] start out at zero before the cycle starts, then we know that for every NO created in step 1, it results in an O_3 in step 2. Given that $[NO]=[O_3]$, the equation becomes

$$[O_3] = \frac{k_1[NO_2]}{k_3[O_3]}$$
$$k_3[O_3]^2 = k_1[NO_2] \tag{3.12}$$

The $[NO_2]$ concentration starts out at some initial state ($[NO_2]_0$) and then decreases. For each NO_2 consumed, one O_3 is produced, so

$$[NO_2] = [NO_2]_0 - [O_3] \tag{3.13}$$

and substituting this into Eq. (3.12) results in the following.

$$k_3[O_3]^2 = k_1([NO_2]_0 - [O_3]) = k_1[NO_2]_0 - k_1[O_3]$$
$$k_3[O_3]^2 + k_1[O_3] - k_1[NO_2]_0 = 0 \tag{3.14}$$

This is a quadratic equation, which gives the following valid solution.

$$[O_3] = \frac{-k_1 + \sqrt{k_1^2 + 4k_3k_1[NO_2]_0}}{2k_3} \tag{3.15}$$

Zenith angle (°)	k_1 (s^{-1})
0	9.61×10^{-3}
10	9.56×10^{-3}
20	9.42×10^{-3}
30	9.17×10^{-3}
40	8.75×10^{-3}
50	8.11×10^{-3}
60	7.08×10^{-3}

When the Sun is directly overhead, the zenith angle is 0°.

Table 3.13 Rate constant values for Eq. (3.11) as a function of the angle of the Sun (zenith).

Assuming that the Sun is directly overhead (zenith angle of 0°), an initial NO_2 concentration of 100 ppbv (which is equivalent to 2.46×10^{12} mcs/cm^3 at 1 atm of pressure), and a k_3 rate constant of 1.9×10^{-14} cm^3/mcs^{-1}/s^{-1} (see Table 3.13 for k_1), we can calculate a steady-state concentration of ozone.

$$[O_3] = \frac{-k_1 + \sqrt{k_1^2 + 4k_3k_1[NO_2]_0}}{2k_3}$$
$$= \frac{-9.61\times10^{-3}s^{-1} + \sqrt{(9.61\times10^{-3}s^{-1})^2 + 4\left(1.9\times10^{-14}\frac{cm^3}{mcs\cdot s}\right)(9.61\times10^{-3}s^{-1})\left(2.5\times10^{12}\frac{mcs}{cm^3}\right)}}{2\left(1.9\times10^{-14}\frac{cm^3}{mcs\cdot s}\right)}$$
$$= 9.0 \times 10^{11}\frac{mcs}{cm^3}$$

A concentration of 9.0×10^{11} mcs/cm^3 can be converted to ppbv at 1 atm of pressure by dividing it by the number density of molecules at that pressure ($2.46\times10^{19}\frac{mcs}{cm^3}$) and then multiplying by a billion.

$$\left(9.0 \times 10^{11}\frac{mcs}{cm^3}\right) \div \left(2.46 \times 10^{19}\frac{mcs}{cm^3}\right) \times \left(1 \times 10^9\right) = 37 \text{ ppbv}$$

This level of ground-level ozone is in the range of the average levels, around 20–60 ppbv. The EPA's Air Quality Index states that an exposure of NO_2 to a range of 85–100 ppbv for 8 h or more is unhealthy for individuals who are sensitive to air pollution, so a 100 ppbv range assumed in the aforementioned problem is a reasonable level for a pollution event in a large city. Note that sunlight is required to start this cycle, and the value of the rate constant k_1 peaks at noon when the Sun is at its highest point in the sky.

3.7.3 The Hydroxyl Radical

The presence of ozone introduces a very reactive oxidant into the troposphere, but it is not enough to explain some of the very short lifetimes for VOCs. Studies in the early 1970s linked the presence of NO_x and ozone to the formation of the hydroxyl radical (·OH or just OH). Because it is a radical and highly reactive, VOCs susceptible to attack by OH are readily oxidized to aldehydes and eventually to CO_2.

You learned earlier that when ozone absorbs radiation, it can photolyze to produce atomic and molecular oxygen in a variety of energy states (Figure 3.19). Each set of products results from different photon energies.

☞ A *radical* or *free radical* is a chemical species with an odd number of electrons or unpaired electrons. An unpaired electron makes the species particularly reactive. Some radicals are printed explicitly, such as Cl·, and sometimes, it is left up to the reader to know, as in Cl.

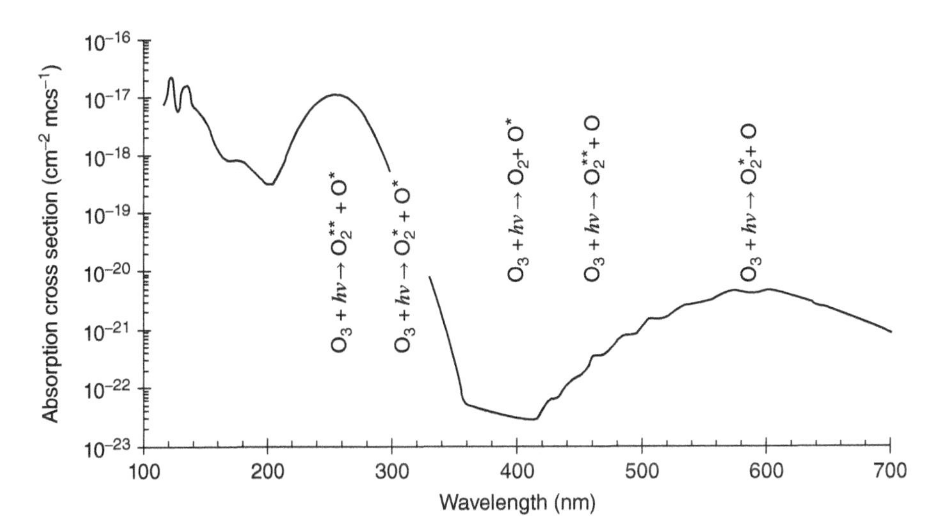

Figure 3.19 The absorption spectrum of ozone showing what photolytic reactions are produced from the different absorption regions. The excited states are more accurately described with the equations to the left and were first introduced in Table 3.5 on page 100. Source: Data from Ackerman (1971).

$O_3 + h\nu \rightarrow O + O_2$
$O_3 + h\nu \rightarrow O + O_2^*$
$O_3 + h\nu \rightarrow O + O_2^{**}$
$O_3 + h\nu \rightarrow O^* + O_2$
$O_3 + h\nu \rightarrow O^* + O_2^*$
$O_3 + h\nu \rightarrow O^* + O_2^{**}$

O represents $O(^3P)$
O^* represents $O(^1D)$
O^{**} represents $O(^1S)$
O_2^* represents $O_2(^1\Delta)$
O_2^{**} represents $O_2(^1\Sigma)$

It is the production of O^* that is of most interest here. While some of the O^* will lose its energy to a chaperone, chances are high in the troposphere that it will react with water.

$$O^* + M \rightarrow O + M \qquad (R3.37)$$
$$O^* + H_2O \rightarrow 2OH \qquad (R3.38)$$

Obviously, Reaction R3.38 is more probable in air that is humid. Kinetic estimates obtain the efficiency of this reaction at 5% for air with a relative humidity of about 10% to as high as 40% efficient for a relative humidity of 80%, and rates maximize when the hydroxyl radical concentration peaks at noon.

Example 3.14: Identifying Radicals

Which of the following species are radicals?

Cl_2	OH	NO_2	BrCl
ClO	OH^-	NO_2^-	SO_2
ClO_2	NO	N_2O_4	O_2

Solution: The easiest way to spot a radical is to count the number of valence electrons. If it is an odd number, then it is a radical.

Species	Valence electrons	A radical?
Cl_2	7+7=14	No
ClO	7+6=13	Yes
ClO_2	7+6+6=19	Yes
OH	6+1=7	Yes
OH^-	6+1+1=8	No
NO	5+6=11	Yes
NO_2	5+6+6=17	Yes
NO_2^-	5+6+6+1=18	No
N_2O_4	5+5+4×6=34	No
BrCl	7+7=14	No
SO_2	6+6+6=18	No
O_2	6+6=12	Yes

Table 3.14 Determining whether a species is a radical.

Table 3.14 lists the species and their valence electrons. The only radical that is tricky to identify is O_2. It has an even number of electrons, but Molecular Orbital Theory predicts that two of its electrons exist as unpaired in separate orbitals. Empirical evidence of this is the fact that O_2 is paramagnetic (reacts to a magnetic field).

Let us examine some of the important molecules and their reaction with the hydroxyl radical and ozone.

3.7.3.1 Carbon Monoxide and Hydroperoxyl Radical Carbon monoxide is a primary pollutant originating from the incomplete combustion of organic fuels, such as fossil fuels and wood. This colorless and odorless gas is toxic because it binds strongly to hemoglobin, represented by Hb.

$$HbO_2 + CO \rightleftharpoons HbCO + O_2 \quad \text{(R3.39)}$$

While CO can be fatal in enclosed spaces, its effects can be reversed by breathing pure oxygen and shifting the equilibrium back to the left.

Carbon monoxide has an atmospheric lifetime of about 0.1 years and can be attacked by the hydroxyl radical.

$$CO + OH \rightarrow CO_2 + H \quad \text{(R3.40)}$$
$$H + O_2 + M \rightarrow HO_2 + M \quad \text{(R3.41)}$$

The HO_2 species is known as the hydroperoxyl radical, which is a reservoir of the hydroxyl radical. Both are often referred to as HO_x (much as NO_x) and can assist in the conversion of NO to NO_2.

$$HO_2 + NO \rightarrow NO_2 + OH \quad \text{(R3.42)}$$

This results in a larger cycle where the loss of HO_2 produces NO_2, which leads to ozone and eventually to the hydroxyl radical. You can think of this as a cycle where carbon monoxide goes in, carbon dioxide comes out, and in between NO_x, ozone, and HO_x, all cycle between the various forms. Also note that in Reaction R3.41, molecular oxygen is consumed when H forms HO_2. This eventually leads to **an additional ozone molecule** because one O replenishes OH when reacting with H, and the other O goes to NO_2 and eventually ozone via Reactions R3.34–R3.36. The cycle ends when OH and NO_2 react.

$$OH + NO_2 \rightarrow HNO_3 \quad \text{(R3.43)}$$

Nitric acid (HNO_3) is relatively stable, very soluble in water, and is one of the main components of acid rain. It can undergo photolysis if it remains in the atmosphere during the day and jumps back into the aforementioned cycles.

$$HNO_3 + h\nu(180\text{ nm}) \rightarrow OH + NO_2 \quad \text{(R3.44)}$$

Mini Summary: NO_x emissions from combustion form ozone and eventually nitric acid. Additional pollutants, such as CO, increase the level of ozone as CO is oxidized to CO_2 by OH.

3.7.3.2 ***Alkanes*** Many of the petroleum fuels used in modern societies consist of hydrocarbons in the alkane class – those containing only C, H, and single bonds fitting the formula C_nH_{2n+2}. Examples would include methane (CH_4), propane (C_3H_8), octane (C_8H_{18}), and cetane ($C_{16}H_{34}$). Inefficiencies in engine design and combustion processes, evaporation at refineries and from gas tanks, and emissions from drilling rigs all contribute to the alkane emission budget. Biogenic sources of alkanes are very low, with ethane (C_2H_6) and butane (C_4H_{10}) having some biogenic source in soils.

Example 3.15: Radical Reactions

Draw the Lewis structure for propane (C_3H_8) knowing that alkanes have a carbon chain as their backbone. Next, draw the skeletal structure of hexane (C_6H_{14}) by analogy without going through the process of counting valence electrons.

Solution: For the propane structure, we will follow the same Lewis structure procedure as before. Follow the duet rule for the hydrogen atoms.

Electrons for (C_3H_8)
(h): $3C+8H = 3\times4+8\times1 = 20\ e^-$
(n): $3\times8+8\times2 = 40\ e^-$
(s): $40-20 = 20\ e^-$
(b): $b = \frac{1}{2}\times20 = 10$ bonds
$FC_C = 4-0-4 = 0$
$FC_H = 1-0-1 = 0$
$\Delta FC = 0$

```
  H H H
  | | |
H-C-C-C-H
  | | |
  H H H
```

There is no possibility for hypervalency and no need for double or triple bonds.

You can see that alkanes are just long chains of carbons surrounded by hydrogen atoms with each carbon bonded to four different atoms, either H or C. Next to the Lewis structure of propane is a simplified structure, which highlights only the carbon backbone. At the vertices of each line segment is a carbon atom, and the hydrogen atoms are all implicitly assumed.

Extending this to hexane, a C_6 chain gives the following.

```
  H H H H H H
  | | | | | |
H-C-C-C-C-C-C-H
  | | | | | |
  H H H H H H
```

While alkanes possess much potential chemical energy, they are quite stable, which makes them useful for fuels since they will remain unreactive in the absence of high temperatures. They are very susceptible to attack by the hydroxyl radical in the following four-step mechanism.

$$RCH_3 + OH\cdot \longrightarrow H_2O + R\text{-}\dot{C}H_2 \quad \text{(R3.45)}$$

$$R\text{-}\dot{C}H_2 + O_2 + M \longrightarrow M + R\text{-}CH_2\text{-}O\text{-}O^\bullet \quad \text{(R3.46)}$$

$$R\text{-}CH_2\text{-}O\text{-}O^\bullet + NO \longrightarrow NO_2 + R\text{-}CH_2\text{-}O^\bullet \quad \text{(R3.47)}$$

$$R{-}CH_2{-}O^{\bullet} + O_2 \longrightarrow HOO^{\bullet} + R{-}CHO \quad (R3.48)$$

Note that the "R" at the beginning of the alkane is used to represent a chain of carbon atoms that is indeterminately long. For example, propane (C_3H_8) can be represented as R-CH_3, where R is CH_3CH_2.

In the first step, the hydroxyl radical pulls off a terminal H from the alkane to form water. In the second step, the alkane radical collides with molecular oxygen (a highly likely event given that O_2 is 21% of the gases in the atmosphere) to form a peroxy radical. In the third step, the peroxy radical reacts with nitric oxide (also highly probably since the alkane pollution often occurs with NO_x production in combustion processes), removing an oxygen atom and forming NO_2. The fourth step is a collision with O_2 again (highly likely), but this time, another terminal H is removed to form the hydroperoxyl radical and an aldehyde (the CHO functional group), which are fairly stable compounds with short atmospheric lives. Notice that two molecules of O_2 have been consumed generating NO_2 and HO_2, which can go on to produce **two ozone molecules**. Analogous to the case of carbon monoxide, a carbon compound goes into the cycle and out comes an aldehyde with an addition of ozone and an interconversion of OH, HO_2, NO, and NO_2.

Before moving on to the next chemical class, let us calculate the lifetime of alkanes against oxidation by the hydroxyl radical. Use the rate constants in Appendix I and Table I.9 on page 327.

Example 3.16: The Lifetime of Propane

Calculate the lifetime of propane against oxidation from the hydroxyl radical at 30 °C, assuming [OH]=$1.5 \times 10^6 \frac{\text{mcs}}{\text{cm}^3}$.

Solution: Remember that the lifetime (τ) of a species can be calculated by Eq. (3.3). We need to determine the rate law of the reaction in question and then derive the lifetime from it.

$$C_3H_8 + OH \rightarrow \text{products}$$

The rate law is given by $\textit{Rate} = k[C_3H_8][OH]$. The lifetime is always equal to the ratio of the species concentration to the rate.

$$\tau_{C_3H_8} = \frac{[C_3H_8]}{\textit{Rate}} = \frac{1}{k[OH]}$$

The propane rate constant value at 30 °C is $1.16 \times 10^{-12} \frac{\text{cm}^3}{\text{mcs·s}}$. Plug this into the lifetime equation along with the concentration of the hydroxyl radical.

$$\tau_{C_3H_8} = \frac{1}{\left(1.16 \times 10^{-12} \frac{\text{cm}^3}{\text{mcs·s}}\right)\left(1.5 \times 10^6 \frac{\text{mcs}}{\text{cm}^3}\right)} = 5.7 \times 10^5 \text{ s}$$

If this τ is converted to days, we get $\tau = 6.6$ days.

The formation of the aldehyde is an intermediate step half-way to forming carbon dioxide. Aldehydes are relatively stable with an atmospheric lifetime of about 1 day, and there are three different routes that aldehydes follow as they further degrade.

Route 1 involves photolysis of the aldehyde to form carbon monoxide and an alkane that is one carbon shorter than it started (C-3 propane would become C-2 ethane with the third carbon forming CO).

$$R{-}C(=O){-}H + h\nu\,(290\text{nm}) \rightarrow R - H + CO \quad \text{(R3.49)}$$

The remaining alkane reenters the oxidation sequence (Reaction R3.45) when it is attacked by OH. Carbon monoxide re-enters by reacting with OH in Reaction R3.40.

Routes 2 and 3 start by repeating steps 1 and 2 from the aldehyde formation.

$$R{-}C(=O){-}H + OH\cdot \longrightarrow H_2O + R{-}\dot{C}{=}O \quad \text{(R3.50)}$$

$$R{-}\dot{C}{=}O + O_2 + M \longrightarrow M + R{-}C(=O){-}OO^\bullet \quad \text{(R3.51)}$$

Route 2 includes NO and results in CO_2 and an alkane radical with one less carbon than it started.

$$R{-}C(=O){-}OO^\bullet + NO \longrightarrow NO_2 + R{-}C(=O){-}O^\bullet \quad \text{(R3.52)}$$

$$R{-}C(=O){-}O^\bullet \longrightarrow R\cdot + O{=}C{=}O \quad \text{(R3.53)}$$

Route 3 incorporates NO_2 to form a stable species known as PAN: peroxylacetyl nitrate.

$$R{-}C(=O){-}OO^\bullet + NO_2 \longrightarrow R{-}C(=O){-}OONO_2 \quad \text{(R3.54)}$$

Notice that Route 1 produces CO, which can eventually produce HO_2 (Reactions R3.40 and R3.41) and ozone. Routes 2 and 3 consume an O_2, introducing another net oxygen atom into the cycle and eventually an ozone molecule.

3.7.3.3 Alkenes Alkenes are hydrocarbons that contain one or more double bonds. Common alkenes with one double bond follow the formula of C_nH_{2n}, such as ethene (C_2H_4) and butene (C_4H_8). Biogenic sources are significant, with emissions of ethene and butene exceeding anthropogenic sources (Kesselmeier and Staudt, 1999). Ethene is emitted from some plants, especially agricultural varieties, during periods of stress such as injury, temperature, or drought. Ethene, also known as ethylene and the aging hormone of plants, is used as a ripening agent for commercial fruits. Over 100 million tons of ethene are produced industrially from petroleum for the agricultural and chemical industries.

Example 3.17: Radical Reaction of an Unsaturated Species

Draw the Lewis structure of propene (C_3H_6).

Solution: For the propene structure, we will follow the same Lewis structure procedure as before. Remember that alkenes are hydrocarbon chains.

Electrons for C_3H_6
(h): $3C+6H = 3\times4+6\times1 = 18\ e^-$
(n): $3\times8+6\times2 = 36\ e^-$
(s): $36-18 = 18\ e^-$
(b): $b = \frac{1}{2}\times18 = 9$ bonds
$FC_{C_{db}} = 4-0-4 = 0$
$FC_{C_{sb}} = 4-0-4 = 0$
$FC_{H} = 1-0-1 = 0$
$\Delta FC = 0$

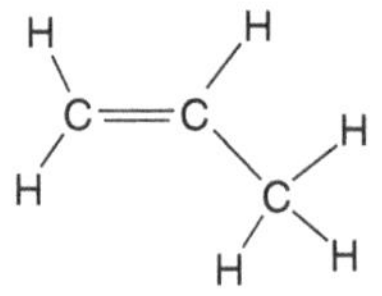

There is no possibility for hypervalency, and one double bond is placed between two carbon atoms. Because there are only three carbon atoms, it does not matter where the double bond is placed.

The double bond of an alkene is attacked by the hydroxyl radical, and the hydroxyl radical becomes an alcohol group.

$$CH_2 = CHCH_3 + \cdot OH \rightarrow HOCH_2 - \dot{C}HCH_3 \quad (R3.55)$$
$$HOCH_2 - \dot{C}HCH_3 + O_2 \rightarrow HOCH_2 - CHO\dot{O}CH_3 \quad (R3.56)$$
$$HOCH_2 - CHO\dot{O}CH_3 + NO \rightarrow NO_2 + HOCH_2 - CH\dot{O}CH_3 \quad (R3.57)$$

The OH can add at either carbon atom sharing the double bond. Thus far, the reaction follows the schema of alkanes (OH attack, followed by a reaction with O_2 and NO). In the next portion of the mechanism, the molecule fractures eventually form two aldehydes.

$$HOCH_2 - CH\dot{O}CH_3 \rightarrow HOCH_2 \cdot + CHOCH_3 \quad (R3.58)$$
$$HOCH_2 \cdot + O_2 \rightarrow HCHO + HO_2\cdot \quad (R3.59)$$

The aldehydes formed ($CHOCH_3$ and HCHO) persist for a day or less and then get oxidized by one of the three routes discussed in the Alkane section. Note, once again, that two molecules of oxygen are consumed in the first set of reactions producing NO_2 and HO_2, which will go on to generate two additional ozone molecules. Once the two aldehydes are oxidized, as described in the previous section, there will be an additional ozone molecule formed from each aldehyde.

This example with propene is the simplest reaction pathway assuming that the OH adds to the terminal carbon atom. Variations of this reaction abound when alkenes containing more than one double bond are oxidized. Complications notwithstanding, hopefully you are beginning to see that radical oxidation of VOCs, whatever their form, in the troposphere is an occasion for generating ozone.

Example 3.18: The Lifetime of Propene

Calculate the lifetime of propene against oxidation from the hydroxyl radical at 10 °C, assuming $[OH]=1.5\times10^6\ \frac{mcs}{cm^3}$.

Solution: See the Section C.1 on page 265.

3.7.3.4 ***Terpenes*** If you have ever hiked in a forest populated with pines (or other conifers) and taken in a deep breath filled with their wonderful aroma, you have inhaled terpenes. Terpenes represent a large class of VOCs that include isoprene, limonene, and pinene (see Figure 3.20). Terpenes are often produced by plants as a direct or indirect defense against herbivores, and they seem to be released in the largest quantities during hot days. They could serve as a bulk climate control since their presence in large forests can seed clouds and block sunlight.

Examining the structures in Figure 3.20, you will notice that this class of compounds is similar to alkenes with at least one C–C double bond. You should not be surprised to find out that the oxidation of terpenes is similar in mechanism with a wide number of variants since the OH can attack at a number of sites on the terpene.

The combination of biogenic and anthropogenic VOC emissions can compromise the air of cities and, surprisingly, national parks. You may have heard about an "Ozone Action Day", which is an alert declared by a municipality to warn residents of high levels of ground-level ozone and photochemical smog. These days often occur during the summer months in large urban centers, but they can also occur far away from cities.

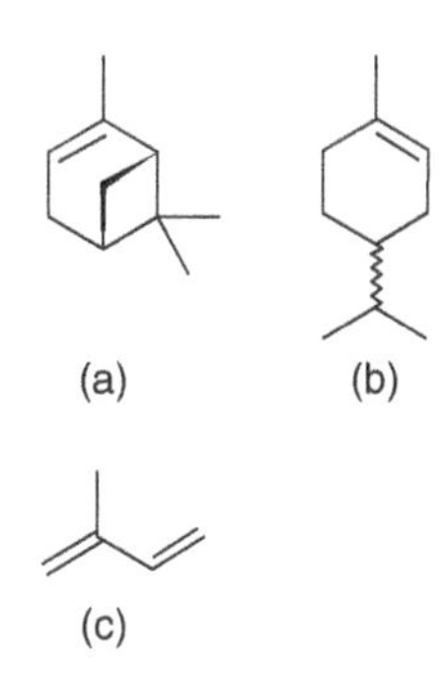

Figure 3.20 Terpenes are a common class of biogenically produced VOCs. Their presence produces the characteristic fragrance of many plants. α-pinene (a) is produced by conifers, limonene (b) is present in the rind of citrus fruits, and isoprene (c) is emitted by several classes of trees such as palm, oak, eucalyptus, and aspen.

3.7.3.5 ***Nitrogen-Containing Compounds*** Besides the nitrogen oxide compounds that have been described previously, there is another major emission of nitrogen to the atmosphere. Ammonia is produced naturally by the decay of proteins, so major anthropogenic contributions of it come from agriculture (livestock and fertilizer) as well as decay processes in landfills and sewage treatment plants. One of the major sources of ammonia is urea ($(NH_2)_2CO$), which is hydrolyzed by microbes using the enzyme urease.

$$(NH_2)_2CO(aq) + 2H_2O(l) \xrightarrow{\text{urease}} 2NH_4^+(aq) + CO_3^{2-}(aq) \quad (R3.60)$$

$$NH_4^+(aq) + OH^-(aq) \rightleftharpoons NH_3(aq) + H_2O(l) \quad (R3.61)$$

$$NH_4^+(aq) + H_2O(l) \rightleftharpoons NH_3(aq) + H_3O^+(aq) \quad (R3.62)$$

Whether acidic or basic, aqueous concentrations of ammonium ion (NH_4^+) are always in equilibrium with ammonia, which can especially become volatile under basic conditions.

Ammonia, as with most other trace gases, reacts with the hydroxyl radical.

$$NH_3 + \cdot OH \rightarrow \cdot NH_2 + H_2O \quad (R3.63)$$

$$\cdot NH_2 + \cdot NO_2 \rightarrow N_2O + H_2O \quad (R3.64)$$

3.7.3.6 ***Sulfur-Containing Compounds*** The final class of tropospheric compounds that are important to understanding tropospheric pollution are sulfur-containing compounds. Major anthropogenic sources of sulfur emissions come from the combustion of coal and petroleum, metal smelting, natural gas drilling, and paper mills. Across the United States, the sulfur content of coal varies, but any small amount of sulfur in coal will lead to significant amounts of SO_x (SO_2 and SO_3) pollution since a typical coal-fired power plant has a voracious appetite for coal. The Kingston Fossil Plant, run by the Tennessee Valley Authority outside of Knoxville, consumes just under 10 tons of coal per minute (TVA, 2013). Even a small percentage of sulfur in coal would lead to a large amount released into the air at that rate.

Sulfur can be found in coal in many forms, one of which is the mineral pyrite (FeS_2), which gets oxidized along with the coal.

$$4FeS_2(s) + 11O_2(g) \rightarrow 2Fe_2O_3(s) + 8SO_2(g) \quad (R3.65)$$

Smelting, another source of sulfur emissions, is a process where metals are extracted from minerals. When the mineral is a sulfide, such as ZnS, the mineral is first roasted in an oven with oxygen to convert the sulfide into an oxide.

$$2ZnS(s) + 3O_2(g) \rightarrow 2SO_2(g) + 2ZnO(s) \qquad \text{(R3.66)}$$

The zinc oxide goes on to further processing, leaving SO_2 as a waste product. Paper mills convert sulfur into sulfurous acid (H_2SO_3) in order to extract lignin from wood as part of a process for converting wood to paper. As a result, significant amounts of SO_x gases escape. All of these industrial processes contribute oxidized sulfur in the form of gaseous SO_2 or SO_3.

Once sulfur is in this form, it escapes into the atmosphere where it encounters molecular oxygen.

$$2SO_2 + O_2 \rightarrow 2SO_3 \qquad \text{(R3.67)}$$

This reaction is exceedingly slow since the double bond of molecular oxygen is strong and SO_2 is a fairly stable species. As with most species, oxidation in the troposphere is caused by the hydroxyl radical.

$$SO_2 + OH + M \rightarrow HOSO_2 + M \qquad \text{(R3.68)}$$
$$HOSO_2 + O_2 \rightarrow +HO_2 + SO_3 \qquad \text{(R3.69)}$$
$$SO_3 + H_2O + M \rightarrow +H_2SO_4 + M \qquad \text{(R3.70)}$$

Note that the $HOSO_2$ species is a radical and hence very reactive. The fate of sulfur in this case is sulfuric acid, the second major component of acid deposition (see Section 3.9). It is very water soluble and precipitates out of the atmosphere.

Reduced forms of sulfur come from natural gas drilling. Along with the methane that is harvested from a natural gas well, significant amounts of hydrogen sulfide (H_2S) and thiols (R-CH_3SH) are collected or released. If these sulfur compounds are collected, they are usually burned on site or sold as "sour gas" and used as a low-grade fuel. Either way, the combustion of these sulfur compounds produces SO_2. The reduced sulfur compounds that escape are oxidized in the troposphere by... wait for it... the hydroxyl radical.

$$H_2S + OH \rightarrow SH + H_2O \qquad \text{(3.16)}$$

Unfortunately, the fate of the SH radical is more complicated than other forms of sulfur. Further oxidation by O_2 is extremely slow, and oxidation by O_3 leads to the formation of HSO.

Dimethyl sulfide (DMS or CH_3SCH_3) is the most abundant biogenic sulfur emission, originating from phytoplankton. It also reacts in the same manner as the other hydroxyl radical oxidation mechanisms.

$$CH_3SCH_3 + OH \rightarrow CH_3SCH_2 + H_2O \qquad \text{(R3.71)}$$
$$CH_3SCH_2 + O_2 \rightarrow CH_3SCH_2OO \qquad \text{(R3.72)}$$
$$CH_3SCH_2OO + NO \rightarrow NO_2 + CH_3SCH_2O \qquad \text{(R3.73)}$$

From here, there is a fracturing of the compound similar to the alkene mechanism. Notice the consumption of O_2 and the formation of NO_2 – another molecule of ozone can form.

$$CH_3SCH_2O \rightarrow CH_3S + H_2CO \qquad \text{(R3.74)}$$
$$CH_3S + O_2 \rightarrow CH_3SOO \qquad \text{(R3.75)}$$
$$CH_3SOO \rightarrow CH_3 + SO_2 \qquad \text{(R3.76)}$$

The formaldehyde (H_2CO) formed in Reaction R3.74 gets cycled into the aldehyde oxidation mechanism mentioned earlier (add another molecule of ozone). Sulfur dioxide is produced in Reaction R3.76 and will eventually lead to sulfuric acid. Finally, the methyl radical in the last reaction will go on to react with O_2 and NO to form another aldehyde – add another two ozone molecules.

3.7.3.7 Nighttime Reactions The reactions initiated by the hydroxyl radical are all ultimately driven by solar radiation. The [OH] peaks at noon when the solar intensity reaches its maximum and drives the bulk of the photochemical oxidation reactions. Under these conditions, there is plenty of energy available (via photons) to cause relatively stable species, such as alkanes, to react because this available energy makes the formation of highly reactive species, such as the hydroxyl radical or other radicals, possible. The cycles previously mentioned can keep going and going because the solar energy drives them. What happens during the night when radiation is not available? This is analogous to a windup toy that keeps going as long as it is being continuously wound. Once the windup process stops, the toy just runs down until it stops.

Most chemical reactions have high activation energies because in order for a reaction to start, bonds must break. Reactions involving highly reactive species, such as the hydroxyl radical, often have a very small activation energy because one reactant brings the energy with it. When these reactive species go away and there is no solar radiation for photolysis, species that would often be reactive during the day become stuck in their own potential energy wells and cannot escape when colliding with other species. Reactive species begin an irreversible descent into thermodynamic stability and are not regenerated, such as in the case of NO_x and O_3.

$$NO_2 + O_3 \xrightarrow{k_{NO_2+O_3}} NO_3 + O_2 \quad (R3.77)$$

$$NO_3 + NO_2 + M \rightleftharpoons N_2O_5 + M \quad (R3.78)$$

$$N_2O_5 + H_2O \rightarrow 2HNO_3 \quad (R3.79)$$

This series of reactions begins the process of cleansing the troposphere of NO_x and ozone by converting them to nitric acid, which is water-soluble and precipitates out of the atmosphere. During the day, some of the NO_2 would be photolyzed to form the more reactive NO and atomic oxygen would go on to form ozone with molecular oxygen. This can no longer occur at night. Reactions R3.77–R3.79 represent the winding-down process of these reactants as each one forms more and more stable species.

Example 3.19: The Free Energy of Some Reactions of Nitrogen

Calculate the change in Free Energy for the Reactions R3.77 through R3.79 and compare them to the Free Energy for Reaction R3.34 on page 119 at 298 K.

Solution: To find the Free Energy for each of the reactions, we just need to look up for their Gibbs free energy values since the temperature is at 298 K.

For Reaction R3.77

$$\begin{aligned}\Delta G^{\circ}_{rxn} &= \Sigma\Delta G^{\circ}_{f\,products} - \Sigma\Delta G^{\circ}_{f\,reactants}\\ &= \left[(1\text{ mol NO}_3)(115.9\text{ kJ/mol}) + (1\text{ mol O}_2)(0.0\text{ kJ/mol})\right]\\ &\quad - \left[(1\text{ mol NO}_2)(51.3\text{ kJ/mol}) + (1\text{ mol O}_3)(163.2\text{ kJ/mol})\right]\\ &= -98.6\text{ kJ}\end{aligned}$$

For Reaction R3.78
$\Delta G^{\circ}_{rxn} = -50.1$ kJ
For Reaction R3.79
$\Delta G^{\circ}_{rxn} = -35.5$ kJ
For Reaction R3.34
$\Delta G^{\circ}_{rxn} = +268.0$ kJ

Reactions R3.77 through R3.79 are all spontaneous reactions that form species that are progressively more stable, whereas Reaction R3.34, which is likely to occur during the day, is nonspontaneous and requires a significant input of energy.

Nitric acid can also react with any ammonia in the atmosphere to form ammonium nitrate, a salt that can act as a site for nucleation of water droplets.

$$NH_3 + HNO_3 \rightarrow NH_4NO_3 \quad \text{(R3.80)}$$

Example 3.20: The Lifetime of NO_2

What is the lifetime of NO_2 against O_3 attack during night given that $k_{NO_2+O_3}$ is $2.8\times10^{-17}\frac{cm^3}{mcs\cdot s}$ on a cool summer night when the temperature drops to 20 °C and the ambient $[O_3]$ is $6.8 \times 10^{11}\frac{mcs}{cm^3}$?

Solution: We just need to use the form of the lifetime given in Eq. (3.3) for Reaction R3.77, which has the following rate law.

$$Rate = k_{NO_2+O_3}[NO_2][O_3]$$

Plugging in the given values results in the following.

$$\tau = \frac{1}{k_{NO_2+O_3}[O_3]} = \frac{1}{\left(2.8\times10^{-17}\frac{cm^3}{mcs\cdot s}\right)\left(6.8\times10^{11}\frac{mcs}{cm^3}\right)}$$
$$= 5.3 \times 10^4 \text{ s} = 15 \text{ h}$$

The rate of NO_2 loss does not seem fast enough to remove it all before morning (maybe 8 h away). This means that some residual NO_2 and O_3 will be available for the photochemical cycle the next day. Some of the NO_x reacts with VOCs and forms nitroperoxy species that persist until the next day when the hydroxyl radical restarts the oxidation process.

The conclusion to draw from the previous discussion is that the presence of solar radiation makes it possible for all sorts of high-energy reactants and reactions to flourish. Endothermic (and endoergic) reactions are possible because there is often plenty of energy available to break bonds and lift atmospheric species out of deep potential energy wells. When this energy is unavailable, the only reactions that are probable are those that are exoergic and form more stable species. These exoergic reactions eventually lead to the thermodynamic fate of each of the elements. Examine Table 3.15, and you can see this trend by just looking at the various forms of nitrogen and sulfur. Nitrogen, in the form of N_2, is very unreactive because it contains a triple bond. Once liberated from this deep energy well by the high temperatures in an internal combustion engine (Reaction R3.31 on page 114), it begins its path toward its thermodynamic fate by becoming NO. Further oxidation leads to NO_2, then NO_3 and N_2O_5 concentrations peak overnight. The final fate of nitrogen is the exceptionally stable gaseous and aqueous forms of HNO_3. Each step goes deeper and deeper into a potential energy well to form the most stable species. If you look at the sulfur table, you can see the same progression: SO_2, then SO_3, then $H_2SO_4(g)$, and finally, $H_2SO_4(aq)$. You saw the same progression in the radical reactions of VOCs with the hydroxyl radical, with carbon eventually forming CO_2, its most stable gaseous form, and $H_2CO_3(aq)$. This is analogous to the windup toy running down and stopping.

The free energy values in these tables quantify the energy released or absorbed when forming each species from the default form found in nature ($N_2(g)$ for nitrogen). You can use them as a shortcut to assessing the stability of a species without having to place it in a chemical reaction and calculate the free energy of the reaction. The most stable form of nitrogen is $HNO_3(aq)$. Each of the nitrogen species in Table 3.15 represents some point along the surface of a potential energy well. Analogous to water running down a hill under the force of gravity, the thermodynamic fate of nitrogen is nitric acid dissolved in water droplets. There can be barriers that stop this process, such as the deep well that holds $N_2(g)$. Molecular nitrogen is, by far, the most abundant form of nitrogen, but it is not the form with the lowest energy. Because N_2 has a triple bond, it is energetically difficult to get nitrogen

Species	ΔG_f° (kJ/mol)
$N_2(g)$	0.0
$N(g)$	455.5
$NO(g)$	87.6
$N_2O(g)$	103.7
$NO_2(g)$	51.3
$NO_3(g)$	115.9
$N_2O_4(g)$	99.8
$N_2O_5(g)$	117.1
$NH_3(g)$	−16.4
$NH_3(aq)$	−46.11
$HNO_2(g)$	−46.0
$HNO_2(aq)$	−50.60
$HNO_3(g)$	−73.5
$HNO_3(aq)$	−111.3

Species	ΔG_f° (kJ/mol)
$S(s)$	0.0
$S(g)$	236.48
$SO_2(g)$	−300.1
$SO_3(g)$	−371.1
$HSO_4^-(aq)$	−755.9
$H_2SO_4(aq)$	−744.53
$SO_4^{2-}(aq)$	−744.60
$H_2S(g)$	−33.59
$HS^-(aq)$	12.08
$(CH_3)_2S(g)$	6.95

Table 3.15 The Free Energy values presented for each of the forms of nitrogen and sulfur suggest the thermodynamic fate for each in an oxidizing atmosphere.

out of its local energy well and into the larger energy well. Lightning can do this, and it is a natural source of HNO_3. The energy barrier that keeps N_2 in its well is a kinetic barrier, and the height of the wall is proportional to the activation energy (E_a) that is required to chemically change N_2. A river dam represents a good analogy to the difference between thermodynamic fate and kinetic barriers. Gravity is exerting a relentless force on the water in the river, and it would flow downstream to the lowest elevation if it were unimpeded – this is thermodynamic fate and is characterized by the ΔG_{rxn}. There is a barrier that prevents the flow of water – these are kinetic barriers represented by E_a. You could predict that in the long-term, say 1 million years, the water will eventually flow downstream because eventually, the sluice gates will open or the dam will crumble or be removed by engineers. In the short-term, the water will remain in a high-energy state because of the dam.

In the case of the tropospheric reactions that have been the focus of this section, the presence of solar radiation turns on thermodynamic control. While the radiation is present, forms of nitrogen and sulfur can interconvert since there is enough energy to lift them out of their potential energy well and keep them "afloat" in the larger potential energy well. Once the solar radiation is gone, molecules will settle back into their potential energy wells unless specific collisional reactions provide enough energy to escape. A specific example is nitrogen dioxide. After it is formed, it has a significant kinetic barrier that prevents it from decomposing. It could persist indefinitely at low temperatures, in the absence of ozone or in the absence of light. Once it is exposed to solar radiation and other reactive species, it can leave its potential energy well and convert to other forms of nitrogen. While the solar radiation is present, it can float around above all of the potential energy wells as if it were ping-pong balls floating in a well filled with water. When the solar radiation ceases, it is similar to draining the water out of the well – some species will get trapped in another energy well or could drop to the bottom of the well.

3.7.3.8 ***Summary of Reaction Involving the Hydroxyl Radical*** There are many different types of gases emitted into the atmosphere by anthropogenic, geologic, and biogenic sources. All are susceptible to attack by the hydroxyl radical, undergo radical oxidation, and participate in the NO_x and HO_x cycles that eventually form combinations of PANs, aldehydes, CO_2, ozone, sulfuric acid, and nitric acid. It is the presence of radicals such as NO_x and OH that produce other HC radicals, and the combination of the high reactivity of radicals and the high pressures of the troposphere causes the lifetimes of these species to be relatively short. There are other mechanisms for oxidation involving direct attack of the

Species	Global emissions (Tg/year)	Removal by OH (%)
CO	2800	85
CH_4	530	90
C_2H_6	20	90
Isoprene	570	90
Terpenes	140	50
NO_2	150	50
SO_2	300	30
$(CH_3)_2S$	30	90
$CFCl_3$	0.3	0

Notice that the CFC is not affected.

Table 3.16 The data in this table shows the removal of various trace species in the troposphere by the hydroxyl radical. Source: Monks (2005).

VOC by ozone and a minor species NO_3, but all eventually lead to similar outcomes: CO_2 for the carbon atoms and the cycle of NO_x, HO_x, and ozone. As shown in Table 3.16, the hydroxyl radical really does earn its moniker as the detergent of the atmosphere. It is an integral part of photochemical smog, which is driven by solar radiation. The fate of photochemical smog is the production of toxic gases during the day and a thermodynamic fate of nitric and sulfuric acid. This photochemical smog is a symptom of communities that use a lot of petroleum fuels for industry, transportation, and residential life.

3.8 CLASSICAL SMOG

In the previous section, you were introduced to photochemical smog, which derives from the combustion of relatively clean fossil fuels, such as petroleum and natural gas. You might even consider photochemical smog to be a sign of the advancement of civilization (although not advanced enough to *prevent* smog events), since its appearance implies the existence of oil refineries and an advancement in the energy density of fuel stocks (petroleum has more chemical energy per kg than coal). Classical smog is the result of primarily burning coal, although burning wood on a large scale would also cause a similar smog. It is primarily composed of smoke, soot, and SO_x emissions.

Coal is the result of geologic processes on ancient vegetation stored deep below the surface. High temperatures and pressures convert vegetation into coal through a process of coalification (Figure 3.21). This is a process where plant litter (leaves, wood, and so forth) is converted from chemical forms, such as humic acids, to coal. In this endothermic process, the % carbon is increased and the % oxygen and % hydrogen are decreased. This increases the energy content of the material (see Table 3.17).

Item	Anthracite coal	Bituminous coal	Lignite coal	Peat	Wood	Humic acids
%C	93.5	85.1	85.0	49.9	50.0	51.6
%H	2.9	5.1	5.1	6.6	6.2	4.2
%O	2.1	7.4	7.5	42.7	43.7	38.4
%N	0.5	0.7	0.7	0.8	0.1	1.8
%S	1.0	1.7	1.7	0.0	0.0	4.1
Energy (MJ/kg)	30–32	28–30	20–24	20–23	17–20	18

Source: Data compiled from Lindstroem (1980), Reddy, Leenheerm, and Malcolm (1992), and Ragland, Aerts, and Baker (1991).

Table 3.17 The elemental composition and heat content of the different stages in the process of coalification show a trend of increasing %C, decreasing %O and %H, and increasing heat content.

Sulfur is a constituent element of coal, typically found in organosulfur forms (thioethers, thiophenes, disulfides, and so forth) and inorganic forms (pyrite, FeS_2; gypsum, $CaSO_4$). Coal is and has been mined in many states across the United States, and each state has its own average level of sulfur, with Western coal having less sulfur compared to Eastern coal. When coal is burned as a fuel, the sulfur is oxidized to sulfur dioxide and escapes in the flue gas along with smoke and soot. These are the main ingredients to classical smog.

☞ A *thermal inversion* is a weather event that inverts the normal temperature profile of the troposphere (where temperature decreases with increasing altitude). The air at the surface can cool during the night under clear skies. In the morning, a layer of warmer air can cap the cold air and prevent mixing (since colder air does not rise). The duration of thermal inversions can be extended by surrounding mountains that form a bowl and prevent winds from mixing with the cold air.

It was a form of classical smog that collected over Donora, Pennsylvania, in October of 1948. This small town of 15,000 was the home of a US steel plant and the scene of one of the worst air pollution events in the US history. A *thermal inversion* above the town began trapping the emission from the steel mill on October 27, 1948. The classic smog thickened until a rain event on October 31. By the time the event was over, 20 people and over 800 animals had died from the smog.

A similar event happened in London, England, in 1952, from December 5–9. A thick cloud of classical smog remained over the city for days. Deaths reported by the hospitals and mortuaries spiked abnormally during and shortly after the event. The smog event claimed the lives of at least 4,000 Londoners, with some estimates closer to 12,000.

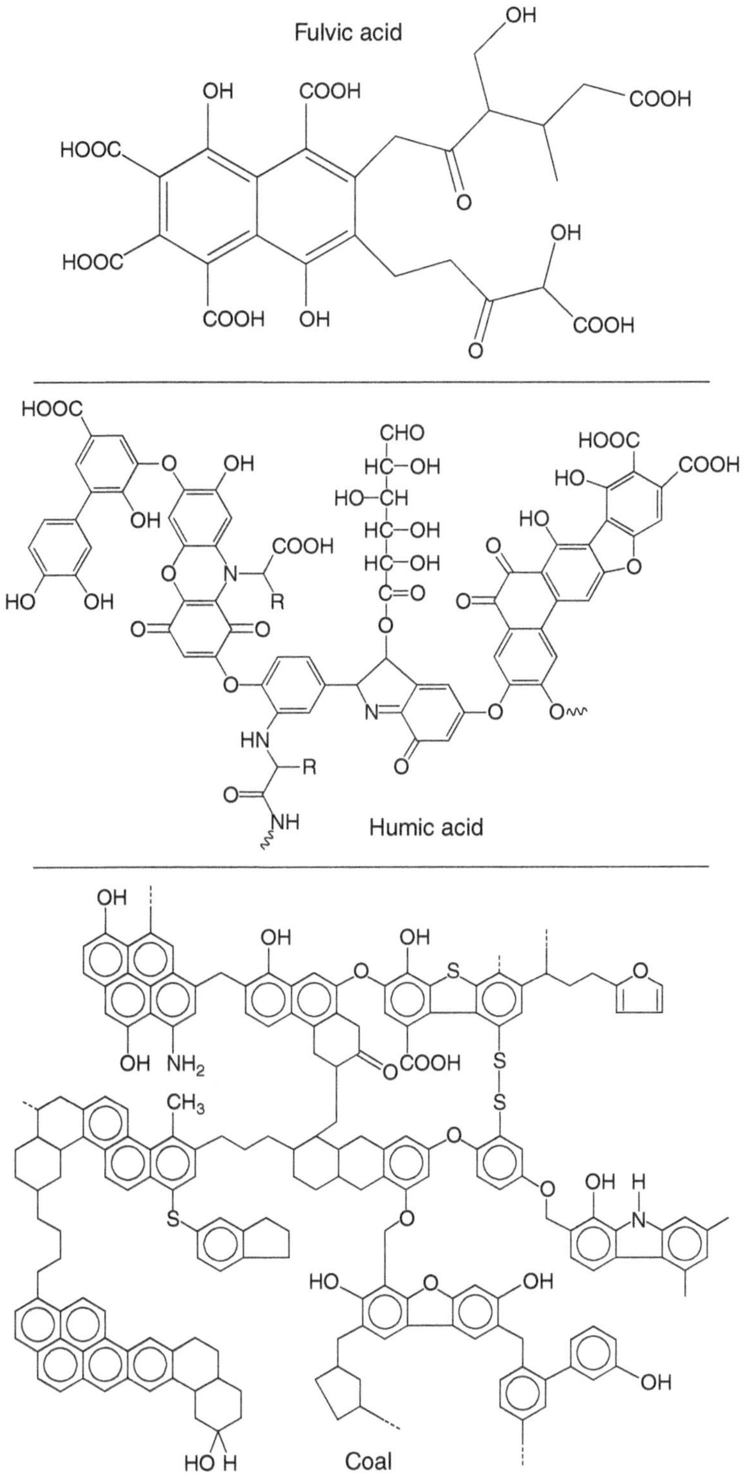

Figure 3.21 The process of coalification converts fulvic and humic acids (decaying plant material) into coal by removing oxygen and increasing aromatic ring structures under intense heat and pressure. Courtesy Michal Sobkowski (2010) and Yikrazuul (2009), respectively.

These two major pollution events were the impetus for major air pollution legislation in the United States. Most of the earlier laws funded research and started to develop emission programs and standards. Arguably, the most significant legislation was the Clean Air Act (CAA) of 1970, which shifted much control over air pollution to the federal government and established comprehensive regulations on emissions and air quality standards. The Environmental Protection Agency was also established in 1970 to carry out the requirements of the CAA. The CAA of 1990 established a program specifically for acid deposition (see Section 3.9) and the phaseout of ozone-depleting substances (see Section 3.12).

☞ *Point sources* refer to specific locations where pollution originates, such as factories, power plants, and drilling rigs. *Nonpoint sources* are those that discharge pollutants from a wide area or from a location that cannot be defined, such as cultivated land, cars, ships, and forests.

The CAA of 1990 was largely successful in reducing SO_x and NO_x emissions despite an expanding economy and increased industrial outputs. One might argue that the emission goals were not very ambitious, but minimal federal standards make sense because industrial chimneys make air pollution a national issue instead of a local issue. Chimneys are some of the tallest structures built by humans because the higher up the pollutants are injected into the troposphere the more dilute the pollution will be when it has a direct impact of the environment very far away from the *point source*.

3.9 ACID DEPOSITION

As the previous sections have illustrated, the thermodynamic fate of the three major elements of photochemical smog (N, S, C) eventually ends up in the highest oxidation state (and the thermodynamic fate of each: HNO_3, H_2SO_4, and CO_2). Sulfuric acid and sulfates, along with nitric acid and other nitrates, can settle out of the atmosphere in a process known as deposition. If water is involved, then the deposition is known as *wet deposition*, commonly called *acid rain*. If the deposition is in the form of dry dust, then it is called *dry deposition*. Both mechanisms are referred to as *acid deposition* since both are highly acidic. While acid rain is the more commonly used moniker, dry deposition actually is often larger than wet deposition. Much depends on meteorological conditions, geography, and local sources of nitrogen and sulfur, but one estimate places wet deposition as equal to or generally less than dry deposition (Seinfeld and Pandis, 2006). Since acid rain is by far easier to measure, it is often the focus of the total deposition.

☞ The process called *acid deposition* includes acidic species settling out of the atmosphere as dry dust (*dry deposition*) or dissolved in rain (*wet deposition*). Wet deposition is commonly known as *acid rain*.

In Chapter 1, you were reintroduced to the pH scale, the logarithmic numerical scale that quantifies the amount of acid or base in a solution. Pure water has a pH of 7.0 at 25 °C due to the autodissociation equilibrium between deprotonated water (hydroxide ion or OH^-) and protonated water (hydronium ion or H_3O^+). The pH of rain is less than this due to the presence of carbon dioxide in the atmosphere, which dissolves in water to form carbonic acid.

☞ For a review of acid and base chemistry, see Review Example 1.21 on page 40.

☞ For a review of chemical equilibrium expressions, see Review Example 1.22 on page 42.

☞ For a review of chemical equilibrium calculations, see Review Example 1.23 on page 44.

Example 3.21: Natural Rain

Determine the pH of natural rain at a temperature of 298 K given that the average concentration of CO_2 is approximately 397 ppmv or 3.97×10^{-4} atm.

Solution: Carbon dioxide is a very soluble gas, as measured by Henry's Law.

$$CO_2(g) \rightleftharpoons CO_2(aq) \qquad K_H = 3.5 \times 10^{-2}\ \mathrm{M/atm} \qquad \text{(R3.81)}$$

Carbon dioxide can react with water in the aqueous phase to form carbonic acid, which is a weak acid, and ionizes in solution to provide a hydronium ion. Because it is a diprotic acid, there are two ionization steps.

$$CO_2(aq) + 2H_2O(l) \rightleftharpoons H_3O^+(aq) + HCO_3^-(aq) \qquad K_{a1} = 4.5 \times 10^{-7} \qquad \text{(R3.82)}$$

$$HCO_3^-(aq) + H_2O(l) \rightleftharpoons H_3O^+(aq) + CO_3^{2-}(aq) \qquad K_{a2} = 4.7 \times 10^{-11} \qquad \text{(R3.83)}$$

Since the K_{a2} is very small, let us ignore it for now and use only K_{a1}. Each of these reactions is part of the greater process of carbon dioxide dissolving in rain droplets. Let us add the two reactions together.

$$CO_2(g) \rightleftharpoons \cancel{CO_2(aq)} \quad (R3.84)$$

$$\cancel{CO_2(aq)} + 2\,H_2O(l) \rightleftharpoons H_3O^+(aq) + HCO_3^-(aq) \quad (R3.85)$$

$$CO_2(g) + 2\,H_2O(l) \rightleftharpoons H_3O^+(aq) + HCO_3^-(aq) \quad (R3.86)$$

Using Hess's Law, the sum of a series of reactions implies the multiplication of each reaction's equilibrium constant.

$$K_{rain} = K_H \times K_{a1} \quad (3.17)$$

We need to take Reaction R3.86 and set up an ICE table with it.

	$CO_2(g)$	+	$2\,H_2O(l)$	$\rightleftharpoons$	$H_3O^+(aq)$	+	$HCO_3^-(aq)$
I	3.97×10^{-4}		- - -		0		0
C	$-x$		- - -		$+x$		$+x$
E	$3.97 \times 10^{-4} - x$		- - -		x		x

The problem we have before us is that the pressure of CO_2 will drop by x, and the concentrations of hydronium and bicarbonate ions will increase by x. Since the pressure of carbon dioxide (P_{CO_2}) is in units of atm, it does not match the molar concentration units of $[H_3O^+]$ and $[HCO_3^-]$. We need to adjust the value of x to convert it to units of atm, using the Ideal Gas Law.

$$PV = nRT$$

$$P = \tfrac{n}{V}(RT)$$

The fraction $\frac{n}{V}$ represents $\frac{mol}{L}$ or Molar, so multiplying it by RT converts it to pressure in atm. Here is the adjusted ICE table.

	$CO_2(g)$	+	$2\,H_2O(l)$	$\rightleftharpoons$	$H_3O^+(aq)$	+	$HCO_3^-(aq)$
I	3.97×10^{-4}		- - -		0		0
C	$-x(RT)$		- - -		$+x$		$+x$
E	$3.97 \times 10^{-4} - x(RT)$		- - -		x		x

We now need to plug in the equilibrium values into the K_{rain} expression.

$$\frac{[H_3O^+][HCO_3^-]}{P_{CO_2}} = K_{rain} = K_H \times K_{a1}$$

$$\frac{(x)(x)}{3.97 \times 10^{-4} - x(RT)} = K_H \times K_{a1}$$

To simplify the math, we can use the 5% rule, then solve for x. Assume $x(RT) \ll 3.97 \times 10^{-4}$ atm.

$$\frac{x^2}{3.97 \times 10^{-4} - \cancel{x(RT)}_{\to 0}} = K_H \times K_{a1}$$

$$x^2 = K_H \times K_{a1}\left(3.97 \times 10^{-4}\ \text{atm}\right)$$

$$x = \sqrt{K_H \times K_{a1}\,(3.97 \times 10^{-4}\ \text{atm})}$$

$$x = \sqrt{\left(3.5 \times 10^{-2}\ \tfrac{M}{atm}\right)(4.5 \times 10^{-7})(3.97 \times 10^{-4}\ \text{atm})}$$

$$x = 2.5 \times 10^{-6}\ \text{M}$$

The values for the equilibrium constants can be found in Appendix I. Checking the assumption, we find the following.

$$5\%\ \text{rule} = \frac{x(RT)}{P_{CO_2}} \times 100\%$$

$$5\%\ \text{rule} = \frac{2.5 \times 10^{-6}\left(0.08206\ \frac{\text{L·atm}}{\text{K·mol}}\right)(298\ \text{K})}{3.97 \times 10^{-4}\ \text{atm}} \times 100\%$$

$$5\%\ \text{rule} = 15\%$$

When the assumption fails, we need to use successive approximations but consider that the atmosphere has a very large volume compared to the volume of rain clouds, so we can consider the concentration of carbon dioxide (P_{CO_2}) to be constant. The little CO_2 that gets absorbed into the water is not significant compared to the vast reservoir of carbon dioxide in the atmosphere. This is an example of an *open system*, introduced in Section 5.2.

Looking back at the equilibrium ICE table, x represents the concentration of hydronium ion, so finding the pH is easy.

$$\text{pH} = -\log[H_3O^+] = -\log(2.5 \times 10^{-6}\ \text{M}) = 5.60$$

The pH of natural rain water, saturated with dissolved CO_2 at a mixing ratio of 397 ppmv, is 5.60.

Now let us check to see if the second acid dissociation (governed by K_{a2}) would have contributed to the pH. We need to consider Reaction R3.83 and set up an ICE table with it. We know from the previous calculation the initial values for $[HCO_3^-]$ and $[H_3O^+]$.

	$HCO_3^-(aq)$	+	$H_2O(l)$	$\rightleftharpoons$	$H_3O^+(aq)$	+	$CO_3^{2-}(aq)$
I	2.5×10^{-6}		---		2.5×10^{-6}		0
C	$-x$		---		$+x$		$+x$
E	$2.5 \times 10^{-6} - x$		---		$2.5 \times 10^{-6} + x$		x

We now need to plug in the equilibrium values into the K_{a2} expression.

$$\frac{[H_3O^+][CO_3^{2-}]}{[HCO_3^-]} = K_{a2}$$

$$\frac{(2.5 \times 10^{-6} + x)(x)}{2.5 \times 10^{-6} - x} = K_{a2}$$

To simplify the math, we can use the 5% rule, then solve for x. Assume $x \ll 2.5 \times 10^{-6}$ M.

$$\frac{(2.5 \times 10^{-6} + \cancel{x}_0)(x)}{2.5 \times 10^{-6} - \cancel{x}_0} = K_{a2}$$

$$\frac{(2.5 \times 10^{-6})(x)}{2.5 \times 10^{-6}} = K_{a2}$$

$$\frac{\cancel{(2.5 \times 10^{-6})}(x)}{\cancel{2.5 \times 10^{-6}}} = K_{a2}$$

$$x = K_{a2} = 4.7 \times 10^{-11}\ \text{M}$$

Checking the assumption, we find that the assumption was good (0.0019%). While the first dissociation of carbonic acid (K_{a1}) contributed 2.5×10^{-6} M H_3O^+, the second (K_{a2}) only contributed an additional 4.7×10^{-11} M, which is insignificant. Conclusion: only the first dissociation was important.

The aforementioned example shows that rain, unaffected by acids such as nitric and sulfuric acid, already starts out acidic. With additional acidic contributions from biogenic and anthropogenic sources, acid rain can reach pH levels below 4. Remember that the pH scale is a logarithmic one, so the difference between pH 4 and pH 5 implies a factor of 10 difference in the concentration of hydronium ion.

Example 3.22: Acid Rain

If natural rain has a pH of 5.60 and a particular sample of acid rain with nitric acid has a pH of 4.20, what is the hydronium ion contribution from the carbonic acid and from nitric acid, and what is the ratio between them?

Solution: In the previous example, the contribution of hydronium from carbonic acid was 2.5×10^{-6} M. A solution with a pH of 4.20 has the following hydronium ion concentration.

$$[H_3O^+] = 10^{-pH} = 10^{-4.20} = 6.31 \times 10^{-5}\ M$$

Some of this $[H_3O^+]$ is from carbonic acid, but most is from nitric acid. The ratio between them is

$$\frac{[H_3O^+]_{nitric}}{[H_3O^+]_{carbonic}} = \frac{6.31 \times 10^{-5} - 2.5 \times 10^{-6}}{2.5 \times 10^{-6}} = 24$$

About 24 times more hydronium ions come from the nitric acid compared to the hydronium ions from carbonic acid.

Acid deposition has wide-ranging and varied impacts on the biosphere and human civilization. Weathering of rocks is accelerated. Acid attacks the chemical structure of various rocks and dissolves some more than others. Lakes and streams that have access to limestone rocks are usually much less affected by acid deposition compared to those that are surrounded by granite. Limestone buffers against acid much more effectively, and regions of the world where limestone geology is abundant tend to feel muted effects of acid deposition compared to those regions with granitic and basaltic geology. For more details on the effects of acid deposition on soil, surface water, and flora, see Chapter 4.

Acidification of lakes and streams places stress on aquatic life and leads to reduced biodiversity and, eventually, sterilization. Healthy lakes maintain a pH of at least 6.5. Lakes with a pH below 5 can only support a few species of fish that are acid tolerant, such as brook trout and eels. At a pH of 4 and below, a body of water supports very little biota and is considered dead.

Terrestrial life does not escape damage from acid deposition. Plants suffer damage to their leaves and needles due to exposure to acid. Soil exposed to acid degrades nutrient levels, reduces microbial processes, and causes toxic metals to leach out of the soil and into the watershed. This causes damage to roots that absorb this water.

3.10 OZONE DESTRUCTION IN THE STRATOSPHERE

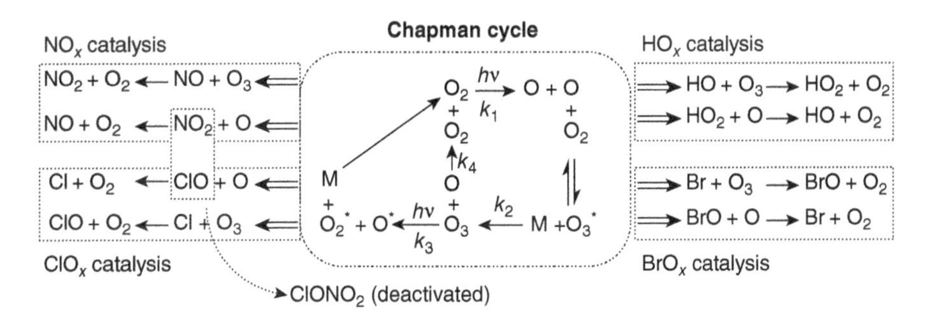

Figure 3.22 The Chapman cycle and several of the catalytic reactions that interfere with ozone production.

$$O_2 + h\nu \xrightarrow{k_1} O + O$$

$$O + O_2 + M \xrightarrow{k_2} O_3 + M$$

$$O_3 + h\nu \xrightarrow{k_3} O^* + O_2^*$$

$$O + O_3 \xrightarrow{k_4} 2O_2$$

Table 3.18 The four reactions of the Chapman cycle.

The Chapman cycle, summarized by the reactions in Table 3.18 and in Figure 3.22, is responsible for the "ozone layer" that lies within the stratosphere from about 15–30 km in altitude. The Chapman cycle absorbs a <240 nm photon in the photolysis of molecular oxygen in step 1 and another 200–300 nm photon in step 3. For organisms on the Earth's surface, these steps provide the important UV shield. Other trace components of the atmosphere can interfere with the Chapman cycle and expedite the destruction of ozone. These species follow the same general mechanism.

$$\begin{array}{r} X + O_3 \rightarrow XO + O_2 \\ XO + O \rightarrow X + O_2 \\ \hline O + O_3 \rightarrow 2O_2 \end{array}$$

When the two reactions involving the "X" species are added, step 4 of the Chapman cycle results and no "X" is involved in the overall reaction. This is an indication that "X" is a catalyst – a species that regenerates itself, is not part of the overall reaction, changes the mechanism by stabilizing the activated complex, and makes the overall reaction proceed much faster. Here are some examples of these catalytic cycles using common trace species such as NO, Cl, Br, HO, and H.

$$\begin{array}{r} NO + O_3 \rightarrow NO_2 + O_2 \\ NO_2 + O \rightarrow NO + O_2 \\ \hline O + O_3 \rightarrow 2O_2 \end{array} \quad (R3.87)$$

$$\begin{array}{r} OH + O_3 \rightarrow HOO + O_2 \\ HOO + O \rightarrow HO + O_2 \\ \hline O + O_3 \rightarrow 2O_2 \end{array} \quad (R3.88)$$

$$\begin{array}{r} Cl + O_3 \rightarrow ClO + O_2 \\ ClO + O \rightarrow Cl + O_2 \\ \hline O + O_3 \rightarrow 2O_2 \end{array} \quad (R3.89)$$

$$\begin{array}{r} H + O_3 \rightarrow HO + O_2 \\ HO + O \rightarrow H + O_2 \\ \hline O + O_3 \rightarrow 2O_2 \end{array} \quad (R3.90)$$

☞ While Cl catalysis will be the focus of the discussion of ozone destruction, Br catalysis is about 50× more efficient. The amount of chlorine species in the atmosphere is about 170× higher than that of bromine, so Cl catalysis still does more damage.

$$\begin{array}{r} Br + O_3 \rightarrow BrO + O_2 \\ BrO + O \rightarrow Br + O_2 \\ \hline O + O_3 \rightarrow 2O_2 \end{array} \quad (R3.91)$$

$$\begin{array}{r} Br + O_3 \rightarrow BrO + O_2 \\ Cl + O_3 \rightarrow ClO + O_2 \\ BrO + ClO \rightarrow BrCl + O_2 \\ BrCl + h\nu \rightarrow Br + Cl \\ \hline 2O_3 \rightarrow 3O_2 \end{array} \quad (R3.92)$$

These catalytic cycles can rapidly destroy ozone and regenerate the catalyst so that it can participate in many cycles. Both X and XO reduce ozone levels – X by removing the third O from O_3, which would normally require a 200–300 nm photon, and XO by reacting with atomic oxygen (generated by step 1) and preventing it from forming O_3. These cycles not only destroy ozone but also prevent its formation.

Contributions to the concentration of NO_x in the stratosphere come mostly from nitrous oxide (N_2O), a by-product of microbial metabolism of nitrogen-based fertilizers in soil. It maintains a concentration of around 300 ppbv in the lower stratosphere and rapidly decreases with increasing altitude due to photolysis. Nitrous oxide requires UV-C radiation (160–200 nm) to initiate a photolysis reaction, so it is a stable species in the troposphere. When it reaches the stratosphere, it participates in the following set of reactions.

$$N_2O + h\nu \rightarrow N_2 + O^* \quad (R3.93)$$

$$N_2O + O^* \rightarrow 2NO \quad (R3.94)$$

$$N_2O + O^* \rightarrow N_2 + O_2 \quad (R3.95)$$

A little more than half of the O^* proceeds via Reaction R3.94, generating the NO to start the catalytic cycle.

Example 3.23: N_2O Photolysis

Draw the Lewis structure for nitrous oxide and determine the formal charge of each atom. Can you rationalize why half of the time Reaction R3.94 occurs and Reaction R3.95 occurs the other half of the time?

Solution: See the Section C.1 on page 265.

The HO_x catalyst is a product of an excited-state oxygen atom reacting with either water or methane. While most of the water in the atmosphere is in the troposphere, some water does make it past the cold temperatures of the tropopause, giving water a stratospheric concentration around 5 ppmv (0.0005%). Methane is also a product of microbial metabolism of biogenic and anthropogenic sources of carbon and maintains a concentration profile that starts near 1.5 ppmv at the tropopause and drops off precipitously at 30 km to about 0.2 ppmv or less. An excited-state oxygen atom is required to produce the catalyst, which comes from the photolysis of ozone or Reaction R3.93.

$$O_3 + h\nu \rightarrow O^* + O_2 \quad \text{(R3.96)}$$
$$H_2O + O^* \rightarrow 2OH \quad \text{(R3.97)}$$
$$CH_4 + O^* \rightarrow OH + CH_3 \quad \text{(R3.98)}$$

The hydroxyl radical, which was described more thoroughly in Section 3.7.3, can now start the ozone destructive catalytic cycle.

Finally, the other major contributors to catalytic ozone destruction are chlorofluorocarbons (CFCs) and bromocarbons (halons) (Figure 3.23). These chemicals are mostly anthropogenic in origin and come from industrial and agricultural sources. CFCs were developed as commercial and residential refrigeration was booming in the late 1920s. The existing refrigeration used mechanical compression of certain gases that, when expanded to a larger chamber, would absorb heat. Gases that have strong dipole moments do this because in the process of expansion, the intermolecular forces between the gas molecules must be broken, and energy is absorbed from the surroundings in order to break these attractions. Gases that were employed at the time were highly toxic ammonia (still used by the international space station!), methyl chloride, corrosive sulfur dioxide, and flammable butane. CFCs made refrigeration in living spaces safe, and halons produced very effective fire extinguishers. Eventually, CFCs were used in aerosol spray cans, as propellants to produce foam and for cleaning electronic circuit boards. In most of the applications, the CFCs were vented to the atmosphere and became part of the tropospheric mix.

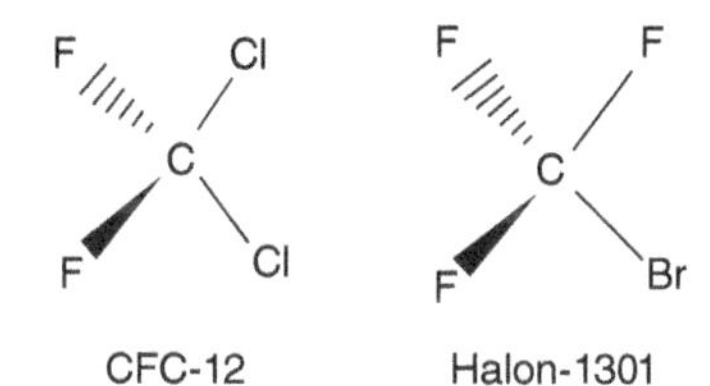

Figure 3.23 CFCs and halons are hydrocarbons with many or all of the hydrogens replaced by a halogen. They are usually very stable, inert, and revolutionized refrigeration by indoor climate control, degreasing, and fire extinguishing.

CFCs and halons are very stable in the troposphere, unlike other hydrocarbons, which can be oxidized by ozone and the hydroxyl radical. Given the thorough mixing of the troposphere, most of the CFCs and halons will eventually find their way into the stratosphere where UV photons from about 160 to 200 nm photolyze the C–Cl bond, and produce atomic chlorine (and bromine in the case of halons).

The rates of these catalytic cycles depend on the altitude, temperature, and concentrations of the various species. The HO_x catalytic cycle dominates the ozone destruction at the top and bottom of the stratosphere, which is not where the ozone layer lies, so the HO_x catalytic cycle does not destroy much ozone. The NO_x cycle dominates at the altitude of the ozone layer (15–30 km) but is responsible for 70% of the loss.

Figure 3.24 tells a different story. Atmospheric scientists, using ground and satellite measurements, had been continuously measuring the density of the ozone column above the South Pole and over the entire globe since the 1970s, although intermittent measurements date back to 1957. At the time, there were about 200 ground stations spread around the globe, with 17 of those stations on the Antarctic continent. It is obvious from the bottom graph in Figure 3.24 that total ozone levels above the pole had been dropping precipitously since the late 1970s, and yet it was not until 1985 that scientists first published the trend. It also seems contradictory that the upper graph indicates that the ozone levels above the Northern Hemisphere were fairly stable. How can they be so different?

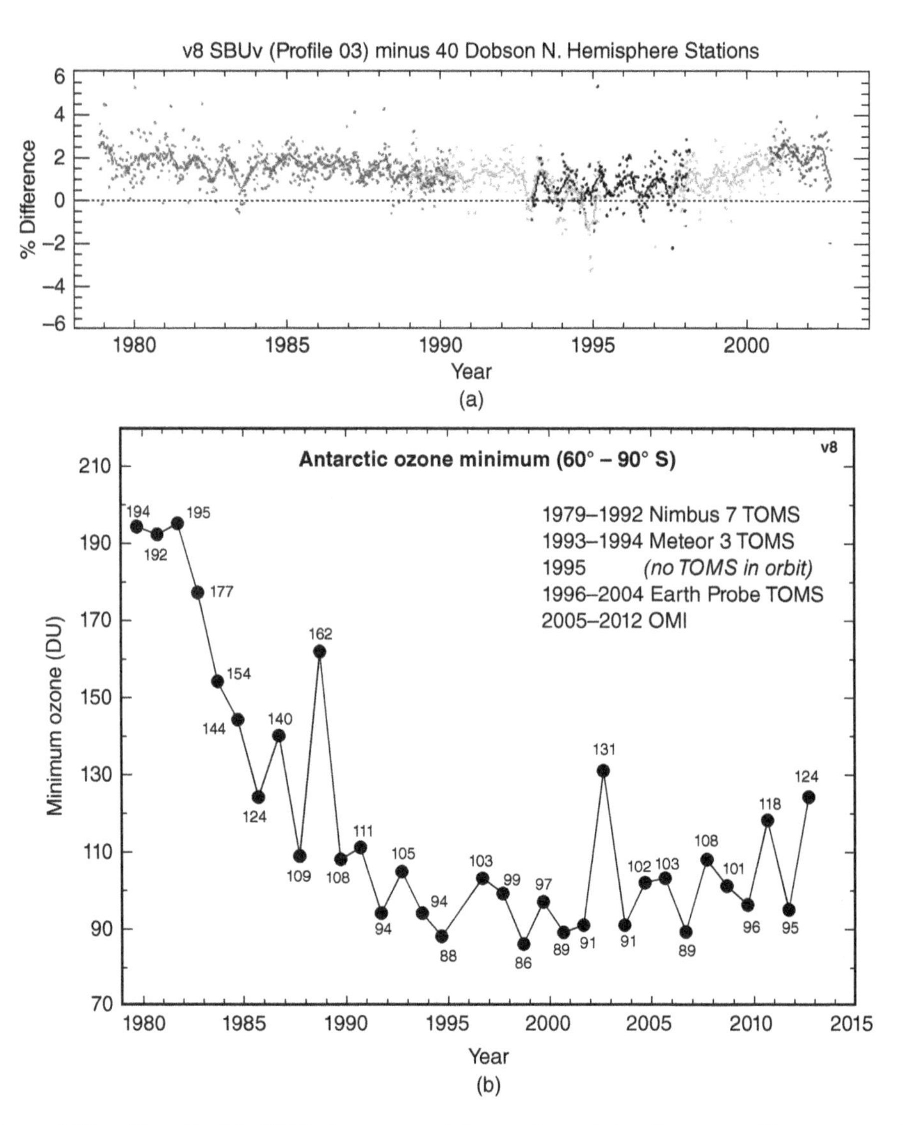

Figure 3.24 Since the late 1970s the ozone minimum above Antarctica was getting worse until around 2005 when ozone thickness was starting to trend upward (b). Above the Northern Hemisphere, ozone levels appeared to be normal (a). Note that the Montreal Protocol was signed in 1987 and went into effect in 1989. Ozone levels are just now returning to pre-1987 levels. Source: NASA.

Satellites were collecting over 100,000 data points per day, but data processing technology was slow back then and most of the data was archived for a time when someone was interested in studying it. NASA scientists had been studying this data all along but thought that the measurements were a result of a calibration error because they were dipping below the "normal" ozone levels. There was a calibration error in some ground-based measurements, resulting in ozone levels reading within normal levels in 1983, so the NASA team had to contend with two measurements giving different results. There was also a latent belief among some scientists that autumn measurements over the pole were a better indicator of ozone loss because there was continual solar radiation – why would ozone loss occur in the early spring after months of darkness? It did not make sense.

In 1985, Joseph Farman published his analysis of the ozone loss in the journal *Nature*, much to the surprise of many... except those NASA scientists who may have wished they had pursued the calibration anomaly a little further.

3.11 THE OZONE HOLE

Figure 3.24 helps to frame the initial confusion over the ozone hole. How is it forming over the South Pole during the transition from winter to spring while ozone levels seem to be pretty stable in other locations? Due to the turbulent mixing of the troposphere, the gases that spawned the catalytic cycles would have been uniformly spread throughout the troposphere and stratosphere, so no one expected an obvious "hot spot" for catalytic activity. The other strange feature of timing was the transition from darkness to light during the ozone hole event – since the catalytic cycles all require some radiation to produce the primary catalysts. Antarctica has a set of properties that produce a synergistic and potent combination for ozone destruction.

3.11.1 Polar Stratospheric Clouds

The bottom of the stratosphere is a very cold place, reaching temperatures as low as −90 °C. The small amount of water vapor that mixes into the stratosphere forms tiny ice crystals at these very cold temperatures. These ice crystals are a combination of nitric acid, from a reaction between NO_2 and H_2O, and water in the form of nitric acid trihydrate (NAT – $HNO_3 \cdot 3H_2O$). It is the chemical composition and phase (solid) that make the polar stratospheric clouds (PSCs) important for the ozone hole.

Chlorine in the stratosphere exists in many forms, some of which are more catalytically active than others. You saw from Reaction Cycles R3.89 and R3.92 that the Cl and ClO forms are catalytically active. These forms are also very reactive and normally exist in less reactive forms (reservoirs) such as HCl, HOCl, and $ClONO_2$ (chlorine nitrate), which are not catalytically active. These species can react on the surface of the NAT ice to release chlorine.

$$HCl(s) + ClONO_2(g) \rightarrow Cl_2(g) + HNO_3(s) \qquad (R3.99)$$

$$HCl(s) + HOCl(g) \rightarrow Cl_2(g) + H_2O(s) \qquad (R3.100)$$

Notice what is happening here – three relatively stable reservoirs of chlorine (HCl is dissolved in the ice, so it is especially inactivated) are being turned into a very reactive form of chlorine. HOCl, Cl_2, and $ClONO_2$ can be photolyzed to form atomic Cl, but the energy requirements are much different, and two atoms of Cl are liberated with Cl_2 instead of just one.

$$ClONO_2 + h\nu(\leq 215\text{ nm}) \rightarrow Cl + NO_3 \qquad (R3.101)$$

$$HOCl + h\nu(\leq 254\text{ nm}) \rightarrow Cl + OH \qquad (R3.102)$$

$$Cl_2 + h\nu(\leq 330\text{ nm}) \rightarrow 2Cl \qquad (R3.103)$$

Photons at 330 nm are much more abundant than those of <254 nm, especially in the lower stratosphere.

Higher levels of NO_2 in the stratosphere tend to reduce chlorine catalysis by binding chlorine in the form of chlorine nitrate.

$$ClO + NO_2 \rightarrow ClONO_2 \qquad (R3.104)$$

The PSCs work to remove NO_2 (as NAT) from the gas phase as well as liberate a reactive form of chlorine. All this happens during the period leading up to the vernal equinox in late September in the Southern Hemisphere. It is still dark prior to this period, so no photochemistry is possible. PSCs are quietly denitrifying the stratosphere above the South Pole and converting inactive chlorine reservoirs into active ones. The "spring" of the ozone hole "mouse trap" is being set. Two more characteristics of Antarctica are also important to this process.

3.11.2 The Polar Vortex

During the winter months a weather pattern develops above the Antarctic pole that is called the polar vortex. This clockwise spinning vortex isolates the stratosphere above the pole and keeps outside, warmer air from increasing the temperature. This promotes the formation of the polar stratospheric clouds (PSCs) and the nitric acid ice crystals (NATs) that form under the extreme cold. These NATs provide the means to produce highly reactive Cl_2 during the winter while removing moderating species, such as NO_2. The vortex also prevents the accumulating chlorine from being diluted by outside air. This combination of

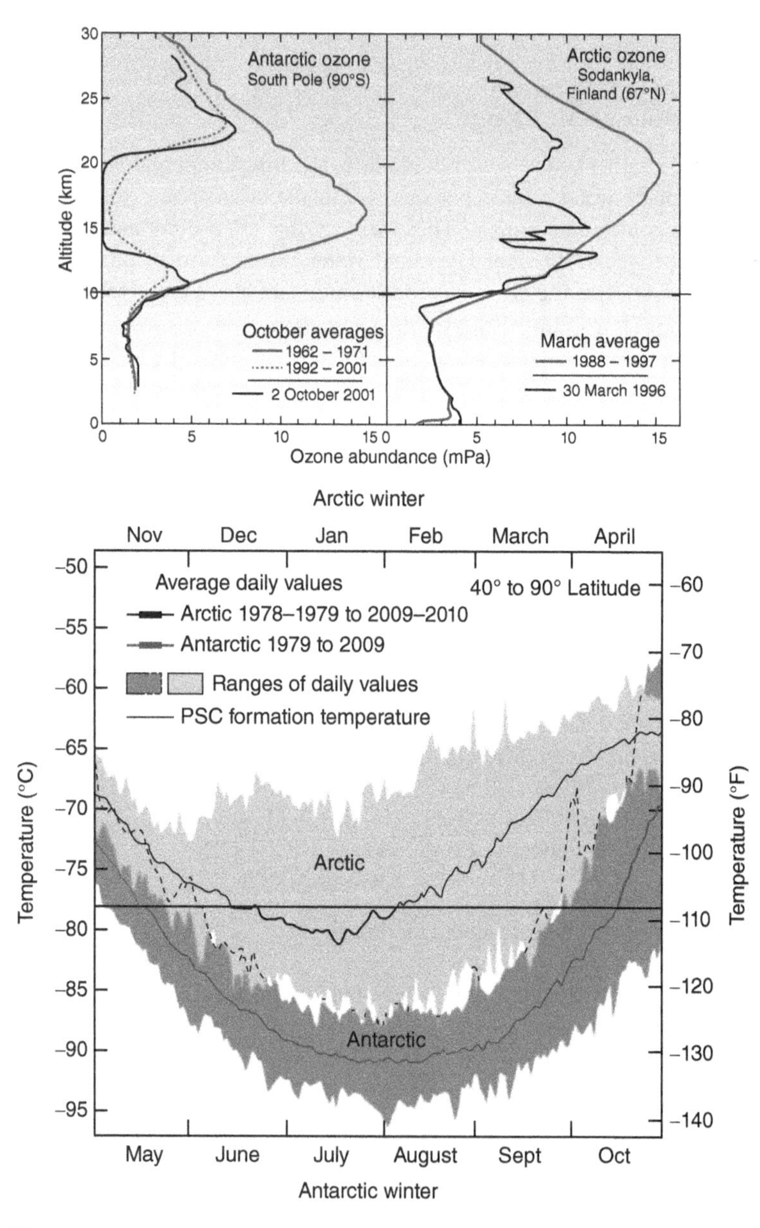

Figure 3.25 The bottom graph compares the winter temperatures above each of the poles. PSCs are not as common above the Arctic, so ozone destruction is much less but still not zero as the upper graph suggests. Courtesy FaheyD.W., and M.I. Hegglin (Coordinating Lead Authors), Twenty Questions and Answers About the Ozone Layer: 2010 Update, Scientific Assessment of Ozone Depletion: 2010, 72 pp., World Meteorological Organization, Geneva, Switzerland, 2011. Source: Reproduced with permission of World Meteorological Organization (WMO).

factors synergistically prepares the gaseous soup for massive ozone destruction as soon as solar radiation arrives with the onset of spring.

3.11.3 The Dark Winter

The final characteristic of Antarctica is not so unique since both poles experience weeks of darkness during the winter months. The absence of solar radiation over the Antarctic pole allows temperatures to plummet, generating PSCs, and allows chlorine levels to build up in the absence of photolysis, which would drain away the concentration.

The three special conditions above the Antarctic pole (PSCs, polar vortex, dark winter) provide the conditions for massive ozone loss in the presence of Cl and Br catalysts when the spring radiation arrives. To make matters worse, because there is so little ozone available to absorb solar radiation, the vortex takes longer than usual to warm up, prolonging the dissipation of the PSCs. Further, the angle of the Sun (zenith) in early spring reduces the efficiency of the Chapman cycle and prevents it from replacing the ozone that is lost through catalysis.

These conditions exist, to a lesser extent, at the North Pole. As Figure 3.25 shows, temperatures do not get as low there and PSCs do not form extensively.

3.12 CFC REPLACEMENTS

The role of chlorine catalysis in ozone destruction was first presented to the world in 1974 by the chemists Sherwood Rowland and Mario Molina. Their work started a global effort to measure the scope of the problem. They shared the 1995 Nobel Prize in Chemistry with Paul Crutzen, who discovered NO_x catalysis. As mentioned before, the ozone hole first came to the attention of the world in 1985 when Joseph Farman reported his analysis of ground and satellite measurements. Shortly after atmospheric scientists recognized the cause of the ozone hole, the international community developed and adopted the Montreal Protocol in 1987, which sought to implement a global ban on the use of CFCs and halons, and the phaseout of HCFCs (CFC replacements) by 2030.

The relative destructiveness of a gas is measured by the Ozone Depleting Potential (ODP). It compares the ability of a gas to decrease the steady-state concentration of ozone compared to CFC-12 (the benchmark substance). An ODP that is greater than 1.0 means that the gas is better at destroying ozone than CFC-12, whereas a fractional number means that the gas is less harmful. Note that all of the CFC replacements, with C–H bonds (such as HCFC-21), have small ODPs. While the HCFCs still contain chlorine, they are attacked in the troposphere by the hydroxyl radical and very little is left to escape into the stratosphere. The GWP column in Table 3.19 is the Global Warming Potential described in Section 3.13. As it happens, the CFCs and their replacements are potent GHGs, so while the CFC replacements are reducing the catalytic destruction of ozone in the stratosphere, they are still enhancing greenhouse warming.

Trade name	Formula	Lifetime (years)	ODP	GWP
CFC-11	CCl_3F	45	1	4,750
CFC-12	CCl_2F_2	100	1	10,890
CFC-113	$C_2F_3Cl_3$	85	0.8	6,130
CFC-114	$C_2F_4Cl_2$	300	1	10,040
CFC-115	C_2F_5Cl	1700	0.6	7,370
Halon 1211	CF_2ClBr	16	3	1,890
Halon 1301	CF_3Br	65	10	7,140
Halon 2402	$C_2F_4Br_2$	20	6	1,640
HCFC-21	$CHFCl_2$	1.7	0.04	151
HCFC-22	CHF_2Cl	12	0.055	1,810
HCFC-123	$C_2HF_3Cl_2$	1.3	0.02	77
HCFC-124	C_2HF_4Cl	5.8	0.022	609
HCFC-142b	$C_2H_3F_2Cl$	17.9	0.065	2,310
HFC-23	CHF_3	270	0	12,240
HFC-32	CH_2F_2	4.9	0	543
HFC-41	CH_3F	2.4	0	90
HFC-134	$C_2H_2F_4$	9.6	0	1,090
HFC-134a	CH_2FCF_3	14	0	1,320
HFC-245ca	$C_3H_3F_5$	6.2	0	682
HFC-152a	$C_2H_4F_2$	1.4	0	122

The lifetime indicates the average residence time in the atmosphere. The ODP column indicates the ozone-depleting potential relative to CFC-12, and the GWP column indicates the global-warming potential relative to CO_2. (Data from the EPA)

Table 3.19 The properties of some selected Class I and Class II ozone-depleting substances.

A replacement for CFCs must have the required properties for substitution (use as a refrigerant, foaming/blowing agent, nonconductivity, low reactivity), but it needs to have zero ozone-depleting potential or stay within the troposphere so that it will not get involved in catalytic ozone destruction. The successful replacements were similar compounds (alkanes) that contained less chlorine and bromine and more fluorine and hydrogen. Let us examine HCFC-21 and HFC-23 from Table 3.19.

In the case of HCFC-21 (seen in Figure 3.26), one of the Cl atoms has been replaced by an H atom (compared to CFC-11). In the case of HFC-23, one Cl atom from CFC-12 has been replaced by an F atom and the other by an H atom. In order for the compounds to have refrigerant properties, they must be polar, which they are. HCFC-21 still has some ODP because it has a Cl atom, but because its atmospheric lifetime is so small compared to CFC-11 (see the third column of Table 3.19), it does not make it to the stratosphere in significant quantities. HCFC-21, however, is on a schedule to be phased out completely by 2030 because it still has a small ODP. HFC-23 has no ODP because it has no Cl or Br atoms, but it does have a significant global-warming potential (GWP) because it is a greenhouse gas. While HFCs are replacing HCFCs as the Montreal Protocol phaseout progresses, we are trading one problem (ODP) for another (GWP). HFC-134a, currently a commonly used replacement for refrigerant CFCs, has an ODP of 0 but a GWP of 1300. The European Union banned it for use in automotive air-conditioning units in 2011 because of its high GWP. Some of the CFC replacements are flammable or toxic, making them an inferior product to the original CFCs.

HCFCs and HFCs usually have much shorter atmospheric lifetimes compared to CFCs. This is because having a C–H bond in the compound makes it susceptible to attack from the hydroxyl radical in the atmosphere. The following example problem explains why.

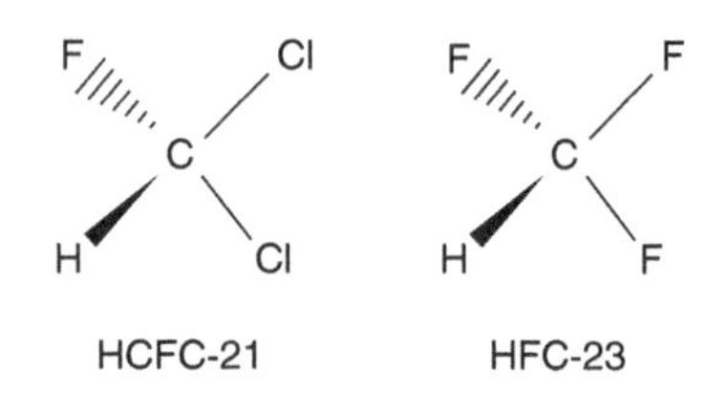

Figure 3.26 The chemical structure of two of the molecules that have replaced CFCs.

Example 3.24: CFC Replacements

CFCs are very inert, but their replacements, HCFCs, are more reactive and especially susceptible to atmospheric oxidation by the hydroxyl radical. Why? Use B.E. and enthalpies to justify your reasoning for the comparison between CFC-12 and HCFC-22 and their reaction with the hydroxyl radical.

Bond	Bond energy (kJ/mol)
C–H	414
C–F	485
C–Cl	339
C–Br	285
O–H	464
O–F	190
O–Cl	203

Table 3.20 Important bond energies for HCFC analysis.

Species	ΔH_f° (kJ/mol)
HOCl	−78.7
OH	39.0
H_2O	−241.8
CCl_2F_2	−477.4
$CHClF_2$	−482.6
$CClF_2$	−279.1

Table 3.21 Important heats of formation for HCFC analysis.

Solution: Since this chapter has been focused on atmospheric chemistry and photolysis, it is natural to postulate what sort of energies would be required to photolyze the C–H, C–Cl, and C–F bonds. Using thermodynamic values (such as B.E.) gives an underestimate, but it is a good starting place. From Table 3.20, it is obvious that the C–F bond is the strongest at 485 kJ/mol, the C–H bond is second at 414 kJ/mol, and the C–Cl bond is the weakest by 75 kJ. This would imply that if individual B.E. were the most important factor, CFCs would be more reactive compared to HCFCs because the C–Cl bond is weak compared to the other. We know that this is in opposition to the empirical data that tells us that CFCs are very unreactive and make it to the stratosphere intact. Bond energies are a dead end.

Empirical studies have shown that the hydroxyl radical is the most important initiator of atmospheric oxidation, so let us look at the first step in the oxidation of CFC-12 (CCl_2F_2) and HCFC-22 ($CHClF_2$) using data from Table 3.21.

Rxn:	CCl_2F_2	+	OH	→	HOCl	+	$CClF_2$
ΔH°_{rxn}(kJ/mol):	(−477.4)		(39)	→	(−78.7)		(−279.1)

$$\Delta H^\circ_{rxn} = \Sigma\Delta H^\circ_{f\,prod} - \Sigma\Delta H^\circ_{f\,react}$$
$$= [(1)(-78.7) + (1)(-279.1)] - [(1)(-477.4) + (1)(39)]$$
$$= 80.6\text{ kJ}$$

Rxn:	$CHClF_2$	+	OH	→	H_2O	+	$CClF_2$
ΔH°_{rxn}(kJ/mol):	(−482.6)		(39)	→	(−241.8)		(−279.1)

$$\Delta H^\circ_{rxn} = \Sigma\Delta H^\circ_{f\,prod} - \Sigma\Delta H^\circ_{f\,react}$$
$$= [(1)(-241.8) + (1)(-279.1)] - [(1)(-482.6) + (1)(39)]$$
$$= -77.3 \text{ kJ}$$

From the aforementioned calculations, you can see that the reaction between the CFC and OH is endothermic. This is a strong indication that it is also nonspontaneous since there is very little entropy change because both sides have two gas equivalents (if free energy values for $CClF_2$ were available, it would have been better to use them rather than enthalpies). The reaction between the HCFC and OH is exothermic and favorable.

Thus, the C–Cl and C–H B.E. must be considered with the O–Cl and O–H bonds that are formed (as part of HOCl and H_2O), which is directly related to the calculated heat of reaction in each case.

The Montreal Protocol and the international efforts that followed it stand as a model for international environmentalism and cooperation. Figure 3.24 is a testament to this because it demonstrates the slowdown and reversal of ozone destruction during the southern vernal equinox. Figure 3.27 shows that the CFC levels are dropping as the HCFC and HFC levels are rising. Estimates place a full recovery by 2050. Why? Because the Chapman cycle, driven by solar radiation, will eventually bring ozone levels back to their former state. It is fortunate for us humans that many of the environmental systems upon which we depend are resilient in this way.

Thomas Midgley, Jr. (1889–1944) was an industrial chemist who worked for General Motors Chemical Company. He helped develop tetraethyl lead (TEL), an anti-knock agent for gasoline cars, and the first CFC's. When workers at the TEL factory began to die and suffer mental illness (in 1924), he attended a press conference where he washed his hands in TEL to demonstrate that it was perfectly safe. Shortly after, Midgley took a long leave from his work while he sought treatment for lead poisoning. TEL was not phased out of general use until the 1970s despite its known neurotoxicity. Midgley invented the first CFC as a new refrigerant, which replaced dangerous gases that were in use. At the 1930 American Chemical Society national conference he demonstrated the non-toxicity and fire extinguishing properties of CFC-11 by breathing in the gas and blowing out a candle.
Midgley received many well deserved awards from the American Chemical Society, but may have been indirectly responsible for the two greatest atmospheric environmental disasters (atmospheric lead poisoning and ozone destruction) of the 20th Century. Blum (2010, pp. 121–123) and Giunta (2006).

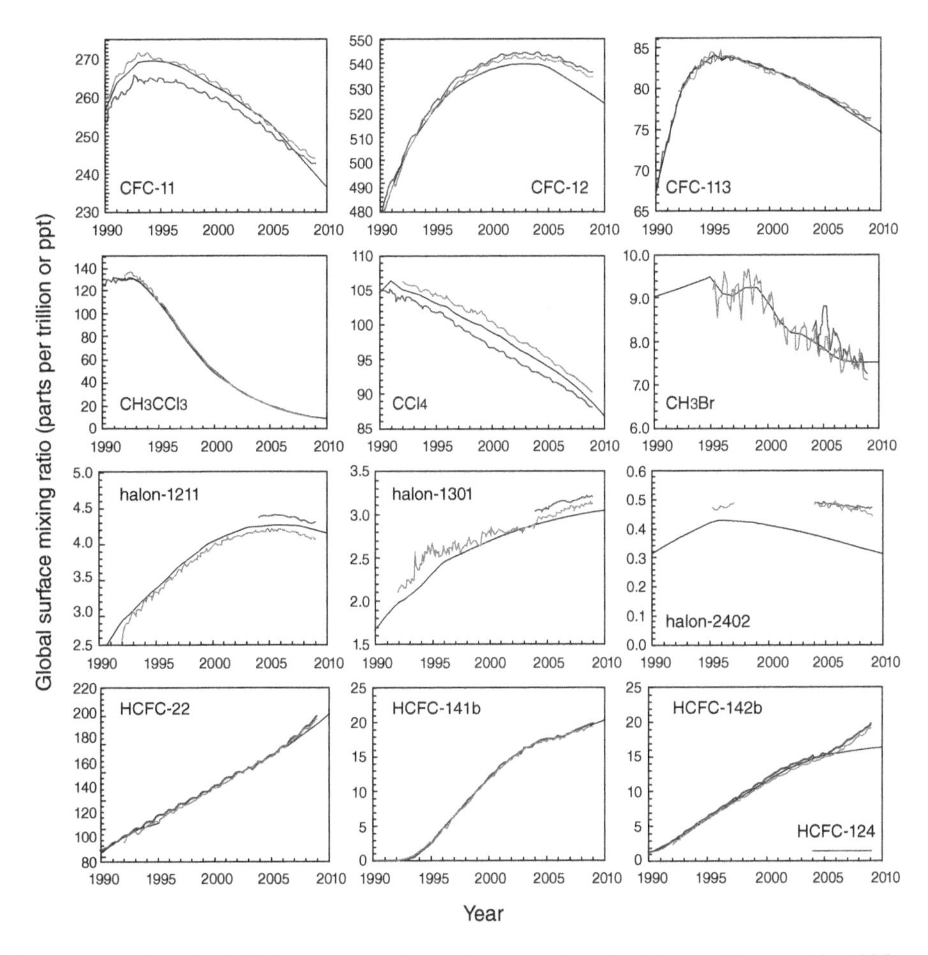

Figure 3.27 Levels of CFCs are beginning to decrease since the Montreal Protocol in 1985, according to the data from World Meteorological Organization. Source: Reproduced with permission of World Meteorological Organization (WMO).

3.13 CLIMATE CHANGE

Natural forcings are factors, such as solar radiation and biosphere activity, that force or drive the local or global climate in one direction or another (warmer or cooler, dryer or wetter). *Anthropogenic forcings* are factors that are produced by humans that can also influence local or global climates. These forcings include greenhouse gas emissions and sulfate aerosol emissions.

Now that you are familiar with GHGs and the role they play in the troposphere, we are ready to tackle the issue of Climate Change, often referred to as Global Warming. You learned in Chapter 1 that the Earth is a dynamic environment with a long legacy of drastic changes in atmospheric composition, ocean pH, composition of the crust, and average global temperatures. These changes were brought about by chemical and physical forces such as accretion, cooling, density differentiation, oxidation, solar radiation, volcanism, and the occasional collision with an asteroid or meteor. The field of climatology incorporates all of these factors, called *natural forcings*. Since the climate can be affected by a myriad of factors, it is a difficult task. You were introduced to one category of forcings in Section 3.6.2 when you learned about GHGs. GHGs absorb IR radiation emitted from the surface of the Earth and cause an increase in the heat budget of the atmosphere. This is one dimension of GHG forcing, and other dimensions are equally as important. These factors in the GHG forcing are used to calculate a GWP.

1. **Concentration:** Gases with higher concentrations will have a larger impact compared to gases with lower concentrations (all other factors being equal).
2. **Spectral window:** Gases that absorb in an empty spectral window will have a larger effect per molecule compared to a gas that absorbs where another gas already absorbs.
3. **Atmospheric lifetime:** Gases that persist longer in the atmosphere will absorb more heat over time than a short-lived gas.
4. **Absorption efficiency:** Gases that absorb more strongly than others do or in more spectral locations will have a larger impact compared to gases that have a weak absorption.

Gas	Lifetime (years)	GWP 20 year	GWP 100 year	GWP 500 year
CO_2	∞	1	1	1
CH_4	12	72	25	7.6
N_2O	114	289	298	153
CFC-12	100	11,000	10,900	5,200
Halon 1301	65	8,480	7,140	2,760
HCFC-22	12	5,160	1,810	549
HCFC-123	1.3	273	77	24
SF_6	3,200	16,300	22,800	32,600

While some of the CO_2 may be pulled from the atmosphere after a century or so, a significant portion could be in the atmosphere for 2–20 centuries Archer *et al.* (2009).

Table 3.22 The lifetime of carbon dioxide is difficult to estimate and has become a controversial topic.

Each of these factors must be evaluated in order to determine the GWP. The actual GWP values shown in Table 3.22 are all relative to CO_2. Notice that the GWP for CO_2 has a value of 1 for each of the GWP time frames. Methane, for example, is 72 times more potent compared to carbon dioxide over a 20 year period, but since it has a short atmospheric lifetime, its impact over longer periods decreases. Notice how incredibly potent the HCFCs are. While they have short lifetimes, they absorb in open spectral windows and their concentrations are increasing (Figure 3.27).

In Figure 3.28, you can see the IR spectrum of the common GHGs. Each absorbs IR radiation specific to the molecular structure. There are two labeled spectral windows in this graph. These windows show that none of the usual GHGs absorb in these wave number ranges. This means that in the atmosphere, radiation in these windows (produced by the surface of the Earth) would escape into space because there are no gases to absorb it. This is the reason why a new GHG that absorbs in these open spectral windows can have such

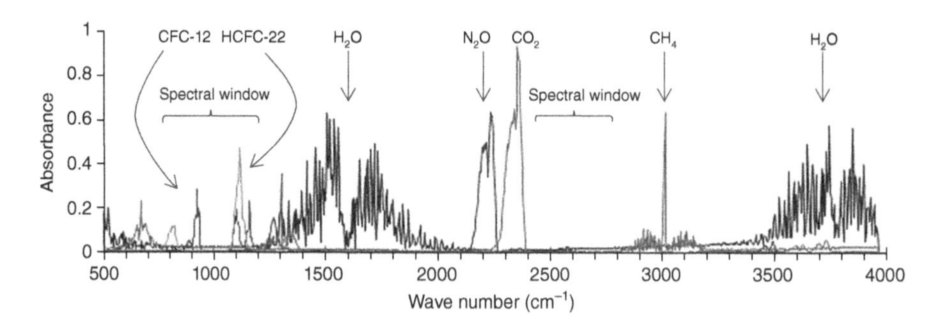

Figure 3.28 The addition of a CFC and an HCFC fills one of the open spectral windows where no GHG currently absorbs. Source: Data from NIST.

a high GWP – a small change in the concentration of such a gas would cause a very large change in the amount of this radiation that escapes into space.

The spectral window near 1000 cm^{-1} in Figure 3.28 is occupied by CFC-12 and HCFC-22. If either of these gases absorbed only at 2350 cm^{-1}, one of the absorption bands of CO_2, then its GWP would be much lower because CO_2 is already absorbing most of the radiation in this area of the spectrum, so an additional absorption would add insignificantly to the total absorption.

Other forcings that climatologists incorporate into their climate models:

- Ozone (stratospheric loss, tropospheric formation)
- Land use
- Water vapor
- Albedo (land use, clouds formation, snow cover)
- Aerosols (sulfate emissions, biomass burning, carbon aerosols)
- Solar irradiance

Global Warming (or Climate Change) is either the "greatest hoax ever perpetrated on the American people," (Inhofe, 2005) according to Senator James Inhofe, or "is arguably the biggest threat we are facing today," (Biddle, 2011) according to the British Foreign Secretary William Hague, or probably something in between. It has occupied countless hours of international negotiations since the Rio de Janeiro Earth Summit in 1992 (and countless gallons of jet fuel to transport representatives of nearly 200 countries to the annual conferences, further contributing to the problem of GHG levels!). The Paris meeting of 2015 yielded an agreement between over 190 countries to begin meaningful emission reductions. Time will tell if these countries hold to their pledges.

The United Nations has established the Intergovernmental Panel on Climate Change (IPCC) and has issued five assessment reports that can be downloaded from their website. The IPCC gathers information from socioeconomic and research studies that inform the issue of climate change. Here is what seems clear and uncontroversial about the issue (at least according to scientists – politicians are another matter!).

1. *On average, the global temperatures are rising.* Since the "Little Ice Age" in the 17th century, temperatures have risen, and much more dramatically since the Industrial Revolution. Other than a slight cooling trend in the 1960s and 1970s, global mean temperatures have been trending upward – about 0.7 °C since the mid-19th century. This trend has been observed from sea surface temperatures (buoys, ships, satellites), as well as thousands of meteorological monitoring stations around the world. Figure 3.29 shows the temperature trend for the past 1000 years. The below average

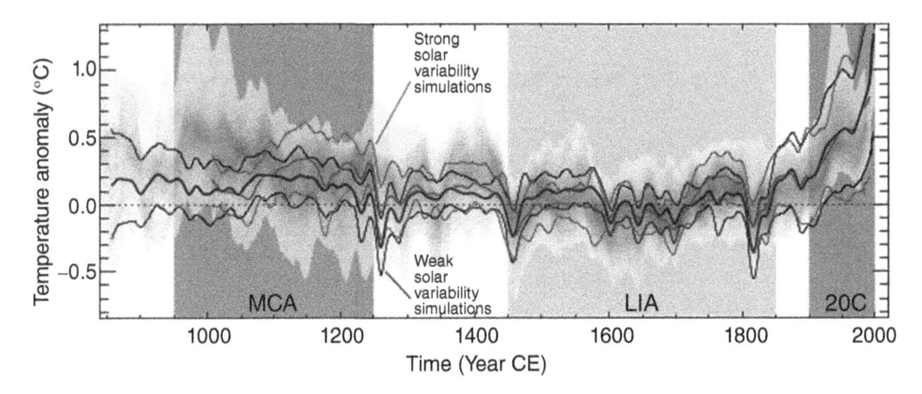

Figure 3.29 Temperature reconstruction over the last millennium, highlighting the warmer Medieval Climate Anomaly (MCA), the cooler Little Ice Age (LIA), and the 20th century (20C). Source: Reproduced with permission of IPCC.

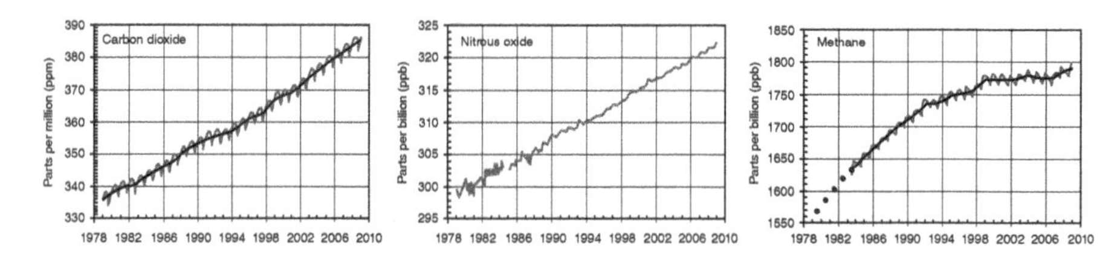

Figure 3.30 Concentration trends for the three most important greenhouse gases. The data was obtained from the National Oceanic and Atmospheric Administration. Source: Data from NOAA ESRL Global Monitoring Division, Boulder, Colorado, USA (http://esrl.noaa.gov/gmd/).

temperatures that persisted from about 1400 to 1900 were followed by a dramatic, although small in absolute terms, rise in temperatures since the Industrial Revolution. Coincidence?

2. *GHG concentrations are on the rise.* Monitoring stations around the world have shown increasing levels of GHGs, the three most important of which are CO_2, CH_4, and N_2O. Since these gases absorb IR radiation emitted by the planet's surface, their increasing concentrations must contribute, in theory, to increasing tropospheric temperatures. Figure 3.30 shows concentration trends for the most important GHGs over the past 40 years.

There is also strong evidence that the current levels of GHGs are unprecedented in the most recent 650 kyrs. Figure 3.31 shows measurements of GHGs taken from tiny gas bubbles trapped in ice cores. While these values only represent concentration over Antarctica, they provide global trends. The δD value represents the changing ratio of heavy water (D_2O, where D is deuterium) to light water (H_2O), which is proportional to changes in temperature. Such evidence is further justification for calling the current geological period, the Anthropocene.

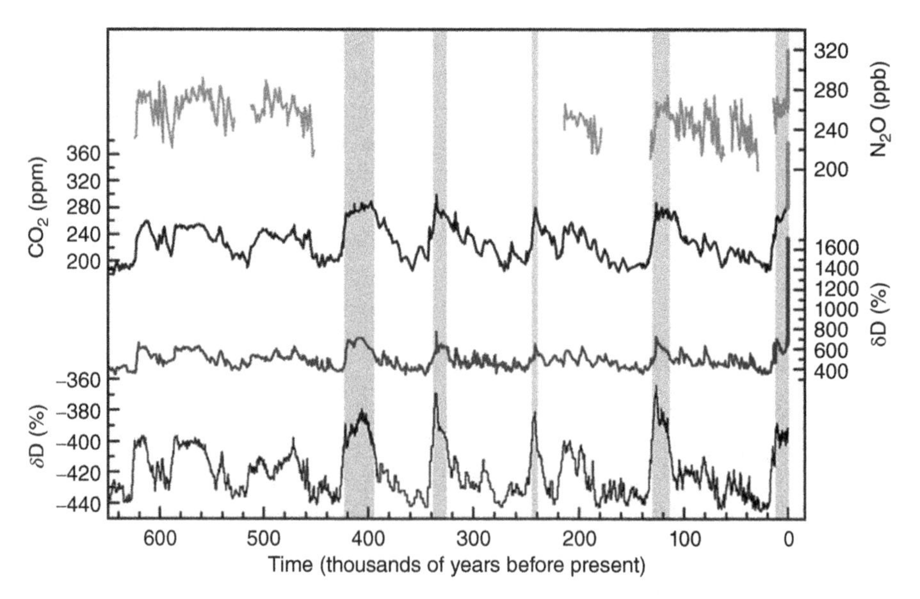

Figure 3.31 Variations of deuterium (δD) in Antarctic ice, which are a proxy for local temperature, and the atmospheric concentrations of the greenhouse gases carbon dioxide (CO_2), methane (CH_4), and nitrous oxide (N_2O) in the air trapped within the ice cores and from recent atmospheric measurements. Data cover 650,000 years, and the shaded bands indicate the current and previous interglacial warm periods. Source: Reproduced with permission of IPCC.

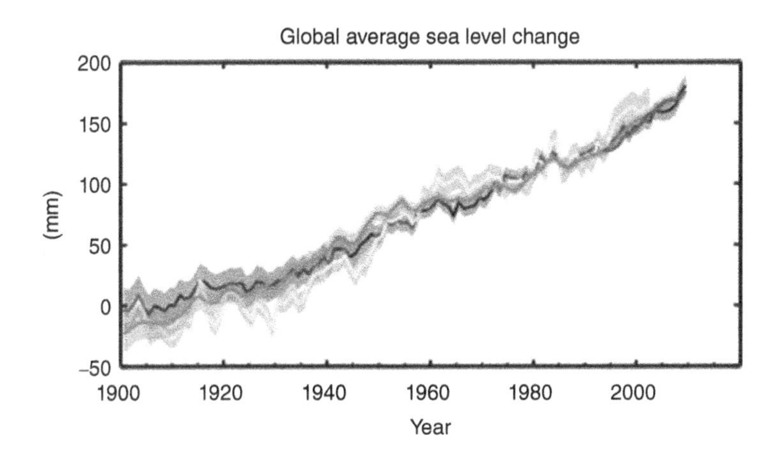

Figure 3.32 Sea level rise over the past 20 years has been steady according to a few different measuring techniques. Source: Reproduced with permission of IPCC.

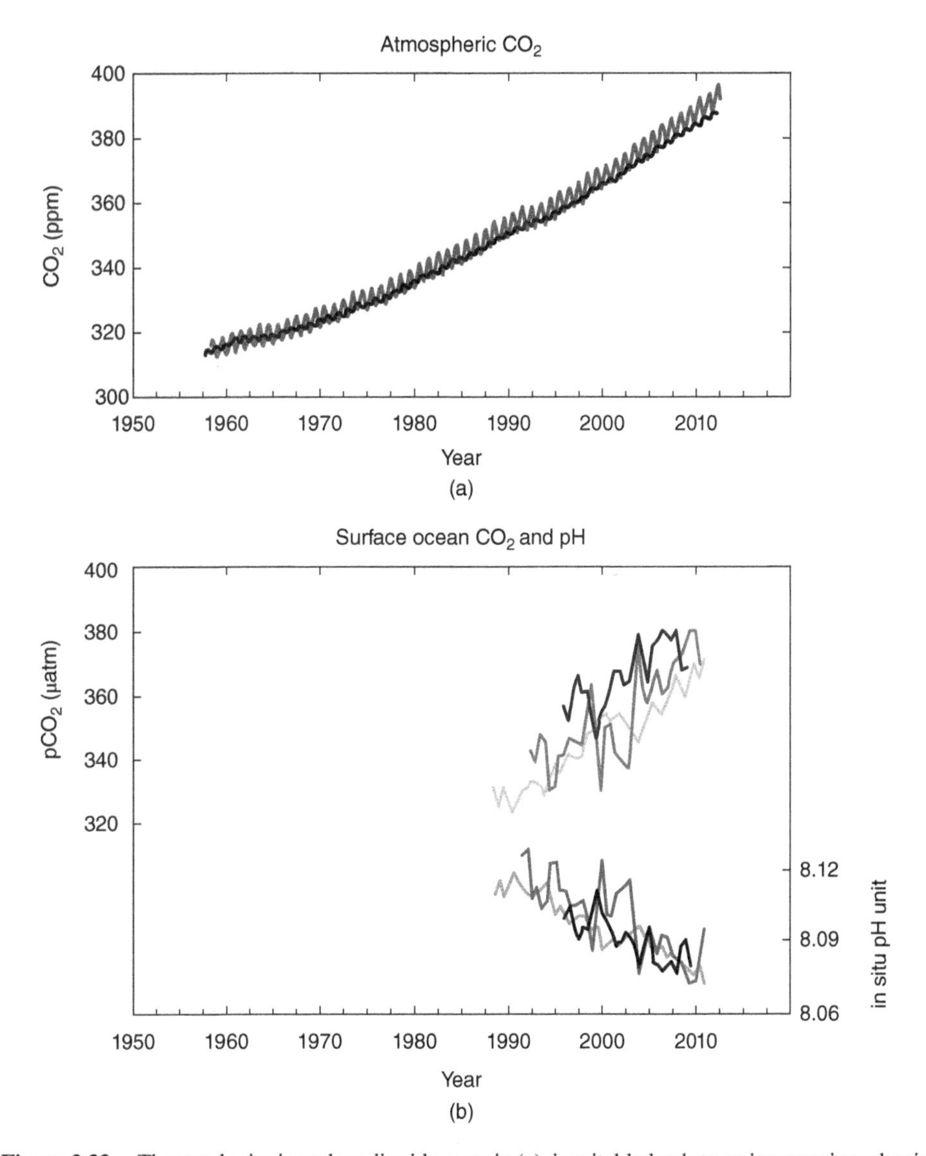

Figure 3.33 The steady rise in carbon dioxide, seen in (a), inevitably leads to an increase in carbonic acid and a decrease in pH as more of the atmospheric CO_2 is dissolved in the oceans, seen in (b). Source: Reproduced with permission of IPCC. Dorea *et al.* (2009).

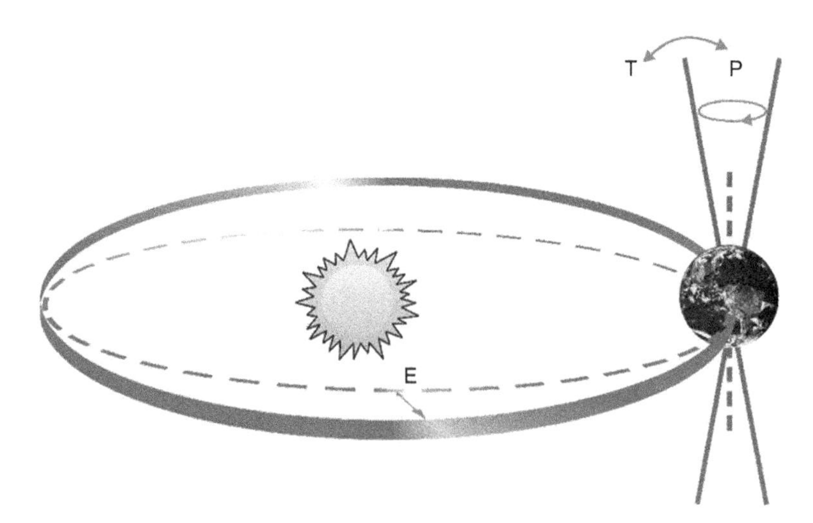

Figure 3.34 Milankovitch Cycles encompass several different irregular patterns in the orbit of the Earth. These irregularities result in predictable changes in the distance between the Earth and the Sun, which lead to changes in climate. The E cycles occur every 100,000 and 400,000 years. The T cycle and P cycle occur at intervals of 41,000 and 20,000 years, respectively. Source: Reproduced with permission of IPCC.

3. *Sea levels are rising.* While sea level rise is not uniform across the globe, satellite-based measurements show an increase of about 3.4 mm/year from 1993 to 2009, obvious from Figure 3.32. Much of the sea level rise is due to the expansion of water with rising global temperatures. It is unfortunate that Figure 3.32 covers such a short time period, but reliable sea level measurements have only become available recently.

4. *Glaciers are receding.* Most glaciers have lost significant mass in the past several decades.

5. *Oceans are acidifying.* It is inevitable that as carbon dioxide levels rise, there will be a concomitant increase in the amount of carbonic acid in the oceans due to dissolved CO_2. This decreases the pH, as shown in Figure 3.33. This area of anthropogenic change has not been extensively studied, but one recent study (Reyes-Nivia *et al.*, 2013) suggests that the "do nothing" scenario for energy policy would lead to increased dissolution of coral reefs.

6. *Milankovitch cycles predict a gradual cooling.* The Earth moves about the Sun with cyclical eccentricities that deviate from a perfect elliptical orbit. These cycles, depicted in Figure 3.34, are collectively called Milankovitch cycles. During each cycle, the amount of solar radiation changes and thus changes the global climate on the Earth.

   You can see the footprint of these cycles when looking at the temperature record taken from Antarctic ice core data that reflect the global climate from the present to 800,000 years before the present (see Figure 3.35) The temperature data comes from Luethi *et al.* (2008), the carbon dioxide data come from Luethi *et al.* (2008), the methane data come from Loulergue *et al.* (2008), and the dust data come from Lambert *et al.* (2008). This figure is similar to Figure 3.31, except an attempt was made to point out the Milankovitch cycles in the temperature record.

   The bottom trace represents a change from the average temperature of the current age (so a negative ΔT means a temperature cooler than the current average, not a temperature below 0 °C). You can see that there have been four periods in the past 800,000 years that have been warmer than the present (the most recent warm period

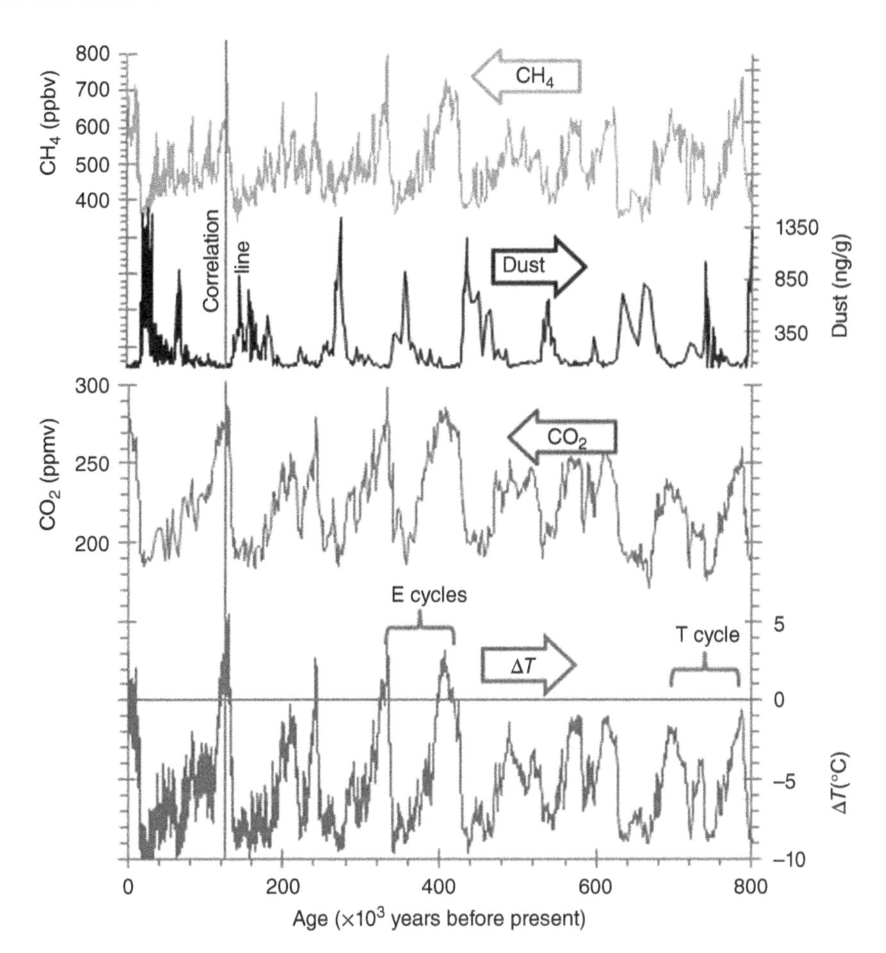

Figure 3.35 Data obtained from the DOME C Antarctic ice cores reveal a very dynamic climate over the past 800,000 years. The traces show (starting from the top) methane, dust, carbon dioxide, and temperature levels. The temperature scale is relative to the current average temperature and the others are absolute scales. Source: Data from NOAA ESRL Global Monitoring Division, Boulder, Colorado, USA (http://esrl.noaa.gov/gmd/).

was 5 °C warmer than present). These line up with the E cycles of the Milankovitch cycles. There are smaller cycles that are about 40,000 years apart and could represent the T cycles. The next three traces are measures of the carbon dioxide, dust, and methane levels in bubbles (CO_2 and CH_4) or dirt trapped in the ice core. Examine the "correlation line" on the graph. You can see that temperature, CO_2, and CH_4 are positively correlated, while these are negatively correlated with dust levels. This should make sense since methane and carbon dioxide are GHGs and would, in theory, trap more heat in the troposphere while dust in the atmosphere would block sunlight, increase the Earth's albedo, and reduce global temperatures. A common source of atmospheric dust is volcanic eruptions. The dust spikes could represent major eruptions, which sent the climate into a cold period. A modern example of this was Mt. Pinatubo in the Philippines, which erupted in 1991 and depressed global temperatures for about 2 years.

Examining the previous Milankovitch E-cycle shows a gradual cooling after the temperature maximum, but this has a resolution no better than thousands of years. Thus, these cycles do not have a significant impact on climate on a timescale that would concern humans (decades to centuries). In fact, the maximum of the most recent E-cycle was 10,000 years ago in the Holocene, so presumably the Earth should be on a slow and uncertain cooling trend. Is the documented warming trend in the past century being attenuated by an E-cycle cooling? Would it have been worse without the Milankovitch cycles?

Given these certainties, there have been informed and uninformed claims in the media about specific weather, drought, and the frequency of cyclones. Some of these uncertainties can be stated directly.

1. *The connection between frequency of hurricanes and climate change has not been established.* Hurricane Katrina captured the attention of Americans for several years after 2005, and recent hurricanes to hit the north-east United States (Hurricane Irene and Sandy) have extended this focus. Is hurricane frequency increasing because of rising temperatures? Data from the National Hurricane Center, seen in Figure 3.36a), shows only those storms that have reached the status of a hurricane (>75 mph winds). Reliable records of wind speeds have only been available via satellite measurements since the early 1960s, even though aircraft measurements were made as far back as the 1940s. You can see from the graph below that on a multidecadal time scale, there appears to be no strong trend for hurricane frequency, although there is an upward slope that is statistically significant. In 2005, the year of Hurricane Katrina, there were 15 named hurricanes – an aberration in the overall trend.

2. *The intensity of hurricanes appears to be increasing.* Dr. Kerry Emanuel, a climatologist from the Massachusetts Institute of Technology and arguably one of the most influential hurricane researchers in the U.S., has reported a trend describing an increase in the power dissipation of hurricanes. The power dissipation is a function of the cube of the wind velocity, so small changes in average wind are magnified. While there is much variability, the correlation between the power dissipation and the sea surface temperature shows an increase over time. When only the maximum wind speed is examined, without taking into account the other factors in Emanuel's power dissipation calculation, no trend in hurricane strength is obvious as a function of time, as seen in Figure 3.36b).

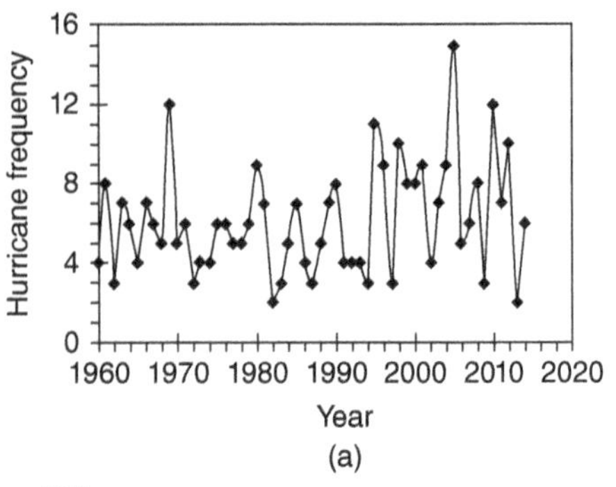

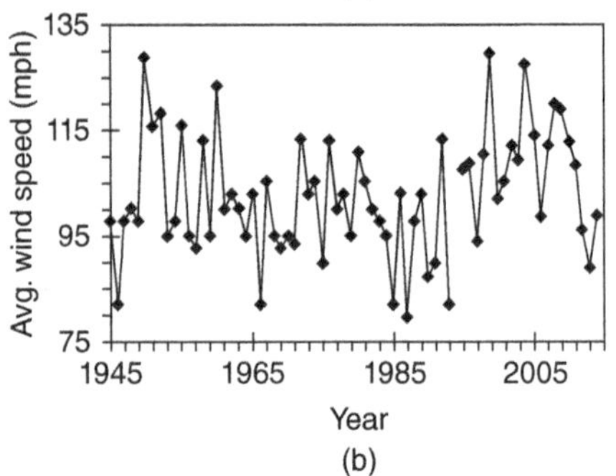

Figure 3.36 Graph (a) shows only the number of named hurricanes (Frequency) as a function of year. The slope of the data is positive (increasing frequency) but does not rise above 3σ. Graph (b) shows the average maximum wind speed for all of the hurricanes in a given year. A trend here is also hard to see. Source: Data from National Hurricane centre.

The complication in the ice core graph is that the correlation between temperature and carbon dioxide can be explained in both ways. If something causes the global temperature to increase, CO_2 will become less soluble in the warmer oceans, which will result in higher atmospheric CO_2 levels. Also, higher CO_2 levels means more heat trapped in the troposphere, which means global temperatures will increase. This relationship represents a *positive feedback cycle* with cause and effect reinforcing each other. One of the major difficulties with historical records is that time resolution can range from 20 to 500 years. Researchers (Marcott *et al.*, 2013; Shakun *et al.*, 2012) report that a rise in carbon dioxide *preceded* a global temperature rise at the end of the last deglaciation some 16,000 years ago. While the relationship between CO_2 levels and temperature is part of a positive feedback cycle, their analysis of the temperature record shows that temperature changes lag CO_2 level changes 90% of the time.

☞ A *positive feedback cycle* is a cyclic relationship that results in an amplification of the cause and effect. Ocean temperatures and CO_2 levels in the atmosphere represent this kind of feedback cycle.

☞ A *negative feedback cycle* is a cyclic relationship that results in an attenuation of the cause and effect. The relationship between temperature and cloud formation represents a negative feedback: when temperatures increase, there is more water evaporation, which leads to more clouds, which reflect sunlight and lead to cooler temperatures.

Example 3.25: Feedback Cycles

What are some other examples of positive and negative feedback cycles?

Solution:

Negative: As temperatures increase, the vegetative growing season lengthens, which means that plants and trees grow more, plants absorb CO_2, so this could lead to lower CO_2 levels, which would lead to less trapped heat in the troposphere, which would lead to lower temperatures.

Positive: As temperatures increase, the permafrost in the Arctic tundra will thaw; the permafrost contains trapped methane, and it would be released to the atmosphere; methane is a GHG, so it would trap more heat, which would lead to higher temperatures.

Milankovitch cycles (and other natural events such as volcanic eruptions and intensity cycles of the Sun) do not exclude human influence on the climate, and they only point to other factors that have influenced the climate in the past. The IPCC uses the following analogy as a reason for continued concern over anthropogenic influence over the climate (IPCC, 2007b).

> By analogy, the fact that forest fires have long been caused naturally by lightning strikes does not mean that fires cannot also be caused by a careless camper.

The evidence that the uncontroversial facts of climate change are caused by anthropogenic pollution comes from Global Circulation Models (GCMs), which demonstrate that natural forcings alone are not sufficient to fit the observed warming trends (see Figure 3.37).

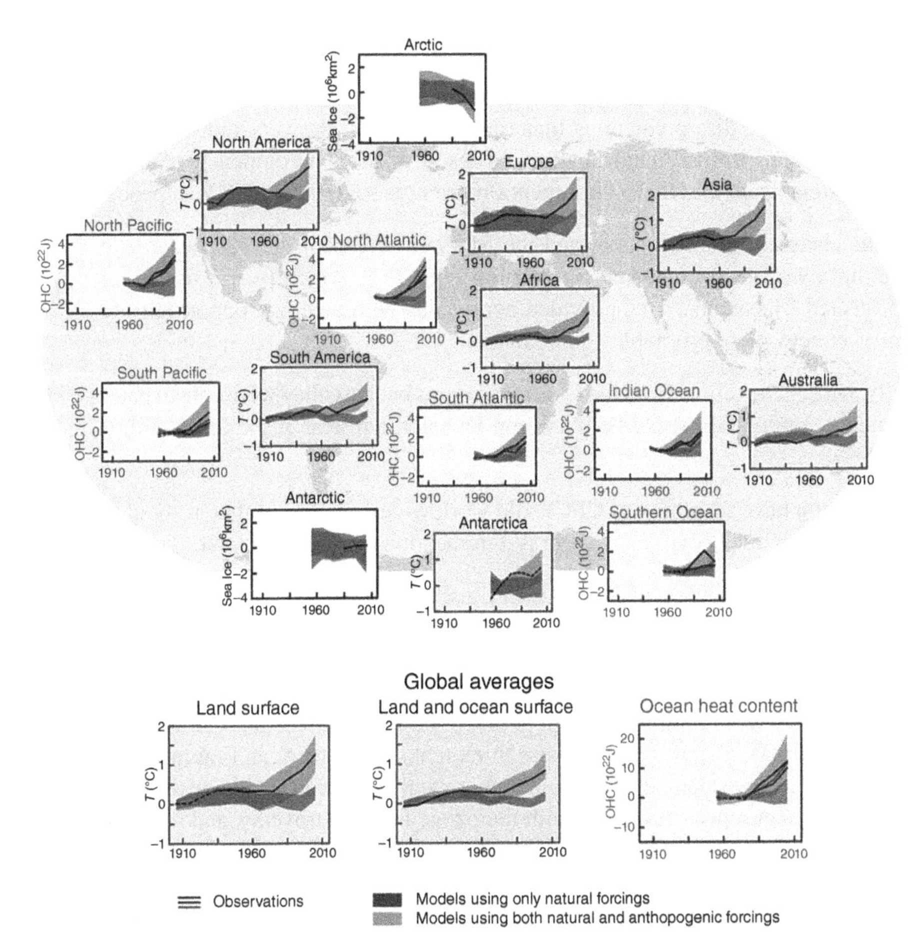

Figure 3.37 Given the temperature trends observed in various parts of the globe (black line), GCMs are unable to model the data using only natural forcings (darker shading). When anthropogenic forcings (such as anthropogenic greenhouse gas emissions) are included in the calculation, then GCMs are able to model the data sufficiently (lighter shading). Source: Reproduced with permission of IPCC.

☞ *The Precautionary Principle* is an approach that shifts the burden of proof to those who would argue that no harm is done. Thus, if a course of action could have widespread and significant deleterious effects, then the best option would be to change course in order to avoid harm, unless it can be sufficiently proven that no harm will be done.

It is impossible to unequivocally demonstrate that anthropogenic forcings are the cause of the observed climate change because there are thousands of factors and no way to run a controlled experiment where climate is observed in the absence of human influence.

Many have appealed to *The Precautionary Principle* and say we should change our national and global policies on the use of fossil fuels and energy consumption. Others fear that our national competitiveness in the global marketplace will be harmed and our economy will slow if the United States adopts GHG controls, such as an energy or carbon tax. Others deny any anthropogenic causes. The 2015 Paris climate change agreement pitted these two political forces against each other as the United States has been trying to implement the provisions of the agreement.

Strangely, statements being made by those denying anthropogenic climate change are reminiscent of the "controversy" over the destruction of the ozone by CFCs (Masters, 2007). In 1975, the chairman of DuPont, under criticism that the CFCs produced by DuPont were destroying ozone, said that the ozone depletion theory was "a science fiction tale...a load of rubbish...utter nonsense." This was after Rowland and Molina reported their findings on the catalytic mechanism of CFCs. In 1979, a DuPont official stated that,

> No ozone depletion has ever been detected...all ozone depletion figures to date are based on a series of uncertain projections.

DuPont further warned that if CFC production were curtailed "entire industries could fold." The Association of European Chemical Companies cautioned that regulations would affect inflation and employment nationally and internationally.

None of these dire predictions materialized. In fact, the very industries that were warning of disaster profited handsomely as a result of the environmental regulations on CFCs. An official from the UN Environment Programme said, in 1990, (Greenpeace, 1997)

> ...the chemical industry supported the Montreal Protocol in 1987 because it set up a worldwide schedule for phasing out CFCs, which [were] no longer protected by patents. This provided companies with an equal opportunity to market new, more profitable compounds.

Ironically, DuPont benefited from the CFC regulations because they were able to patent CFC replacements. Joseph Glass, Du Pont's Freon Division Director at the time, said, (Greenpeace, 1997)

> When you have $3 billion of CFCs sold worldwide and 70% of that is about to be regulated out of existence, there is a tremendous market potential.

A more recent example of corporate intrigue is the case of Exxon-Mobil and its complicated public and private positions on climate change. A 2015 investigation by the New York Attorney General revealed that for decades Exxon-Mobil was using climate change predictions developed by its corporate scientists to make decisions about Arctic drilling while the company was funding nonprofit public policy think tanks to promote doubt about the science of climate change. One of these think tanks, the American Enterprise Institute, offered $10,000 to any scientist or economist that would publically criticize the findings of an IPCC assessment report. As it was with the ozone hole controversy, and reminiscent of the corporate denial of the link between smoking and lung cancer, corporate interests can effect public policy in complicated ways.

3.14 MEASUREMENTS OF ATMOSPHERIC CONSTITUENTS

In your experience with chemistry and other sciences thus far, you have likely been asked to make measurements or observations about samples that are in the lab. Most chemistry labs require the measurement of items that you can either make or test under controlled circumstances. Measurements of atmospheric species are quite difficult in comparison for the following reasons (think "three R's").

1. **Remote:** Unless the sample of interest is at ground level, most samples of interest lie far out of reach and possibly in a different layer of the atmosphere that could be miles away. These gases are only accessible by aircraft (troposphere or stratosphere)

or satellites. Satellite-and ground-based techniques use optics to measure a column of gas directly above (ground-based) or below (satellite-based) the instrument. This column of gas can be many kilometers thick. For obvious reasons, unless you have access to such sophisticated and expensive equipment, you are not likely to be able to make measurements of most of the gases discussed in this chapter.

2. **Reactive:** The gases that are often of interest are reactive species, such as ozone, NO_x, SO_x, or atomic oxygen. These samples would likely not survive long enough for you to capture them outside and bring them into the lab. Because they are so remote and require an aircraft to get a sample, it is easier to bring the measurement equipment to the sample than to bring the sample to the lab. It is also very easy to contaminate gases since they can never be stored in open containers similar to liquids or solids. Transferring the sample from a sealed container to a measuring device also requires special equipment involving vacuum pumps, valves, and transfer lines.

3. **Rare:** Nearly all of the interesting atmospheric gases are very low in concentration. This means that the instrumentation must be very sensitive to detect gases at the ppmv level or smaller. Most of the measurements of atmospheric gases use the interaction of light and the target gas, so often Beer's Law ($A = \epsilon bC$) is used. If the concentration (C) of the analyte gas is low, then the most obvious way to increase the magnitude of the signal so that it is measurable is to increase the path length (b). Satellite- and ground-based measurements often measure gases in a column of air that can be over 100 km long. This requires expensive optics and leaves the measurement sensitive to errors that can be caused by heterogeneities in the sample, such as changes in humidity, pressure, temperature, and composition across several kilometers of sample.

Conclusion: Atmospheric measurements are difficult!!

Some of the methods mentioned here are impossible for you to measure in an undergraduate setting because the cost of the specialized equipment is prohibitive. It is worth knowing these methods in some detail along with the instrumentation you might be using in laboratory for this course.

3.14.1 Satellite-Based Measurements

The focus in the first half of this chapter was on the absorption of solar radiation of gases. Gases also emit specific photons when they are placed into an excited state either through reactions, heat or absorption. Either of these two phenomena can be used to measure the presence and concentration of gases in the atmosphere.

Satellite spectrometers can use light generated directly from the Sun or reflected off the Moon, Earth's surface, or the atmosphere to measure the concentration of gases. In a backscatter mode, called nadir mode, the spectrometer looks at solar radiation that is scattered by gases in the atmosphere. The spectrometer uses one or more wavelengths where the analyte gas absorbs strongly and one or more where the analyte does not absorb strongly. The wavelength that is not absorbed by the analyte is used to determine the blank since it represents the amount of light backscattered from the Sun assuming no analyte is present (no absorption). The absorbed wavelength should be less intense because, while also being backscattered to some extent, some of this radiation is absorbed by the analyte. The difference between the two intensities can be used to calculate the concentration of the analyte gas.

In the case of ozone, for example, the spectrometer might measure across a wavelength range of 300 to 340 nm. Ozone absorbs light at 300 nm very well (look back to Figure 3.19 on page 121) and absorbs near 340 nm poorly. In the nadir mode, the satellite can look straight down through all of the layers of the atmosphere and take a measurement of the light intensity at 340 nm (the blank), and then it compares this to the light levels at 300, 310, 320, and 330 nm for example. We know from Section 3.5 that shorter wavelengths (such as 300 nm) are absorbed at higher altitudes and longer wavelengths (such as 320 nm) are able to penetrate deeper into the atmosphere (lower stratosphere or upper troposphere) before they are absorbed. This means that the spectrometer can use the 300 nm measurement to measure the ozone concentration at the top or middle of the stratosphere and the 320 nm measurement to measure ozone at the bottom of the stratosphere.

In a different mode, commonly called limb mode, the spectrometer looks at the emission from the analyte gas after it has been excited by solar radiation. Ultraviolet radiation excites the gas molecules, some of which eventually get rid of this energy by emitting photons at a lower wavelength (others convert the energy to heat). For ozone and other species, the emission is in the microwave region. The spectrometer can measure at different angles, which translates to different depths in the atmosphere. This mode can also measure backscattering. The spectrometer would compare intensities of absorbed and reference wavelengths that are generated by the Sun and backscattered. Because this technique scans the atmosphere at an angle in either mode, it yields a better vertical resolution compared to the backscattering mode.

In a third mode commonly called occultation, the satellite positions itself such that the atmosphere is between the spectrometer and a light source, such as the Sun or Moon. This mode is similarly to watching the Sun set in the evening. From the satellite's perspective, it can track the light of a star as it moves from above the horizon of the atmosphere, then into the atmosphere, and finally below the horizon. As the star or other object dips into the atmosphere, portions of its spectrum begin to be absorbed by the gases. The angle of the star's descent is related to its depth in the atmosphere. If the spectrometer is set to a characteristic absorption of ozone, it can set its blank measurement when the star is high above the atmosphere (no absorption) and then compare this to measurements of the same wavelength as the star dips into the atmosphere. The intensity should weaken as the light travels through more atmosphere and more ozone. In this manner, the measurement can estimate the ozone levels as a function of altitude.

3.14.2 Ground-Based Measurements

3.14.2.1 ***LIDAR*** Common ground-based measurements involve measuring backscatter, similarly to the nadir satellite method. Powerful lasers are shot straight up into the atmosphere. One wavelength that is strongly absorbed is compared to another that is not absorbed (the reference or blank). A large telescope focuses on a certain layer of the atmosphere and measures the backscatter intensity, comparing the absorbed wavelength to the reference beam. For ozone, these wavelengths are typically 308 nm (absorbed) and 351 nm (reference).

3.14.2.2 ***Cavity Ring-Down Spectroscopy*** Cavity Ring-Down Spectroscopy is a method used to facilitate small-scale monitoring of various GHGs. These systems work by placing a sample gas inside of a chamber that is surrounded by mirrors. Laser light is introduced into the chamber. The laser reflects off the mirrored surfaces repeatedly, with only a small amount leaking out of one of the mirrors and to a detector. Once the source laser is turned off, any empty chamber will cause the light intensity to decay over time as the laser pulse slowly loses intensity. If there is a gas in the chamber that absorbs the laser light, then the intensity will decay much faster. A comparison of the "ring-down" time between the sample and the reference (empty chamber) is proportional to the concentration of the gas. Table 3.23 lists the types of gases that can be measured and their detection limits. These spectrometers were priced between $50k and $150k at the date of this printing.

Gas	λ (μm)	LOD (ppbv)
CH_4	1.65	52
CO_2	1.57	2500
CO	1.57	2000
NH_3	1.5	19
NO	5.2	0.7
NO_2	0.410	0.4
NO_3	0.662	0.002
N_2O_5	0.662	0.0012
OH	0.309	ppmv

(Data from the EPA)

Table 3.23 Gases that can be measured using a Cavity Ring-Down Spectroscopy, along with the characteristic wavelength that gets absorbed by each gas, and the Limit Of Detection (LOD).

3.14.3 Ambient Monitoring

Other small-scale and less expensive option for benchtop or remote monitoring are ambient monitoring instruments that specialize in a specific gas. Each of these systems usually contains an internal calibration mechanism, remote control, and data transfer capabilities to allow these boxes to be placed on a lab bench or at a remote monitoring station. Each instrument uses a characteristic absorption wavelength (UV or IR) or chemiluminescence for NO_x and SO_x.[4] Table 3.24 contains a list of the gases that can be measured along with their detection limits. These spectrometers were priced between \$9k and \$23k at the date of this printing.

Gas	LOD (ppbv)	Range (ppmv)
CH_4	50	0.05–5000
CO	40	(0.04–10,000)
H_2S	1.5	1.5–10
NH_3	1.0	0.05–20
NO_x	0.02	0.02–50
O_3	1.0	0.001–200
SO_2	0.05	0.05–1000

They also can be interfaced with the Internet to provide remote control operation.

Table 3.24 Small-scale ambient spectrometers are available to quantitate gases in the lab or atop monitoring stations. These instruments have comparable detection limits to the satellite and ground-based measurements.

3.14.4 Infrared Spectroscopy

The equipment available at your institution will limit you to the measurement of certain atmospheric species. Nearly all undergraduate institutions have a Fourier Transform InfraRed (FTIR) spectrometer, and many of the gases you might be interested in measuring are GHGs. These gases absorb in the IR region of the spectrum, making IR spectroscopy an appropriate measuring technique. These instruments are meant for use in an organic chemistry lab and, unfortunately, have detection limits on the order of parts per thousand. A more thorough description of this instrument can be found in Appendix F on page 304.

3.15 IMPORTANT TERMS

- ★ troposphere
- ★ stratosphere
- ★ mesosphere
- ★ thermosphere
- ★ exosphere
- ★ tropopause
- ★ stratopause
- ★ mesopause
- ★ comparisons to the Venusian atmosphere
- ★ photolytic reactions
- ★ chaperone molecules
- ★ actinic flux
- ★ lifetime (τ)
- ★ UV spectrum
  - · UV-A
  - · UV-B
  - · UV-C
- ★ ozone
  - · ground-level production
  - · stratospheric
  - · the Chapman Cycle
  - · catalytic destruction
  - · CFCs, halons
  - · HCFCs, HFCs
- ★ Earth's energy budget
  - · albedo
  - · solar irradiance
  - · solar constant
- ★ the Greenhouse Effect
  - · greenhouse gases
  - · vibrational modes
  - · oscillating dipole moment
  - · IR-active modes
  - · IR-inactive modes
- ★ engine types
  - · four-stroke
  - · two-stroke
  - · diesel
  - · emission comparison
  - · fuel additives and alternatives
    - ▪ tetraethyl lead
    - ▪ MTBE
    - ▪ ethanol
- ★ smog
  - · classical
  - · photochemical
- ★ the hydroxyl radical
  - · reaction with
    - ▪ alkanes, alkenes
    - ▪ terpenes
    - ▪ sulfur compounds
    - ▪ nitrogen compounds
- ★ free radicals
- ★ homolytic cleavage
- ★ heterolytic cleavage
- ★ thermal inversion

[4]Chemiluminescence is the phenomenon where a chemical reaction results in an excited-state molecule that emits light. In the case of the NO_x detector, NO_x is converted to NO and reacted with O_3 generated to form NO_2, O_2, and a 1200 nm photon of light. In the case of SO_x, SO_x is converted to SO and reacted with O_3 to form SO_2, O_2 and a 360 nm photon.

- ★ acid deposition
- ★ pH of natural rain
- ★ point source
- ★ nonpoint source
- ★ coalification
- ★ Donora, PA, smog event
- ★ London smog event
- ★ Antarctic Ozone Hole
  - · polar vortex
  - · polar stratospheric clouds
  - · nitric acid trihydrates
  - · polar vortex
  - · Montreal Protocol
- ★ global-warming potential
- ★ ozone-destruction potential
- ★ Climate Change
  - · natural forcings
  - · anthropogenic forcings
  - · radiative forcings
    - ■ concentration
    - ■ absorption
    - ■ spectral window
    - ■ atmospheric lifetime
  - · signs of climate change
    - ■ global temperatures
    - ■ sea level
    - ■ ocean acidification
    - ■ receding glaciers
    - ■ Milankovitch cycles
    - ■ ice cores
  - · positive feedback cycles
  - · negative feedback cycles
  - · the Precautionary Principle
- ★ measurements
  - · nadir
  - · limbs
  - · occultation
  - · LIDAR
  - · cavity ring-down spectroscopy.
  - · chemiluminescence
  - · IR spectroscopy

EXERCISES

1. Why does the thermosphere have the temperature trend that it has? What range of electromagnetic radiation and what gases are responsible for this trend?
2. Why does the mesosphere have the temperature trend that it has? What range of electromagnetic radiation and what gases are responsible for this trend?
3. Why does the stratosphere have the temperature trend that it has? What range of electromagnetic radiation and what gases are responsible for this trend?
4. Why does the troposphere have the temperature trend that it has? What range of electromagnetic radiation and what gases are responsible for this trend?
5. Describe the role that a chaperone molecules plays in atmospheric reactions. Give an example.
6. Summarize the key reason why ozone persists at significant levels in the stratosphere and not the mesosphere or thermosphere.
7. Describe, in your own words, what a species lifetime (represented by τ) says about a species or related reaction.
8. Write the four reactions of the Chapman Cycle.
9. Describe the reactions in the Chapman Cycle. What are the two main species that are formed and destroyed?
10. Describe the usefulness and limitation of using the enthalpy (ΔH) of an endothermic reaction to estimate the photon energy necessary to initiate a reaction.
11. Derive the lifetime formulas for the following reactions.

    (a) the lifetime of NO in $NO + O_3 \xrightarrow{k} NO_2 + O_2$

(b) the lifetime of O_3 in $NO_2 + O_3 \xrightarrow{k} NO_3 + O_2$

(c) the lifetime of NO_2 in $NO_2 + OH + M \xrightarrow{k} HNO_3 + M$

12. Calculate the lifetime of atomic oxygen in Reaction R3.21 at an altitude of 90 km. Refer to the values in Table 3.6.

13. Calculate the lifetime of formaldehyde against OH oxidation at 15 °C assuming that the concentration of OH is $3.67 \times 10^6 \frac{mcs}{cm^3}$. Convert the lifetime to days or years.

14. What does the albedo describe?

15. If the Earth were covered entirely by deciduous trees (and no clouds), would the albedo increase or decrease? Would this lead to an increase or a decrease in the Earth's surface temperature?

16. If the Earth were covered entirely by snow (and no clouds), would the albedo increase or decrease? Would this lead to an increase or decrease in the Earth's surface temperature? Calculate the expected temperature of the Earth using the albedo of snow from Table 3.7.

17. Ignoring any greenhouse effect, calculate the maximum distance from our Sun that a planet can orbit while barely maintaining a global surface layer of snow. Compare this distance to the orbits of the inner planets and draw a conclusion about the necessity of a greenhouse effect.

18. Would the solar constant (E_{sc}) be larger or smaller if the Earth's orbit was closer to the Sun? If the orbit decreased by a factor of 2, how would the solar constant change?

19. If our Sun were replaced by a small, white dwarf star (smaller diameter, all else being the same) and the Earth had the same orbit, how would it affect the climate on the Earth?

20. Why do two-stroke and four-stroke gasoline engines differ on the a) HC and b) NO_x emissions?

21. Why do four-stroke gasoline and diesel engines differ on the a) PM and b) NO_x emissions?

22. Why do fuel additives, such as ethanol and MTBE, and biodiesel reduce CO and PM emissions?

23. How could you design a car with an internal combustion engine, without worrying about expense, so that it would have zero NO_x emissions?

24. Methane and nitrous oxide emissions from anthropogenic sources are on the increase. Explain how each affects ozone chemistry and the greenhouse effect.

25. Describe the anthropogenic pollutants that ozone destruction and the enhanced greenhouse effect have in common. Describe how the chemicals act in each context.

26. Knowing that the following reaction occurs

$$H_2O + N_2O_5 \rightarrow 2\ HNO_3$$

draw the Lewis structure for N_2O_5 next to a water molecule in such a way that you can clearly circle the two HNO_3 molecules that result from the collision. HINT: there is a bridging oxygen atom.

27. Knowing that the following reaction occurs

$$N_2O_4 \rightleftharpoons 2\ NO_2$$

draw the Lewis structure for N_2O_4.

28. Which biogenic species is the least stable against oxidation by the hydroxyl radical at 20 °C – isoprene or α-pinene?

29. Describe the conditions that can overcome kinetic barriers and drive a species to its thermodynamic fate?

30. Propose a mechanism for Reaction R3.33 that uses NO_3 as an intermediate.

31. While nitric acid is very soluble in water, in the gas phase, it can undergo the following reaction.

$$HNO_3 \rightarrow H + NO_3 \qquad (R3.105)$$

Is this reaction likely to occur during the day or night? Give a reason for your answer and explain what reactant is missing from the reaction.

32. Describe the difference between photochemical smog and classical smog. Where can they be found, what are the constituents, and what human activities can cause each?

33. What is the difference between O and O* or O_2, O_2^*, and O_2^{**}? What can happen to these species?

34. Which photolysis reaction will require a photon with the smallest wavelength? Explain.

    (a) $O_3 + h\nu \rightarrow O_2^{**} + O$
    (b) $O_3 + h\nu \rightarrow O_2 + O$

35. Describe a photolytic reaction. How is it different than a radical reaction?

36. How can radical reactions degrade stable species in the absence of light, whereas photolytic degradation of the same stable species requires light?

37. Write the radical reaction mechanism between butane and the hydroxyl radical, which yields carbon dioxide (stop after the first CO_2 formed). How many molecules of ozone would be formed if all of the carbons in butane were to be converted to CO_2?

38. What is the difference between a biogenic, an anthropogenic, and a geologic source?

39. Name one anthropogenic, one geologic, and one biogenic source for sulfur emissions and the specific form of sulfur that is the primary pollutant.

40. Name one anthropogenic, one geologic, and one biogenic source for nitrogen emissions and the specific form of nitrogen that is the primary pollutant.

41. Name one anthropogenic, one geologic, and one biogenic source for HC or VOC and the specific chemical formula of the primary pollutant.

42. What is the pH of a solution that contains 1.43×10^{-5} M HNO_3?

43. If a sample of acid rain has a pH of 4.81, what is the hydronium ion concentration?

44. Determine the pH of natural rain when the average concentration of CO_2 reaches 550 ppmv or 5.50×10^{-4} atm, the level expected in the year 2050 if the industrialized countries follow the "business as usual" model and do not enact a carbon tax or any CO_2 sequestration initiative.

45. Calculate the pH of the rain droplets in a cloud that is exposed to a SO_2 concentration of 2.0 ppbv (at a total pressure of 1.0 atm).

46. How can you recognize a catalyst in the reactions of a mechanism?

47. What does the catalyst do to the rate of the reaction? How does the catalyst manage to do this (discuss the answer in terms of energy)?

48. What are the three major factors that lead to significant ozone loss over the Antarctic pole? Describe their synergy. How is this different from the Arctic pole?

49. GHGs trap heat in the troposphere and result in colder temperatures in the tropopause and lower stratosphere. How might increases in GHG emissions affect ozone destruction over the Arctic pole?

50. Describe one possible positive feedback cycle and one negative feedback cycle related to Climate Change.

51. Barium sulfate is frequently used to enhance X-rays and CAT scans when imaging of the gastrointestinal tract is necessary. Soft tissue is transparent to X-rays, and therefore, usually only bones and other dense tissue are visible. Barium sulfate absorbs X-rays, so when it is ingested as an aqueous suspension, it temporarily coats the inside of the intestines and allows them to be visible in an X-ray photograph. The barium ion, however, is toxic. Barium sulfate is quite insoluble in pure water but becomes more soluble in acidic solutions because the sulfate ion acts as a weak base. Determine the concentration of Ba^{2+} in the human stomach given the HCl concentration is approximately 0.10 M (so the initial $[H_3O^+]$ or $[H^+]$=0.10 M). Consider the K_{sp} of barium sulfate and the first K_b of the sulfate ion (or the second K_a of sulfuric acid). Also, remember that the stomach continually produces acid, so the concentration of acid remains relatively constant despite any reaction that may occur as a result of the barium sulfate. HINT: the resulting reaction should be similar to barium sulfate dissolving in acid, and you may need to use the autodissociation reaction of water in your calculation.

BIBLIOGRAPHY

Ackerman, M. In *Mesospheric Models and Related Experiments*; Fiocco, G., Ed.; D. Reidel Publishing Company: Dordrecht, 1971; pp. 149–159.

Ahlvik, P. *A Comparison of Two Gasoline and Two Diesel Cars with Varying Emission Control Technologies*; 2002; Directions in Engine-Efficiency and Emissions Research (DEER) Conference, US DOE.

Anderson, J. G.; Brune, W. H.; Proffitt, M. H. *Journal of Geophysical Research* **1989**, *94(D9)*, 11465–11479.

Anderson, J. G.; Toohey, D. W.; Brune, W. H. *Science* **1991**, *251(4989)*, 39–46.

Archer, D.; Eby, M.; Brovkin, V.; Ridgwell, A.; *et al. Annual Review of Earth and Planetary Sciences* **2009**, *37*, 117–134.

Baker, K. D.; Nagy, A. F.; Olsen, R. O.; Oran, E. S.; Randhawa, J.; Strobel, D. F.; Tohmatsu, T. *Journal of Geophysical Research* **1977**, *82(22)*, 3281–3286.

Batista, P. P.; Takahashi, H.; Clemesha, B. R.; Llewellyn, E. J. *Journal of Geophysical Research Space Physics* **1996**, *101(A4)*, 7917–7921.

Biddle, B. J. *Comprehending the Climate Crisis*; iUniverse.com, 2011.

Blum, D. *The Poisoner's Handbook*; Penguin Books: New York, 2010.

Bowman, C. T. *Symposium (International) on Combustion* **1973**, *14*, 729–738.

Brion, C.; Tan, K.; van der Wiel, M.; van der Leeuw, P. *Journal of Electron Spectroscopy and Related Phenomena* **1979**, *17*, 101–119.

Bychev, R. M.; Petrova, G. I.; Bychev, M. I. *Journal of Mining Sciences* **2004**, *40(2)*, 210.

Canakci, M.; Gerpen, J. H. V. *Transaction of the American Society of Agricultural Engineers* **2003**, *46(4)*, 937–944.

CBD, Human Population Growth and Climate Change. Electronic, http://www.biologicaldiversity.org/campaigns/overpopulation/climate/.

Chan, W.; Cooper, G.; Brion, C. *Chemical Physics* **1993a**, *178*, 387–401.

Chan, W.; Cooper, G.; Brion, C. *Chemical Physics* **1993b**, *178*, 401–413.

Chan, W.; Cooper, G.; Sodhi, R.; Brion, C. *Chemical Physics* **1993c**, *170*, 81–97.

Chordiya, M. A.; Senthilkumaran, K.; Gangurde, H. H.; Tamizharasi, S. *International Journal of Pharmaceutical Fronteir Research* **2011**, *1(3)*, 96–112.

Christie, M. *Proceedings of the International Commission on History of Meteorology* **2004**, *1.1*, 99–105.

Dorea, J. E.; Lukas, R.; Sadler, D. W.; Church, M. J.; Karl, D. M. *Proceedings of the National Academy of Sciences of the United States of America* **2009**, *106(30)*, 12235–12240.

Dwyer, D. After Backlash, Ethanol Industry Thriving. Electronically, 2012; http://www.npr.org/2012/04/26/151417943/checking-in-on-eurozone-economies.

Ehhalt, D. H. *Physical Chemistry Chemical Physics* **1999**, *1*, 5401–5408.

Emanuel, K. *Nature* **2005**, *436(4)*, 686–688.

Emanuel, K. *Bulletin of the American Meteorological Society* **2008**, *89*, ES10–ES20.

Emanuel, K. *American Meteorological Society* **2008**, *89*, ES10–ES20.

Emmert, J. T.; Stevens, M. H.; Bernath, P. F.; Drob, D. P.; Boone, C. D. *Nature Geoscience* **2012**, *5*, 868–871.

EPA, Effects of Acid Rain - Materials. Electronic, 2012a.

EPA, Environmental Effects of Acid Rain. Electronic, 2012b; http://www.epa.gov/region1/eco/acidrain/enveffects.html.

EPA, History of the Clean Air Act. Electronic, 2012c; http://epa.gov/air/caa/caa_history.html.

EPA, Ozone Layer Protection - Regulatory Programs: HCFC Phaseout Schedule. Electronic, 2013; http://www.epa.gov/ozone/title6/phaseout/hcfc.html.

Friedl, R. R.; Brune, W. H.; Anderson, J. G. *Journal of Physical Chemistry* **1985**, *89*, 5505–5510.

Galloway, J. N.; Knap, A. H.; Church, T. M. *Journal of Geophysical Research* **1983**, *88(C15)*, 10859–10864.

Gillis, J.; Krauss, C. Exxon Mobil Investigated for Possible Climate Change Lies by New York Attorney General. Electronic, 2015; http://www.nytimes.com/2015/11/06/science/exxon-mobil-under-investigation-in-new-york-over-climate-statements.html?_r=0.

Girard, J. E. *Principles of Environmental Chemistry*; Jones & Bartlett Learning: Burlington, MA, 2013.

Giunta, C. J. *Bulletin for the History of Chemistry* **2006**, *2*, 66–74.

Glinton, S. How A Little Lab in West Virginia Caught Volkswagen's Big Cheat. Electronic, 2015; http://www.npr.org/2015/09/24/443053672/how-a-little-lab-in-west-virginia-caught-volkswagens-big-cheat.

Graham, L. A.; Belisle, S. L.; Baas, C.-L. *Atmospheric Environment* **2008**, *42(19)*, 4498–4516.

Greenpeace, DuPont: A Case Study in 3D Corporate Strategy. Electronic, 1997; https://courses.seas.harvard.edu/climate/eli/Courses/EPS281r/Sources/Ozone-hole/more/Greenpeace-on-DuPont.pdf.

Gurvich, L. V.; Veyts, I. V.; Alcock, C. B. *Thermodynamic Properties of Individual Substances*, 4th ed.; Hemisphere Publishing Corp.: New York, 1991; Vol. 2.

Hargreaves, S. Exxon Linked to Climate Change Pay Out. Electronic, 2007; http://money.cnn.com/2007/02/02/news/companies/exxon_science/index.htm?cnn=yes.

Hasekamp, O. P. Ozone profile retrieval from satellite measurements of backscattered sunlight. PhD thesis, Vrije Universiteit Amsterdam, 2002.

Haynes, W. *CRC Handbook of Chemistry and Physics*, 93rd ed.; CRC Handbook of Chemistry and Physics; Taylor & Francis: Boca Raton, FL, 2012.

Hedin, J.; Rapp, M.; Khaplanov, M.; Stegman, J.; Witt, G. *Annales Geophysicae* **2012**, *30*, 1611–1621.

Hegelund, F.; Larsen, R. W.; Nicolaisen, F.; Palmer, M. *Journal of Molecular Spectroscopy* **2005**, *229(2)*, 238–242.

Hendrick, F.; Rozanov, A.; Johnston, P. V.; *et al. Atmospheric Measurement Techniques* **2009**, *2*, 273–285.

Hochschild, G.; Berg, H.; Kopp, G.; Krupa, R.; Kuntz, M. Profiles of stratospheric ClO and O_3 simultaneously retrieved from millimeter wave radiometry at Ny-Alesund (Svalbard) during March 1997. Geoscience and Remote Sensing Symposium Proceedings, IGARSS '98. 1998 IEEE International. July 1998; pp. 2618–2620. Vol. 5.

Iida, Y.; Carnovale, F.; Daviel, S.; Brion, C. *Chemical Physics* **1986**, *105*, 211–225.

Inhofe, S. J. The Greatest Hoax: How the Global Warming Conspiracy Threatens Your Future, 1st ed.; WND: Washington, DC, 2005. ISBN: 1936488493.

IPCC, IPCC Fourth Assessment Report: Climate Change 2007: 2.10.2 Direct Global Warming Potentials. Electronic, 2007a.

IPCC, IPCC Fourth Assessment Report: Climate Change 2007. Electronic, 2007b.

Iverach, D.; Basden, K. S.; Kirov, N. Y. *Symposium (International) on Combustion* **1973**, *14(1)*, 767–775.

Jacobs, D. J. *Introduction to Atmospheric Chemistry*; Princeton University Press: Princeton, NJ, 1999.

Jia, L.-W.; Shen, M.-Q.; Wang, J.; Lin, M.-Q. *Journal of Hazardous Materials* **2005**, *123(1-3)*, 29–34.

Jouzel, J.; Masson-Delmotte, V.; Cattani, O.; *et al. Science* **2007**, *317(5839)*, 793–797.

Jüttner, F.; Backhaus, D.; Matthias, U.; Essers, U.; Greiner, R.; Mahr, B. *Water Research* **1995**, *29(8)*, 1976–1982.

Keller-Rudek, H.; Moortgat, G. K.; Sander, R.; Sorensen, R. *Earth System Science Data* **2013**, *5(2)*, 365–373.

Kesselmeier, J.; Staudt, M. *Journal of Atmospheric Chemistry* **1999**, *33*, 23–88.

Kostsov, V. S.; Timofeyev, Y. M. *Atmospheric and Oceanic Physics* **2003**, *39(3)*, 322–332.

Kulkarni, P.; Bortoli, D.; Costa, M. J.; Silva, A. M.; Ravegnani, F.; Giovanelli, G. Comparison of NO2 vertical profiles from satellite and ground based measurements over Antarctica. General Assembly and Scientific Symposium, XXXth URSI, August 2011; pp. 1–4.

Lambert, F.; Delmonte, B.; Petit, J.; *et al. Nature* **2008**, *452*, 616–619.

Lewis, B.; Carver, J. *Journal of Quantitative Spectroscopy and Radiative Transfer* **1983**, *30*, 287–309.

Lindstroem, O. *Ambio* **1980**, *9*, 309–313.

Lissauer, J. I.; Pater, I. *Fundamental Planetary Science: Physics, Chemistry and Habitability*. Cambridge University Press: New York, **2013**.

Loulergue, L.; Schilt, A.; Spahni, R.; *et al. Nature* **2008**, *453*, 383–386.

Luethi, D.; Floch, M. L.; Bereiter, B.; *et al. Nature* **2008**, *453*, 379–382.

Madronich, S.; Hastie, D R.; Ridley, B. A.; Schiff, H. I. *Journal of Atmospheric Chemistry* **1983**, *1(1)*, 3–25.

Mann, M. E.; Hughes, M. K. *Geophysical Research Letters* **1999**, *26(6)*, 759.

Marcott, S. A.; Shakun, J. D.; Clark, P. U.; Mix, A. C. *Science* **2013**, *339(6124)*, 1198–1201.

Masters, J. M. The Skeptics vs. the Ozone Hole. Electronic, 2007; http://www.wunderground.com/resources/climate/ozone_skeptics.asp.

Matsumi, Y.; Kawasaki, M. *Chemical Reviews* **2003**, *103*, 4767–4781.

Merrill, R.; Mikel, D.; Colby, J.; Footer, T.; Crawford, P.; Alvarez-Aviles, L. In *EPA Handbook: Optical Remote Sensing for Measurement and Monitoring of Emissions Flux*; Mikel, D. K., Merrill, R., Eds.; Environmental Protection Agency, 2011.

Milloy, S. *NASCAR Knocked By Lead-Heads*; FoxNews.com, 2005; http://www.foxnews.com/story/2005/03/04/nascar-knocked-by-lead-heads.html.

Mlynczak, M. G.; Hunt, L. A.; Mast, J. C.; *et al. Journal of Geophysical Research: Atmospheres* **2013**, *118*, 5724–5735.

Monks, P. S. *Chemical Society Reviews* **2005**, *34*, 376–395.

Moore, C. E. *Atomic Energy Levels*; NSRDS: Washington, D.C., 1971; Vol. 1.

Murtaugh, P. Family Planning: A Major Environmental Emphasis. Electronic, 2009; http:

//oregonstate.edu/ua/ncs/archives/2009/jul/family-planning-major-environmental-emphasis.

NADP, Acid Rain. Electronic, 2012; http://nadp.sws.uiuc.edu/educ/acidrain.aspx.

Nee, J. B.; Lee, P. C. *Journal of Physical Chemistry A* **1997**, *101*, 6653–6657.

Needleman, H. *Annual Review of Medicine* **2004**, *55*, 209–0222.

Nerem, R. S.; Chamber, D. P.; Choe, C.; Mitchum, G. T. *Marine Geodesy* **2010**, *33(S1)*, 435–466.

O'Neil, J.; Steele, G.; McNair, C. S.; Matusiak, M. M.; Madlem, J. *Journal of Occupational and Environmental Hygiene* **2006**, *3(2)*, 67–71.

Ogawa, M.; Cook, G. *Chemical Physics* **1958**, *28*, 173–174.

Parrish, D. D.; Murphy, P. C.; Albritton, D. L.; Fehsenfeld, F. C. *Atmospheric Environment* **1967**, *17(7)*, 1365–1379.

Ragland, K. W.; Aerts, D. J.; Baker, A. J. *Bioresource Teachnology* **1991**, *37*, 161–168.

Reddy, M. M.; Leenheerm, J. A.; Malcolm, R. L. Chapter 1: Elemental Analysis and Heat of Combustion of Fulvic Acid from Suwannee River, In *Assessment of Hydrological Conditions with Emphasis on Water Quality and Wastewater Injection, Southwest Sarasota and West Charlotte Counties, Florida*; Hutchinson, C. B., Ed.; USGS, 1992; p. 81.

Reisel, J. R.; Carter, C. D.; Laurendeau, N. M. *Energy and Fuels* **1997**, *11*, 1092–1100.

Reyes-Nivia, C.; Diaz-Pulido, G.; Kline, D.; Guldberg, O.; Dove, S. *Global Change Biology* **2013**, *19(6)*, 1919–1929.

Schwartz, S. E. *Science* **1989**, *243(4892)*, 753–763.

Seinfeld, J. H.; Pandis, S. N. *Atmospheric Chemistry and Physics: From Air Pollution to Climate Change*; Wiley-Interscience: Hoboken, NJ, 2006.

Shakun, J. D.; Clark, P. U.; He, F.; *et al. Nature* **2012**, *484*, 49–54.

Shapiro, A. V.; Rozanov, E.; Shapiro, A. I.; Wang, S.; Egorova, T.; Schmutz, W.; Peter, T. *Atmospheric Chemistry and Physics* **2012**, *12*, 3181–3188.

Shyn, T. W.; Sweeney, C. J.; Grafe, A.; Sharp, W. E. *Physical Review A* **1994**, *50(6)*, 4794–4801.

Torr, D. G.; Torr, M. R. *Journal of Atmospheric and Terrestrial Physics* **1979**, *41*, 797–839.

Tsukahara, H.; Ishidab, T.; Mayumia, M. *Nitric Oxide* **1999**, *3(3)*, 191–198.

TVA, Coal-Fired Power Plant. Electronic, 2013; http://www.tva.com/power/coalart.htm.

U.S. Department of Energy, Ethanol. Electronic, 2013; http://www.fueleconomy.gov/feg/ethanol.shtml.

Velazco, V.; Wood, S. W.; Sinnhuber, M.; *et al. Atmospheric Chemistry and Physics* **2007**, *7*, 1305–1312.

Wallace, D. F. David Foster Wallace on Life and Work. Electronic, 2008; http://www.wsj.com/articles/SB122178211966454607, 2005 Kenyon College commencement speech

Wayne, R. P. *Research on Chemical Intermediates* **1994**, *20(3)*, 395–422.

et al., C. A. E. *Scientific Assessment of Ozone Depletion*; 1998.

Wong, M. H. *Young Scientists Journal* **2012**, *5(11)*, 23–30.

4

THE LITHOSPHERE

A nation that destroys its soils destroys itself. Forests are the lungs of our land, purifying the air and giving fresh strength to our people.

—Franklin D. Roosevelt

I like terra firma; the more firma, the less terra.

—George S. Kaufman

4.1 INTRODUCTION

Humans are terrestrial creatures. It has been about 360 Ma since our evolutionary ancestors first emerged from the ocean to live on land. Ironically, the concept of seasteading (floating cities with sovereign governments) might lead some in our species back to the sea, but for the foreseeable future, most of us will remain on land. The lithosphere has, similarly to the hydrosphere and atmosphere, changed in dramatic ways since the formation of the Earth. The activity of plate tectonics makes this change ongoing. As with the other spheres of the Earth, it has been affected by humans in important ways (deforestation and desertification, large-scale agriculture, contamination, acid deposition, and so on). The purpose of this chapter is to describe the chemistry of soil and some of the environmental issues facing our beloved terra firma.

4.2 SOIL FORMATION

Soil is an exceedingly complex, heterogeneous mixture of inorganic components (minerals), organic components (*humus*), and voids between solid materials that can be filled with water or air. The proportion of these ingredients varies by geographic region, time, and depth. Describing soil by its individual components misses the intricate interplay between them but is the best approximation available. The inorganic portion originates from the physical and chemical weathering of rocks. These weathering processes not only break the rocks into the small pieces called gravel, clay, silt, and sand but can also chemically alter the minerals via processes of dissolution, ion exchange, or oxidation. The organic portion of the soil derives from the decomposition of plant and animal litter, which is so complex that it defies a simple list of constituents. Organic constituents are often referred to as soil humus or humic substances, which are microbially active and subject to aerobic and anaerobic decomposition.

☞ Soil *horizons* are roughly parallel layers of soil that are distinctly different in composition and/or grain size. Most soils contain common horizons, such as organic matter, surface soil, subsoil, parent rock, and bedrock.

If you have ever played in a sandbox and excavated a little too deeply you have likely observed a transition from a layer of one type of soil to a layer that is abruptly different. These layers are referred to as *horizons* (see Figure 4.1) and contain increasingly less organic matter as depth increases. The number of horizons varies by soil type, with a trend of decreasing organic content, and an underlying layer of sandy/silty clay or metal oxides. Finally, weathered bedrock is common.

Environmental Chemistry: An Analytical Approach, First Edition. Kenneth S. Overway.

Companion website: www.wiley.com/go/overway/environmental_chemistry

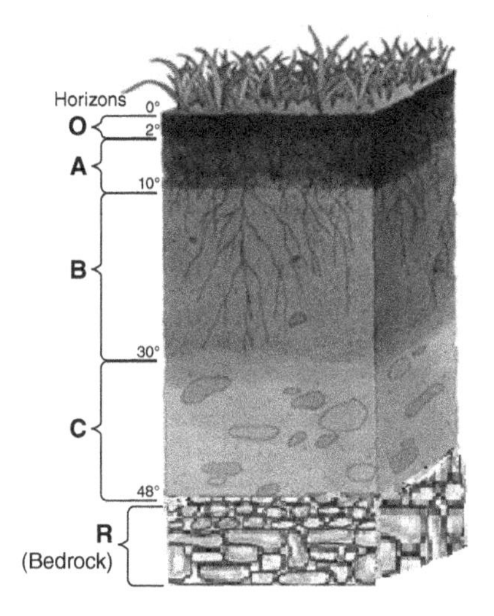

Figure 4.1 A vertical soil profile usually has several horizons. Organic matter tends to be most abundant in the top layer (horizon O). Horizon A tends to be a mixture of mineral and organic components and contains much of the biological activity. Horizon B contains dramatically more inorganic material that consists of clays and other small particles. Horizon C tends to be strictly inorganic material with larger weathered particles from the parent rock. Finally, a layer of bedrock undergirds the soil. Other intermediate layers are possible. Source: http://en.wikipedia.org/wiki/File:SOIL_PROFILE.png. Used under BY-SA 3.0 //creativecommons.org/licenses/by-sa/3.0/deed.en.

When you examine Tables 4.1 and 4.2, the top few elements of the two soils are similar in both lists. These elements should also be familiar to you because of your knowledge of nucleosynthesis (Chapter 1). They are abundant on the Earth because they are the favorable end products of the fusion cycles that evolve in the core of massive stars. Section 1.2.4 (p. 5) described in detail the onion-like layers that form in the core of these stars as lighter elements fuse into heavier elements. The nebular cloud that formed our solar system was a remnant of one or more of these stars. The composition of the Earth's surface is directly related to the abundance of these elements. Differentiation of the elements and compounds, caused by density differences, drove much of the iron and nickel into the core of the Earth and left the lighter silicates in the crust. Initially, the rocks solidified from the magma and lava flows, combined with the ash that settled on the Earth from volcanic eruptions, to form the first soils. Physical and chemical processes continue to shape the variety of soils found around the globe. Soils that are newly formed from geologic material, as in Table 4.1, contain mostly inorganic compounds. Soils that are located near dense vegetation or have been used in agriculture will have a higher organic matter content, such as in Table 4.2.

45-year old volcanic soil

Compound	% by Mass	Element	% by Mass
SiO_2	61.13	O	48.79
Al_2O_3	17.24	Si	28.57
H_2O	4.78	Al	9.12
Na_2O	3.9	Fe	3.80
CaO	3.61	Na	2.89
FeO	2.59	Ca	2.58
Fe_2O_3	2.56	K	1.48
K_2O	1.78	H	0.56
Humus	0.45	C	0.26
CO_2	0.04	N	0.05
pH	6.0	S	0.00

The humus is assumed to have an approximate chemical formula of humic acid ($C_{187}H_{186}O_{89}N_9S$).

Table 4.1 The composition and elemental analysis of a volcanic soil from the British West Indies (van Baren, 1931).

Limestone soil

Compound	% by Mass	Element	% by Mass
SiO_2	73.38	O	52.40
Al_2O_3	7.59	Si	34.30
H_2O	6.05	Al	4.02
Humus	4.42	Fe	2.48
Fe_2O_3	2.63	C(org.)	2.47
K_2O	1.38	K	1.15
CaO	1.01	H	0.88
TiO_2	0.93	Ca	0.72
FeO	0.82	Ti	0.56
CO_2	0.74	Na	0.41
Na_2O	0.55	Mg	0.23
MgO	0.38	C(inorg.)	0.20
P_2O_5	0.10	N	0.14
MnO	0.08	Mn	0.06
Cl	0.05	S	0.06
SO_3	0.05	Cl	0.05
N_2O_5	0.001	P	0.04
pH	7.3		

Organic (humus) and inorganic carbons (carbonates) are calculated separately.

Table 4.2 The composition and elemental analysis of the top layer (horizon A) of a limestone soil from the Netherlands (van Baren, 1930).

☞ *Physical weathering* processes involve physical changes that turn rocks into soil while not chemically altering the composition, such as *abrasion*, *freeze/thaw cycling*, *expansion* and *contraction*, and weathering from biological sources.

4.2.1 Physical Weathering

Synonymous with mechanical weathering, physical weathering is the process involving the conversion of large rocks to small grains of minerals. The chemical composition remains unchanged in these processes – only the size changes. *Abrasion* is a process by which the force of wind, water, or ice grinds a rock surface into smaller particles. This can occur as a result of the massive force of a glacier grinding the rocks on a mountainside as it advances or the minuscule, yet incessant, force of wind-borne dust particles sandblasting the rock away. *Freeze/thaw cycles* occur when moisture that has infiltrated the cracks and pores of a rock freezes and enlarges cracks as a result of the expansion. Repeated freeze/thaw cycles cause the rock to flake and chip, exposing more cracks and pores to further weathering. In the same manner, the minerals themselves experience *expansion and contraction* due to various heat sources or repeated cycles of wetting and drying, resulting in cracking and flaking. Plant roots also infiltrate cracks in rocks and exert enough pressure to cause further disintegration. The end result of physical weathering is to increase the surface area and decrease the particle size of the rocky material. Increased surface area favors chemical weathering processes, which further degrade the original mineral.

☞ *Abrasion* is a physical weathering process where rock particles are scraped away by the force of other particles through wind, water, and glacier action. *Erosion* is the removal of soil by means of wind and water.

4.2.2 Chemical Weathering

Chemical weathering processes result in a change in the chemical composition of the mineral that the rock is originally composed of. Certain minerals, called primary minerals, retain the chemical composition of the parent rock. Several of these primary minerals are listed in Table 4.3. When these minerals are exposed to water, they can be altered through a process of *hydration* in which molecules of water form bonds to the mineral to form a hydrate. These hydrates have different physical properties, and the dehydration of the mineral can require baking at a few hundred degrees centigrade. *Hydrolysis* is a chemical process where water dissociates into H^+ or OH^- and displaces some of the elements in the mineral, such as the transformation of the mineral albite ($NaAlSi_3O_8$) to kaolinite ($Si_4Al_4O_{10}(OH)_8$) – a secondary mineral.

Mineral group	Specific mineral	Chemical formula
Amphibole	Edenite	$NaCa_2Mg_5AlSi_7O_{22}(OH)_2$
Feldspar	Albite	$NaAlSi_3O_8$
Mica	Biotite	$K(Mg,Fe)_3(AlSi_3O_{10})(F,OH)_2$
Pyroxene	Jadeite	$NaAlSi_2O_6$
Quartz	Quartz	SiO_2
Olivine	Olivine	$(Mg,Fe)_2SiO_4$

Table 4.3 A few examples of primary minerals and their chemical formulas.

$$4NaAlSi_3O_8(s) + 4H^+(aq) + 18H_2O(l) \rightarrow Si_4Al_4O_{10}(OH)_8(s) + 8H_4SiO_4(aq) + 4Na^+(aq) \quad \text{(R4.1)}$$

Chemical weathering through *oxidation* or *reduction* causes the oxidation state of an element in the mineral to change by transferring an electron from one species to another, such as in the transformation of magnetite ($FeO \cdot Fe_2O_3$) to hematite (Fe_2O_3).

$$4FeO{\cdot}Fe_2O_3(s) + O_2(g) \rightarrow 6Fe_2O_3(s) \quad \text{(R4.2)}$$

Review Example 4.1: Oxidation States

Identify the oxidation states of the elements that were oxidized or reduced in Reaction R4.2, and label the oxidizing agent and reducing agent.

Solution: See Section D.1 on page 277.

☞ *Chemical weathering* processes involve chemical changes to minerals that alter the chemical formula of the mineral such as *hydration*, *hydrolysis*, *oxidation* and *reduction*, and *dissolution*.

☞ For a review of redox terms and determining oxidation states, see Review Example 1.13 on page 30.

Finally, *dissolution* is another chemical weathering process where components of minerals become soluble under changing pH conditions. Since natural rain contains carbonic acid (pH of 5.6; see the derivation on page 134) and acid rain can have a pH much lower, minerals can become soluble and change composition, such as the dissolution of limestone.

$$CaCO_3(s) + H_2CO_3(aq) \rightarrow Ca^{2+}(aq) + 2HCO_3^-(aq) \quad \text{(R4.3)}$$

4.2.3 Minerals

Silicate and aluminosilicate minerals are the most abundant inorganic components of soil. They are characterized by layers of tetrahedral SiO_4 units and octahedral AlO_6 units. These units are connected to other units via shared oxygen atoms to form tetrahedral or octahedral sheets, as seen in Figure 4.2. Because each oxygen atom is shared by two silicon atoms, the empirical formula is SiO_2, with silicon in the +4 oxidation state and each oxygen in the −2 oxidation state.

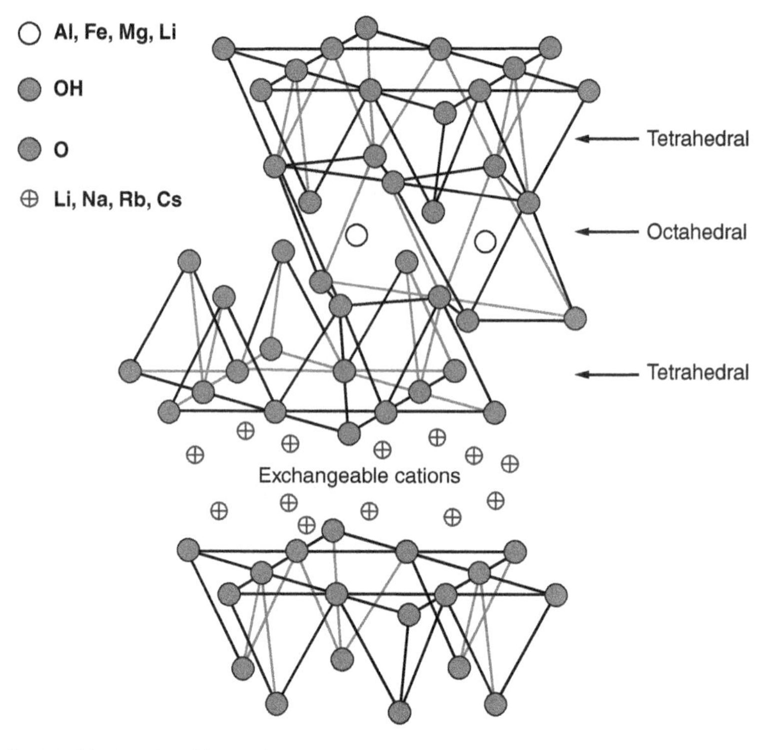

Figure 4.2 In this aluminosilicate sheet, you can see the two layers made from silicate tetrahedrons that form the top and bottom of a sandwich layer with an octahedral sheet of aluminum in between. The aluminum atoms are explicitly shown in the center of the octahedral structures. The silicon atoms are not explicitly shown but reside at the centers of the tetrahedral oxygen structures. Spaces between these layers can contain exchangeable cations. Source: http://wwww.intechopen.com/books/nanocomposites-new-trends-and-developments/polymer-nanocomposite-hydrogels-for-water-purification.

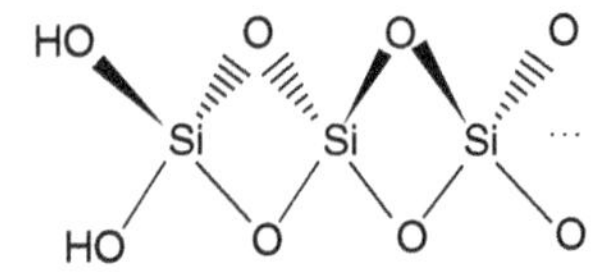

Figure 4.3 Silica, with an empirical formula of SiO_2, really is a network covalent structure of SiO_4 units. Each O atom is shared by two Si atoms, giving the empirical formula. At the end of the polymer chain (left side), the SiO group terminates with an H atom to balance the charge. If the polymer continues (right side) then the O atoms form a neutral polar surface.

☞ *Adsorption* is the process of atoms and molecules attaching themselves to the surface of a solid.

The oxygens at the surface of a sheet are neutral. At the edges of a sheet where the polymer stops, as in Figure 4.3, the oxygen atoms would not be shared by another Si cell and are either negatively charged $Si–O^-$ groups or silanol groups (Si–OH) depending on the pH. In either form, the Si–O bond is highly polar and each oxygen atom has two or more lone pairs of electrons ready to interact with any polar species or cation. The surface of a silicate sheet can *adsorb* cations using an ion–dipole force. When ionized, the silanol groups can adsorb cations using ionic bonds. In either case, the mineral retains these cations and can exchange them with other ions that enter the space, such as hydronium ions (H_3O^+) from acid rain or K^+ and Na^+ from salt water solutions. These cations play an important role in delivering mineral nutrients to plant roots and in buffering the effects of acid deposition.

Other common mineral types are combinations of carbonates, sulfides, sulfates, phosphates, nitrates, halides, oxides, and hydroxides of common cations in a variety of combinations. They result in simple formulas, such as halite (NaCl) and fluorite (CaF_2), or nonspecific formulas such as apatite ($Ca_5(PO_4)_3(F,Cl,OH)$) where variability occurs in the amount of fluorine, chlorine, and hydroxide present in the sample.

4.2.4 Organic Matter and Decay

Soil humus is mainly composed of *biopolymers* in various states of decay, driven by a legion of microorganisms. A vast majority of the organic material in soils is derived from leaf litter and other plant debris. The resulting humus is a complex mixture of biopolymers degraded by plant *senescence* and decomposition brought about by a host of microbes.

Polymer etymologically means "many parts." The "parts" from which polymers are assembled are *monomers*, which means "one part." You are likely familiar with plastics, which are polymeric forms of different organic molecules such as ethylene in high-density polyethylene) (HDPE or #2 plastic) or vinyl chloride in polyvinyl chloride (PVC or #3 plastic).

Protein, saccharides, DNA, RNA, and lipids are *biopolymers* since they are composed of smaller units.

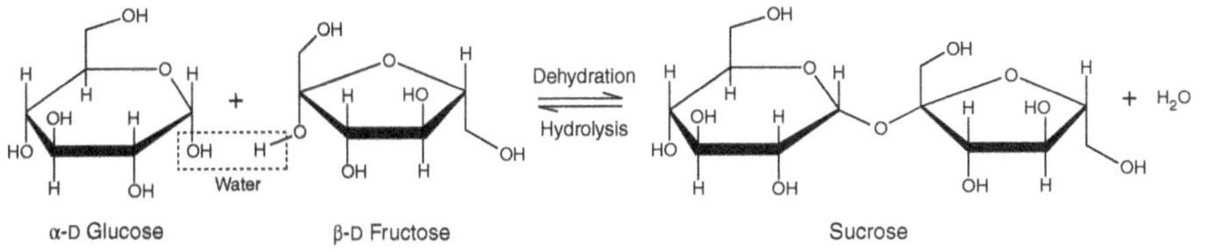

4.2.4.1 Biopolymers Carbohydrate *polymers*, such as cellulose and other polysaccharides, are mostly composed of carbon, hydrogen, and oxygen. Proteins, which are polymers of amino acids, comprise most of the rest of the initial bulk of soil organic matter. These biopolymers consist of *monomeric* units (monosaccharides, amino acids, fatty acids) that are linked by removing a hydrogen from one monomeric unit and a hydroxyl from the other monomeric unit, forming a molecule of water and a covalent linkage between the two monomers, as shown in Reaction R4.4. Polymers linked in this manner result from a dehydration reaction because the linkage results in release of water.

α-D Glucose + β-D Fructose ⇌ (Dehydration / Hydrolysis) Sucrose + H_2O (Water)

(R4.4)

Dehydration (also called condensation) reactions usually occur within an organism as it attempts to assemble cellular machinery from monomeric units. The decay of organic matter in soil usually proceeds in the opposite direction, dismantling biopolymers into smaller units. This process is called hydrolysis, and it is the opposite of a dehydration reaction. Hydrolysis breaks the linkage bond between two monomers and caps the ends of the linkage bonds with either H or OH from water. Example of the hydrolysis of a protein dipeptide bond and lipid ester bond can be seen in Reactions R4.5 and R4.6 below.

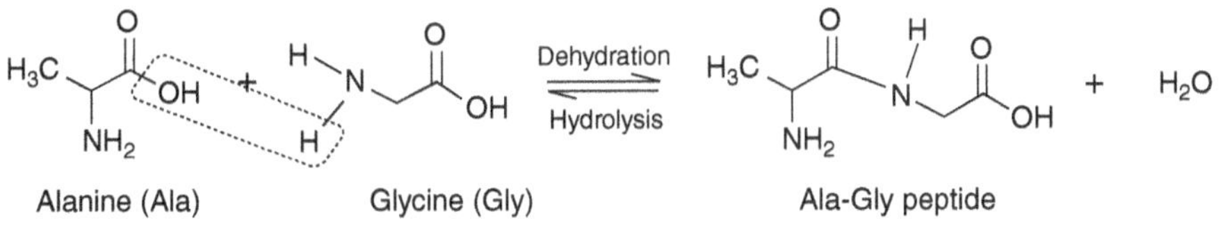

Alanine (Ala) + Glycine (Gly) ⇌ (Dehydration / Hydrolysis) Ala-Gly peptide + H_2O

(R4.5)

Glycerol + Fatty acids ⇌ (Dehydration / Hydrolysis) A triglyceride + 3 H_2O (Water)

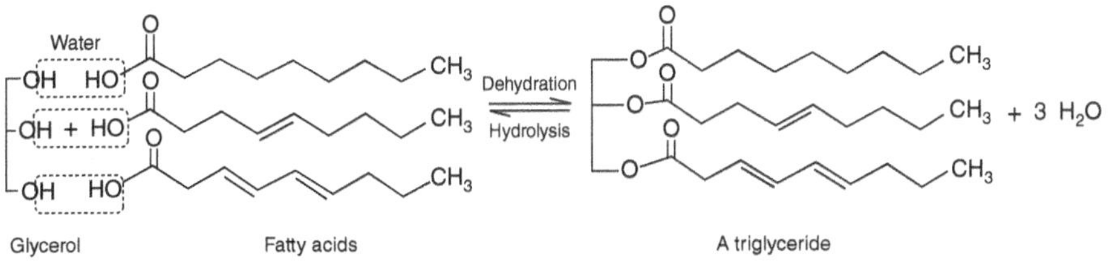

(R4.6)

The hydrolysis process converts the mostly insoluble polymeric biomolecules into monomer units. These smaller molecules are much more soluble in water and that makes them more accessible to microbes as nutrients, starting a succession of reactions that begin as aerobic and end as anaerobic, assuming that oxygen levels are not replenished. These reactions represent overall reactions, where the actual mechanisms are mediated by enzymes and energy-laden biomolecules, such as nicotinamide adenine dinucleotide (NADH) and adenosine triphosphate (ATP).

4.2.4.2 Leaf Senescence Leaf senescence is a programmed process by which plants reclaim some of the nutrients from leaves before they are shed. It is a carefully regulated period when photosynthesis ceases and biopolymers are dismantled. The monomeric units, along with minerals, are reabsorbed into the plant as resources for seeds or overwintering. Senescence is the beginning of the conversion of plant matter into soil organic matter. The plant secretes enzymes into the leaf, which depolymerize the cellulose and proteins, generating monosaccharides and amino acids.

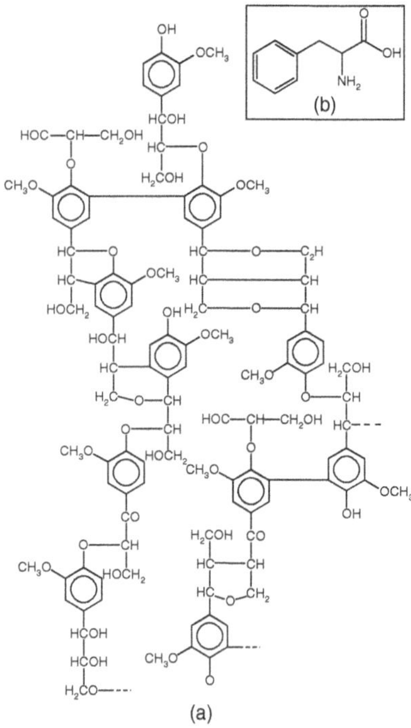

Much of the organic matter that is not reclaimed during senescence is *lignin*, which is a hydrophobic polymer that provides mechanical strength for cell walls. When combined with polysaccharides in the vascular system of plants, the hydrophobicity of lignin prevents water from diffusing throughout the plant, channeling the vascular transport of water. Much of its hydrophobicity comes from the phenyl group of the amino acid phenylalanine, which plants use to synthesize lignin. It is highly insoluble in water and, as a result, is resistant to degradation. Currently, not much is known of the degradation process that converts lignin into soil humus and *humic acids* (Figure 4.4), which form the bulk of soil organic material. The oxidation of lignin is a result of photosynthesis during senescence. As chloroplasts are dismantled, the minor pigments in leaves become obvious, providing the yellow and red colors we associate with autumn. The degradation of the chloroplasts in leaves alters photosynthetic mechanisms and leads to the production of reactive oxygen species, such as superoxide (O_2^-) during the usual process of oxidizing water in photosynthesis. These reactive species destroy cell walls, begin to oxidize lignin and other biopolymers, increase the hydrophilicity of the material, and accelerate the conversion of the leaf into humus.

Figure 4.4 Lignin (a) provides much of the rigidity that allows woody plants to grow extensive structures. Its hydrophobicity (derived from phenylalanine (b)) makes it resistant to degradation, but some fungi and bacteria are able to convert it into humic and fulvic acids, as seen in Figure 3.21 found on page 133. The total combination of all of the organic material in soil is collectively called *humus*. Source: http://en.wikipedia.org/wiki/File:Lignin_structure.svg. Used under BY-SA 3.0 //creativecommons.org/licenses/by-sa/3.0/deed.en.

☞ *Senescence* is the active degradation of a biological structure and the reclamation of the resources.

☞ Respiration in organisms can be accomplished by *aerobic oxidation* or *anaerobic oxidation*. Aerobic oxidation occurs under *aerobic* conditions, which imply the presence of available molecular oxygen. Anaerobic oxidation occurs under *anoxic* conditions, which imply the very low or negligible concentrations of molecular oxygen.

4.2.4.3 Microbial Degradation The decomposition of organic matter in soil is the result of a plethora of microbes extracting chemical energy from the material. The large, complex macromolecular structures shed from flora and fauna become food for the armies of microorganism that are classified in more detail in Section 4.2.5. The details of decomposition are complex given the large array of biomolecules that are involved. Our purpose here is to understand the conditions and classes of chemicals that form. One such condition that occurs when the organic matter has sufficient access to aerated moisture is aerobic oxidation. Once a series of extracellular hydrolysis reactions reduce polysaccharides and proteins to monomeric units, such as glucose and alanine using excreted enzymes, microbes can consume these smaller compounds and use *aerobic oxidation* of these monosaccharides and amino acids for energy production. Ultimately, this process produces the most thermodynamically stable products, such as in this overall reaction involving glucose.

$$C_6H_{12}O_6 + 6O_2 \rightarrow 6CO_2 + 6H_2O \quad \text{(R4.7)}$$

Aerobic oxidation consumes a great deal of molecular oxygen. In many environments where aerobic oxidation occurs, oxygen levels drop precipitously to produce *anoxic* conditions, under which *anaerobic oxidation* begins. These anaerobic reaction mechanisms can be described as *acidogenesis* and *acetogenesis*, where monosaccharides and amino acids are converted into simple carboxylic acids such as butyric acid and acetic acid. Since many of the larger organic acids are eventually converted to acetic acid, acidogenesis often becomes acetogenesis.

$$C_6H_{12}O_6(aq) + 2H_2O(l) \rightarrow 2CH_3COO^-(aq) + 2H^+(aq) + 2CO_2(g) + 4H_2(g) \quad \text{(R4.8)}$$

$$C_6H_{12}O_6(aq) \rightarrow 2CH_3CH_2OH(aq) + 2CO_2(g) \quad \text{(R4.9)}$$

$$C_6H_{12}O_6(aq) + 2H_2(g) \rightarrow 2CH_3CH_2COO^-(aq) + 2H_3O^+(aq) \quad \text{(R4.10)}$$

$$NH_3^+CH(CH_3)COO^-(aq) + 2NH_3^+CH_2COO^-(aq) \rightarrow 3CH_3COO^- + 3NH_4^+(aq) + CO_2(g) \quad \text{(R4.11)}$$

The products from the acidogenesis stage often degrade further to become acetic acid (or the acetate ion under neutral conditions).

$$CH_3CH_2COO^-(aq) + 3H_2O(l) \rightarrow CH_3COO^-(aq) + H^+(aq) + HCO_3^-(aq) + 3H_2(g) \quad (R4.12)$$

$$CH_3CH_2OH(aq) + H_2O(l) \rightarrow CH_3COO^-(aq) + H^+(aq) + 2H_2(g) \quad (R4.13)$$

$$CH_3CH_2CH_2COO^-(aq) + 2H_2O(l) \rightarrow 2CH_3COO^-(aq) + 2H_2(g) + H^+(aq) \quad (R4.14)$$

Organic functional groups

Substituent class	Formula	Example	Condensed and line structures
alcohol	ROH	Ethanol	H_3C-CH_2-OH
amine	RNH_2	Ethyl amine	$H_3C-CH_2-NH_2$
carboxylic acid	RCOOH	Acetic acid	$H_3C-C(=O)-OH$
ether	ROR′	Diethyl ether	$H_3C-CH_2-O-CH_2-CH_3$
ester	RCOOR′	Ethyl acetate	$H_3C-C(=O)-O-CH_2-CH_3$
aldehyde	RCHO	Ethanal	$H_3C-C(=O)-H$
phenyl	RPh	Ethyl benzene	$H_3C-CH_2-C_6H_5$
ketone	RCOR′	Methyl ethyl ketone	$H_3C-CH_2-C(=O)-CH_3$
thiol	RSH	Ethanethiol	H_3C-CH_2-SH

Table 4.4 Lewis structure and line-angle formulas of some of the common organic functional groups.

☞ *Acidogenesis* is an anaerobic, microbial degradation stage that immediately follows hydrolysis and is characterized by the conversion of monosaccharides and fatty acids into short-chain carboxylic acids such as propanoic, butyric, and acetic acid (Table 4.4). *Acetogenesis* is a similar and often concomitant anaerobic process, in which carboxylic acids and saccharides are converted into acetic acid. These two processes eventually lead to the final stage of decomposition of *methanogenesis*, where microbes convert acetic acid (along with carbonate and molecular hydrogen) to CO_2 and CH_4.

The production of these organic acids can contribute to the reduction of the pH, sometimes to levels around pH 4 or below. In many cases, however, there is a healthy population of *methanogenic* microbes, which use acetic acid and carbon dioxide (in the form of hydrogen carbonate) to form methane.

$$HCO_3^-(aq) + 4H_2(g) + H^+(aq) \rightarrow CH_4(g) + 3H_2O(l) \quad (R4.15)$$

$$SO_4^{2-}(aq) + 4H_2(g) + H^+(aq) \rightarrow HS^-(aq) + 4H_2O(l) \quad (R4.16)$$

$$SO_4^{2-}(aq) + CH_3COO^-(aq) + H^+(aq) \rightarrow H_2S(aq) + 2HCO_3^-(aq) \quad (R4.17)$$

$$CH_3COO^-(aq) + H_2O(l) \rightarrow CH_4(g) + HCO_3^-(aq) \quad (R4.18)$$

$$2CH_3CH_2OH(aq) + CO_2(g) \rightarrow CH_4(g) + 2CH_3COO^-(aq) + 2H^+(aq) \quad (R4.19)$$

These processes of hydrolysis, acidogenesis, acetogenesis, and methanogenesis lead to the formation of small, water-soluble compounds that have a dramatic impact on water quality, as discussed in Chapter 5. Each of these processes occurs to varying degrees depending upon environmental conditions of the organic matter of the soil. The force behind this decay is a host of microorganisms that inhabit the soil. Microorganisms use catalytic enzymes to break the stable linkage bonds in biopolymers and consume the monomers as nutrients. The microorganisms consist of bacteria, fungi, actinomycetes, algae, and protozoa. Ten grams of fertile soil contains more than 7 billion bacteria – as much as all of the people on the planet (Sposito, 2008, p. 65). For our purposes, it is not important to understand all of the species of microorganisms in soil, but understanding a few classifications will help us to be descriptive about the conditions under which microorganisms function.

Review Example 4.2: Oxidation States in Methanogenesis

Determine the oxidation state of carbon before and after the following reaction.

$$HCO_3^-(aq) + 4H_2(g) + H^+(aq) \rightarrow CH_4(g) + 3H_2O(l) \quad (R4.20)$$

Solution: See Section D.1 on page 277.

4.2.5 Microorganism Classifications

There are a bewildering number of microorganisms in any sample of soil, so memorizing their genus and species would be difficult and provide diminishing returns. A more useful way to describe these microbes, in the context of soil and water quality, is to describe various aspects of their function and the environment in which they thrive. We will look at three different classifications that are most appropriate for the study of the environment from a chemical perspective.

1. **Food source**

   (a) *Autotrophs* produce their own food from inorganic sources. *Photoautotrophs* use solar radiation to drive photosynthetic processes. Algae are a good representative of this classification. *Chemoautotrophs* use energy from redox reactions of inorganic compounds to drive their synthetic processes. The extremophiles that live near the hydrothermal vents at the bottom of the ocean are an example of a chemoautotrophic organism.

(b) *Heterotrophs* consume presynthesized nutrients produced by other organisms. The many varieties of fungi and bacteria that inhabit the soil would be considered heterotrophs.

2. **Respiration**

   (a) *Aerobes* are organisms that use molecular oxygen as the oxidizing agent in metabolic reactions that produce energy.

   (b) *Anaerobes* are organisms that use other oxidizing agents (NO_3^-, SO_4^{2-}, metals, or carbon) in energy-producing metabolic reactions. Anaerobes thrive under anoxic conditions.

   (c) There are *obligate* aerobes that cannot survive without oxygen and obligate anaerobes that cannot survive in the presence of oxygen.

   (d) *Facultative* anaerobes are microbes that have evolved to use oxygen when it is available for respiration but use other oxidizing agents when conditions become anoxic.

3. **Temperature preference**

   (a) *Thermophiles* are organisms that prefer very warm temperatures ($T > 45$ °C).

   (b) *Mesophiles* are organisms that prefer moderate temperatures (20 °C $\leq T \leq$ 45 °C).

   (c) *Psychrophiles* are organisms that prefer cold temperatures ($T < 20$ °C).

Example 4.1: Categorizing Organisms

Using the three categories listed earlier, describe the following organisms by food source, respiration, and temperature preference.

1. Humans
2. Microbes that produce methane from carbohydrates in the absence of oxygen, find oxygen toxic, and prefer temperatures around 55 °C
3. Algae that thrive near the surface of the ocean at temperatures near 25 °C
4. Archaebacteria that thrive in the isolated lakes below the ice sheets of Antarctica and feed on reduced sulfur compounds

Solution: See Section D.1 on page 278.

4.2.6 Respiration and Redox Chemistry

Many soils possess several properties that promote microbial activity. The high surface area of minerals and humus affords microbes ample access to organic and inorganic nutrients. The mixture of liquid and gaseous phase resources, such as water and molecular oxygen, further promotes biological activity. Soil that is closer to the surface, or has been overturned recently, has access to the atmospheric reservoir of molecular oxygen, but soil that exists far below the surface or is covered by a sufficiently thick layer of water can have little or no oxygen. These extremes provide environments where microbes with diverse respiratory machinery can thrive. As in all organisms, microbes extract chemical energy from organic and inorganic nutrients to drive their metabolism. Since photoautotrophs, such as trees and herbaceous plants, synthesize high-energy polysaccharides, lipids, lignin, and proteins from inorganic material and solar energy from the Sun, their litter contains plenty of chemical energy to supply the needs of the army of microbes that inhabit soil.

Redox reactions represent the default method of harvesting chemical energy for life on the Earth. Recall from Review Example 1.13 in Chapter 1 (page 30) that redox reactions are characterized by a transfer of electrons between an oxidizing agent and a reducing agent. If a redox reaction has a free energy less than zero, then the reaction is spontaneous and can be used by a microbe as a source of chemical energy. Redox reactions are the force behind *electrochemical cells* (batteries). Their chemical potential is usually measured by their electromotive force or cell potential (E), measured in voltage, and can be derived from the free energy of a reaction.

☞ An *electrochemical cell* is a device that contains a reduction half-cell and an oxidation half-cell between which electrons flow. The voltage between these two half-cells is called the electromotive force or the cell potential.

Example 4.2: Cell Potential and the Nernst Equation

In your general chemistry course, you likely learned about the relationship between the free energy of a reaction and its equilibrium constant.

$$\Delta G^\circ = -RT \ln K_{eq} \quad (4.1)$$

This equation states that the standard-state free energy of a reaction is directly related to the equilibrium constant of a chemical reaction. The free energy term describes how much free energy is available when all of the reactants, starting at concentrations of 1.0 M (or 1.0 atm for gases), turn into products (with concentrations of 1.0 M or 1.0 atm) – this is considered the standard state. Rarely are environmentally important chemicals at 1.0 M concentrations. Even if a reaction starts with standard-state concentrations, as it moves toward equilibrium, the concentrations of reactants and products will change. Thus, Eq. (4.1) is not practically useful for real environmental situations. A more useful relationship is

$$\Delta G = \Delta G^\circ + RT \ln Q \quad (4.2)$$

where the equilibrium constant (K_{eq}) is replaced by the reaction quotient (Q), which describes the state of a reaction at some point in its progress toward equilibrium (R is the gas law constant and T is the absolute temperature). Equation (4.2) can provide the free energy of a reaction when reactant and product concentrations are nonstandard and temperatures are other than 298 K.

☞ Thermodynamic constants at *standard conditions* are denoted by a ° defined as 1.0 M concentrations, 1.0 atm pressures, and 298 K.

The cell potential of a chemical reaction is related to the free energy by

$$\Delta G = -nFE \quad (4.3)$$

where n is the number of moles of electrons transferred in the redox reaction, F is Faraday's constant (96,485.309 coulombs per mole of electrons), and E is the electromotive force (in volts). Note that since a spontaneous reaction has $\Delta G < 0$, a spontaneous cell potential is greater than zero. Substituting Eq. (4.3) into Eq. (4.2) results in

$$-nFE = -nFE^\circ + RT \ln Q$$

$$E = E^\circ - \frac{RT}{nF} \ln Q \quad (4.4)$$

where E is the cell potential at nonstandard conditions and E° is the cell potential at standard conditions. Equation (4.4) is known as the Nernst equation, after German chemist/physicist Walther Nernst (1864–1941). The Nernst equation can be used to estimate the electrochemical potential for redox reactions that exist under environmental (nonstandard) conditions. When these conditions allow for an $E > 0$, then the redox reaction is spontaneous, and it is likely, if nutrients are involved, that microbes could use the reaction to harvest energy.

Exercises

1. Determine the standard free energy for the oxidation of glucose at 298 K, given by the following reaction.

$$C_6H_{12}O_6(s) + 6O_2(g) \rightarrow 6CO_2(g) + 6H_2O(l) \quad (R4.21)$$

2. Find the free energy for Reaction R4.21 under reasonable environmental conditions of $P_{O_2} = 0.2095$ atm, $P_{CO_2} = 3.97 \times 10^{-4}$ atm, and $T = 16.5$ °C.

3. The two half-reactions for Reaction R4.21 can be found in Table I.8 (on page 326) are

$$6CO_2(g) + 24H^+(aq) + 24e^- \rightarrow C_6H_{12}O_6(s) + 6H_2O(g) \quad (R4.22)$$

$$O_2(g) + 4H^+(aq) + 4e^- \rightarrow 2H_2O(l) \quad (R4.23)$$

These two reactions can be combined by finding the least common multiple of the number of electrons and then flipping one reaction (reactants become products and vice versa) such that the sum of their half-cell potentials is positive. It is important to realize that the cell potential is an *intensive property* and does not change when the coefficients of the reactions are increased in order to match the number of electrons.

4. Determine the standard-state cell potential as well as the cell potential at the environmental conditions of $P_{O_2} = 0.2095$ atm, $P_{CO_2} = 3.97 \times 10^{-4}$ atm, and $T = 16.5$ °C.

Solution: See Section D.1 on page 278.

An *intensive property* does not depend on the amount of the material, such as density and hardness (gold has the same density whether you have got one ounce or one pound). An *extensive property* does depend on the amount of the material, such as mass or volume. In the case of electrochemical cells, think of a AAA battery and a D battery – one is much smaller than the other but both have the same cell potential (1.5 V for a nonrechargeable alkaline dry cell). Scaling up the concentration of reactants does not increase the voltage but does increase the current capacity (1200 mAh for a AAA cell and 12,000 mAh for a D cell, where mAh is milliAmpere·hours).

You may be most familiar with respiration that involves molecular oxygen since this is the reaction we use to harvest the chemical energy from glucose and store it in molecules such as ATP. As it turns out, this method of respiration provides the most energy per molecule of glucose. It is often the case that regions within soil or bodies of water can become anoxic when aerobic microbes consume all of the available oxygen. Microbes are very flexible in their ability to use other oxidizing agents to harvest energy from glucose and other carbon-based nutrients.

Table 4.5 shows the various redox reactions that life on the Earth uses to produce energy from carbohydrates (glucose in this case). These reactions are overall reactions that compare the energy available under certain conditions. The first reaction involves molecular oxygen and aerobic respiration. Using the standard free energy of formation values, the free energy produced per 1 mole of glucose is −2874.8 kJ. Using Eq. (4.3) to convert the free energy into a hypothetical cell potential yields 1.24 V.

Reaction	$\Delta G°$ (kJ)	$E°$ (V)	$E°(w)^a$ (V)
$C_6H_{12}O_6(aq) + 6O_2(g) \rightarrow 6CO_2(g) + 6H_2O(l)$	−2874.8	1.24	1.28
$C_6H_{12}O_6(aq) + \frac{24}{5}NO_3^-(aq) + \frac{24}{5}H^+(aq) \rightarrow 6CO_2(g) + \frac{42}{5}H_2O(l) + \frac{12}{5}N_2(g)$	−2922.0	1.26	1.20
$C_6H_{12}O_6(aq) + 12MnO_2(s) + 6HCO_3^-(aq) + 6H^+(aq) \rightarrow 12MnCO_3(s) + 12H_2O(l)$	−2645.0	1.14	1.20
$C_6H_{12}O_6(aq) + 24FeO(OH)(s) + 18HCO_3^-(aq) + 18H^+(aq) \rightarrow 24FeCO_3(s) + 36H_2O(l)$	−1262.9	0.55	0.32
$SO_4^{2-}(aq) + CH_3COO^-(aq) + 2H^+(aq) \rightarrow H_2S(g) + CO_2(g) + HCO_3^-(aq) + H_2O(l)$	−138.13	0.18	0.49
$C_6H_{12}O_6(aq) + 3SO_4^{2-}(aq) + 6H^+(aq) \rightarrow 6CO_2(aq) + 3H_2S(g) + 6H_2O(l)$	−741.4	0.32	0.30
$CH_3COO^-(aq) + H_2O(l) \rightarrow HCO_3^-(aq) + CH_4(g)$	−75.9	0.10	0.07

[a] $E°(w)$ refers to the cell potential under average environmental conditions (pH = 7, $[HCO_3^-] = 1$ mM, $P_{CH_4} = 1.80 \times 10^{-6}$ atm, $P_{CO_2} = 3.97 \times 10^{-4}$ atm, $[CH_3COO^-] = 1$ μM, $[C_6H_{12}O_6] = 1$ μM, $P_{H_2S} = 3.0 \times 10^{-8}$ atm, $[SO_4^{2-}] = 1$ mM).

Table 4.5 Microbial respiration reactions for aerobic, anaerobic, and methanogenic organisms. When these reactions are placed on a cell potential scale, they are commonly referred to as the redox ladder.

The standard-state cell potential does not represent usual environmental conditions. Normal environmental conditions, designated by $E°(w)$, are listed below the table in the footnote. Factoring these conditions into Eq. (4.4) results in an adjusted cell potential of 1.28 V. Remember that positive cell potentials imply spontaneous reactions, and the larger the cell potential, the more energy that is derived from it.

Anaerobic oxidation of glucose using the nitrate ion results in a lower cell potential and less energy compared to aerobic oxidation, given environmental conditions. The nitrate ion is, however, a very common species and the thermodynamically most stable form of nitrogen in the aerobic aquatic and terrestrial environments. In fact, the nitrate reaction in

Table 4.5 represents, simultaneously, one of the major routes of anthropogenic fertilizer loss in agricultural soils (arguably a bad thing) and the loss of nutritive nitrogen pollution in natural and anthropogenic wastewater treatment (a good thing).

Anaerobic microbes can also use manganese and iron mineral deposits, where available, as oxidizing agents in an increasingly less energy dense respiration. Further down in Table 4.5 is a reaction involving the acetate and sulfate ions, the acetate having been derived from acetogenesis. Finally, the least energy-efficient route for respiration is methanogenesis, which involves a *disproportionation reaction* involving the carbon atoms of the acetate ion.

☞ A *disproportionation reaction* is a type of redox reaction where the same species is simultaneously oxidized and reduced. In the case of methanogenesis of the acetate ion, an electron is transferred from the carboxylate carbon (it is oxidized) to the methyl carbon (it is reduced).

Review Example 4.3: The Disproportionation of the Acetate Ion

What are the oxidation states of the two carbon atoms in the acetate ion and how do they change when they become HCO_3^- and CH_4 according to Reaction R4.18 (on page 172)? How many electrons are exchanged in this reaction and in which direction do the electrons flow? HINT: start by determining the oxidation state for the carbons in ethane (C_2H_6) and assume that the methyl carbon in the acetate ion has the same oxidation number as the carbons in ethane.

Solution: See Section D.1 on page 280.

Victims of manure pit accidents succumb to the toxic effects of H_2S and asphyxiation. While these accidents are rare, when they do occur, they usually claim multiple victims. In 2007, four members of a Mennonite family in Virginia died when three of the members, one by one, went into a manure pit to save the first victim and were subsequently overcome by the products of anaerobic respiration (mainly H_2S, NH_3, and a lack of oxygen) (Hallam, Liao, and Choi, 2011; Spencer, 2007).

The value of the redox ladder in Table 4.5 is twofold: (1) it shows some of the essential C, N, and O nutrients that make microbial respiration possible, and (2) it shows the fate of these nutrients. For carbon nutrients, aerobic respiration results in the formation of carbon dioxide, whereas anaerobic respiration results in carbon dioxide and methane. Anoxic conditions can result in the formation of hydrogen sulfide (H_2S), a toxic gas that smells of rotten eggs and is associated with swamps, or methane, a potential source of energy that can be reclaimed from landfills or sewage treatment.

At the beginning of this chapter, the organic component of soil was described as a complex mixture of plant and animal litter that defies a simple compositional description. Hopefully, you now appreciate the dynamic and complex nature of organic matter. It is a vitally important component of soil, which is the central focus of agriculture – large and small scale.

In the early hours of December 22, 2008, a retention pond at the Kingston coal power plant in Tennessee ruptured and spilled over a billion gallons of fly ash sludge into the Cinch and Emory rivers and inundated 15 homes just west of Knoxville, TN. Fly ash is a byproduct of coal combustion, and it is stored in a water slurry at many coal-fired power plants. Once the organic components of coal are burned away, the remaining fly ash is a concentrate containing high levels of toxic metals and radioactivity from uranium-laden minerals. The 300-acre spill site at the Kingston power plant contained high levels of arsenic, beryllium, cadmium, and chromium. (Flory, 2008; McBride *et al.*, 1977). New evidence suggests that even the wet storage of the fly ash is contaminating groundwater around power plants (Schaeffer, 2013).

4.3 METALS AND COMPLEXATION

You should have surmised by now that the atmosphere, lithosphere, and hydrosphere are so interconnected that it is difficult to describe processes in isolation of the others. The topic of metals in the environment is one such topic. The lithosphere is the primary source of metals in the environment, but a discussion of hydrosphere is necessary to develop a complete understanding of their effects. Tables 4.1 and 4.2 contain the most abundant metals in typical soils, but there are many trace-level metals that play a significant role in the health of the soil, flora, and fauna of a particular biogeological system.

Many metals you may be familiar with are considered macronutrients for most organisms. These include calcium, associated with bones, sodium and potassium, associated with nerve signals, and magnesium, associated with chlorophyll. There are also several metals that are micronutrients – required at low levels for various biological functions but toxic at high concentrations. Iron is essential for the formation of hemoglobin, but in doses above 10–20 mg Fe per kg of body weight, abdominal pains, stomach ulceration, and liver damage may occur. Zinc is important for enzymatic activity, DNA transcription, and the immune system but at toxic levels can cause abdominal pain and diarrhea. Other metals serve no beneficial biological functions and are toxic even at very low levels. These metals, such as lead, mercury, and cadmium, are often referred to as *heavy metals* and are disproportionately important considering their trace levels in the environment. Clearly, the environmental significance of a metal is tied to its chemical properties and not its physical properties.

The term *heavy metal* is a vague moniker since it is not clearly linked to the atomic mass, bulk mass, density, or some other physical property of the metal. Even if it seems to apply to toxic metals such as Pb and Hg, it is not useful in differentiating between macronutrients, micronutrients, and toxic metals since these definitions imply something about the metal's chemical reactivity. Further, what sense is to be made of the differing toxicities of Pb(IV) and Pb(II), Cr(III), and Cr(VI) species, and what is to be made of toxic semimetals such as arsenic? Clearly, it is not just the element that is of importance but also its charge state.

A more environmentally useful classification system involves separating metals by their ability to covalently bond with *ligands*. This measure, referred to as the *covalent index*, is calculated by

$$\text{Covalent index} = \frac{(EN_{\text{metal}} - EN_{\text{ligand}})^2}{\text{ionic radius}} \tag{4.5}$$

where EN is the electronegativity of the element.

On the low end of the covalent index are Class A metals (Table 4.6). They form ionic bonds with ligands, tend to have low polarizability, and can be identified by their lack of valence d and p orbital electrons. Metals that belong to the group are in the alkali and alkaline earth families (the s-block of the periodic table) or transition and *p*-block metals that have lost all of their valence d and p electrons to achieve a noble gas electron configuration. These metals tend to have low electronegativity values and prefer to form ionic bonds with ligand elements that have high electronegativity values. Thus, Class A metals prefer to bind to ligands with oxygen instead of nitrogen and sulfur ($EN_O > EN_S, EN_N$).

On the high end of the covalent index scale are Class B metals. They can be identified by their electron configurations, usually having an $(n-1)\text{d}^{10}$ or $(n-1)\text{d}^{10}\,\text{np}^2$ configuration (such as Cd(II) with its $5\text{s}^0\,4\text{d}^{10}$ configuration or Pb(II) with its $6\text{s}^0\,4\text{f}^{14}\,5\text{d}^{10}$ configuration). These metals tend to occupy the 4d and 5d rows of the transition block and rows 5 and 6 of the p-block in the periodic table. Because of their position on the periodic table, these metals have moderate electronegativity values and when they bind to ligands containing moderately electronegative electron donors, the resulting bonds are very covalent in nature. Class B metals prefer to bind to ligands with sulfur, carbon, and nitrogen instead of oxygen. The tendency of some of these Class B metals to easily form metal–carbon bonds makes them especially toxic because of their volatility (and subsequent ingestion in the lungs) and tendency to bioaccumulate in fatty tissues.

The toxicity of Class B metals is due to their ability to generate reactive oxygen species (peroxides and superoxides) and their affinity to covalently bond to sulfur, nitrogen, and carbon, three very essential elements of proteins and DNA/RNA. When metals bind to these critical molecules, the three-dimensional shape changes and the biomolecular function is inhibited. Class A metals prefer forming ionic complexes with ligands containing oxygen (and sometimes nitrogen), and since one of the most common oxygen-containing species is water, organisms can easily eliminate the metal before levels become toxic.

Borderline metals are those metals with a covalent index on the spectrum between Class A metals and Class B metals. These metals are needed in biomolecules at a low concentration, and thus, they are often considered micronutrients. Borderline metals tend to be 3d transition metals, with a few exceptions. Two examples of micronutrients from this group include zinc and iron. Zinc(II) cations are employed in a variety of proteins to stabilize the folding of protein structures. Several such structures, commonly called a zinc finger, bind to DNA and are used by nature and biochemists to control the expression of genes. You are undoubtedly familiar with iron's role in oxygen transport where it is placed at the center of each heme group in hemoglobin. The transformation of iron between its +2 and +3 oxidation states controls the binding of molecular oxygen.

☞ A *ligand* is a chemical species that uses its nonbonding electrons to bind to metals. The electron-rich ligand and the electron-poor metal form a *complex*.

☞ For a review of electronegativity, see Review Example 1.19 on page 37.

Class A metals: Metals of low covalent index, small ionic radius, and low polarizability (or "hard" metals). Examples are the alkali and alkaline earth metals, and Sc^{3+}, Ti^{4+}, and Al^{3+}. These metals are usually nontoxic, and the more abundant elements are often macronutrients.

Class B metals: Metals with large ionic radius and high polarizability (or "soft" metals). Examples are Cu^+, Pd^+, Ag^+, Cd^{2+}, Ir^{2+}, Pt^{2+}, Au^+, Hg^{2+}, Ti^{2+}, Pb^{2+}. These metals are usually toxic at all concentrations and contain no nutritional value.

Borderline metals: Metals that have properties intermediate to Class A and B metals. Examples are V, Cr, Mn, Fe, Co, Ni, Cu^{2+}, Zn, Rh, Pb^{4+}, Sn. These metals are often micronutrients but exhibit toxicity at high levels.

Source: Duffus (2002) and Nieboer and Richardson (1980).
One major caveat to this schema involves the classification of Fe^{3+} and Fe^{2+}. Duffus places Fe^{3+} in the Class A category because of its low polarizability, whereas Nieboer *et al.* places it in the Class B category.

Table 4.6 This metal classification schema, currently preferred by environmental scientists, more naturally separates metals by their toxicity than the meaningless term "heavy metals."

Review Example 4.4: Metal Electron Configuration

Electron configurations are important in identifying metals from each of the categories outlined earlier. Use your knowledge of electron configurations and the periodic table to answer the following questions.

Exercises

1. What is the electron configuration of Zn(II)?
2. What is the electron configuration of Cu(II)?
3. What is the electron configuration of Sn(II)?
4. What is the electron configuration of Sn(IV)?
5. What is the electron configuration of Ca(II)?

Solution: See Section D.1 on page 280.

The ligands that bind to metals in soil come from the functional groups found in soil organic matter, as seen in the structure of lignin in Figure 4.4 on page 170 and humic acids in Figure 3.21 on page 133. For the metals that are bound to smaller organic units, such as humic and fulvic acids, these ligands can increase the solubility of the metals in water if the metals would normally form insoluble carbonates, sulfates, or hydroxides (when those anions are present). If the metals are bound to larger, more insoluble ligands (humus or large humic acids), then the solubility of the metals in water is decreased. Thus, the degree to which metals are retained in soil and sediments is related to the solubility of the ligands to which they bind.

4.3.1 Phytoremediation

Common Name	Pollutant(s) Accumulated
Alpine Pennycress	Cd, Ni, Pb, Zn
Indian Mustard	Pb, Zn
Chinese Cabbage	Zn, Pb
Indigo Bush	Pb
Field Chickweed	Cd
Sunflower	As , Cd, Cr
Bermuda Grass	hydrocarbons
White Clover	PCBs[a]

Notice that metals are not the only target of phytoremediation.
[a] PCBs = polychlorinated biphenyls.

Table 4.7 Several plants, referred to as *hyperaccumulators*, can be used in *phytoremediation* processes for the removal of some toxic contaminant. *Source:* Puget Sound (1992).

Toxic metal contamination of soil as a result of natural or anthropogenic factors can negatively impact the ecological and agricultural health of the soil. It is serendipitous that certain plants have evolved mechanisms for preferential absorption (*hyperaccumulation*) (Table 4.7) and sequestration of certain toxic metals. The metals are stored in the plant in such a way that the toxic effects on the cellular machinery are muted. This accumulation of the toxic metals in the stems and leaves of the plant results in the removal of the metals from the soil. This process, called *phytoremediation*, is a cost-effective (albeit slow) method of detoxifying a contaminated site. A contaminated site can be seeded with the proper hyperaccumulator and allowed to sit fallow for a several years. The plant material that is harvested each season contains toxic metal concentration levels as high as 5% of the dry weight of the plant mass. This plant mass can be properly disposed of, eventually leaving the soil with greatly reduced levels of contaminant metals. Phytoremediation is limited to the depth of the hyperaccumulator's root system.

Toxic or nutritive effects of metals in soil are largely due to their concentration and mobility. Rain events can leach metals from the soil if the metal forms a soluble complex or if the soil's pH is affected. The next section will focus on the pH effects in soil as a result of internal and external factors.

4.4 ACID DEPOSITION AND SOIL

Contributions to a soil's pH can come from internal and external sources. The external sources include the mildly acidic contribution of carbonic acid, derived from carbon dioxide levels in the air. With global atmospheric levels of CO_2 around 400 ppm, "natural" rain falling on soil has a pH of 5.6 as derived in Example 3.21 in Chapter 3 (page 134). Also described in Chapter 3 was acid deposition derived from natural and anthropogenic sources, often with a pH well below 5. Sources affecting pH that are internal to soils include decomposition of organic matter (acidogenesis and acetogenesis) and various buffering mechanisms that are the focus of this section. Predicting the pH of a soil sample is highly dependent on the inorganic constituents and their ability to maintain the pH despite the contributions of acidic sources. As we shall see, the geologic origin of the inorganic soil components is vitally important.

4.4.1 Limestone Buffering

Limestone is a sedimentary rock that derives from biotic and abiotic sources. As mentioned in Chapter 1, one of the principal gases in the Earth's early atmosphere, formed as a result of vulcanism, was carbon dioxide. When the early atmosphere cooled and formed the first oceans, most of the carbon dioxide dissolved to form carbonic acid and subsequently precipitated as metal carbonates (such as $CaCO_3$ and $MgCO_3$). Biotic forces, such as the use of calcium carbonate to form coral reefs and the shells of other marine invertebrates, also contribute to limestone rocks. Carbonate minerals are sparingly soluble in water, so their interaction with surface water sets up a carbonate buffering system that stabilizes the pH near 8.2 (for calcium carbonates).

Example 4.3: Limestone Buffering

Determine the pH of a water and soil mixture containing significant amounts of limestone ($CaCO_3$). Assume that the atmospheric concentration of CO_2 is approximately 397 ppm and the temperature is 20.0 °C.

Solution: Carbon dioxide is a very soluble gas, as measured by Henry's Law (see Section 5.3.2).

$$CO_2(g) \rightleftharpoons CO_2(aq) \qquad K_H = 3.5 \times 10^{-2}\ \mathrm{M/atm} \tag{4.6}$$

Carbon dioxide can react with water in the aqueous phase to form carbonic acid. We will use only the first ionization reaction (it was proved that the second was insignificant in the calculation for the pH of pure rain water).

$$CO_2(aq) + 2\,H_2O(l) \rightleftharpoons H_2CO_3(aq) + H_2O(l) \rightleftharpoons H_3O^+(aq) + HCO_3^-(aq) \quad K_{a1} = 4.5 \times 10^{-7}$$

Limestone is slightly soluble in water.

$$CaCO_3(s) \rightleftharpoons Ca^{2+}(aq) + CO_3^{2-}(aq) \qquad K_{sp} = 3.36 \times 10^{-9}$$

The calcium ion has no acid/base activity, but the carbonate ion is a weak base.

$$CO_3^{2-}(aq) + H_2O(l) \rightleftharpoons HCO_3^-(aq) + OH^-(aq) \qquad K_{b1} = 2.1 \times 10^{-4}$$

The K_{b2} for the carbonate ion is related to the K_{a1} reaction for carbonic acid, so it is already represented. Adding all of these reactions together yields the following.

$$\begin{aligned}
CO_2(g) &\rightleftharpoons \cancel{CO_2(aq)} \\
\cancel{CO_2(aq)} + 2\,H_2O(l) &\rightleftharpoons H_3O^+(aq) + HCO_3^-(aq) \\
CaCO_3(s) &\rightleftharpoons Ca^{2+}(aq) + \cancel{CO_3^{2-}(aq)} \\
\cancel{CO_3^{2-}(aq)} + H_2O(l) &\rightleftharpoons HCO_3^-(aq) + OH^-(aq) \\
\hline
CO_2(g) + CaCO_3(s) + 3\,H_2O(l) &\rightleftharpoons H_3O^+(aq) + 2\,HCO_3^-(aq) + Ca^{2+}(aq) + OH^-(aq)
\end{aligned}$$

Since the hydronium and hydroxide ions are both on the same side of the reaction, we can use the autodissociation of water to simplify the reaction.

$$\begin{aligned}
CO_2(g) + CaCO_3(s) + \cancel{3}\,H_2O(l) &\rightleftharpoons \cancel{H_3O^+(aq)} + 2\,HCO_3^-(aq) + Ca^{2+}(aq) + \cancel{OH^-(aq)} \\
\cancel{H_3O^+(aq)} + \cancel{OH^-(aq)} &\rightleftharpoons \cancel{2\,H_2O(l)} \qquad \text{(the } 1/K_w \text{ reaction)} \\
\hline
CO_2(g) + CaCO_3(s) + H_2O(l) &\rightleftharpoons 2\,HCO_3^-(aq) + Ca^{2+}(aq)
\end{aligned}$$

Using Hess's Law, the sum of a series of reactions implies the multiplication of each reaction's equilibrium constant.

$$K = \frac{K_H \times K_{a1} \times K_{sp} \times K_{b1}}{K_w} \tag{4.7}$$

Now we set up an ICE table. Remember that the RT factor comes from the Ideal Gas law equation, allowing the conversion of a molar concentration to a partial pressure.

	$CO_2(g)$	+ $CaCO_3(s)$	+ $H_2O(l)$	⇌ $2\ HCO_3^-(aq)$	+ $Ca^{2+}(aq)$
I	3.97×10^{-4}	---	---	0	0
C	$-x(RT)$	---	---	$+2x$	$+x$
E	$3.97\times10^{-4} - x(RT)$	---	---	$2x$	x

We now need to plug in the equilibrium values into the K expression.

$$\frac{[HCO_3^-]^2[Ca^{2+}]}{P_{CO_2}} = \frac{K_H \times K_{a1} \times K_{sp} \times K_{b1}}{K_w}$$

$$\frac{(2x)^2(x)}{3.97 \times 10^{-4} - x(RT)} = \frac{K_H \times K_{a1} \times K_{sp} \times K_{b1}}{K_w}$$

To simplify the math, we can use the 5% rule, then solve for x. Assume that $x(RT) \ll 3.97\times10^{-4}$ atm knowing full well that P_{CO_2} will not change because the atmosphere is a large reservoir.

$$\frac{(2x)^2(x)}{3.97 \times 10^{-4} - \cancel{x(RT)}_{\to 0}} = \frac{K_H \times K_{a1} \times K_{sp} \times K_{b1}}{K_w}$$

$$4x^3 = \frac{K_H \times K_{a1} \times K_{sp} \times K_{b1} \left(3.97 \times 10^{-4}\right)}{K_w}$$

$$x = \sqrt[3]{\frac{K_H \times K_{a1} \times K_{sp} \times K_{b1}\,(3.97 \times 10^{-4})}{4K_w}}$$

$$x = \sqrt[3]{\frac{\left(3.5 \times 10^{-2}\right)\left(4.5 \times 10^{-7}\right)\left(3.36 \times 10^{-9}\right) \times \left(2.1 \times 10^{-4}\right)\left(3.97 \times 10^{-4}\right)}{4\,(1.00 \times 10^{-14})}}$$

$$x = \sqrt[3]{1.103 \times 10^{-10}}$$

$$x = 4.796 \times 10^{-4}$$

The assumption fails, but we know P_{CO_2} will not change because of the large atmospheric reservoir, so we can ignore this problem. Looking back at the equilibrium ICE table, x does not represent the concentration of hydronium ion, so we need to do some more work. Examining the reactions we have used, HCO_3^- is produced when carbon dioxide dissolves in water and CO_3^{2-} is produced when calcium carbonate dissolves. This is a recipe for a buffer, so we will use the buffer equation (Henderson–Hasselbalch equation).

$$\text{pH} = \text{p}K_a + \log\left(\frac{[\text{base}]}{[\text{acid}]}\right)$$

We know $[HCO_3^-] = 2x$, and this would be the conjugate acid of the buffer. The conjugate base of the buffer is CO_3^{2-}. We can go back to the K_{sp} reaction and determine $[CO_3^{2-}]$ since we know that $[Ca^{2+}] = x$.

$$K_{sp} = [Ca^{2+}][CO_3^{2-}]$$

$$\frac{K_{sp}}{[Ca^{2+}]} = [CO_3^{2-}]$$

$$\frac{3.36 \times 10^{-9}}{4.796 \times 10^{-4}} = [CO_3^{2-}]$$

$$7.006 \times 10^{-6} = [CO_3^{2-}]$$

The final term in the buffer equation we need is the pK_a, and this must be the pK_a related to the acid reaction for HCO_3^-. We know that p$K_a = -\log(K_a)$, where K_a in this case is K_{a2} for the carbonic acid series of reactions.

$$\text{pH} = \text{p}K_a + \log\left(\frac{[\text{CO}_3^{2-}]}{[\text{HCO}_3^-]}\right)$$

$$\text{pH} = 10.33 + \log\left(\frac{7.006 \times 10^{-6}}{2 \times 4.796 \times 10^{-4}}\right)$$

$$\text{pH} = 8.1936 = 8.19$$

The final calculation shows that freshwaters that have access to geologic limestone will maintain an alkaline pH of 8.19.

The difficulty in specifying an exact pH that the limestone buffer maintains is complicated by other carbonates that are found in the environment (see Table 4.8) that have a range of K_{sp} values and therefore would slightly modify the calculation in Example 4.3, yielding a range of pH set points. Since $CaCO_3$ is the most common crustal carbonate, we can consider pH 8.2 to be the most commonly expected buffer set point, but the existence of other carbonates means that the limestone buffering mechanism covers a range of alkaline values.

Compound	K_{sp}	Equil. pH
$CaCO_3$ [a]	3.36×10^{-9}	8.19
$CaCO_3$ [b]	6.00×10^{-9}	8.28
$FeCO_3$	3.31×10^{-11}	7.52
$MgCO_3$	6.82×10^{-6}	9.30
$MgCO_3 \cdot 3H_2O$	2.38×10^{-6}	9.14
$MgCO_3 \cdot 5H_2O$	3.79×10^{-6}	9.21
$MnCO_3$	2.24×10^{-11}	7.47

[a] Calcite polymorph.
[b] Aragonite polymorph.

Table 4.8 Equilibrium pH values, as calculated from Example 4.3, for water that is saturated with CO_2 and these carbonate minerals.

Indeed, agricultural soils are often limed in order to increase the pH. Liming involves the addition of carbonate- and hydroxide-rich minerals to add acid-neutralizing compounds to the soil. The amount of limestone available contributes to the soil's *acid neutralizing capacity* (ANC). Acid is neutralized by limestone with the action of the carbonate buffer,

$$MCO_3(s) + 2H^+(aq) \rightleftharpoons M^{2+}(aq) + H_2O(l) + CO_2(g) \quad \text{(R4.24)}$$

where the metal M is most likely calcium or magnesium. This overall reaction includes the intermediate formation of the hydrogen carbonate ion and the buffering equilibrium derived in Example 4.3. The net effect of this limestone buffering is to remove the H^+ from solution, thus raising the pH, mobilizing calcium and magnesium from the soil and into the watershed, and a release of carbon dioxide into the atmosphere. The 2:1 stoichiometry of Reaction R4.24 makes this buffering mechanism a significant factor in the ANC of a soil. The limestone must be homogeneously incorporated into the soil for the buffering to be effective. A heterogeneous distribution of chunks of limestone in the soil will not buffer soil even in close proximity (Ulrich, 1983, p. 132). Liming an agricultural soil is effective because the added minerals are finely divided and mixed into the top soil horizons.

☞ *Acid-neutralizing capacity* is a measure of a soil's ability to neutralize the effects of an acid attack. It is proportional to the amount of each of the buffering materials described in this section.

4.4.2 Cation-Exchange Buffering

Silicate minerals compose over 90% of the Earth's crust by mass. As a result, the most common acid buffering mechanism in soil comes from the structure of these silicate minerals. As mentioned in Section 4.2.3, the tetrahedral sheets formed by silicate minerals produce a negatively charged surface that adsorbs cations. These cations can be exchanged with the cationic hydronium ion, which effectively neutralizes acidic contributions to the soil. Since silicate minerals can adsorb a variety of cations (Ca, Mg, Fe, K, Na, Hg, and others), the identity of the metal cation is represented by M in the cation-exchange reaction.

$$\text{soil} - M(s) + nH^+(aq) \rightleftharpoons \text{soil} - H_n(s) + M^{n+}(aq) \quad \text{(R4.25)}$$

$$\text{soil} - Ca(s) + 2H^+(aq) \rightleftharpoons \text{soil} - H_2(s) + Ca^{2+}(aq) \quad \text{(R4.26)}$$

In this neutralization process, the mobile hydronium ion adsorbs to the silicate sheet and replaces a metal cation, which becomes mobile and dissolves into the acidified water. Thus, in the process of neutralizing acid, metallic cations leach out of the soil.

Buffering also occurs in the chemical weathering mechanism of dissolution, as described in Reactions R4.1 and R4.3 on page 167. This type of buffering is sometimes considered distinct from the cation-exchange buffering system because it involves a chemical dissolution process. In these reactions, acid alters the structure of the mineral, converting it from a primary mineral to a secondary mineral as these two reactions demonstrate. The effect on the soil pH is the same – hydronium ions are absorbed by the mineral and prevent the pH of the soil from decreasing.

Just as there were complicating factors in the example of the limestone buffer involving different mineral types, the cation-exchange buffering mechanism also has similar complications. Ulrich (1983) reports that the combination of the cation-exchange buffer and the silicate chemical weathering mechanisms results in a pH range of 6.2 to 4.2, meaning that as long as there are cations to exchange or silicate minerals for acids to attack, the pH of the soil will maintain a value in this range until soil reactants are exhausted. The efficiency of the silicate buffering mechanism is also dependent on the granular size of the silicate minerals, with smaller particles providing the largest ANC in the short term (more surface area means a faster neutralization) and in the long term if the hydronium ion is incapable of exchanging with the metals buried in the center of a large particle.

4.4.3 Aluminum Buffering

The final inorganic buffering mechanism involves the aluminum oxide/hydroxide octahedra that are part of the alumino mineral structure. Since aluminosilicates take a range of empirical formulas of Al_2O_3 in some minerals, $Al(OH)_3$, and several in between, such as $AlO(OH)$, the range of solubility of the aluminum octahedra in the presence of acid can maintain a range of buffered pH values. In this buffering mechanism, the aluminum mineral is dissolving in the presence of acid, which is another form of chemical weathering. Table 4.9 shows several examples of minerals and their equilibrium dissolution constants along with the pH that the particular mineral will maintain as long as the mineral is not exhausted by the amount of acid to which it is exposed.

Mineral	Reaction	$\log(K_{eq})$	Buffered pH
Kaolinite	$Al_2Si_2O_5(OH)_4(s) + 6H^+(aq) \rightleftharpoons 2Al^{3+}(aq) + 2H_4SiO_4(aq) + H_2O(l)$	5.45	5.3
Corundum	$0.5Al_2O_3(s) + 3H^+(aq) \rightleftharpoons Al^{3+}(aq) + 1.5H_2O(l)$	9.73	5.2
Amorphous	$Al(OH)_3(s) + 3H^+(aq) \rightleftharpoons Al^{3+}(aq) + 3H_2O(l)$	9.66	5.1
Bayerite	$Al(OH)_3(s) + 3H^+(aq) \rightleftharpoons Al^{3+}(aq) + 3H_2O(l)$	8.51	4.4
Boehmite	$AlO(OH)(s) + 3H^+(aq) \rightleftharpoons Al^{3+}(aq) + 2H_2O(l)$	8.13	4.4
Gibbsite	$Al(OH)_3(s) + 3H^+(aq) \rightleftharpoons Al^{3+}(aq) + 3H_2O(l)$	8.04	4.3
Diaspore	$AlO(OH)(s) + 3H^+(aq) \rightleftharpoons Al^{3+}(aq) + 2H_2O(l)$	7.92	4.3

K_{eq} values from Sparks (2003, p. 129).

Table 4.9 A select set of minerals, which represent the range of aluminosilicate minerals participating in the aluminum soil buffer, with the pH they maintain in the presence of aqueous acid, as calculated by successive approximations.

Again, the complication of this soil buffer is that any of the reactions in Table 4.9 can represent the aluminum buffer, causing a range of buffering points between 5.3 and 4.3 until the aluminosilicate minerals have been exhausted by the acid. Mineral grain size also affects the ANC of the soil.

4.4.4 Biotic Buffering Systems

If the Earth were a sterile planet, then the description of soil buffering would be complete with inorganic buffering systems. Careful, controlled studies of the acidification of lakes and soil have revealed that biotic systems are often more significant compared to the mineral systems in certain environments. One such study, performed on Lake 223 in Ontario, Canada, showed that anaerobic microbes played the most significant role in buffering the lake's pH (Havas, 1989).This isolated lake was studied for 2 years in its pristine state in 1974 and 1975, and then a regimen of acidification using sulfuric acid lasted from 1976 to 1981, during which the pH dropped from 6.6 to 5.0. Most of the acidity was neutralized by the microbes in the lake sediment where anoxic conditions favored the reduction of sulfur to hydrogen sulfide, describe earlier in Table 4.5.

$$SO_4^{2-}(aq) + CH_3COO^-(aq) + 2H^+(aq) \rightarrow H_2S(g) + CO_2(g) + HCO_3^-(aq) + H_2O(l) \quad (R4.27)$$

$$C_6H_{12}O_6(s) + 3SO_4^{2-}(aq) + 6H^+(aq) \rightarrow 6CO_2(aq) + 3H_2S(g) + 6H_2O(l) \quad (R4.28)$$

Notice in the aforementioned reactions the consumption of H^+ ions as a consequence of anaerobic respiration of carbon nutrients.

Another significant buffering mechanism is performed by tree roots during the uptake of nutrients.

$$\text{root} : CO_3^{2-} + 2NO_3^- + H_3O^+(aq) \rightarrow \text{root} : (NO_3^-)_2 + HCO_3^-(aq) + H_2O(l) \quad (R4.29)$$

In this process, tree roots maintain charge neutrality by exchanging a carbonate ion with two nitrate ions, which could have been derived from acidic deposition (thus the presence of hydronium). Because the carbonate ion has a small pK_{b1}, carbonate ions neutralize the acid at any pH below 10.33. This buffering mechanism can be significant in a forest where roots are innumerable and nitrate fertilizer is in abundance. Ammonium fertilizers, however, add acid to soil because microbial nitrification produces acid, so the source of the nitrate ion (eventually derived from NO_2^-) is an important consideration.

$$2NH_4^+(aq) + 3O_2(g) + 2H_2O(l) \rightarrow NO_2^-(aq) + 4H_3O^+(aq) \quad (R4.30)$$

4.4.5 Buffering Summary

In summary, there are up to three important inorganic buffering mechanisms (assuming that the silicate buffering and cation exchange are the same) that can mitigate the effects of acid deposition. Some soils with carbonate minerals, such as limestone, maintain a pH around 8.2 (but in the range of 7.5–9.3) as long as the limestone buffering minerals are present. Once this buffering mechanism has been depleted or if was never present, then it is highly likely that a cation-exchange buffering mechanism will be present and will maintain a pH of around 6 (but in the range of 4.2–6.2) as long as the silicate minerals are still present and contain exchangeable cations. Finally, it is very likely that the soil will contain an aluminum buffering system and will maintain a pH of around 5 (but in the range of 5.3–4.3) as long as there are aluminum-based minerals present in the soil.

These buffering mechanisms can be roughly represented by the graph in Figure 4.5. The variations in mineral solubility, crystal structure, metal binding affinity, and composition make the description of these inorganic buffering systems less than precise. Generally, the limestone buffering system, expressed by various carbonate minerals, maintains a soil pH of 7–9, with the most common limestone ($CaCO_3$) centered at a pH of 8.2. Due to the sparsity of specific cation-exchange equilibria in the literature, the silicate/cation-exchange buffering system covers a pH range of 6.2–4.2. Finally, the aluminum buffering mechanism, also due to variations in mineral reactivity, covers a pH range of 5.2–4.2. Ulrich (1983) suggests the aluminum buffer range goes as low as a pH of 2.8.

There are also significant biotic mechanisms that buffer soils where trees are present and aquatic systems where anoxic conditions and sufficient sulfate ions prevail.

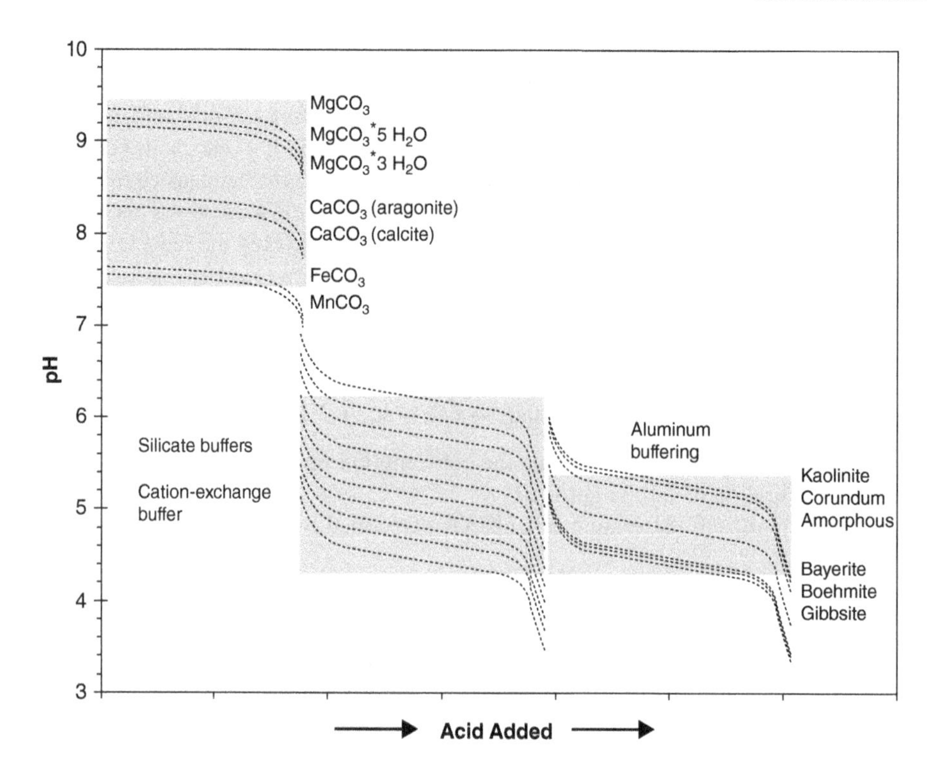

Figure 4.5 A pictorial summary of the inorganic buffering mechanisms found in soil. Variations in the limestone (top left), silicate and cation exchange (middle), and aluminum buffering mechanisms (bottom right) generate three different ranges of buffer center points, with a large amount of overlap in the lower two buffers. The graph simulates the titration of a soil sample assuming that all three inorganic buffering mechanisms are present.

4.4.6 Aluminum Toxicity

Plants exhibit a wide variety of preferences for soil pH. High bush blueberries (*Vaccinium corymbosum*) and Frasier firs (*Abies fraseri*), for example, prefer soils with a pH between 4.5 and 5.5. The American plum (*Prunus americana*) prefers soils with a pH between 6.5 and 8.5. As soil is exposed to chronic acid deposition cycles, the paramount concern with soil acidification is the mobilization of aluminum at pHs below 5.0. Aluminum toxicity can manifest itself in a variety of ways. Phosphate nutrients, for example, bind strongly to free aluminum cations, resulting in an insoluble form of phosphate that is unavailable to plants. The Al^{3+} cation can also bind to DNA, curtailing cell division in roots, and bind to certain enzymes involved in carbohydrate metabolism and polysaccharide assembly. Given the abundance of aluminum in the Earth's crust, the dangers of aluminum toxicity can be severe and widespread in areas with natural acidity and anthropogenic acid deposition. This is not to mention the toxicity of aluminum to aquatic organisms as the mobilized aluminum ions enter the watershed from the soil and concentrate in various bodies of water.

Certain organisms have developed mechanisms for limiting the toxicity of aluminum. The Al^{3+} cation can be detoxified by forming a complex with organic matter that is not important for cellular function. The Common Beach (*Fagus sylvatica*), Norway spruce (*Picea abies*), and Scots pine (*Pinus sylvestris*) employ mycorrhiza, a symbiotic fungal infestation of the root system, which excretes oxalic acid. The electron lone pairs on the oxygens of this organic acid effectively complexes the aluminum cation, detoxifying the soil for the host tree (Ulrich, 1983, p. 141). As this suggests, the more organic content in the soil horizon, the less likely there will be aluminum toxicity owing to the high degree of aluminum complexation and the decrease in free aluminum concentration as a result of the lower aluminosilicate content in the soil.

4.5 MEASUREMENTS

Soil contains a bewildering amount of compounds and elements, so in order to choose an instrument to make a feasible measurement, one has to define a narrow range of analytes to study. One could study the structures of the organic matter or the elemental composition of it. The mineral content of soil is copious and diverse, so a study of the mineral type or of the metal content of the mineral could be initiated. Metals can also be found in the organic matter in the form of metal–ligand complexes, so an extraction and study of the metal content of organic matter could be accomplished. Finally, there are a host of nutrients, both natural and anthropogenic, whose concentration could be a cause for concern when very low or very high. These nutrients are typically classified by their major element, such as N, C, S, and P. When in the environment, these nutrients are often seen as polyatomic anions, such as nitrate, carbonate, sulfate, and phosphate. These nutrients will be the focus of Chapter 5, and their analysis can be found at the end of that chapter.

4.5.1 Metals

Metals can be found in both the organic and inorganic portions of soil. In the organic material, metals are usually trace components and found to be adsorbed to organic functional groups that are highly polarized and electron-rich. These metals can usually be extracted into an aqueous solution by digesting the organic material with a strong acid or ashed in an oven (to remove the organic material) and extracted. Acid digestion involves the addition of a strong acid, such as nitric or hydrochloric acid, to the sample and then heating the mixture to increase the extraction kinetics. Microwave bombs (Figure 4.6) are often used in conjunction with acid digestion. A microwave bomb usually consists of a polytetrafluoroethylene (PTFE) canister into which the sample and the digesting acid are placed. The lid to the canister forms a seal to withstand the increase in pressure that comes with the heating process. The canister is then placed inside of a larger plastic vessel that has a top that screws onto the vessel to generate a pressurized seal that will prevent the sample from erupting from the canister. The vessel is then placed in a microwave oven and heated for a period of time. Since the microwave radiation is only absorbed by the solution inside the vessel, heating of the sample is rapid because the rest of the vessel does not absorb any radiation. High temperatures and pressures in the bomb encourage thorough extraction of metals from the sample.

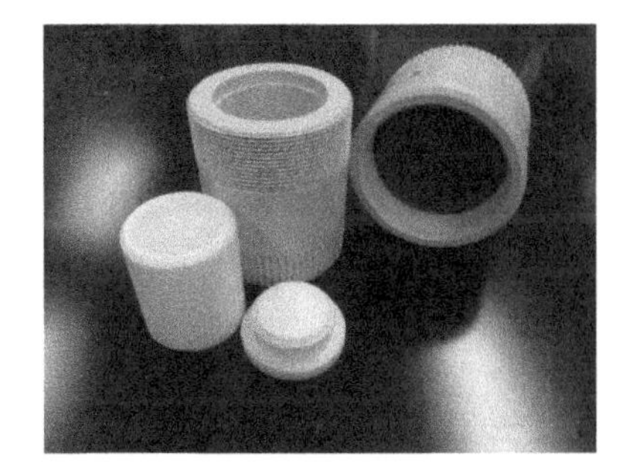

Figure 4.6 Microwave bombs provide a fast and efficient means of acid digestion. A sample and a digesting acid are placed in the PTFE canister, which is inserted into the bomb vessel. High temperatures and pressures can be safely achieved in minutes. Courtesy Ken Overway, 2013.

When the sample matrix is organic matter, typically a strong acid with some oxidizing power, such as nitric acid, is used. Sulfuric acid can also be used in a mixture with HNO_3, but because its boiling point is higher than the melting point of the PTFE canister, the use of H_2SO_4 should be done carefully and with limited heating times. If the sample matrix is a nonsilicate mineral, such as a carbonate or phosphate, then hydrochloric acid or aqua regia (1:3 vol/vol HNO_3:HCl) is sufficient for digestion. For silicate minerals, only hydrofluoric acid (HF) will dissolve the matrix. If the analysis requires only the metals on the surface of a finely ground silicate mineral, then HF is not needed and HNO_3 or HCl is sufficient. Once heated, the vessel must be allowed to cool and depressurized before the vessel is unscrewed and the canister opened. Extreme caution should be exercised when using any of these acids, especially HF. Because of its explosive nature, perchloric acid $HClO_4$ should be avoided. If you are going to use a microwave bomb, carefully read and follow the manufacturer's instructions and guidelines.

The acid digestion process should liberate most of the available metals from the matrix. The sample must now be filtered to remove any undissolved material and placed in a volumetric flask (avoid using glass if HF was used as the digesting acid). Depending on the sensitivity of the instrumentation, a dilution of the sample may be necessary to prevent a highly concentrated sample from overloading the detector and causing the analytical signal to be nonlinear.

Once in the solution phase, the metals can be quantitatively analyzed by the use of either atomic absorption spectroscopy (AAS) or an inductively coupled plasma (ICP) atomizer in tandem with either a mass spectrometer (MS) or an optical emission spectroscope (OES) in the form of an ICP-MS or an ICP-OES instrument. These instruments, described in more detail in Appendix F, introduce the sample to a flame where it is atomized and detected either by the absorption of light (AAS), by the photonic emission of the analyte (ICP-OES), or by the selective detection of the analyte's mass (ICP-MS). AAS is the least sensitive of the three instruments, so only samples with analytes greater than a few mg/L, depending on the analyte and the soil pH, should be used with it. Both instruments have the selectivity to measure single or multiple analytes in a mixture, so separating the analyte from the other soluble components of the mixture is usually not necessary. There are some interfering components that may degrade the signal, so you should consult the manufacturer's guidelines for the particular analyte. These manuals often list the interferents of concern and suggest ways to minimize their effects.

Another common, powerful, and nonspectroscopic method for the analysis of metallic cations is ion chromatography (IC). This instrument, also described in more detail in Appendix F, separates the metal cations as a function of time based on their affinity for the stationary phase. Each metal elutes from the instrument at a unique time and is detected based on an electrochemical detector.

A less common instrument found in some well-equipped colleges and universities is an X-ray fluorescence spectroscope (XRF). In XRF, the core electrons of metals are excited via absorbed *x*-rays. When the core electrons fall back to their ground states, they emit photons whose wavelengths are characteristic of the metal. One of the major advantages of XRF is that it does not require acid digestion and the extensive sample preparation that the solution-phase measurements require. This reduces analysis time and does not destroy the sample. It is not a very sensitive method and is comparable to AAS.

4.5.2 pH and the Equilibrium Soil Solution

Since pH is an aqueous phenomenon involving the hydronium ion (H_3O^+), soil itself does not have a measurable pH. Soil pH is determined by adding neutral, deionized water to the soil and allowing it to equilibrate. This is sometimes referred to as the equilibrium soil solution (ESS). Some procedures suggest that minimal water should be added to the soil sample in order to form a paste (Reeve, 2002) or slurry, enough moisture to wet the bulb of a pH meter. The Virginia Tech Soil Testing Laboratory combines the soil and distilled water in a 1:1 vol/vol ratio (Maguire and Heckendorn, 2009). Using more water than a 1:1 ratio often produces a solution with low salt content, making the pH calibration standards very dissimilar (and introducing a systematic error) and also generates a dilute solution with very little buffering capacity in which it is difficult to measure a stable pH (Vaughan, 1998).To avoid this problem, a 0.01 M $CaCl_2$ or 1.0 M KCl solution is added to the soil instead of water, producing a high ionic strength solution that renders a stable pH measurement. Using a salt solution also makes comparison of the pH of samples easier because it masks the differences in salt content in each soil sample, which can affect the pH measured.

Once the ESS is made, it must be allowed to equilibrate, and this is aided by stirring. Equilibration times vary from 1 hour (Reeve, 2002) to 20 s (Vaughan, 1998).While the solution is equilibrating, a standard pH meter should be calibrated using buffered calibration solutions. If a two-point calibration is used and if the soil will have a pH less than 7, then calibration solutions with a pH of 4 and 7 should be used. If the pH of the soil is greater than 7, pH 7 and 10 calibration solutions would be preferred in order to improve accuracy. Once the pH probe is calibrated, it is advisable to turn off the stirring in the soil sample, gently stir the pH probe in the soil for a few seconds, and then let the probe equilibrate. Because pH measurements of the ESS using pure water can vary from the ESS using a salt solution, it is important to record, along with the pH, exactly which solution was used.

Temperature can be a very large source of error in pH measurements. Meters that include an automatic temperature compensator (ATC) probe help to mitigate changes in temperature. If an ATC is not available, it is important that the pH meter is calibrated at the same temperature as the sample in situations where measurements are made in the field. If it is not possible or convenient to calibrate the pH electrode at the same temperature as the samples, then a manual correction can be made. The temperature compensation factor is 0.003 pH/°C/pH unit away from pH 7.

4.6 IMPORTANT TERMS

- ★ soil horizon
- ★ physical weathering
  - · abrasion
  - · freeze/thaw
  - · expansion/contraction
  - · biological forces
- ★ chemical weathering
  - · hydration
  - · hydrolysis
  - · oxidation/reduction
  - · dissolution
- ★ oxidation state
- ★ adsorption
- ★ silicate sheets
- ★ aluminosilicates
- ★ cation exchange
- ★ biopolymers
  - · polysaccharides→simple sugars
  - · protein→amino acids
  - · lipids→fatty acids
- ★ leaf senescence
- ★ humus
- ★ humic acids
- ★ lignin
- ★ microbial degradation
  - · aerobic oxidation
  - · anaerobic oxidation
  - · acidogenesis
  - · acetogenesis
  - · methanogenesis
- ★ aerobic conditions
- ★ anoxic conditions
- ★ microorganism classes
  - · autotrophs
  - · chemoautotrophs
  - · photoautotrophs
  - · heterotrophs
  - · aerobes
  - · anaerobes
  - · obligate/facultative
  - · thermophiles
  - · mesophiles
  - · psychrophiles
- ★ Nernst equation
- ★ intensive properties
- ★ extensive properties
- ★ redox ladder
  - · oxygen respiration
  - · nitrate respiration
  - · sulfate respiration
  - · disproportionation respiration
- ★ ligand
- ★ metal complex
- ★ heavy metals
- ★ Class A metal
- ★ borderline metal
- ★ Class B metal
- ★ phytoremediation
- ★ inorganic soil buffers
  - · limestone/carbonate
  - · cation exchange
  - · aluminum exchange
- ★ biotic soil buffering
- ★ acid-neutralizing capacity

EXERCISES

1. Describe the purpose of dividing soil into horizons. Why are they useful?
2. Which soil horizon is likely to be the most effective (short term) at buffering soil: Horizon B (containing small particle sizes and high surface area) or Horizon C (containing some small particles and some larger particles with less surface area)?
3. Separate the items in the following list into chemical and physical weathering processes: hydration, abrasion, freeze–thaw, dissolution, oxidation/reduction, hydrolysis, expansion/contraction.

4. Which type of mineral is most abundant on the Earth's surface? Why? Consider what you know about the formation of the Earth.

5. What role does water play in the formation of biopolymers (saccharides, proteins, and fats)?

6. How are lignins made hydrophilic as a result of leaf senescence?

7. Organic matter is further degraded into what chemical forms by what processes?

8. Classify the microbes that are used in the tertiary treatment stage of wastewater, by their metabolic pathway, temperature preference, and food source. These "bugs" thrive in the tanks with alternating stages of vigorous mixing with air and stages where dissolved oxygen drops to near zero, and they consume the organic nutrients at temperatures between 0 and 40.

9. List the important chemical species that replace O_2 in anaerobic respiration reactions in the order of their energetic potential under environmental conditions. What chemicals do these species (C, N, S elements) produce in the reactions?

10. In the limestone buffer system involving minerals such as $CaCO_3$ and $MgCO_3$, which ionic species does the actual buffering and how does the other species make a difference in the buffer?

11. Write a representative reaction for each inorganic buffering mechanism, listing the pH range where each buffers the soil.

12. Explain which of the soils in Table 4.10 would maintain the highest pH.

Soil 1		Soil 2		Soil 3	
Compound	Percent composition	Compound	Percent Composition	Compound	Percent Composition
SiO_2	43.38	$MgCO_3$	37.68	SiO_2	57.29
$NaAlSi_3O_8$	28.91	SiO_2	32.17	Al_2O_3	27.35
Fe_2SiO_4	14.55	Al_2O_3	26.84	Fe_2O_3	10.32
Al_2O_3	13.16	Humus	3.31	TiO_2	5.04

Table 4.10 Simulated compositional analysis of some soil samples.

13. Which class of metal would have a stronger affinity to humic acids with only carboxylate groups?

14. Which class of metal would be more likely to bond with proteins containing sulfur?

15. Given the data in Table 4.11, write a short narrative describing the level and type of microbial activities (identified by respiration preferences) that are thriving at each stage of the measurement.

Days	O_2(aq) (mg/L)	CO_2(aq) (mg/L)	CH_4(aq) (mg/L)	H_2S(aq) (mg/L)
1	6.2	0.6	–	–
2	1.5	1.3	–	–
5	0.3	1.8	–	–
10	–	2.0	0.6	0.01
15	–	2.1	0.8	0.16

Table 4.11 Simulated water analysis after a contamination of organic waste.

16. What deleterious effects does soluble aluminum have on plants?

17. List two major instruments that are used to measure the elemental analysis of metal in various samples?

BIBLIOGRAPHY

Agency for Toxic Substances and Disease Registry (ATSDR), ToxGuide of Hydrgen Sulfide. Electronic, 2013; http://www.atsdr.cdc.gov/toxguides/toxguide-114.pdf.

Blaylock, M. J.; Salt, D. E.; Dushenkov, S.; Gussman, C.; Kapulnik, Y.; Ensley, B. D.; Raskin, I. *Environmental Science and Technology* **1997**, *31*, 860–865.

Buchanan-Wollaston, V. *Journal of Experimental Botany* **1997**, *48(307)*, 181–199.

Duffus, J. H. *Pure and Applied Chemistry* **2002**, *74(5)*, 793–807.

Evangelou, V. P. *Environmental Soil and Water Chemistry*, 1st ed.; Wiley Interscience: New York, 1998.

Famulari, S. Phytoremediation. Electronic, 2013; http://www.ndsu.edu/pubweb/famulari_research/index.php.

Fazeli, A. R.; Tareen, J. A. K.; Basavalingu, B.; Bhandage, G. T. *Proceedings of the Indian Academy of Sciences - Earth and Planetary Sciences* **1991**, *100(1)*, 37–39.

Federer, A. A.; Hornbeck, J. W. *Water, Air, & Soil Pollution* **1985**, *26*, 163–173.

Flory, J. EPA Found High Arsenic Levels Day After Ash Spill, Knoxville News Sentinel, January 3, 2009; http://www.knoxnews.com/news/2009/jan/03/epa-found-high-arsenic-levels-day-after-ash-spill/.

Fosbol, P. L.; Thomsen, K.; Stenby, E. H. *Corrosion Engineering Science and Technology* **2010**, *45(2)*, 115–135.

Grispen, V. M.; Nelissen, H. J.; Verkleij, J. A. *Environmental Pollution* **2006**, *144*, 77–83.

Gurvich, L. V.; Veyts, I. V.; Alcock, C. B. *Thermodynamic Properties of Individual Substances*, 4th ed.; Hemisphere Publishing Corp.: New York, 1991; Vol. 2.

Hallam, D. M.; Liao, J.; Choi, K. *Manure Pit Injuries: Rare, Deadly, and Preventable*; 2011.

Hancock, L. M. S.; Ernst, C. L.; Charneskie, R.; Ruane, L. G. *American Journal of Botany* **2012**, *99(9)*, 1445–1452.

Hanson, B.; Garifullina, G. F.; Wangeline, S. D. L. A.; Ackley, A.; Kramer, K.; Norton, A. P.; Lawrence, C. B.; Pilo-Smits, E. A. H. *New Phytologist* **2003**, *159*, 461–469.

Havas, M. In *Acid Precipitation: Soils, Aquatic Processes, and Lake Acidification*; Norton, S. A., Lindberg, S. E., Page, A. L., Eds.; *Advances in Environmental Science*; Springer-Verlag: New York, 1989; Vol. 4; pp. 189–193.

Haynes, W. *CRC Handbook of Chemistry and Physics*, 93rd ed.; *CRC Handbook of Chemistry and Physics*; Taylor & Francis Group: Boca Raton, FL, 2012.

Inglett, P. W.; Reddy, K. R.; Corstanje, R. Anaerobic Soils, In *Encyclopaedia of Soils in the Environment*, 1st ed.; Hillel, D., Ed.; Academic Press, University of Virginia: Charlottesville, VA, 2004; Vol. 2; p. 72.

Kim, M.; Gomec, C. Y.; Ahn, Y.; Speece, R. E. *Environmental Technology* **2003**, *24(9)*, 1183–1190.

Lepedus, H.; Stolfa, I.; Radic, S.; Perica, M. C.; Pevalek-Kozlina, B.; Cesar, V. *Croatica Chemica Acta* **2008**, *81(1)*, 97–103.

Lui, W.; Zhou, Q.; Zhang, Z.; Hua, T.; Cai, Z. *Journal of Agricultural and Food Chemistry* **2011**, *59*, 8324–8330.

McBride, J. P.; Moore, R. E.; Witherspoon, J. P.; Blanco, R. E. *Radiological Impacts of Airborne Effluents of Coal-Fired Power Plants and Nuclear Power Plants*; Rep. ORNL-5315, 1977.

Maguire, R.; Heckendorn, S. Soil Test Note 1 - Explanation of Soil Tests. Electronic, 2009.

Manousaki, E.; Kalogerakis, N. *Industrial and Engineering Chemistry Research* **2011**, *50*, 656–660.

Milner, M. J.; Kochian, L. V. *Annals of Botany* **2008**, *102*, 3–13.

Myers, C. R.; Nealson, K. H. *Science* **1988**, *240(4857)*, 1319–1321.

Nieboer, E.; Richardson, D. *Environmental Pollution (Series B)* **1980**, *1*, 3–26.

Nisman, B. *Bacteriology Reviews* **1954**, *18(1)*, 16–42.

Ostrem, K. Greening waste: anaerobic digestion for treating the organic fraction of municipal solid wastes. PhD thesis, Department of Earth and Environmental Engineering, Columbia University, 2004.

Puget Sound. LID Technical Guidance Manual for Puget Sound. *Sampling of Plant Species Studied for Phytoremediation*. Puget Sound Partnership: Washington, **2005**; pp. 209–214.

Rastogi, A.; Ghosh, A. K.; Suresh, S. In *Thermodynamics - Physical Chemistry of Aqueous Systems*; Moreno-Pirajan, J. C., Ed.; InTech, 2011, http://www.intechopen.com/books/thermodynamics-physical-chemistry-of-aqueous-systems.

Reeve, R. N. *Introduction to Environmental Analysis*; John Wiley & Sons, Ltd: West Sussex, 2002.

Reuss, J. O.; Walthall, P. M. In *Acid Precipitation: Soils, Aquatic Processes, and Lake Acidification*; Norton, S. A., Lindberg, S. E., Page, A. L., Eds.; *Advances in Environmental Science*; Springer-Verlag: New York, 1989; Vol. 4; pp. 1–31.

Roue, R. A.; Haselton, H. T. Jr.; Hemingway, B. S. *American Mineralogist* **1984**, *69*, 340–357.

Rout, G.; Samantaray, S.; Das, P. *Agronomie* **2001**, *21(1)*, 3–21.

Schaeffer, E. *TVA's Toxic Legacy: Groundwater Contaminated by Tennessee Valley Authority Coal Ash*; 2013.

Shah, K.; Kumar, R. G.; Verma, S.; Dubey, R. *Plant Science* **2001**, *161*, 1135–1144.

Sparks, D. *Environmental Soil Chemistry*, 2nd ed.; Academic Press: San Diego, CA, 2003.

Spencer, H. Manure Pit Tragedy Not Unprecedented. Electronic, 2007; http://www.readthehook.com/76922/manure-pit-tragedy-not-unprecedented.

Sposito, G. *The Chemistry of Soils*, 1st ed.; Oxford University Press: Oxford, 2008.

Szczygłowska, M.; Piekarska, A.; Konieczka, P.; Namieśnik, J. *International Journal of Molecular Science* **2011**, *12(11)*, 7760–7771.

Ulrich, B. In *Effects of Accumulation of air pollutants in forest ecosystems.*; Ulrich, B., Pankrath, J., Eds.; Reidel: Boston, MA, 1983; pp. 127–146.

van Assche, F.; Clijsters, H. *Plant, Cell and Environment* **1990**, *13*, 195–206.

van Baren, J. *Mitteilungen des Geologischen Instituts der Landbouwshogeschool in Wageningen*; H. Veenman & Zonen, 1930.

van Baren, J. *Mededelingen: Landbouwhogeschool*; H. Veeman & Zonen: Wageningen, Netherlands, 1931.

Vaughan, B. Soil pH Troubleshooting. Electronic, 1998; http://www.naptprogram.org/files/napt/publications/method-papers/1998-soil-ph-troubleshooting.pdf.

Wershaw, R. L.; Leenheer, J. A.; Kennedy, K. R. In *Understanding Humic Substances: Advanced Methods, Properties and Applications*; Davies, G., Ghabbour, E. A., Eds.; The Royal Society of Chemistry: Cambridge, 1999; pp. 19–30.

Ye, R.; Jin, Q.; Bohannan, B.; Keller, J. K.; McAllister, S. A.; Bridgham, S. D. *Soil Biology and Biochemistry* **2012**, *54*, 36–47.

Yeh, J.-M.; Chang, K.-C. *Journal of Industrial and Engineering Chemistry* **2008**, *14(3)*, 275–291.

Zen, E.-A.; Chernosky, J., Joseph V. *American Mineralogist* **1976**, *61*, 1156–1166.

5

THE HYDROSPHERE

When the well's dry, we know the worth of water.

—Benjamin Franklin

I never drink water because of the disgusting things that fish do in it.

—W. C. Fields

5.1 INTRODUCTION

Of all of the substances on the planet, water holds the most prominent place in our conscience. Over half of our body weight is due to water, and it covers about 70% of the Earth's surface. These oceanic reservoirs hold a prominent position in our literature and contain the last remaining frontier of the unknown on the planet. It is also one of the most precious resources we share with the biosphere. Healthy human populations and productive croplands depend on the amount and quality of the available water. Yet, the global watershed has become the intentional and unintentional receptacle for residential and industrial waste. W. C. Fields, and all of us, should worry less about what the fish are doing in the water and more about what we do to the water.

Still, the properties and function of water in the geobiosphere are a wonder of nature and will require a wide range of chemistry to understand.

5.2 THE UNUSUAL PROPERTIES OF WATER

In chemical terms, water is a very unusual substance. It is a light molecule, and yet, compared to others of the same molar mass, it is liquid instead of gaseous at room temperature. It is less dense in its solid phase, whereas a vast majority of other substances are more dense as a solid. It is also an effective solvent, into which nonpolar gases, liquids, and solids dissolve to a small extent and polar substances to a large extent. Finally, its bonding structure is labile enough to enable proton exchange between solute species, creating the whole phenomenon of Arrhenius acid–base chemistry.

The electronic structure of water holds the key to all of these properties. The Lewis structure, seen in Figure 5.1, shows two hydrogen atoms attached to an oxygen atom with a bent molecular geometry. This arrangement results in a highly polarized structure owing to the electronegativity differences between the two atoms ($\Delta EN = 3.5 - 2.1 = 1.4$). The polarity of the molecule generates a strong intermolecular force, which causes water to reside in a liquid form, while other heavier substances remain gaseous (such as N_2, O_2, CO_2, and Cl_2). This dipolar force is so ubiquitous in chemistry and biochemistry that it is given a distinct name – hydrogen bonding.

Figure 5.1 The Lewis structure of water shows a molecule in the tetrahedral orbital group with a bent molecular shape having an average bond angle of 104.5°.

Environmental Chemistry: An Analytical Approach, First Edition. Kenneth S. Overway.

Companion website: www.wiley.com/go/overway/environmental_chemistry

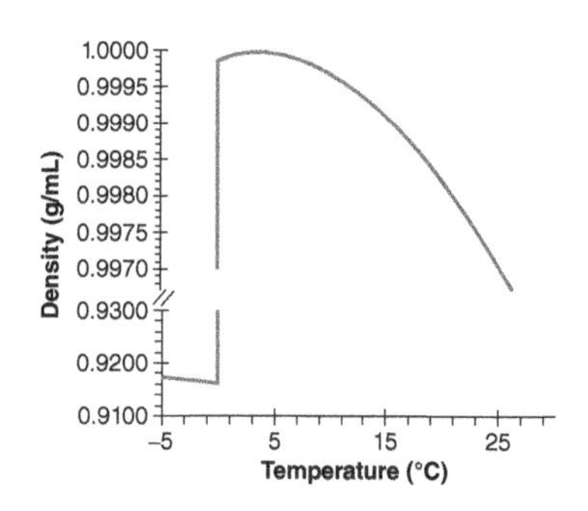

Figure 5.2 The density of liquid water reaches its maximum at about 4 °C. As it warms, the density continues to decrease. Notice the large change in density when it freezes. Source: Data from Eric W. Peterson, Illinois State University.

The Lewis structure also shows four potential points of hydrogen bonding between other molecules. On average, the liquid phase of water exhibits 2–3 hydrogen bonds per molecule. This allows the molecules to remain condensed and yet move extensively throughout the solution. In contrast, the solid phase maximizes the number of hydrogen bonds, resulting in a solid-phase crystal structure that takes up more space compared to the liquid phase – about 8% less dense at 0 °C. The transition between the liquid and solid phases must necessarily reflect this drastic change in density. Figure 5.2 shows the density of water as a function of temperature. Water behaves similarly to other liquids in that its density increases as temperature decreases, but at about 4 °C, the trend reverses and the density decreases. This represents the point where the molecule is thermodynamically forced to adopt its lowest energy intermolecular orientation by forming the fourth hydrogen bond. The resulting macromolecular orientation takes up more volume, reflecting the density decrease and leading to our common experience – ice floats on water.

Water's polarity is also responsible for its effectiveness as a solvent to polar solutes. The molecular dipole of water orients itself around ions or other polar molecules to form macromolecular cages. This ability to form interactions with other species often results in a net decrease in enthalpy of the system, evolving heat into the solution. The dissolution of some solutes is endothermic, and in these cases, the solubility is driven by the increased entropy of dissolution.

Review Example 5.1: The thermodynamics of solubility

When potassium chloride dissolves in water, the resulting solution temperature decreases. When methanol dissolves in water, the solution temperature increases.

Exercises

1. Calculate the change in enthalpy, entropy, and free energy for the dissolution of potassium chloride in water at 25 °C.
2. Calculate the change in enthalpy, entropy, and free energy for the dissolution of methanol in water at 25 °C.
3. Explain the temperature observations.

Solution: See Section E.1 on page 285.

Water's role as a solvent in environmental systems means that no thorough understanding of the hydrosphere is possible without studying the many chemical equilibria that are present in an aqueous system. The various equilibrium constants and reactions shown in Figure 5.3 are critical to understanding the distribution of environmentally significant chemicals. Your previous exposure to chemical equilibria in general chemistry assumed a *closed system* where only heat is exchanged between the system and surroundings while no chemicals are lost to or added from the surroundings. In a closed system, reactions eventually reach equilibrium, and straightforward equilibrium calculations are sufficient to predict the concentrations of products and reactants. The actual environment is an *open system* where very few chemical reactions reach equilibrium because matter is constantly being exchanged between the system and surroundings. Fortunately, a model does not have to be strictly true for it to be useful. While real environmental chemical reactions take place under dynamic conditions, the simplified model presented here will allow you to make general descriptions of aquatic systems that are still useful.

On a small scale, you may have learned that aqueous solutions are homogeneous, but in the natural environment, this rarely is the case. Besides regional differences in dissolved organic and inorganic components, heat and salinity play a major role in the heterogeneity of waterbodies, both small and large, and cause water circulation on local and global scales.

5.2.1 Freshwater Stratification

Figure 5.2 shows clearly that changes in temperature cause changes in water density. The inverse relationship between temperature and density results in a stratification of most

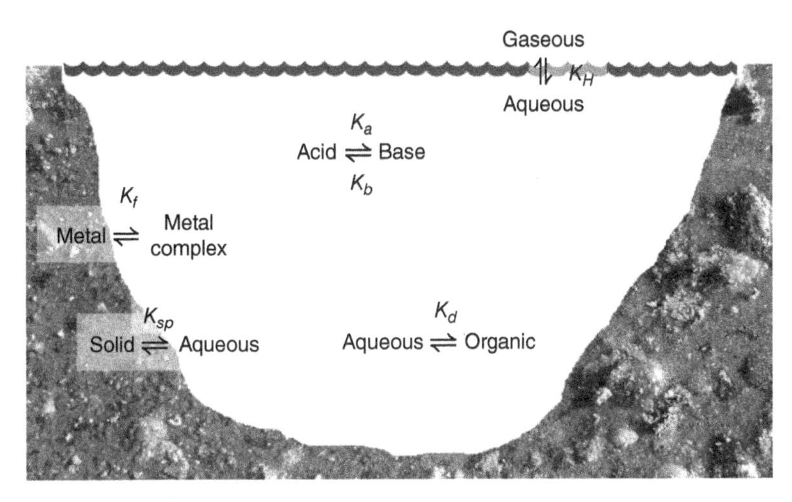

Figure 5.3 Understanding the hydrosphere requires an understanding of chemical equilibria of several types, some of which were shown earlier. The exchange of gases between air and water is described by Henry's Law (K_H). Acid–base reactions are integral to water chemistry (K_a and K_b). The exchange of salts and metals between the surrounding soil and the aqueous phase is a function of solubility products (K_{sp}) and complex ion formations (K_f). Finally, solutes distribute themselves between the water and hydrophobic solutes and sediment as described by a distribution coefficient (K_d).

☞ An *open system* is one where matter in the system can exchange with the surroundings, such as water evaporating from a lake (system) into the air (surroundings). Heat can also flow from the system to the surroundings. In a *closed system*, heat can be exchanged, but no matter is exchanged with the surroundings. In a closed system, there is a fixed total mass. This is an ideal situation that is rarely observed in nature.

waterbodies. In freshwater lakes, the stratification occurs in the warmer months when the upper layer of water, called the *epilimnion*, receives daily heat from the Sun and can exchange gases with the atmosphere. As a result, the epilimnion is warmer, has higher levels of dissolved oxygen (DO), and is less dense compared to the layers below it. At the bottom of a lake, the *hypolimnion* maintains the coldest and densest water, and without access to the atmosphere, DO levels are usually low. Separating these two layers is the *metalimnion* or *thermocline*, which contains a rapid vertical change in density and prevents the epilimnion and hypolimnion from mixing. The higher oxygen levels in the epilimnion support a more diverse aerobic ecology, while the hypolimnion contains high levels of nutrients as a result of *detritus* from the epilimnion sinking to the hypolimnion and supports an anaerobic ecology.

This stratification remains relatively stable as long as the epilimnion maintains warm temperatures and low water density. In colder months, winter temperatures cool the epilimnion and decrease the density difference between the layers. As a result, surface water sinks and deeper water is pushed to the surface. This *overturn* leads to an increase in the DO levels of deeper waters and an increase in the nutrient levels in surface waters. Stratification returns when rates of heating at the surface exceed the rate of heat exchange during the mixing process.

☞ *Freshwater stratification*, driven by heat from the Sun and atmosphere, causes the formation of an *epilimnion* at the top of a freshwater lake and a *hypolimnion* at the bottom, separated by a *thermocline* or *metalimnion*. Stratification in the ocean, resulting from heat and salinity, produces large-scale *thermohaline conveyors* that circulate water and nutrients throughout the globe.

5.2.2 The Thermohaline Circulation

A related stratification occurs in the oceans, where the upper layer contains water of a relatively uniform temperature. Ocean stratification, however, is much more dynamic than in freshwater, forming vast conveyor systems that circulate water across the globe. The Gulf Stream, emerging out of the Gulf of Mexico, contains relatively warm water (10–20 °C) and moves northward along the surface of the northern Atlantic Ocean toward northern Europe. Over this distance, the water temperature drops to around 10 °C and evaporates more moisture into the atmosphere than it gains from precipitation. When this stream reaches Greenland, it has become cold and salty compared to the surrounding waters. This combination increases the water density and causes the stream to sink to the ocean floor and flow southward along the Greenland continental shelf, continuing all the way to Antarctica,

☞ *Detritus* is a particulate organic material formed as a result of biological activity. In aquatic systems, it is often called "marine snow" since it falls from the epilimnion and piles up on the ocean floor. In terrestrial environments, detritus is mostly composed of leaf litter but can contain dead organisms and animal feces. In either environment, detritus is home to legions of microorganisms and is the source for much of the nutrients for other organisms.

where it joins a deep ocean, circumpolar stream that encircles the polar continent. Branches of this circumpolar stream eventually rise to the surface in various locations, resulting in a redistribution of nutrients that is similar to the overturn of freshwater lakes.

This *thermohaline circulation*, driven by heat and salinity, is responsible for maintaining a climate in northern Europe which is much warmer than its latitude suggests, and for delivering vast quantities of nutrients that maintain very productive fisheries around the globe. Some climatologists are speculating that a warmer globe due to anthropogenic climate change may attenuate some of these thermohaline conveyors. A weakened Atlantic thermohaline conveyor would bring a colder climate to northern Europe because it would receive less heat from the Gulf Stream.

5.2.3 Salinity

Salinity is one of the most obvious and essential properties of seawater that makes it different from freshwater. It is defined as the mass of dissolved inorganic matter in 1 kg of seawater after all fluoride, bromide, and iodide are replaced by an equivalent of chloride ion. The halides, other than chloride, typically compose no more than 0.2% of the solute mass, so ignoring the contribution of the other halides results in a small error.

Most of the planet's seawater contains a narrow range of salinity, from 33‰ to 37‰. Relatively isolated seawater systems, such as the Baltic Sea (<10‰) and the Great Salt Lake (between 50‰ and 270‰), can vary widely from the norm, but small differences in salinity and temperature are responsible for the thermohaline circulation described earlier.

The measurement of salinity, described in Section 5.2.3, is part of a measurement of total dissolved solids and can be accomplished by means of evaporation or conductivity.

☞ The permil symbol ‰ is often used to represent parts per thousand in order to avoid confusion with parts per trillion (ppt). Just as % represents parts per hundred, the ‰ represents parts per thousand. The calculation in Example 5.1 is set up for ‰ since the units are g per kg (a ratio of 1 to 1000). The Mediterranean sea, for example, has an average salinity of 38‰, which is equivalent to 3.8%.

Example 5.1: Saliniity

What is the salinity of a 100.0-g seawater sample if a preweighed 250-mL beaker (58.7254 g) containing the sample was boiled to dryness, baked in a drying oven overnight at 180 °C, cooled in a desiccator, and reweighed to give a mass of 62.0541 g?

Solution: Since salinity is defined as the mass of solute that is dissolved in 1 kg of seawater, the difference between the beaker after boiling down the sample and the clean beaker was 3.3287 g. Since this represents mostly the mass of inorganic solute (there is very little dissolved organic matter in seawater, and much of organic solute would likely evaporate with the temperatures required to boil down the sample), this would amount to the following salinity.

$$Salinity = \left(\frac{3.3287 \text{ g solute}}{0.1000 \text{ kg seawater}}\right) = 33.29‰ \quad (5.1)$$

5.3 WATER AS A SOLVENT

As surface dwellers, humans interact with two major fluid solutions regularly. One is our atmosphere, discussed in Chapter 3 extensively, which is composed of molecular nitrogen (usually considered the "solvent"), molecular oxygen, argon, and several trace components that vary greatly from region to region. Our hydrosphere is the other solution, where water is the solvent and a host of other species are the solutes. While water is often called the universal solvent, its polarity makes the solubility of polar solutes favorable. Because the hydrosphere is in intimate contact with both the lithosphere and the atmosphere, we find both dissolved gases and solids present in surface, ground, and rain waters.

5.3.1 Dissolved Solids

Table 5.1 shows the wide range of dissolved constituents in natural waters. The most prominent salt in all sources of water is sodium chloride, with sulfate, magnesium, bicarbonate, and potassium ions completing the list of major components. Most of these dissolved ionic salts (and silica) come from hydrospheric contact with geologic sources, with freshwater watersheds dissolving minerals as they drain precipitation from landmasses to the oceans. The exception is the bicarbonate ion, which can also come from dissolved CO_2 from the atmosphere. Curiously, the calcium and bicarbonate ions are sometimes the dominant solutes in freshwater systems, but not in the oceans, where the pH is more alkaline. This is the result of chemical equilibrium processes that limit the concentration of certain species as the result of pH and solubility. In fact, since ocean salinity is relatively stable, these processes must be responsible for the removal of solute species as fast as rivers deliver more dissolved solids.

Species	Average seawater[a] (mg/L)	Amazon River[b] (mg/L)	Mississippi River[b] (mg/L)	Great Salt Lake[c] (mg/L)	California Rain Water[c] (mg/L)	New Mexico groundwater[c] (mg/L)
Cl^-	19,000	1.9	1.9	140,000	17	17
Na^+	10,500	1.8	20	83,600	9.4	37
SO_4^{2-}	2,700	3	7.9	16,400	7.6	15
Mg^{2+}	1,350	1.1	10	7,200	1.2	1.1
Ca^{2+}	410	4.3	38	241	0.8	6.5
K^+	390		2.9	4,070		3
HCO_3^-	142	19	113	251	4	77
Br^-	67					
Sr^{2+}	8					
SiO_2	6.4	7		48	0.3	103
B	4.5					
F^-	1.3	0.2	0.3			
pH	7.5–8.4	6.5	7.4	7.4	5.5	6.7

[a] From Hem (1985, p. 7) Table 2.
[b] From Hem (1985, p. 9) Table 3.
[c] From Nelson (2013).

Table 5.1 Major constituents of various waterbodies.

The rejection of the steady-state concentrations of the solutes in seawater was used by creationists to challenge the deep age of the Earth. Assuming that their god had made the oceans slightly salty at the creation and the net annual additions of salt from terrestrial watersheds, Austin and Humphreys estimated that the Earth's oceans are a maximum of 62 million years old given the current salinity of the oceans. Since they claim that a massive addition of salt from Noah's flood greatly increased salinity levels as a result of "fountains of the great deep" (Genesis 7:11), they claim that the likely age is closer to the biblically supported age of 6000 years. Confirmation bias? (Austin and Humphreys, 1990)

Sparingly soluble salts, such as calcium and magnesium carbonates, precipitate as limestone sedimentary rocks. Biological processes contribute to the loss of these carbonates through the formation of exoskeletons and coral reefs. As you learned in Chapter 4, the sulfate ion is used by some anaerobic organisms as an electron acceptor in respiration reactions. These sulfur-reducing microbes contribute to the production of the sulfide ion and the formation of the mineral pyrite (FeS_2) and keep the iron and sulfur concentrations in check. Deep-sea hydrothermal vents also contribute to the oceanic homeostasis. Seawater that percolates into the oceanic crust eventually comes into contact with hot basalt rocks, causing the formation of insoluble magnesium and calcium minerals. The water that emerges from hydrothermal vents on the ocean floor has a very low concentration of these cations and contributes to the steady-state concentrations in the larger ocean. Sea spray introduces millions of metric tonnes of salt, mostly sodium chloride, into the atmosphere annually and redistributes this salt over the entire planet.

The salinity of the oceans has a number of effects on the properties of the solution. The freezing point of water is depressed by the interaction with the solutes, according to the colligative properties of water, resulting in a freezing point around −2 °C. Seawater is about 2.5% denser than freshwater, affecting the mixing rates of freshwater and seawater in *estuaries*. Finally, the salinity decreases the solubility of gases in solution, affecting the extremely important parameter known as dissolved oxygen.

☞ An *estuary* is a partially enclosed coastal body of water at the terminus of a river where freshwater and salt water mix.

5.3.2 Dissolved Oxygen

One of the most important indicators of water quality is the amount of molecular oxygen dissolved in solution. Gases can be dissolved into surface waters as a result of the air–water interface. The degree to which this occurs is governed by Henry's Law, named after the English chemist William Henry (1774–1836),

$$K_H = \frac{[O_2]}{P_{O_2}} \tag{5.2}$$

where $[O_2]$ is the molar concentration of oxygen and P_{O_2} is the partial pressure of oxygen as derived from the following reaction.

$$O_2(g) \rightleftharpoons O_2(aq) \qquad \text{(R5.1)}$$

Example 5.2: Dissolved Oxygen in Pure Water

Determine the maximum amount of DO in pure water at standard temperature (25 °C), in the standard units of mg O_2/L. Assume a barometric pressure of 1.00 atm and 21% oxygen.

Solution: Starting with Eq. (5.2), solve for the aqueous concentration term.

$$\begin{aligned}[O_2] &= K_H P_{O_2}\\ &= \left(1.2 \times 10^{-3}\ \tfrac{\text{M}}{\text{atm}}\right)(1.00\ \text{atm} \cdot 0.21)\\ &= 2.52 \times 10^{-4}\ \text{M O}_2\end{aligned}$$

Convert the molar unit to mg O_2/L.

$$\begin{aligned}[O_2] &= \left(2.52 \times 10^{-4}\ \tfrac{\text{mol O}_2}{\text{L}}\right)\left(\frac{2 * 15.9994\ \text{g O}_2}{1\ \text{mol O}_2}\right)\left(\frac{1000\ \text{mg}}{1\ \text{g}}\right)\\ &= 8.0637 = 8.1\ \tfrac{\text{mg O}_2}{\text{L}}\end{aligned}$$

This represents a solution saturated with O_2. Environmental samples that contain contaminants would have a DO level less than this.

Henry's Law can be found defined differently depending on the source. Some texts define it as in Reaction R5.1, setting the gaseous component as the reactant and the dissolved component as the product. Other sources reverse the reaction, setting the dissolved component as the reactant and the gaseous component as the product. This results in the Henry's Law constant being the reciprocal of the one defined in Eq. (5.2). Make sure that you clearly identify the definition of Henry's Law when using other sources of constants.

The Henry's Law constant (K_H) is just a specific form of a general equilibrium constant. A list of selected Henry's Law constants for various gases can be seen in the Appendix I on page 321. Notice that the solubility of a gas increases as its partial pressure increases, according to Eq. (5.2). While the partial pressure of molecular oxygen is relatively constant at sea level, Henry's Law explains why mountain climbers have difficulty breathing at the top of mountains where the partial pressure of oxygen decreases to about 0.07 atm at the summit of Mt. Everest from 0.21 atm at sea level, causing the solubility of $[O_2]$ to decrease proportionally. Given Henry's Law and the increasing partial pressure of carbon dioxide in the atmosphere due to anthropogenic fossil fuel emissions, oceanic dissolved carbon dioxide is on the rise, leading to an increase in the concentration of carbonic acid and a decrease in the ocean's pH.

*5.3.2.1 **Temperature Effects*** The solubility of gases in water is also affected by the temperature of the solution. Since the Henry's Law constant is just an equilibrium constant, it is related to the free energy of a reaction.

$$\Delta G^\circ = -RT \ln K_H \tag{5.3}$$

Since ΔG° is itself temperature dependent, it must be replaced by two less temperature-dependent terms, enthalpy and entropy.

$$\Delta H^\circ - T\Delta S^\circ = -RT \ln K_H \tag{5.4}$$

Since Table I.3 contains Henry's Law constants that were measured at standard temperature (25 °C), we need to find K_H at a nonstandard temperature with respect to K_H°, Henry's Law constant at standard temperature. Rearranging Eq. (5.4) and making a ratio between K_H and K_H°, we get

$$\begin{aligned}
\ln K_H &= -\frac{\Delta H^\circ}{R}\left(\frac{1}{T}\right) + \frac{\Delta S^\circ}{R} \\
K_H &= e^{\left(-\frac{\Delta H^\circ}{R}\left(\frac{1}{T}\right)+\frac{\Delta S^\circ}{R}\right)} \\
\frac{K_H}{K_H^\circ} &= \frac{e^{\left(-\frac{\Delta H^\circ}{R}\left(\frac{1}{T}\right)+\frac{\Delta S^\circ}{R}\right)}}{e^{\left(-\frac{\Delta H^\circ}{R}\left(\frac{1}{T^\circ}\right)+\frac{\Delta S^\circ}{R}\right)}} \\
&= e^{\left(-\frac{\Delta H^\circ}{R}\left(\frac{1}{T}\right)+\frac{\Delta S^\circ}{R}-\left(-\frac{\Delta H^\circ}{R}\left(\frac{1}{T^\circ}\right)+\frac{\Delta S^\circ}{R}\right)\right)} \\
&= e^{\left(-\frac{\Delta H^\circ}{R}\left(\frac{1}{T}\right)+\frac{\Delta S^\circ}{R}+\frac{\Delta H^\circ}{R}\left(\frac{1}{T^\circ}\right)-\frac{\Delta S^\circ}{R}\right)} \\
&= e^{\left(\frac{\Delta H^\circ}{R}\left(\frac{1}{T^\circ}-\frac{1}{T}\right)\right)} \\
K_H &= K_H^\circ \cdot e^{\left(\delta_H\left(\frac{1}{T^\circ}-\frac{1}{T}\right)\right)}
\end{aligned} \tag{5.5}$$

where δ_H is the Henry's Law temperature correction factor (found in Table I.3), T is the actual temperature of the solution, and T° is standard temperature (298 K). The correction factor is a convenient substitution for the enthalpy of the Henry's Law reaction and the gas law constant.

☞ *Estuaries* are transitional habitats between rivers and the ocean. The salinity of these waterbodies depends on the tidal forces and water temperatures. Warm, less dense freshwater can float above cooler, saltier ocean tides and generate a stratified aquatic system. If temperature differences are smaller and tidal flows are dominant, a high degree of mixing occurs to form a nearly homogeneous salinity.

Example 5.3: Temperature Dependence of Henry's Law

Using Eq. (5.5), plot Henry's Law constant (K_H) for molecular oxygen from 5 °C to 45 °C in intervals of 5 °C, then calculate the DO level in units for mg O_2/L for this range, and plot this data set in the same graph that contains K_H. Assume a barometric pressure of 1.00 atm of pressure and 21% oxygen.

Solution: See Section E.1 on page 286.

As Example 5.3 shows, as the temperature of the solution increases, the solubility of the gas decreases. This has important implications for natural and anthropogenic warming. Anthropogenic warming of natural waters is called *thermal pollution*, which is usually associated with industrial processes that require large quantities of cooling water. Electric power plants and other industrial facilities are built near large waterbodies in order to provide a reservoir for dumping excess heat. Electric utilities that employ a once-through cooling process (pulling water into the facility and flushing it out after a single cooling exposure) consume over 30,000 gal $MW^{-1}h^{-1}$ (gallons of water per megawatt per hour) and can return water that is up to 30 °C warmer than the source. Given that a typical power plant's capacity is on the order of 100 MW to a few GW, a day of operation could raise the temperature of 80 million to 2 billion gallons of water and cut oxygen levels by around 40%. This can have severe consequences for aerobic aquatic life that depends on the DO levels.

☞ *Thermal pollution* is caused by the use of water from an aquatic environment as a reservoir for waste heat. The heat lowers the solubility of gases, most importantly DO. The lowered DO and elevated temperatures negatively impact the aerobic aquatic organisms.

Lake Anna in Virginia is the site of the North Anna nuclear power plant, operated by Dominion Power. Its first two 940 MW reactors went online in 1978 and 1980, and it has received approval from the U.S. Nuclear Regulatory Commission to construct a third reactor on the site. The plant was built on Lake Anna, an artificial lake that was created by damming the North Anna River. In order to mitigate the effects of thermal pollution, Lake Anna was divided into a public 9600-acre reservoir and a 3400-acre private waste heat reservoir for the power plant. Three stone dikes separate the "hot" side from the "cold" side. The average temperature of the lake at its head waters is 65 °F (18 °C), and the average temperature at the discharge point of the power plant is 79 °F (26 °C) when the plant is running at 70% capacity, but temperatures at specific points in the waste heat treatment reservoir can rise above 90 °F (32 °C). Dominion Power claims that their discharge temperature has never gone above 103.6 °F, coming close to the Clean Water Act limit of 104 °F. The dikes prevent the hot water from mixing with the cold water until the far end of the lake, where the hot side flows into the main lake through a submerged opening in the dike, at which point the water temperature is presumably close to the cold side (the coldest water mixes with the public lake because of thermal stratification and the submerged discharge pipe). In this way, the North Anna nuclear power plant spares the majority of the lake's ecology from the warm temperatures and loss of DO.

5.3.2.2 ***Salinity Effects*** The solubility of gases is also affected by the salinity of the water. In your general chemistry course, you learned that the concentration terms in an equilibrium expression were unitless, which is true, but it is likely that the reason for this was never explained. The terms in the equilibrium expression should actually be *activities* and not concentrations. For the reaction $A \longrightarrow B$, the equilibrium expression is

$$K = \frac{\{B\}}{\{A\}} = \frac{\gamma_B \dfrac{[B]}{[B]_0}}{\gamma_A \dfrac{[A]}{[A]_0}} \tag{5.6}$$

where $\{A\}$ and $\{B\}$ are the *activities* of solutes A and B in solution, $[A]_0$ and $[B]_0$ are the standard-state concentrations (1 M for aqueous components and 1 atm for gases), and γ_A and γ_B are called the *activity coefficients* of each species. The activity is the effective concentration of these species, not the actual concentration. In dilute solutions, the calculated concentration is approximately the same as the activity, and each γ term is effectively unity. In solutions having a high salinity, the species is no longer surrounded by only water molecules but has become surrounded by a cage of ions from another ionic solute. These ionic cages prevent species A and B from interacting with each other and other species in solution, so they are effectively absent from the solution. Thus, their activity is somewhat less than their calculated concentration would predict. In the case of dissolved gases, the highly ionic environment in saline solutions accentuates the polarity of water and makes nonpolar gases, such as O_2 and N_2, less soluble.

Deriving a relatively simple adjustment to DO levels in saline water is not possible, and a couple of models have been developed that employ a more rigorous mathematical relationship than is useful to present here (Benson and Krause, 1984; Weiss, 1970). The result, however, is a relationship between DO, salinity, and temperature that comes within 0.02% of empirical measurements. This relationship drives the form that is presented on a website by the U.S. Geological Survey.[1] This form simply requires the entry of a range of

[1]The USGS DO table website can be found at http://water.usgs.gov/software/DOTABLES/

temperatures and/or salinity in order to estimate the DO levels in any freshwater or marine environment, and it clearly shows that DO levels decrease as salinity increases.

In order to make use of the USGS DO table form, the salinity of the sample needs to be measured. A discussion of the instrumental methods to determine salinity can be found in Section 5.10.2. Empirical measurements of actual DO levels using the Winkler Method or ion-selective electrodes (ISEs) are also discussed later in this chapter in Section 5.10.7.

☞ Solutions with high salinity have a high *ionic strength*. This causes important solute species to be surrounded by a cage of ions and prevents the solute from interacting with other aqueous species. This lowers the *activity* of the solute, which represents the effective concentration of the species in solution. The activity is calculated by multiplying an *activity coefficient* (γ) by the actual concentration of the solute. The activity coefficient approaches a value of 1 for dilute solutions. Ionic strength is calculated in a particular way, so it is not the same as salinity, though salinity will often be used nontechnically to refer to solutions with high ionic strength.

Review Example 5.2: Dissolved Oxygen in Saline Waters

Use the USGS DO table website to plot the DO levels for a freshwater system (salinity = 0‰) where the water temperature may range from 5 to 40 °C in intervals of 5 °C, and then retrieve similar data for a seawater system (salinity = 34‰) where the water temperature may range from 5 to 40 °C in intervals of 5 °C. Plot both of these data sets on the same graph to demonstrate the effect temperature and salinity have on DO. Assume that the atmospheric pressure ranges from 1.0 atm to 1.1 atm in increments of 0.1 atm. Plot only the data sets for the atmospheric pressure condition of 1.0 atm. Once you locate the website, go to Section B – Oxygen Solubility Tables – and use the web form to complete this problem.

Solution: See Section E.1 on page 286.

A good rule of thumb to remember is that the cleanest freshwater at the average ground temperature (60 °F) saturated with oxygen will have a DO level of about 10 mg/L O_2. Anything warmer, saltier, or dirtier will have a lower DO level.

5.4 THE CARBON CYCLE

When life on the Earth is described in science-fiction movies, it is often described as "carbon-based" life and for a good reason. A glance at Table 5.2 shows carbon as the second most abundant element in the human body. Carbon is central to the biosphere and a key element in a larger geochemical cycle that involves inorganic carbon moving through the atmosphere, hydrosphere, and lithosphere. In Chapter 3, you learned how hydrocarbons help to drive the production of photochemical smog and carbon's role in the two most important greenhouse gases. Carbon is essential to the microbial respiration reactions that are responsible for the formation of the organic component of soil from leaf litter. In this chapter, I hope to highlight the many forms of carbon that you have already studied and put them in the context of the carbon cycle with an emphasis on the hydrosphere.

Element	Percent By mass	Element	Percent By mass
O	61	P	1.1
C	23	K	0.20
H	10	S	0.20
N	2.5	Na	0.14
Ca	1.4	Cl	0.14

Source: Data from Emsley (1998).

Table 5.2 Approximate elemental composition of the human body.

In general, the nutrient cycles have reservoirs in each of the compartments of the environment that can be described as relatively stable or dynamic and as either inorganic or organic. The carbon reservoir in the atmosphere is mostly inorganic in the form of carbon dioxide, which is stable and has a relatively long life. Other forms of organic carbon do exist in the atmosphere, but given their relatively low concentrations and the presence of oxidizing agents (such as the hydroxyl radical and ozone), organic carbon in the atmosphere is transient. In the hydrosphere, the reservoir of carbon is mostly in the inorganic forms of carbonate species (HCO_3^- and CO_3^{2-}). Current estimates suggest that the oceans contain about 50 times more carbon by mass compared to the atmosphere, with about 3% of the oceanic carbon mass in organic forms. On the surface of the lithosphere, there is a very dynamic reservoir of organic carbon originating from the biosphere. Below the surface, the main lithospheric reservoir of inorganic carbon is in the sedimentary carbonate minerals, representing over 99% of the total carbon on the planet. The subsurface, lithospheric, organic carbon reservoirs consist mainly of the coal beds, petroleum and methane deposits, and shale formations, which were surface deposits buried long ago and have achieved relatively static chemical forms. Let us examine some of the conversion processes that affect the hydrosphere.

5.4.1 Anthropogenic Contributions

The most obvious anthropogenic contribution to the global carbon cycle is through atmospheric emissions, principally carbon dioxide and other hydrocarbons that eventually form CO_2. Much of this comes from energy production, cement manufacturing, and agriculture. Concerning CO_2, 40–45% of human emissions remain in the atmosphere, 20–35% are absorbed by the oceans, and 20–40% are absorbed by the biosphere. In the ancient past, increases in dissolved CO_2 concentrations were regulated by the precipitation of carbonate minerals as a result of dissolved cations from weathered rocks – a process that is slower than the rate of atmospheric inputs of CO_2. Dissolved carbon exists predominately as the bicarbonate ion, which is in equilibrium with carbonate sediments and atmospheric carbon dioxide. This is a gross simplification, however, because stratification in the oceans causes the deep water to retain carbon for thousands of years.

A more localized, anthropogenic contribution to the carbon cycle is the discharge of raw sewage. While Europe and North America treat most of the sewage they produce, much of the rest of the world does not. Besides being a reservoir for many deadly waterborne diseases, sewage contributes to reduced water quality and increased sedimentation. Some of the carbon is returned to the atmosphere via microbial respiration, and some of the carbon stays locked in sediment.

5.4.2 Biotic Processes

In the oceans, about 25% of the inorganic carbon that is converted to organic carbon by photoautotrophs sinks as detritus to the ocean floor where it is consumed by heterotrophic bacteria and converted back to inorganic carbon. It is estimated that this process of transporting carbon to the deep ocean has suppressed the atmospheric carbon dioxide concentration by 150–200 ppmv, which is a 38–50% difference. Organic carbon takes on a bewildering number of forms, such as carbohydrates and protein. This organic carbon is consumed by heterotrophic species, returning some of the carbon to the atmosphere and retaining the rest in a condensed form.

Inorganic carbon in the oceans is utilized by some organisms in the production of carbonate exoskeletons, which eventually transport this carbon to the deep ocean. This process actually causes some of the CO_2 to leave the ocean and enter the atmosphere because of the calcification reaction, where one bicarbonate ion enters the solid phase and another becomes gaseous.

$$2HCO_3^-(aq) + Ca^{2+}(aq) \rightleftharpoons CaCO_3(s) + CO_2(g) + H_2O(l) \qquad (R5.2)$$

Photosynthetic organisms on land convert inorganic carbon to organic carbon, storing some in various condensed forms and releasing some back to the atmosphere. Some of the organic carbon is buried and becomes the peat, petroleum, and natural gas reservoirs after long stretches of geologic time.

5.4.3 Summary

Carbon's primary biogeochemical reservoirs store carbon as CO_2 in the atmosphere, bicarbonate ion and organic carbon in the hydrosphere, and carbonate minerals and organic carbon in the lithosphere. The oxidative nature of the atmosphere drives all gaseous carbon to the thermodynamically stable form of CO_2, carbon's highest oxidation state of +4. Photosynthetic organisms reduce carbon to oxidation states from −4 to −2 when it forms organic carbon. This generates a steady supply of energy-rich chemicals, which supply nutrients and energy to much of the remaining biosphere.

5.5 THE NITROGEN CYCLE

Nitrogen is an essential element for biological activity since it is a central element in amino acids, which are the building blocks of proteins. The Earth's atmosphere is a tremendous reservoir of nitrogen in the form of dinitrogen gas, but the nitrogen triple bond makes it very stable and effectively useless as a nutrient to most biological systems. This inert behavior of nitrogen gas limits the bioavailability of nitrogen and thus makes it a limiting reactant for biological activity in many environmental systems. In Chapter 3, you learned about the abiotic portion of the nitrogen cycle whereby combustion processes and photolytic reactions in the atmosphere form gaseous oxides of nitrogen (NO_x) and eventually lead to nitric acid. Lightning strikes also provide enough activation energy to produce NO_x from atmospheric nitrogen and oxygen. These alternative forms of nitrogen eventually oxidize to the nitrate ion, which is a bioavailable form of nitrogen.

As you read this section, you should keep the thermodynamics of the nitrogen cycle in mind. From the fixation of nitrogen to the formation of the nitrate ion, most of the transformations of the nitrogen cycle represent a downhill fall in which the form of nitrogen gets more and more stable, all the while microbes are harvesting the energy lost at each step. From a redox perspective, the nitrogen cycle is a progression of increasing oxidation states of nitrogen while the species is in an aerobic environment. In the presence of oxygen, which is more electronegative, free energy can be released by the oxidation of nitrogen and the reduction of oxygen. This fact is exploited by the army of microbes that drive the nitrogen cycle.

Review Example 5.3: Oxidation of States of Nitrogen

In the following nitrogen cycle progression, calculate the oxidation state of nitrogen for each species.

$$N_2 \rightarrow NH_3/NH_4^+ \rightarrow NO_2^- \rightarrow NO_3^- \tag{5.7}$$

Solution: See Section E.1 on page 287.

Human population had been limited by the amount of bioavailable nitrogen in agricultural production. Cultivated crops deplete nitrogen and other nutrients from soil, and the only readily available preindustrial fertilizer was livestock manure. In 1909, the German scientist Fritz Haber demonstrated a conversion of gaseous nitrogen to ammonia through the following reaction.

$$N_2(g) + 3\,H_2(g) \rightleftharpoons 2\,NH_3(g) \tag{R5.3}$$

This reaction, known as the Haber synthesis and later scaled up by another German scientist named Carl Bosch (Haber–Bosch process), made the production of synthetic fertilizer possible. This reaction is spontaneous at room temperature but kinetically slow. Raising the temperature above 300 °C and increasing the reaction pressures to over 300 atm in the presence of an iron catalyst result in an acceptable reaction rate and 10–20% yield. While gaseous nitrogen can be easily extracted from the air, hydrogen gas must be synthesized in order to produce ammonia. Methane can produce hydrogen gas in a process called steam reformation.

$$CH_4(g) + H_2O(g) \rightleftharpoons CO(g) + 3\,H_2(g) \tag{R5.4}$$

> The Haber–Bosch process was one of several technological advances that began what is now called the Green Revolution, which took place after World War II. The availability of artificial fertilizer, synthetic pesticides, high-yield crops, and irrigation dramatically increased global crop yields. The human population began an exponential increase from about 2 billion people to the current total of 7 billion, a response that biologists see in most other species when the limits to ecological resources are removed.

Fritz Haber (1868–1934) was a German chemist who won the Nobel Prize in chemistry in 1918. He was simultaneously responsible for indirectly saving the lives of billions of people as a result of the Haber synthesis and responsible for the ghastly deaths of thousands of soldiers in World War I because of the poisonous gases he developed for the German army. For more details on Haber's interesting life, go to RadioLab.org and search for "Haber." (Courtesy Wikimedia Commons, 1919.)

Carbon monoxide can produce more hydrogen when reacted with water.

$$CO(g) + H_2O(g) \rightleftharpoons CO_2(g) + H_2(g) \quad (R5.5)$$

These reactions form the basis of the industrial production of artificial fertilizer ammonium nitrate (NH_4NO_3), which has revolutionized food production and led to a tremendous amount of pollution that will be the focus of later sections.

While anthropogenic sources of nutritive nitrogen have become a significant source of nitrogen pollution, other abiotic sources (such as lightning) have an insignificant effect on the biosphere.

5.5.1 Nitrogen Fixation and Assimilation

Diazotrophic bacteria possess the enzyme nitrogenase that permits them to convert nitrogen gas to ammonia in a process called *nitrogen fixation*. Nitrogenase is a metalloprotein that contains iron and molybdenum–iron active centers that catalyze the reduction of nitrogen gas to ammonia. The whole process is driven by ATP.

$$N_2 + 8H^+ + 8e^- + 16ATP \rightarrow 2NH_3 + H_2 + 16ADP + 16PO_4^{3-} \quad (R5.6)$$

The reduction occurs in a stepwise manner in which N≡N is reduced to H−N=N−H, then to $H_2N{-}NH_2$, and then finally to $2NH_3$ molecules – all the while the molecule is being stabilized by the protein. Ammoniacal nitrogen can be used by most other organisms to synthesize amino acids and other organic-nitrogen biomolecules in a process called *assimilation*. In soil, diazotrophic bacteria often form symbiotic colonies in the root structures of plants and produce natural nitrogen-based fertilizer for their host. In marine environments, cyanobacteria can fix nitrogen in the same manner and provide a source for nutritive nitrogen to phytoplankton and other organisms. The nitrate ion, however, is usually the most abundant form of nitrogen, so phytoplankton employ enzymatic reduction of the nitrate ion to form the ammonium ion and then assimilate it into amino acids. Phytoplankton that occupy the epilimnion of the ocean can assimilate the ammonium ion without a costly reduction of nitrogen.

☞ *Nitrogen fixation* is a process by which gaseous N_2 is converted into ammoniacal nitrogen (NH_3 or NH_4^+), a nutritive form of nitrogen.

☞ The term *diazo* comes from Greek and means "two nitrogens." Gaseous nitrogen is a diazo compound. You will see the same term used later to describe colorful derivatives of nitrogen compounds that are used in quantitative methods.

5.5.2 Ammonification

Proteins can undergo hydrolysis to produce smaller peptides and amino acids, which are a nutrient source for many organisms. Some organisms extract energy from amino acids through a catabolic mechanism, resulting in the expulsion of the ammonium ion. The driving force in these reactions is the energy contained in the organic carbon that is released upon oxidation, such as in the oxidation of alanine to the pyruvate and ammonium ions.

$$CH_3CHNH_2COOH(aq) + \tfrac{1}{2}O_2(g) \longrightarrow CH_3COCOO^-(aq) + NH_4^+(aq) \quad (R5.7)$$

Humans excrete ammonia by converting it into urea ($CO(NH_2)_2$). Bacteria consume the amino acids and excrete the ammonium ion directly. This process of ammoniacal nitrogen production from organic nitrogen is called *ammonification* and is the reverse assimilation process.

☞ Nitrogen *assimilation* is a process by which ammoniacal nitrogen is used to synthesize amino acids. Nitrogen *ammonification* is a process by which ammoniacal nitrogen is produced from amino acids.

Before the development of artificial fertilizer, rare geologic deposits of nitrate minerals and guano (bird poop – very rich in N, P, and K) were valuable resources. One of the richest deposits of both was located along the west coast of Peru. A war was fought between Peru, Bolivia, and Chile from 1879 to 1883 to control these valuable resources.

5.5.3 Nitrification

While ammonia is highly toxic to some organisms, in an aerobic environment, it is an energy-rich molecule that can be used by some microbes to produce energy. When combined with molecular oxygen, there is a great deal of free energy produced when the ammonium ion is oxidized.

$$2\,NH_4^+(aq) + 3\,O_2(g) \longrightarrow 2\,NO_2^-(aq) + 2\,H_2O(l) + 4\,H^+(aq) \quad (R5.8)$$

Other microbes oxidize the nitrite ion to the nitrate ion.

$$2\,NO_2^-(aq) + O_2(aq) \longrightarrow 2\,NO_3^-(aq) \qquad (R5.9)$$

Review Example 5.4: Oxidation of Ammonium

Calculate the change in free energy for Reactions R5.8 and R5.9.

Solution: See Section E.1 on page 288.

The nitrate ion is the form of nitrogen that is the most oxidized and has the lowest free energy of formation of the nitrogen species. This means that there is no more energy to be harvested from the further oxidation of nitrogen. The nitrate ion is the most stable form of nitrogen and the thermodynamic fate of nitrogen in an aerobic environment. It tends to be present in higher concentrations compared to the other forms in the long term. As mentioned in the Assimilation section, phytoplankton can convert it to the ammonium ion when ammoniacal nitrogen is in low concentration. Since available nitrogen limits biological productivity, accumulation of the nitrate ion can lead to algal blooms and eutrophication, which is discussed later. Because the nitrate ion is very soluble in water, it tends to wash out of soils and the atmosphere and into the watershed. This leads to more frequent biologically limiting concentrations of nitrogen in soil compared to aquatic environments.

5.5.4 Denitrification

Completing the nitrogen cycle requires the various forms of nitrogen to revert back to the major reservoir of nitrogen in the biogeosphere, which is N_2. Since the nitrate ion is the terminal form of nitrogen in an aerobic environment, conversion of it back to molecular nitrogen occurs under anoxic conditions. Chapter 4 discussed the various metabolic pathways and the free energy that could be harvested from each in Table 4.5 on page 175. In an aerobic environment, using O_2 as the electron acceptor yields the most energy. The second energy-rich metabolic path involves the use of the nitrate ion.

$$C_6H_{12}O_6(aq) + \tfrac{24}{5}NO_3^-(aq) + \tfrac{24}{5}H^+(aq) \rightarrow 6CO_2(g) + \tfrac{42}{5}H_2O(l) + \tfrac{12}{5}N_2(g) \qquad (R5.10)$$

Thus, anaerobic microbes complete the nitrogen cycle by producing molecular nitrogen. In this reaction, it is the oxidation of carbon that forms the basis for the free energy yielded in the reaction. Some chemoautotrophs can perform a disproportionation of two nitrogen species to harvest energy and complete the nitrogen cycle.

$$NO_2^-(aq) + NH_4^+(aq) \longrightarrow N_2(g) + 2\,H_2O(l) \qquad (R5.11)$$

5.5.5 Summary

The atmospheric and hydrospheric nitrogen cycle describes the interchange of many forms of nitrogen that ultimately reside in the inert atmospheric reservoir of N_2 and the highly oxidized geohydrospheric form of NO_3^-. Each cycle involves its own complicated series of transformations driven by radiation, oxidation, and reduction. Some of these transformations are abiotic and others are mediated by microbial enzymes. In all cases, however, you should recognize the chemical driving force of free energy, from which there is no escape.

5.6 THE PHOSPHORUS CYCLE

Phosphorus is an essential nutrient because of its role in the phosphate backbone of deoxyribonucleic acid (DNA) and ribonucleic acid (RNA) and its central role in the

Phosphorus was discovered most recently by the German chemist Hennig Brandt in 1669 by boiling down 50 gallons of urine in search of the "philosopher's stone" – a substance that could turn inexpensive metals into gold. The procedure he followed is described in Ogilvy (2006, pg. 58).

- Boil urine to reduce it to a thick syrup.
- Heat until a red oil distills up from it, and draw that off.
- Allow the remainder to cool, where it consists of a black spongy upper part and a salty lower part.
- Discard the salt, mix the red oil back into the black material.
- Heat that mixture strongly for 16 h.
- First white fumes come off, then an oil, then phosphorus.
- The phosphorus may be passed into cold water to solidify.

What Brandt produced was white phosphorus, a form of phosphorus that carries on a slow reaction with oxygen and glows continuously – all from several buckets of urine. Do not try this at home – go to someone else's house!

chemistry of adenosine triphosphate (ATP). It is part of the mineral structure in the teeth and bones as hydroxyapatite ($Ca_5(PO_4)_3(OH)$), and all biological membranes are composed of phospholipids. It is also present in vitamins B_6 ($C_8H_{10}NO_6P$) and B_{12} ($C_{63}H_{88}CoN_{14}O_{14}P$). The phosphorus cycle is much simpler compared to the other cycles because the number of forms that phosphorus takes in the environment is very limited. Until recently, scientists thought that the atmosphere played virtually no role in the cycle, but some studies in the late 1990s showed that some environments, such as the Mediterranean Sea, receive up to 10% of their annual input of phosphorus from the atmosphere in the form of wet and dry deposition (Migon, Sandroni, and Bethoux, 2001). The source of this phosphorus is likely from dust (natural phosphorus from phosphate rock weathering; anthropogenic from incinerators and agricultural fertilizers).

Phosphorus has a +5 oxidation state in the phosphate ion and does not usually participate in redox reactions similarly to nitrogen, which is in the same chemical family. In rare anoxic environments, phosphorus can form phosphine (PH_3). It is very toxic as a gas but quickly oxidizes in an aerobic environment.

The major reservoirs of phosphorus are in phosphate mineral deposits that are found in specific places around the world. Guano deposits represent rare forms of biogenic phosphorus deposits. Islands along the coast of South America and other Pacific islands, sites that host large populations of migratory birds, had large reserves of mineralized guano. These sites have been heavily harvested and will eventually become depleted. Bone Valley, in central Florida, is the largest geological phosphate deposit in the United States. Most of the world's reserves of phosphate minerals are found along the Atlantic coast of Morocco. Unlike carbon and nitrogen, for which a large atmospheric reservoir is distributed evenly around the globe, phosphorus is highly localized. Similarly to petroleum, phosphate mining is part of a global market. But unlike fossil fuels, which are an energy source that has potential replacements such as wind or solar, phosphorus reserves have no replacement. Once the reserves are gone, we will have to figure out how to recycle the phosphorus from the crops we produce, harvest it from the relatively low concentration that is found in the environment, or go after some of the deep sea deposits in the Atlantic and Pacific Oceans.

Phosphorus is found in aquatic systems as orthophosphate (PO_4^{3-}) or as polyphosphates. Polyphosphates can hydrolyze in acidic solutions to form orthophosphate, or phosphate can be found attached to organic molecules. This organic phosphate can come in soluble or insoluble forms. Insoluble organic phosphates, combined with the low solubility of orthophosphate in the presence of naturally occurring cations (see Table 5.3), keeps the sediment in aquatic environments relatively rich in phosphorus and the epilimnion relatively poor. Because the phosphate ion is fairly insoluble with commonly occurring cations, it tends to remain fixed in soil compared to the nitrate ion and does not remain in solution at high concentrations in aquatic environments. This leads to more frequent biologically limiting concentrations of phosphorus in aquatic environments compared to soil – just the opposite of nitrogen.

Compound	pK_{sp}
$AlPO_4$	20.007
$Ca_3(PO_4)_2$	32.684
$Ca_{10}(PO_4)_6(F,OH)_2$	118[a]
$Cu_3(PO_4)_2$	36.854
$FePO_4 \cdot 2H_2O$	15.004
$Mg_3(PO_4)_2$	23.983

Source: Data from Haynes (2012).
This also means that phosphate mineral residence time in sediment is long.
[a] Somasundaran, Amankonah, and Ananthapadmabhan (1985).

Table 5.3 Phosphate minerals are quite insoluble, which means that, in typical aquatic environments, where cations such as calcium and magnesium are present, levels of dissolved orthophosphate will be low.

Anthropogenic contributions to excess phosphates in the environment come from agricultural and residential fertilizers as well as detergents. In fertilizers, phosphorus is often included as phosphorus pentoxide (P_2O_5) or as phosphate minerals, such as hydroxyapatite, that have been hydrolyzed with sulfuric or phosphoric acid to make them more water-soluble. The inclusion of phosphates in fertilizer is obvious, as phosphate is a limiting reactant in biological productivity and an essential factor in the Green Revolution. Phosphates also make dishwashing detergents very effective, which deserves a little more explanation.

Soaps and detergents have historically been produced from a reaction between lye and animal fat. The lye would hydrolyze the fatty esters in the fat to form long-chain carbon compounds with carboxylate ends. These amphiphilic compounds are water soluble in alkaline solutions and form micellular structures (see Figure 1.12 on page 19) that envelop oils inside the hydrophobic interior of the micelle, allowing the insoluble organic "dirt" to be washed away. The efficacy of these "natural" soaps suffers greatly when combined with hard water, containing high concentrations of calcium and magnesium cations, because the cations effectively remove the soap from solution by the formation of insoluble soap scum. Industrial chemists corrected this problem by replacing the carboxylate fatty acids with synthetic surfactants such as sodium lauryl sulfate ($NaC_{12}H_{25}SO_4$) or sodium dodecylbenzenesulfonate ($C_{18}H_{29}NaSO_3$), amphiphilic compounds that still form the micellar structures but are more soluble in the presence of hard water. The efficacy of the surfactants is still affected by hard water, so modern synthetic detergents contain surfactants and builders. The builders, such as sodium phosphate (Na_3PO_4) or sodium triphosphate ($Na_5P_3O_{10}$), were added to remove any interfering cations from the solution to maximize the surfactant concentration since phosphate salts are very insoluble. Before phosphate regulations, common detergents contained 30–50% builders by mass and contributed about 50–70% of the phosphate input into the environment. Detergent manufacturers have replaced phosphate builders with sodium aluminum silicate zeolites, which perform a cation exchange between the Na^+ in the zeolite and the Ca^{2+} and Mg^{2+} in the water. This softens the water and reduces surfactant precipitation.

Phosphate's role in the degradation of water quality was proven conclusively in the 1970s when limnologists began adding carbon, nitrogen, or phosphorus nutrients to various isolated lakes. The lakes that were given phosphates showed sustained algal blooms and heavy sedimentation, a condition called *eutrophication*. An *oligotrophic* lake has very low concentrations of dissolved C, N, and P and, consequently, has a moderate-to-low aquatic plant growth and high DO levels. A *eutrophic* lake has an overabundance of nutrients, which causes the algae in the epilimnion to multiply. As the algae die, they form a sediment at the bottom of the lake where aerobic microbes rapidly consume all of the DO as they metabolize the nutrient-rich sediment. This results in anoxic conditions at the bottom of the lake, making it uninhabitable to most of the large, aerobic organisms. The long-term consequence of eutrophication is the conversion of the lake to a swampy shallow wetland, a process that normally takes centuries when driven by natural conditions (*natural eutrophication*) but can take only decades when it is the result of anthropogenic pollution (*cultural eutrophication*).

☞ Artisanal soap is made by combining animal fat with lye produced from wood ash. The combustion of wood produces soda ash (Na_2CO_3) and potash (K_2CO_3). When lime ($Ca(OH)_2$) is added during the ashing process, caustic ash or lye is produced (NaOH or KOH), which is a more effective hydrolyzing agent. The conversion of ash to caustic ash generates insoluble $CaCO_3$ and permits the more soluble hydroxide to be separated from the wood ash when it is combined with water and filtered in the following reaction. $Ca(OH)_2 + K_2CO_3 \rightarrow CaCO_3 + 2KOH$ (R5.12)

☞ *Eutrophication* is a process by which a lake eventually fills as a result of receiving sediment from the watershed and from organic detritus produced by the algae. *Natural eutrophication* occurs over centuries as a result of natural processes. *Cultural eutrophication* is an accelerated form caused by mainly phosphorus pollution (in freshwater) and nitrogen and phosphorus pollution in estuaries and marine habitats. *Eutrophic* water can lead to anoxic conditions due to the aerobic degradation of the organic sediment. These *dead zones* drive out large aerobic organisms and lower biodiversity. An *oligotrophic* lake is the opposite – low nutrient levels, high DO, and low biological productivity.

5.7 THE SULFUR CYCLE

Sulfur is also an essential element, although by mass it is the least significant compared to C, N, or P according to Table 5.2 (p. 199). Sulfur is necessary for the amino acids cysteine and methionine. Cysteine is tremendously important for protein folding since it often forms a disulfide bridge when two cysteine amino acids on the protein chain line up and covalently bond. Iron–sulfur proteins play a central role in energy storage within mitochondria, and sulfur is part of vitamins such as B_1 ($C_{12}H_{17}ClN_4OS$) and B_7 ($C_{10}H_{16}N_2O_3S$).

Sulfur is similar to carbon and nitrogen in that it has a very large and nearly ubiquitous reservoir. According to Table 5.1 (p. 195), seawater contains about 2.7 g/L of the sulfate ion and freshwater contains roughly 3–10 mg/L. Sulfur's geologic reservoir is the relatively common minerals gypsum ($CaSO_4$) and pyrite (FeS_2). It can also be found in its elemental form in deposits near volcanoes and hot springs. The high concentration of molecular oxygen in the atmosphere means that the exposure of sulfide minerals leads to an oxidation of sulfate minerals. Sulfides are exceptionally insoluble in water, whereas sulfates are much more soluble. This leads to a transport cycle of lithospheric sulfur to hydrospheric sulfur.

Other anthropogenic contributions of sulfur place sulfur minerals in combustion processes, in the use of coal for fuel and smelting processes, producing gaseous sulfur dioxide and eventually acid deposition. The smelting of minerals is an industrial process where metals are recovered from their naturally occurring minerals. Sulfide ores of copper, lead, and zinc are roasted in the presence of oxygen to remove the sulfur and leave a more soluble oxide mineral.

$$2ZnS(s) + 3O_2(g) \longrightarrow 2ZnO(s) + 2SO_2(g) \qquad (R5.13)$$

If you have ever had your hair curled at a salon or home using a permanent wave curl (or "perm"), then you have had the disulfide bridges in your hair temporarily decoupled. A solution of ammonium thioglycolate coats the hair and decouples some of the bonds via reduction. Once the hair has been curled in a certain style and rinsed, a dilute solution of hydrogen peroxide is used to reset the disulfide bonds via oxidation. Once reformed, the disulfide bridges hold the hair in the position when it was set.

Zinc oxide is dissolved in a sulfuric acid solution to produce Zn^{2+} and then converted to Zn metal by electrolysis. The SO_2 can be converted to sulfuric acid and sold as a valuable by-product, but it is not always captured and contributes to acid deposition when vented to the atmosphere in the same way as sulfur emissions from coal combustion.

One particularly disruptive sulfur oxidation process is known as *acid mine drainage*. In the mining process, removing valuable minerals causes significant amounts of sulfide minerals to be transported from their subterranean origin and exposed to oxygen in large open-air piles.

☞ *Acid mine drainage* is a pollution event that occurs outside of mines as a result of the oxidation of sulfide minerals and the subsequent production of sulfuric acid.

$$2FeS_2(s) + 7O_2(g) + 2H_2O(l) \longrightarrow 2Fe^{2+}(aq) + 4SO_4^{2-}(aq) + 4H^+(aq) \qquad (R5.14)$$

What results is a colorful and corrosive plume of rusty iron and sulfate ion, imparting a very low pH to the receiving watershed.

Review Example 5.5: Sulfur Oxidation States

What are the oxidation states for the following forms of sulfur?

SO_4^{2-}, S, $(CH_3)_2S$, COS (O=C=S), H_2S

Solution: See Section E.1 on page 288.

Sulfur is assimilated into biological tissue by reduction of sulfur from the sulfate ion and the synthesis of the amino acids cysteine and methionine. From there, sulfur becomes part of many proteins. You may recall from Chapter 4 that one of the anaerobic metabolic pathways used the sulfate ion as the electron acceptor, producing H_2S as a by-product. This is known as dissimilatory sulfate reduction. Anoxygenic photosynthetic bacteria harvest energy using sunlight and H_2S gas, producing elemental sulfur. Other chemotrophic bacteria convert H_2S and elemental sulfur to the sulfate ion.

The most abundant biogenic emission of atmospheric sulfur is dimethyl sulfide ($(CH_3)_2S$ or DMS), which is produced by phytoplankton. Oceanic emissions of DMS are a little less than half of the annual anthropogenic sulfur emissions, so DMS plays a significant role in total atmospheric sulfur chemistry. In fact, the biogenic emission of DMS is suspected to increase as more dissolved CO_2 encourages the productivity of phytoplankton. As the DMS emissions increase and lead to increased levels of sulfate aerosols as a result of atmospheric oxidation, the albedo of the atmosphere increases as a result of the reflectivity of the sulfate aerosols. This hypothetical negative feedback cycle should cool the planet.

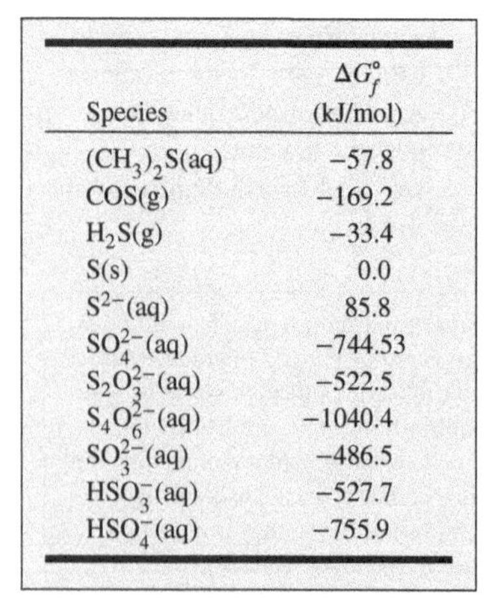

Species	ΔG_f° (kJ/mol)
$(CH_3)_2S(aq)$	−57.8
$COS(g)$	−169.2
$H_2S(g)$	−33.4
$S(s)$	0.0
$S^{2-}(aq)$	85.8
$SO_4^{2-}(aq)$	−744.53
$S_2O_3^{2-}(aq)$	−522.5
$S_4O_6^{2-}(aq)$	−1040.4
$SO_3^{2-}(aq)$	−486.5
$HSO_3^-(aq)$	−527.7
$HSO_4^-(aq)$	−755.9

Table 5.4 The free energy of the common hydrospheric forms of sulfur.

5.7.1 Summary

Sulfur can be found in several different forms and oxidation states. Table 5.4 shows the various forms and their free energy of formation values. The sulfate ion and its analogs (HSO_4^- and H_2SO_4) have the lowest ΔG_f° per sulfur atom, indicating that they are the thermodynamic fate of sulfur in an aerobic environment. Sulfate and sulfide compounds are, by far, the most commonly found forms of sulfur in the environment, with a large sulfate reservoir in the oceans and sulfide and sulfate minerals in the lithosphere. There are, however, several microbial life forms that have found sulfur quite useful in energy production, not to mention sulfur's important role in vitamins and two essential amino acids. In anoxic environments, sulfur can be reduced from the sulfate ion to the noxious and toxic H_2S and elemental sulfur. These sulfur species represent a range of oxidation states from +6 to −2.

5.8 WATER QUALITY

While it is painfully obvious to many that the purity and availability of water are essential for civilization and the biosphere, maintaining clean water supplies under the threat of anthropogenic pollution has been a perennial struggle for humans and federal enforcement of water quality standards did not begin in the United States until the Clean Water Act of 1972. Since the necessary purity of water is different for drinking water, agricultural irrigation, industrial cooling, recreation, and waste removal, the various standards for water quality rely on measurements of chemical parameters (Table 5.5). While these parameters are discussed in greater detail in the following sections, they focus on species that should already be familiar to you: O_2, NO_3^-, PO_4^{3-}, SO_4^{2-}, dissolved organic and inorganic matter, pH, and others. As part of most environmental chemistry laboratory courses, you will be testing for some of these species, and the methods for these studies are described later. More than 100 pollutants are monitored by the EPA as part of various federal legislation. These laws provide a mechanism for the EPA and state authorities to regulate everything from wastewater processing to drinking water production.

Drinking water, sometimes called potable water, is regulated by the EPA in the United States, which requires certified testing of the water supply. Chloride, for example, must not exceed 250 mg/L while manganese cannot exceed 0.05 mg/L. These tests are performed at regular intervals and must be disclosed to the public. You can go to your municipality's website and find an annual report disclosing the contaminant levels.

Many Americans (about 15%) obtain their drinking water by simply drilling a well into the ground and pumping the water into their homes without treatment. Another 15% are connected to public water systems that tap into groundwater. While surface waters are exposed to harmful pathogens and toxic chemicals, groundwater is often clean enough to drink. The average drinking well is drilled about 100–400 feet below the surface and has casing installed that goes down some or most of this depth to prevent surface water from intruding into the well. As long as septic tanks, concentrated animal feeding operations (CAFOs), and industrial storage tanks are far away from the well, the water that is recovered from these drilled wells is potable. This is not always the case, however. Some areas of the country have geologic factors that make well water undrinkable. Some of these problems are the results of industrial drilling and mining operations. Biological contamination is also an issue. One study reported a significant increase in intestinal illness in children after a rain event in areas that relied upon private wells (Bergquist, 2014).

To achieve drinking water standards, municipal water systems use a variety of techniques to purify the water. Suspended solids are removed using *flocculants*, which bind to these small particles, resulting in larger particles that settle out of solution. Filtration can be used in conjunction with, or as an alternative to, flocculants to improve clarity. Filtration can also be used to remove pathogens from the water if the filter pore size is ≤ 1 μm for protozoa or ≤ 0.3 μm for bacteria. Chlorination is a very effective method of disinfection and eliminates viruses as well as microbes. Dissolved contaminants cannot be removed through filtration and require ion exchange for inorganic contaminants and adsorption with activated carbon for organic contaminants.

Municipalities have a number of options for drinking water treatment, depending on the quality of the water source. Private residents have a more limited set of options available for treating their own well water. Water softening systems, which use an ion-exchange mechanism to exchange sodium ions for calcium, magnesium, and iron ions, can be employed to reduce rusty stains in sinks, and calcification in pipe are often caused by untreated well water. In addition, small purifiers can be installed, which use a combination of charcoal filters, UV sterilization, and/or reverse-osmosis membranes in order to remove dissolved organic contaminants, sterilize the water, and reduce the level of dissolved contaminants.

Many Americans who express a heightened concern over health choose to drink bottled water over tap water. While the EPA regulates public drinking water, the Food and Drug Administration (FDA) regulates commercial bottled water. While the FDA has generally adopted the water quality standards of municipal drinking water, the FDA does not require commercial bottlers to disclose the sources of the water, what contaminants it contains, or how it was treated before bottling. The EPA requires municipal tap water to be treated for pathogens that cause intestinal illness and diarrhea, whereas the FDA does not require this

Alkalinity: A measure of the buffering capacity of water to neutralize inputs of acid. It measures dissolved species such as carbonates and sulfates.

BOD/COD: A measure of the nutrient load in water by specifying how much DO would be consumed if dissolved organics, sulfides, ammonia, and so on, were oxidized.

DO: A measure of the amount of the dissolved oxygen.

Nitrogen: A measure of the dissolved and suspended nitrogen species, such as nitrate, nitrite, ammonia, and organic nitrogen.

Phosphorus: A measure of the dissolved and suspended phosphorus species, such as phosphate and organic phosphorus.

Fecal coliform: A measure of bacteria that are indicators of the presence of human and animal waste.

Hardness: A measure of the dissolved calcium and magnesium salts in the water.

Metals: A measure of the dissolved metals in solution, both nutritive and toxic.

Turbidity: A measure of the cloudiness of water, as caused by suspended solids. Turbid water blocks sunlight and increases surface water temperatures.

Temperature: An obvious measure that affects DO, metabolic rates, and solubility.

pH: A measure of the hydrogen ion concentration as expressed in the pH scale from 1 to 14. Ecologically healthy waters range from pH 6 to 9.

Total dissolved solids: A measure of the dissolved ionic and organic compounds. Water hardness is a subset of this measure.

Table 5.5 Common parameters of water quality.

☞ *Coagulants* or *flocculants* are compounds that form charged complexes when dissolved in water. The charged surfaces attract other particles, which *adsorb* onto the charged surface. These particles grow in size and can no longer be suspended in the water, whereupon they sink to the bottom to form sediment. Alum ($KAl(SO_4)_2 \cdot 12\,H_2O$), a natural mineral and spice used in pickling, was used by ancient Egyptians as a coagulant to purify water. Other coagulants are $Al_2(SO_4)_3$, $AlCl_3$, $NaAlO_2$, $Fe_2(SO_4)_3$, $FeSO_4$, $FeCl_3$, $FeClSO_4$, and organic polymers.

for bottled water. While the cost of bottled water is hundreds to thousands of times higher than that of tap water, a National Resources Defense Council study also showed that 35–40% of bottled water comes from public water supplies anyway. So, the next time you drink bottled water, the water may be no cleaner than the water found in the tank of a school toilet.

Speaking of toilets...

At the beginning of the 20th century, a young dentist noticed that many citizens of Colorado Springs had brown teeth that were extremely resistant to tooth decay, eventually identifying high levels of fluoride in the drinking water supply as the cause of the "fluorosis." Decades later, another dentist wondered if it were possible to lower the fluoride level in water enough to avoid fluorosis but keep the tooth decay resistance. In 1945, Grand Rapids, MI, became the first city to add fluoride to its drinking water, and within 11 years, a controlled study showed the children in Grand Rapids had a 60% reduction in tooth decay compared to paired cities without fluoridation. While the U.S. Centers for Disease Control and Prevention (CDC) has listed the fluoridation of the public water supply as one of the 10 great public health achievements of the 20th century, there are still contemporary opponents of fluoridation that claim that it is a government conspiracy to control the population or just an infringement on personal freedom (to have bad teeth?).

5.9 WASTEWATER TREATMENT

One of the primary uses of water, both domestic and commercial, is to remove waste products. In the context of domestic use, this includes showers, laundry, and sewage – all making use of the solvent properties of water to carry away waste. Besides cleaning agents, such as bleach, most domestic waste contains biodegradable organic compounds. Commercial waste encompasses a variety of wastes, from dissolved and suspended organic compounds (both natural and synthetic) to inorganic waste. Some of this waste is biodegradable, and some of it is highly toxic (synthetic organic waste and metal cation waste). If even the toxic components of this waste were to be introduced to an aquatic system, much of it would provide nutrients for microbes and other organisms that inhabit the body of water. This nutrient load can quickly generate anoxic water since aerobic microbes would consume the little DO that exists at environmental temperatures.

The historical approach to wastewater was simply to get rid of it. Ancient methods of wastewater removal, as demonstrated by the Romans, used canals and ditches to move wastewater out of buildings and homes and into the Tiber River. Eventually, tiles and pipes were used to increase the efficiency of removal. No treatment was implemented other than the natural biological processes provided by the waterways into which the raw sewage was sent. The 12th-century Cistercian Abbey in Arnsburg, Germany, redirected a portion of the Wetter River to flow through the sewer system of the abbey and back into the river. Even into the 19th century, most major cities discharged untreated waste into the same groundwater and surface waters that were used for drinking water. Diseases such as cholera and typhoid fever are easily spread when untreated sewage and drinking water mix. The last major outbreak of cholera in the United States occurred in 1911, but cholera outbreaks often appear in other parts of the world whenever there is a breakdown in a city's wastewater treatment system.

A recent major outbreak of cholera occurred in the October following the devastating January 2010 earthquake that hit the island of Haiti. Genetic testing has traced the contamination to a strain of cholera that comes from Asia. Nepalese soldiers, assisting the Haitians as part of the United Nations aid mission, are suspected to have infected the water north of the capital Port-au-Prince. The outbreak had killed more than 7000 people.

While scientists were learning about the role microorganisms played in the decomposition of raw sewage, some municipalities in Europe experimented with the application of raw sewage to irrigation fields and successfully reduced the dissolved organic load in the effluent by over 60% (Wiesmann, Choi, and Dombrowski, 2006). This method of water purification, using sand and gravel, is still used in some countries because of its simplicity in construction and efficacy.

In the following sections, the modern process of wastewater treatment will be described. The purpose of this treatment is to remove the nutrient load and pathogens from the water so that it does not contaminate drinking water supplies and reduce water quality.

5.9.1 Biochemical Oxygen Demand and Chemical Oxygen Demand

One of the primary measures of water quality is its biochemical oxygen demand (BOD), which is a measure of how much DO would be needed if aerobic microbes could consume all of the nutrients in the water. Since DO levels in most aquatic systems are below 10 mg O_2/L, even a small amount of nutrient load can compromise DO levels for aerobic organisms. Raw sewage routinely has a BOD of 300 mg O_2/L or more, so a discharge of raw sewage into a body of water would lead to anoxic conditions if the dilution ratio is not very high.

The measurement of BOD, described in more detail in Section 5.10.5, takes place in a sealed bottle that is incubated for 5 days at 20 °C. This permits the microbes present to consume as much DO as the nutrient load requires and the DO levels allow.

A DO measurement before and after incubation allows an estimate of the amount of oxygen needed to consume the nutrients.

Since metabolism is a redox process, aerobic metabolism employs a reduction half-reaction involving molecular oxygen.

$$O_2 + 4H^+ + 4e^- \longrightarrow 2H_2O \qquad (R5.15)$$

Reaction R5.15 shows that four electrons are consumed for every molecule of oxygen used. In any waste stream, there are a plethora of nutritive molecules with different stoichiometric relationships with oxygen, so BOD must be measured experimentally. Since it is the DO level, which is of paramount concern, the specific molecular identity of the waste is usually not important unless it has toxic effects. It is possible, however, to theoretically predict a BOD value for a waste stream if the complex mixture of waste is simplified. Since carbohydrates are very common in many waste streams, one could assume that all of the waste is equivalent to glucose ($C_6H_{12}O_6$) or some other model organic compound. To estimate the BOD for a particular molecule, an oxidation half-reaction can be written for the substance. Once the half-reaction is balanced for charge, it can be stoichiometrically related to O_2 by equating the electrons given in Reaction R5.15.

Example 5.4: BOD

What is the BOD of a 100.0 mg/L solution of glucose?

Solution: The oxidation of glucose ($C_6H_{12}O_6$) to CO_2 can be written in the following way (via a stepwise process). Remember that oxygen and hydrogen are added to a redox reaction in the balancing process by adding water and H^+.

$$C_6H_{12}O_6 \longrightarrow CO_2 \qquad (R5.16)$$

$$C_6H_{12}O_6 \longrightarrow \mathbf{6}CO_2 \qquad (R5.17)$$

$$C_6H_{12}O_6 + 6H_2O \longrightarrow 6CO_2 \qquad (R5.18)$$

$$C_6H_{12}O_6 + 6H_2O \longrightarrow 6CO_2 + 24H^+ \qquad (R5.19)$$

$$C_6H_{12}O_6 + 6H_2O \longrightarrow 6CO_2 + 24H^+ + \mathbf{24e^-} \qquad (R5.20)$$

☞ For a review of the rules for balancing redox reaction, see Review Example 1.24 on page 47.

Reaction R5.20 asserts that for every molecule of glucose consumed, 24 electrons are produced in the oxidation. Reaction R5.15 shows that for every molecule of O_2 consumed, four electrons are consumed. This gives a 24:4 or 6:1 stoichiometric ratio between glucose and oxygen.

$$\begin{aligned} BOD &= \left(\frac{100.0 \text{ mg } C_6H_{12}O_6}{1 \text{ L}}\right)\left(\frac{1 \text{ mol } C_6H_{12}O_6}{180.1572 \text{ g}}\right) \\ &\times \left(\frac{24\, e^-}{1 \text{ mol } C_6H_{12}O_6}\right)\left(\frac{1 \text{ mol } O_2}{4\, e^-}\right)\left(\frac{31.9988 \text{ g } O_2}{1 \text{ mol } O_2}\right) \\ &= 106.569 = 106.6 \text{ mg } O_2/\text{L} \end{aligned} \qquad (5.8)$$

Thus, a 100.0 mg/L solution of glucose, if allowed to be fully consumed (oxidized) by aerobes, would require 106.6 mg O_2/L. Clearly, this solution would become anoxic without constant aeration.

As mentioned previously, a biochemical oxygen demand test takes at least 5 days to complete. This time constraint is intolerable in many instances since it may be expensive or impossible to store a quantity of industrial or domestic waste while a BOD test is completed. A chemical oxygen demand (COD) test produces an acceptable estimate of BOD in most situations. Unlike the BOD test, a COD test employs a very strong oxidizing agent, instead of a biologically mediated oxidation, to accomplish the oxidation of the waste.

$$Cr_2O_7^{2-} + 14H^+ + 6e^- \longrightarrow 2Cr^{2+} + 7H_2O \qquad (R5.21)$$

The stoichiometric relationship between the electrons consumed in Reaction R5.15 and R5.21 can be used to convert the amount of dichromate ion consumed to the equivalent amount of oxygen consumed. Because of the reactivity of the dichromate ion, a COD test takes a couple of hours to complete, so it is routinely used to estimate BOD. Unfortunately, the dichromate ion is not as discriminating with the waste as microbes are. Some of the waste cannot be digested by the microbes and would not contribute to the BOD in a biological test. In the presence of the dichromate ion, this biologically inert waste gets oxidized, resulting in a COD measurement that is an overestimate of the BOD. Given that the purpose of wastewater treatment is to remove as much of the nutritive waste as possible, a COD test provides a conservative estimate of the BOD, such that a treatment regimen engineered with a COD measurement in mind will certainly eliminate all of the BOD.

5.9.2 Primary Treatment

☞ The *influent* is the untreated waste stream that enters a municipal sewage treatment plant. After that stage, the treated stream (*effluent*) enters the next stage until the final effluent is discharged into a waterway.

Municipal wastewater comes from a variety of sources, from domestic and commercial to storm water runoff. Besides dissolved and suspended organic material, wastewater can contain a bewildering assortment of dirt, wood, rags, plastic, and other random objects. These items must be removed before the BOD can be reduced since they can plug or destroy the equipment that carries out most of the treatment in later stages. The *influent* is passed through coarse and fine screens or grates to remove these objects. Various mechanical devices can be employed to scoop or scrape the debris out of the waste stream and place it in a hopper, which eventually is taken to a landfill for long-term storage or to an incinerator.

Once the large debris has been removed, the influent flow rate is decreased such that suspended particles, such as sand and other grit, settle out of the liquid, and insoluble oils and scum collect at the surface. Sometimes, aeration assists in causing oils and other low-density suspended material to float to the surface. Sometimes, coagulants are used to increase the efficiency of the sedimentation process. The sediment and scum are removed from the liquid, dried, and sent to a landfill. The resulting effluent from primary treatment is mostly free of suspended inorganic material and debris and contains mostly water and dissolved organic material. The BOD of this effluent is still very high (>200 mg O_2/L) and requires biological treatment that comes in the next stage.

5.9.3 Secondary Treatment

The influent to this stage of treatment contains a large amount of dissolved organic and inorganic nutrients containing the elements C, N, P, and S. Most secondary treatment methods start with aerobic digestion of these nutrients using the native microbes found in the sewage. Some methods of aerobic treatment pump the influent into a large tank containing inert material onto which colonies of microbes grow. This material can be rocks, coke or coal, slag, plastic balls, or vertically oriented corrugated plastic sheets. This mode of aerobic digestion, often called *trickle filtering*, allows the influent to trickle through the medium in the presence of pockets of air such that high DO levels are maintained. The other method of aerobic digestion, called the *activated sludge* process, takes place in aerated tanks where microbe colonies are suspended on particulate matter. Oxygen is supplied to the wastewater in these tanks by pressurized air bubbling up from the bottom or by large mechanical beaters that mix the surface water with air.

Whichever method is used, aerobic digestion is the goal of this stage in secondary treatment. Microbes consume the nutrients and form ever-growing colonies either on a trickle surface or on a suspended sludge. In trickle filtering, the biological material sloughs off of the inert medium and is collected at the bottom of the trickle filter and sent to a sediment tank where, as in the activated sludge process, the sludge settles to the bottom tank. Since this sludge contains a healthy population of microbes, a portion of it is pumped back to the beginning of the aeration tank to maintain an optimal population of microorganisms.

Chemically, the aerobic treatment process involves oxidizing carbon-based nutrients and converting them to CO_2 and biosolids (sludge), since the aerobic, mesophilic, and heterotrophic microbes are consuming the nutrients and building proteins, fats, and other biomolecules as they grow and multiply. The effectiveness of aerobic treatment of the wastewater lies in the conversion of the dissolved organic matter into the biosolids, which can be separated from the water. The removal of this nutritive energy (BOD) from the effluent through a chemical conversion and phase change makes aerobic digestion the key step in the treatment.

In this process, the digestion of amino acids results in the production of ammonia. The aerobic environment promotes nitrification, as described in Section 5.5. Much of the nitrogen remains in solution, and it will be the focus of tertiary treatment.

The activated sludge process typically removes around 85–90% of the BOD, leaving the effluent BOD at about 15 mg O_2/L. The trickle filter method is not as efficient, bringing the BOD down to around 60 mg O_2/L, but it is less costly compared to the activated sludge process. The effluent from this process has completed the secondary treatment stage and is sent on to tertiary treatment. The sludge, on the other hand, still contains enough chemical energy and nutrients to be hazardous if discharged into the environment. Anaerobic digestion of the sludge renders it less problematic for the environment.

Commonly used coagulants are combined with inorganic salts, such as lime ($Ca(OH)_2$), to control the pH and generate the floc.

$$Al_2(SO_4)_3 + 3Ca(OH)_2 \rightarrow 2\,Al(OH)_3 + 3\,CaSO_4 \quad (R5.22)$$

$$2FeCl_3 + 3Ca(OH)_2 \rightarrow 2\,Fe(OH)_3 + 3\,CaCl_2 \quad (R5.23)$$

The iron or aluminum hydroxide species that result can be manipulated using the solution pH to form soluble complexes that are either negatively charged (neutral to alkaline pH) or positively charged (acidic pH). In the basic solutions, the hydroxide complexes pick up extra hydroxides to form negative ions, such as $Fe(OH)_4^-$ and $Al(OH)_4^-$. In acidic solutions, hydroxides are removed from the complex to form complexes such as $Al(OH)^{2+}$ or $Fe(OH)_2^+$. These complexes, both positively and negatively charged, are known as floc and adsorb dissolved and suspended organic and inorganic material. The organic material usually contains polar alcohol, amino, and carboxylate substituents, which are attracted to the charged floc surface. As more and more of this material sticks to the floc surface, the floc particles grow larger and eventually precipitate out of solution.

5.9.4 Anaerobic Digestion

Although technically not part of the secondary treatment process, anaerobic digestion of the sludge produced from the aerobic stage is another biologically mediated process that removes nutrient energy. The sludge from secondary treatment must be removed and disposed of since it contains a large number of pathogens, produces a noxious odor, and contains a significant amount of BOD. Under anoxic conditions, however, this waste can produce a significant amount of methane gas, which can minimally be used to offset the cost of the process or, under ideal conditions, can produce surplus energy in the form of biogas.

In Chapter 4, you learned about anaerobic metabolic pathways, one of which was the disproportionation of carbon compounds to produce carbon dioxide and methane. Sludge from secondary treatment can be pumped into a large storage building that contains a floating ceiling or rubber diaphragm that prevents the sludge from mixing with air that might contain oxygen. Under these anoxic conditions, along with elevated temperatures (>45 °C), thermophilic methanogenic microbes metabolize the sludge and produce a biogas that is between 50% and 70% methane (the rest is mostly carbon dioxide and some hydrogen sulfide).

If you compare the free energy of formation for methane ($\Delta G_f^\circ = -50.5$ kJ/mol) and carbon dioxide ($\Delta G_f^\circ = -394.4$ kJ/mol), you will notice that methane's ΔG_f° is much higher. This indicates that there is a lot of chemical energy stored in methane, which should come as no surprise since methane is a common fuel known as natural gas. The effectiveness of anaerobic digestion of sludge lies in this principle – a large amount of the chemical energy stored in the sludge is converted into methane, which bubbles to the surface and easily separates from the remaining biosolids. In comparison, the aerobic digestion in the secondary treatment phase converted the chemical energy in the dissolved organic material to sludge.

According to the U.S. Department of Energy, more than 20% of the municipal wastewater treatment facilities employ anaerobic digestion. In these applications of anaerobic digestion, the methane gas is used to heat the anaerobic digester to maintain the thermophilic conditions necessary for maximum methane production. In about 2% of these facilities, excess methane is used to produce electricity to offset the power used to run the machinery at the plant (Federal Energy Management Program, 2005).

If you ever visit a municipal sewage treatment plant, your first stop will likely be at the solid waste hopper. You will see rags, tampon applicators, condoms, paper money, ID cards, and an interesting array of other objects. At one plant tour in Pennsylvania, I asked one of the engineers to tell me what was the strangest item he found after the primary treatment. His response – a bowling ball. At another plant, it was a human spleen. You will spend hours trying to imagine how they got into the sewage system!

5.9.5 Tertiary Treatment

Tertiary wastewater treatment encompasses several processes that defy a single description other than "the treatment that comes after secondary treatment." The objective in this stage of treatment is to remove additional nutrients, increase the water clarity, and/or disinfect. Nutrient removal can be accomplished chemically or biologically and can be incorporated into the secondary treatment process, so the placement of tertiary treatment is not always distinct from secondary treatment.

Tertiary treatment is often necessary because conventional secondary treatment leaves most of the phosphorus and nitrogen nutrients in the effluent. State and national efforts to reduce pollution in large watersheds, such as the Chesapeake Bay, have forced municipal wastewater treatment facilities to reduce their nitrogen and phosphorus nutrient discharges well below that which can be achieved through secondary treatment alone.

5.9.5.1 Biological Nitrogen Removal Biological nitrogen removal is the process by which denitrification is achieved. The Nitrogen Cycle section of this book described the process of nitrification, where nitrogen species such as ammonia and the nitrite ion are oxidized to form the nitrate ion under aerobic conditions. The process of denitrification occurs under anoxic conditions and converts the nitrate ion into molecular nitrogen via Reactions R5.10 and R5.11. Once in the nonnutritive N_2 form, nitrogen escapes into atmosphere and lowers the nutrient load of the water. Since secondary treatment designs include aerobic digestion, creating the conditions for an anoxic zone immediately after the aerobic zone allows for this sequential treatment. The aerobic zone promotes the conversion of all organic and inorganic nitrogen to the nitrate ion form, and then the anoxic zone promotes the conversion of the nitrate ion to gaseous nitrogen. Practically, this is accomplished by constructing the secondary treatment tanks such that the influent moves into a chamber with an aerator and then flows around a barrier into another chamber with no aerator. Aerobic microbes quickly generate anoxic conditions once they are separated from the aeration zone, which allows facultative microbes to begin the denitrification process. This process can be repeated several times, where immediately following an anoxic zone is another aerobic zone.

As Reaction R5.10 indicates, denitrification is a metabolic reaction that consumes carbon nutrients. In many wastewater treatment plants, the aerobic digestion process is so efficient at removing carbon nutrients that there is often not enough left for denitrification to be efficient. Practically, this is solved by adding a carbon nutrient to the anoxic zones to promote denitrification, such as methanol.

5.9.5.2 Chemical Nitrogen Removal Ammonia stripping is an alternative process to nitrogen removal via denitrification. When nitrogen is present as ammoniacal nitrogen, it participates in an equilibrium between two forms of nitrogen.

$$NH_3(g) + H_2O(l) \rightleftharpoons NH_4^+(aq) + OH^-(aq) \tag{R5.24}$$

It is an application of Le Chatelier's Principle to use the pH of the solution to shift nitrogen from one form or the other. Under high pH conditions, nitrogen exists predominantly in the ammonia form, which is less soluble and can be removed from the solution as a gas. Practically, this is accomplished by adding lime or sodium hydroxide to raise the pH and then spraying the water inside of a sealed chamber. As the liquid droplets collect at the bottom of the chamber, the ammonia becomes gaseous and can pumped out of the top of the chamber.

Once removed, the ammonia can be vented to the atmosphere, sent into a high-temperature chamber where it can be catalytically converted to N_2, or sent into another chamber where it reacts with a sulfuric acid solution to form ammonium sulfate.

$$2NH_3(g) + H_2SO_4(aq) \rightleftharpoons (NH_4)_2SO_4(aq) \tag{R5.25}$$

While venting the ammonia to the atmosphere is the least expensive option, it could result in significant air pollution and noxious odors. The catalytic destruction is energy-intensive because of the high temperatures necessary. The recovery of the nitrogen as ammonium sulfate allows it to be reused as an agricultural fertilizer.

*5.9.5.3 **Chemical Phosphorus Removal*** The predominant forms of phosphorus nutrients in wastewater are orthophosphate or polyphosphate ion forms. The phosphate ion can be precipitated from a solution that has a high calcium or aluminum ion concentration. Adding a mineral, such as lime or alum, will cause most of the inorganic phosphate to precipitate. Conveniently, lime is also an additive used for ammonia stripping, so the chemical removal of ammonia and the phosphate ion can be accomplished in the same process. The calcium phosphate that precipitates can be sent to a landfill or reused as agricultural fertilizer.

☞ *Lime*, the mineral, (not to be confused with the fruit!) has an ionic formula of CaO. It occurs naturally but can also be produced by roasting limestone ($CaCO_3$). When dissolved in water, it forms a very caustic solution of calcium hydroxide.

*5.9.5.4 **Biological Phosphorus Removal*** Certain microbes, known as polyphosphate accumulating organisms (PAOs), are able to absorb significantly more phosphate compared to other microbes. There exists aerobic and denitrifying PAOs, which can be added to the secondary treatment stage and contribute to aerobic digestion or the denitrification process while they are removing phosphate nutrients from the wastewater. As the PAO population grows, the dead PAOs settle out into the sludge and can be removed from the effluent.

5.9.6 Filtration

In the final stages of the wastewater treatment process, filters can be used to provide additional clarification of the effluent before it is discharged. Filtration through sand or anthracite coal traps fine suspended particles that do not settle out of clarifying ponds. An additional function of filtration through anthracite coal is the adsorption of dissolved organic matter. The anthracite acts as a charcoal filter, functioning the same as the filter packs on the popular Brita® water filtration system.

5.9.7 Disinfection

Often, the final stage of wastewater treatment is disinfection. A host of very nasty pathogens survive the wastewater treatment process, such as hepatitis, typhoid, polio, and cholera. The elimination of these pathogens from our drinking water supplies has been one of the greatest achievements of human civilization. If you have ever lived through a flood or other natural disaster that disrupts the water supply, you know that one of the most important things to do is to boil any water that will be used for drinking. The CDC reports that 1.5 million people die every year due to waterborne pathogens.

There are several ways to disinfect the effluent from a wastewater treatment plant. Chlorine gas was one of the first chemicals used to disinfect water. The gas is bubbled into the water, and the highly reactive chlorine will attack most organic molecules, effectively disrupting molecular processes and killing or deactivating the pathogens. Chlorine gas is very dangerous and is stored in large tanks under high pressure, making it doubly dangerous to store and use. Sodium hypochlorite, which is the active ingredient in bleach, is another chlorine-based disinfectant that is dissolved in solution and much safer to store and use. The advantage of these chlorine disinfectants is that they are usually the least expensive means of disinfection. Their major disadvantage is that residual chlorine remains in the water and can be acutely toxic to the ecosystem into which the water is discharged, and they both form stable chlorinated organic compounds, which can bioaccumulate and result in persistent toxins that can cause long-term damage.

Chlorine gas reacts with water to form hypochlorous acid

$$Cl_2(g) + H_2O(l) \rightleftharpoons HOCl(aq) + HCl(aq) \tag{R5.26}$$

Chlorine gas is a very effective and relatively inexpensive means of disinfection. It is also tremendously toxic. Just before 3 a.m. on January 6, 2005, a rail car filled with chlorine ruptured in South Carolina, spilling 60 tons of chlorine. Of the nine people who died, one was a trucker who was sleeping in the cab of his truck. Because of the danger in storing and using chlorine gas, many wastewater treatment facilities have chosen to use the more expensive sodium hypochlorite because of the reduced costs in training and risk when using chlorine gas.

so both disinfection with either Cl_2 or sodium hypochlorite leave the same residual chlorine species. The hypochlorite ion can be destroyed by several compounds such as sulfur dioxide gas, sodium sulfite, sodium metabisulfite ($Na_2S_2O_5$), or sodium bisulfite. All of these compounds produce the sulfite ion in solution, which attacks the hypochlorite ion to render the nontoxic chloride and sulfate ions.

$$SO_3^{2-}(aq) + OCl^-(aq) \rightarrow SO_4^{2-}(aq) + Cl^-(aq) \quad (R5.27)$$

This is accomplished by allowing the disinfected water to mix with a solution of sulfite ion sometime before the water is discharged.

Two alternatives to chlorine disinfection make use of either powerful UV lamps or an ozone generator. Both consume a large amount of electricity but do not leave toxic residual compounds in the water. The UV lamps destroy all microbes and viruses through UV photolysis, and bubbling ozone through the water causes all organic molecules to undergo harsh oxidation and effectively kills everything. There is no residual compound from UV disinfection, and after the excess ozone bubbles out of the solution, only molecular oxygen is left over.

5.9.8 Biosolids

The sludge that is formed as part of the secondary treatment process is the accumulation of the coagulated microbes that have done the work of reducing the nutrient load. This mass of living and dead cells possesses a lot of nutrient energy in the form of biomolecules and must not be discharged into the environment. It is usually pumped to a sludge press, which removes most of the water to form what is called a dewatered cake. This cake is stabilized compost. Piling up this material, or mixing it with wood chips, and allowing it to "compost" at elevated temperatures kills many of the pathogens and renders it fit for garden and agricultural land applications or disposal in a landfill. According to the US EPA, class A biosolids, which are free of harmful pathogens, can be applied to land and home gardens with few restrictions. Class B biosolids, which contain a few pathogens, can be applied to agricultural fields with a controlled frequency of application, depending on the pathogen count.

Since the biosolid cake is mostly carbon compounds, it is combustible and can be sent to an incinerator. If the cake is dry enough, then the thermal energy can be used to generate electricity. If the cake is not dry enough, then much of the heat energy from the incinerator goes into evaporating the water and reduces the energy output.

5.9.9 Septic Tanks and Sewage Fields

Many Americans live outside of an urban or suburban setting and have no access to a municipal sewage system. Some of these homes use sewage tanks to temporarily store sewage, which perform very little treatment to the waste and must be emptied periodically. Other households used a septic field system where the household waste flows into a large concrete tank separated into two compartments. This tank allows solids to settle to the bottom and grease to form a scum layer on top. The first compartment receives the bulk of the solids, and the second compartment receives the small amount of solids that escape the first. Most of the septic waste is water, so the excess water drains out of the other end of the tank through a special pipe that is submerged below the scum but above the sludge. Escaping solids will plug the drain field and clog the system, so the submerged pipe will avoid this. Some aerobic and anaerobic digestion occurs in the tank, but its primary purpose is to retain solids and prevent clogging. Any gases generated in the process diffuse out of the tanks. Solids must be pumped out before they overflow the tank and clog the system. The liquid that escaped enters the drain field, is dispersed over a large area in the yard, and percolates through the soil. The process of percolation is usually sufficient to remove pathogens from the water before it reaches the underlying groundwater aquifer, which is the source of the drinking water. The USGS suggests that any septic field be at least 100 horizontal feet from the nearest well head, and a well should be at least 45 feet deep.

5.10 MEASUREMENTS

Unlike the measurements you read about in the preceding chapters, the area of water quality methods is organized and standardized. There are several governmental and nongovernmental agencies that publish methods of analysis for many environmentally significant chemical species. These methods are compiled in reference books or have been placed on the Internet for subscription or free access. The National Water Quality Monitoring Council, a publicly funded organization, has set up an online database of these methods and provides links to the source of each method. The National Environmental Methods Index (NEMI) is one of the best sources for discovering a method and obtaining a copy of it. Methods from government sources are usually free downloads, such as from the U.S. EPA. The American Public Health Association is a private nonprofit organization that publishes the *Standard Methods for the Examination of Water and Wastewater* (hereafter referred to as *Standard Methods*) compendium of water quality methods and requires a subscription for online downloads or the purchase of the *Standard Methods* book.

In these sources, each method is designated by an alphanumeric label and a title describing the chemical species and technique used for analysis. Methods from different agencies are often very similar, but if your results will be used in a larger study led by a public or private organization, you should make sure that you use a method the organization has approved. In the following methodological descriptions, I have provided at least one or more methods for each measurement. I encourage you to make use of the following resources in order to find the specific methods that you require.

Resources for locating water quality methods of analysis

NEMI database: www.nemi.gov
U.S. EPA list of approved methods:
water.epa.gov/scitech/methods/cwa/methods_ index.cfm
***Standard Methods* database:** www.standardmethods.org/
***Standard Methods* compendium:** Clesceri, L.S., A.E. Greenberg, and A.D. Eaton, (eds). 1998. *Standard Methods for the Examination of Water and Wastewater.* 20th Edition. ISBN: 0875532357. American Public Health Association. Washington, D.C.
Hach Methods: hach.com/epa

Some methods are available from third-party websites, so knowing the method name, number, and agency will help you locate these resources. Use the NEMI database to obtain this information.

5.10.1 Potentiometric pH Measurements

The pH of a solution is most commonly measured using a potentiometric pH meter, which is basically a hydronium-ion-selective electrode. The operational relationship that is used in potentiometric pH measurements is

$$\mathrm{pH}_{sample} = \mathrm{pH}_{std} - \frac{\left(E_{sample} - E_{std}\right)}{RT\ln\left(\frac{10}{F}\right)} \tag{5.9}$$

where pH_{sample} and pH_{std} are the pH of the sample and buffer, respectively, E_{sample} and E_{std} are the measured electrode potentials for the sample and buffer, respectively, R is the gas law constant in Joules, T is the temperature in Kelvin, and F is Faraday's constant.

Component	Molal conc. (mol/kg)	Amount to add to 1 kg H_2O
NaCl	0.40764	23.8233 g
Na_2SO_4	0.02927	4.1576 g
KCl	0.01058	0.7887 g
$MgCl_2$	0.05474	5.2119 g
$CaCl_2$	0.01075	1.1931 g
HCl (conc.)	0.02	1.667 mL
TRIS buffer	0.04	4.8456 g

The salinity is determined by adding the grams of salt added per 1 L. Adding the masses of NaCl, Na_2SO_4, KCl, $MgCl_2$, and $CaCl_2$ gives 35.1746 g in this case, assuming that 1 kg of water is approximately a liter.

Table 5.6 A modified recipe from DelValls and Dickson (1998) for a seawater TRIS buffer having a salinity of 35.2‰ and a pH of 8.09.

The analyst is largely unaware of this relationship because it is established as part of the calibration procedure and handled by the pH meter. With freshwater samples, standard reference buffers can be purchased for pH 4, 7, and 10. These buffers need to be at the same temperature as the sample when they are used to calibrate the pH electrode. Automatic Temperature Compensation (ATC) probes can be connected to many pH meters in order to correct for temperature differences from sample to sample since the measurement is temperature dependent. For more details on a standard method approved for measurements of pH, see method 4500-H in the *Standard Methods* compendium.

Seawater measurements require a standard buffer that contains the same salinity as the sample. Since marine samples can have variable salinity values, a reference buffer must be custom-made for the sample. A recipe has been provided in Table 5.6 (DelValls and Dickson, 1998).

A more accurate measurement of pH in seawater requires adjustments of the electrochemical measurement with other parameters such as temperature, alkalinity, and the total carbonate concentration (Piedrahita, 1995).

5.10.1.1 Spectrophotometric pH Measurements In the spectroscopic method using a 3.5 mL cuvette, 3 mL of the seawater sample is placed into the cuvette and baseline readings at 573 and 433 nm are made. After 5 μL of a cresol red indicator is added to the cuvette, the sample is mixed, and the absorbance at 573 and 433 nm is achieved. The pH is determined by the following equation:

$$\mathrm{pH} = 7.8164 + 0.004(35 - S) + \log\left(\frac{\frac{A_{573}}{A_{433}} - 0.0286}{2.7985 - 0.09025 \cdot \frac{A_{573}}{A_{433}}}\right) \tag{5.10}$$

where A_{573} and A_{433} represent the absorbance (with the baseline subtracted) of the sample at 573 and 433 nm, and S represents the salinity of the sample in units of ‰. The constants in this equation were derived (Byrne and Breland, 1989) for a mixture of cresol red indicator and seawater at 25 °C and would change with a different indicator dye, such as *m*-cresol purple. The resulting pH is within ± 0.004 pH units according to the authors.

Spectroscopic pH measurements have not been adopted as standard methods by the EPA, the USGS, or the organization sponsoring the *Standard Methods*. A more thorough description of this method is available (Easley and Byrne, 2012).

5.10.2 Total Dissolved Solids (TDS)

Determining the amount of solids dissolved in a water sample presumably looks for ionic and molecular compounds. The ionic compounds dissolved will increase the conductivity of the solution proportional to the Total Dissolved Solids (TDS) load in the solution. Conductivity probes and meters are used for this purpose. A typical conductivity probe consists of two platinum wires separated from each other by a fixed space. An AC voltage is applied to the wires and the resistance, according to Ohm's law ($V = IR$), is measured. As the solution salinity increases, the resistance across the space between the wires decreases. Conductivity is measured in Siemens, which is the reciprocal of an ohm (Ω, unit of resistance). This measurement is relatively insensitive to the dissolved organic compounds and is more appropriate for measuring salinity. For more details on an approved method for measurements TDS via conductivity, see method 2540 in the *Standard methods* compendium.

TDS measurements can also be made by filtering the sample to remove suspended solids (2.0 μm pore or smaller), evaporating or boiling a sample to dryness, drying the residue in an oven for a couple of hours, cooling in a desiccator, and weighing the residual solids. Heating the sample would produce a negative bias toward those molecular compounds that might be volatile and evaporate if the temperature is great enough. The temperature used to dry the residual solids has a bearing on the inorganic portion of the TDS as well. Mechanically (trapped in the solid) and chemically bonded water (hydrates) will contribute to the mass if the sample is dried at relatively low temperatures, such as 105 °C. This water can be driven off by drying at 180 °C. Some carbonate salts decompose below 200 °C (such as $MnCO_3$, $NiCO_3$, and $CuCO_3$), but most common carbonates, sulfates, and nitrates will decompose above 200 °C, making drying ovens an acceptable means of drying the sample.

Since most of the organic compounds will vaporize at drying oven temperatures, this process necessarily reduces the contribution of dissolved organic solids to the TDS measurement. A more accurate measure of the dissolved organic compounds is accomplished by Total Organic Carbon (TOC) described in Section 5.10.4.

For more details on some standard methods approved for measurements of TDS via drying oven, see *Standard Methods* 2540 and EPA Method 1601.

5.10.3 Salinity

Salinity is a subset of the TDS measurement, representing the inorganic portion. A conductivity measurement in the TDS method would result in a conductance (μSeimens/cm) and can be converted to a salinity result using a calibration solution of known salinity.

If salinity is measured by the drying method of TDS, as described earlier, then the mass of the residue per 1 kg of sample is a measure of the salinity in units of ‰, as described in Example 5.1 on page 194.

5.10.4 Total Organic Carbon (TOC)

Measurements of TDS via conductivity miss many of the dissolved organic compounds that do not contribute to a solution's conductivity. The measurement of TOC provides a remedy for this by selectively measuring these species. Any environmental, aqueous solution contains inorganic carbon, in the form of dissolved carbon and other carbonates, and organic carbon. Since the inorganic carbon is largely from exposure to carbonate minerals or the atmosphere, these forms of carbon need to be removed before the organic carbon can be measured. This is accomplished by acidifying the solution and driving the carbonic acid reaction toward dissolved carbon dioxide via Le Chatelier's principle.

$$CO_2(aq) + H_2O(l) \rightleftharpoons H^+(aq) + HCO_3^-(aq) \qquad (R5.28)$$

☞ *Sparging* is a process by which an inert gas, such as N_2 or Ar, is bubbled through a solution in order to displace a reactive dissolved gas, such as O_2, from the solution.

With most of the inorganic carbon in the dissolved CO_2 form, it is removed by *sparging* the solution with an inert gas such as nitrogen or helium. Sparging drives the CO_2 out of the solution, leaving only dissolved organic carbon in the solution. If the analyst wishes to measure the amount of inorganic carbon, most TOC instruments have a TIC mode (Total Inorganic Carbon) and use infrared absorption to measure the concentration of the CO_2 that was sparged out of the sample.

The remaining solution contains dissolved organic carbon (DOC) and particulate organic carbon (POC) (insoluble but suspended). Separating the DOC from the POC can be accomplished by filtration. If the POC is left in the sample, then the subsequent digest will measure both DOC and POC. The remaining organic carbon is oxidized to CO_2 using a strong oxidizing agent (such as $K_2S_2O_8$, potassium persulfate), and then the CO_2 can be measured by sparging it out of the solution and into an infrared detector. Adding the TIC to the TOC results in a measure of the TC (total carbon).

For more details on some standard methods approved for measurements of TOC using various oxidants, see EPA Method 415 and the series of methods under Method 5310 in *Standard Methods*.

5.10.5 Biochemical Oxygen Demand (BOD)

Since BOD relies on microbes to oxidize the nutrient load in the sample, the method used for BOD analysis must replicate environmental conditions. The overall method is to fully aerate the sample, immediately measure the DO level, let the sample incubate for 5 days in the absence of light and in a temperature-controlled environment, and then measure the DO level again. In some samples, the pH must be adjusted to a value between 6.5 and 8.5 prior to the trial in order to maximize the microbiological activity. Amendments of mineral nutrients might be required if the sample is deficient in Ca, Mg, Fe, or phosphate ion. The measured BOD that results is the difference between the initial and the final DO level after the incubation period.

In practice, the nutrient load may be so high in some samples that the DO levels will approach zero before all of the nutrients are digested, resulting in an underestimate of the BOD. This requires a series of diluted samples so that about 30% of the oxygen remains to ensure that nutrient load is the limiting reactant. The BOD is calculated according to the following equation.

$$\text{BOD}_5\ (\text{mg O}_2/\text{L}) = \frac{D_1 - D_2}{P} \tag{5.11}$$

For example, examine Table 5.7.

According to the table, 2.00 mL of the sample was diluted to a total volume of 300 mL in a BOD incubation bottle. Its initial BOD reading was given as 8.1 mg/L O_2 (D_1). Two more dilutions were made with 5.00 and 10.00 mL of sample to a total volume of 300 mL.

Aliquot volume (mL)	D_1 (mg O_2/L)	D_2 (mg O_2/L)	Dilution factor (P)	BOD_5 (mg O_2/L)
2.00	8.1	5.6	2/300	375
5.00	8.0	1.7	5/300	378
10.00	8.1	0.0	10/300	243
Blank	8.2	8.0	300/300	0.2

D_1 represents the initial DO reading after sample preparation, D_2 represents the final DO reading after the 5 day incubation, and P represents the dilution factor as it appears in Eq. (5.11). BOD incubation bottles with a total volume of 300 mL were used.

Table 5.7 An example BOD experiment with three dilutions.

A blank was prepared to show that the diluted samples began at near DO saturation. After a 5-day incubation at 20 °C, the DO of each sample was measured again (D_2). Since the dilution for the first trial was 2.00 mL in 300 mL, the calculated DO is given as follows.

$$BOD_5\ (\text{mg } O_2/\text{L}) = \frac{8.1 - 5.6}{\frac{2}{300}} = 375 \text{ mg } O_2/\text{L}$$

As Table 5.7 shows, the first dilution resulted in a successful trial because the final DO measurement did not drop below the suggested 30% of saturation or 2.4 mg/L O_2 in this case. The second dilution was probably a good trial because the BOD_5 result was close to the first dilution. The third dilution was too concentrated because the DO was exhausted and there remains a nutrient load in this sample.

Standard procedures for BOD analysis are provided by EPA Method 405 and *Standard Methods* 5210.

5.10.6 Chemical Oxygen Demand (COD)

While the measurement of the BOD has a physiological meaning for the sample, it requires attention to specific incubation conditions and 5 days before the final determination can be made. These methodological deficits are remedied by the implementation of a COD method. The COD technique makes use of a harsh oxidizing reagent to accomplish in a couple of hours what took 5 days in the BOD method.

In this method, a known amount of the sample is added to a test tube containing a mixture of sulfuric acid and the dichromate ion. The dichromate ion oxidizes nearly all species dissolved in the solution, leading to an overestimate of the BOD since many of these chemical species would not serve as nutrients in a biological test. Whereas in the BOD test, microbes mediate the oxidation of nutrients using DO, the COD method makes use of the dichromate ion as the oxidizing agent.

$$Cr_2O_7^{2-} + 14H^+ + 6e^- \longrightarrow 2Cr^{2+} + 7H_2O \qquad \text{(R5.29)}$$

The dichromate ion is such a strong oxidizing agent that even nonnutritive chemical species (organic and inorganic) in the sample are oxidized, causing the COD measurement, in units of mg/L O_2, to be an overestimate of the BOD measurement.

Reaction R5.21 reveals the two spectroscopic modes of measuring COD. The $Cr_2O_7^{2-}$ ion absorbs strongly at 420 nm, so the loss of the absorbance at 420 nm compared to the reagent blank is proportional to the COD. The Cr^{3+} ion generated by the oxidation absorbs visible light at 620 nm. The increase in the measured absorbance at 620 nm is also proportional to the COD.

The digestion of the sample can be accomplished in about 2 h by placing the sample and the sulfuric acid/dichromate mixture into a heater block to let it digest in a sealed container for 2 h. For very dilute samples with little nutrients (90 mg/L O_2 or less), the loss of absorbance of $Cr_2O_7^{2-}$ is used. This requires that the dichromate ion concentration is standardized.

For higher COD samples (100 to 900 mg O_2/L), the absorbance of Cr^{3+} at 620 nm is proportional to the COD. The advantage of using the chromium ion absorbance is that the dichromate solution used to digest the sample does not need to be standardized for dichromate. Since the concentration of Cr^{3+} should be zero for the reagent blank, this reference point allows dichromate to be in excess and approximate.

5.10.7 Dissolved Oxygen

There are two common methods for measuring DO in water samples – a titrimetric method called the Winkler method and an electrochemical method using ISEs.

5.10.7.1 Titrimetric Method The Winkler method, developed by the Hungarian chemist Lejos Winkler (1863–1939) in 1888, is a titration of iodine produced as a result of a reaction that "fixes" the solution so that further exposure of the sample to atmospheric oxygen does not lead to inaccuracies. In this method, a solution of manganese(II) sulfate and an alkaline solution of potassium iodide are added to the sample, resulting in the following reaction.

$$2Mn^{2+}(aq) + 4OH^-(aq) + O_2(aq) \longrightarrow 2MnO(OH)_2(s) \quad (R5.30)$$

Once the DO has been removed from the solution and converted to the brown manganese precipitate, the solution is acidified with either sulfuric acid or solid sulfamic acid, which causes the manganese precipitate to redissolve and react with the iodide ion.

$$MnO(OH)_2(s) + 2I^-(aq) + 6H^+(aq) \longrightarrow Mn^{2+}(aq) + I_2(aq) + 3H_2O(l) \quad (R5.31)$$

Once these two reactions are complete, the sample has been "fixed." Since the solution is acidic, no more manganese ions can react with DO to form the precipitate, which requires alkaline conditions. With the possibility of contamination gone, the sample can be transported to a laboratory for further analysis or can be titrated on site.

With the aqueous iodine present, the sample will have a light to dark yellow color, proportional to the amount of DO. The color is too light, however, to serve as a good indicator for the end point of the titration that will determine the DO level. Since iodine reacts with dissolved starch to form an inky blue complex, starch is added in order to enhance the color of the solution. Since the iodine concentration is proportional to the original DO level, a stoichiometric removal of the iodine from solution – turning the inky blue solution to clear – will result in a DO measurement. Iodine removal is accomplished by titrimetric addition of sodium thiosulfate, which converts iodine back to the iodide ion.

$$I_2(aq) + 2S_2O_3^{2-}(aq) \longrightarrow 2I^-(aq) + S_4O_6^{2-}(aq) \quad (R5.32)$$

Every four moles of thiosulfate ion reacts with one mole of DO. If the fixed sample is taken to a laboratory, a burette containing a standardized thiosulfate ion solution can be used to titrate the sample and determine the DO level. If 200.0 mL of the fixed sample is titrated, then the volume of a 0.025 M thiosulfate ion solution added to turn the inky blue solution to colorless is equal to the mg O_2/L.

$$\left(1.0\ \tfrac{\text{mg}}{\text{L}}O_2 \times \frac{200.0\ \text{mL}}{1000\ \frac{\text{mL}}{\text{L}}}\right)\left(\frac{1\ \text{mol}\ O_2}{31.9988\ \text{g}\ O_2}\right)\left(\frac{2\ \text{mol}\ MnO(OH)_2}{1\ \text{mol}\ O_2}\right)$$
$$\times\left(\frac{2\ \text{mol}\ I^-}{1\ \text{mol}\ MnO(OH)_2}\right)\left(\frac{2\ \text{mol}\ S_2O_3^{2-}}{2\ \text{mol}\ I^-}\right)\left(\frac{1\ \text{L}}{0.025\ \text{mol}\ S_2O_3^{2-}}\right) = 1.0\ \text{mL} \quad (5.12)$$

Equation (5.12) shows the one-to-one relationship between a milliliter of the 0.025 M thiosulfate ion solution and 1.0 $\frac{\text{mg}}{\text{L}}$ O_2 if 200 mL of fixed sample is titrated. Prepackaged Winkler kits are available that contain all of the appropriate reagents in a small container (sold by brands such as Hach and LaMotte). In order to conserve the amount of titrant and reduce the total volume of the sample and subsequent waste, one popular commercial kit scales down the sample volume to 20 mL and provides a titration syringe that delivers 0.1 mL per major ruled mark on the syringe body. Thus, each major mark on the syringe represents 1 $\frac{\text{mg}}{\text{L}}$ O_2 in a titration.

The limiting reactant in the Winkler method, besides the DO, is the thiosulfate ion solution. It must be standardized before a DO determination can be made. This is achieved by using a potassium iodate primary standard with excess KI. The iodate ion reacts with the iodide ion to form I_3^-.

$$IO_3^-(aq) + 8I^-(aq) + 6H^+(aq) \longrightarrow 3I_3^-(aq) + 3H_2O(l) \qquad (R5.33)$$

This solution can be titrated with the thiosulfate standard.

$$I_3^-(aq) + 2S_2O_3^{2-}(aq) \longrightarrow 3I^-(aq) + S_4O_6^{2-}(l) \qquad (R5.34)$$

In the modified Winkler method, the alkaline iodine solution contains a small amount of sodium azide (NaN_3) to eliminate any interference caused by the nitrite ion present in the sample. The nitrite ion would convert the iodide ion to iodine, giving an overestimate of the DO levels since the concentration of iodine is an indirect result of Reactions R5.30–R5.31. Instead, the azide ion converts the nitrite ion to nitrogen gas and nitrous oxide.

$$NaN_3(aq) + H^+(aq) \longrightarrow HN_3(aq) + Na^+(aq) \qquad (R5.35)$$

$$HN_3(aq) + NO_2^-(aq) + H^+(g) \longrightarrow N_2(g) + N_2O(g) + H_2O(l) \qquad (R5.36)$$

Since the nitrite ion is often found in freshwater systems, the modified Winkler method is typically employed to improve the accuracy of the method.

For more details on standard methods approved for measurement of DO using the Winkler method, both titrimetric and other, see EPA Method 360.2 and Hach Method 8229.

Standardization

If a 0.025 M thiosulfate secondary standard (can be purchased or made by adding 6.205 g of $Na_2S_2O_3 \cdot 5H_2O$ to a 1-L volumetric flask and diluting with deionized water) is to be used for the Winkler method, then the standardizing solution can be made by quantitatively weighing about 0.01 g of a KIO_3 primary standard and mixing it with about 0.2 g of KI in 40 mL of deionized water and 5 mL of 1 M HCl. About 10 drops of a starch indicator are added to the iodate primary standard solution to enhance the color for the titration. The iodate solution is titrated to turn colorless with the thiosulfate solution. The standardization calculation is as follows.

$$?\,M\,S_2O_3^{2-} = (g\,KIO_3)\,(214.01\,g/mol)\,(KIO_3\text{ purity})\,(6)\,(\text{vol in mL }S_2O_3^{2-})\,(1000) \qquad (5.13)$$

The starch indicator can be made by adding approximately 1 g of iodometric starch to 10 mL of deionized water, mixing, and pouring the mixture into 100 mL of boiling deionized water. Boil for an additional few minutes. Allow any precipitate to settle and decant the liquid as the starch indicator.

5.10.7.2 Dissolved Oxygen via ISE ISEs are potentiometric devices that measure the voltage generated by a concentration gradient that develops across a membrane. In the case of a DO ISE, DO diffuses across an organic membrane and is electrochemically reduced at the electrode. This generates a current, which is proportional to the DO level in the solution.

Using ISEs to measure DO is very attractive because of the rapid measurement and the ability to use the ISE in a remote sensor *in situ*. A DO ISE can be attached to a multisenor monitoring device (along with electrochemical sensors for pH, calcium, temperature, and others) and lowered into a deep well to provide measurements for a variety of water quality parameters without the need for removing a sample, storing it, and transporting it to a lab for analysis. The downside to ISEs is that they tend to be less accurate compared to the Winkler method, yielding an inaccuracy of around of 2–10%. The measurement is temperature-sensitive, and the membranes are not very rugged and can foul if exposed to excessive contaminants. They are subject to interference from other ions, which can result in poor precision. Since ISEs operate on the principle of electrochemistry, they actually measure

☞ *In situ* refers to a measurement made in the natural environment. *In vitro* measurements are usually accomplished by removing the sample from its environment and processing it in glassware (thus the *vitro* part of the name) in a laboratory. While *in situ* measurements are preferred because the environmental context is not disturbed, they are often harder to make because of limitations in equipment and environmental conditions.

activity and not concentration, so they are susceptible to inaccuracies resulting from ionic strength differences. EPA Method 360.1 contains more details on the standard method using a DO ISE.

5.10.8 The Nitrate Ion

The measurement of nitrate ion levels is vitally important to water quality. Consequently, there are several different methods for measuring the nitrate ion. Most environmental samples will contain a variety of other chemical species that might interfere with a nitrate analysis. While the nitrate ion does absorb UV radiation below 220 nm, so do many organic molecules and all of the halides, which are all common constituents of environmental samples. Thus, any method for detecting the nitrate ion must either be highly specific or manage to separate out the nitrate ion either spectroscopically or physically.

A spectroscopic separation usually requires a *derivatization*, which chemically changes the analyte into a different species that absorbs in a region of the spectrum with less spectral interferences. Examples of these methods are described in the Spectroscopy section. Physical separation involves the use of methods such as filtration or extraction. Most spectroscopic and chromatographic methods require some sort of prefiltration to remove suspended solids from the sample. Chromatography, described in more detail in Appendix F on page 302, physically separates analytes so that they are relatively pure when they pass through the detector, thus eliminating most interferences. An example of this type of method is described in more detail in the Ion Chromatography section.

☞ *Derivatization* is a process by which the target analyte is chemically altered to form a compound that is analytically different in some way. For spectroscopic measurements, derivatization results in a shift in the absorbance or emission of the analyte in order to improve detection and quantitation. In other methods, such as esterification, the analyte is converted to an ester, which usually has a lower boiling point compared to the analyte. This allows the analyte to be analyzed by a gas chromatograph, which usually requires compounds to have boiling points below 300 °C.

5.10.8.1 Spectroscopy Spectroscopic separation of the nitrate ion from other substances that would interfere with its detection is accomplished by shifting its absorbance from the UV to the visible region of the spectrum. Looking at a filtered water sample, it should be obvious that there is very little absorbance in the visible region of the spectrum (since it is colorless) and thus very little interference from absorbing species. Since the nitrate ion is very stable, it does not easily react with species that are *chromophoric*. Standard methods have been developed that reduce the nitrate ion to the nitrite ion, which then reacts with a chromophoric species.

☞ A *chromophore* is a chemical species or substituent of a molecule that absorbs visible radiation and gives the molecule its color, just as a *fluorophore* is responsible for giving a molecule the ability to fluoresce. Chromophores and fluorophores are routinely part of derivatization processes to improve the detection of analytes.

EPA Method 353 and *Standard Methods* 4500-NO3 achieve the reduction of the nitrate ion using cadmium.

$$\mathrm{Cd(s) + NO_3^-(aq) + 2H^+(aq) \longrightarrow Cd^{2+}(aq) + NO_2^-(aq) + H_2O(l)} \qquad \text{(R5.37)}$$

Reduction using cadmium produces toxic waste that must be disposed of properly. An alternative method uses the enzyme nitrate reductase to reduce the nitrate ion to the nitrite ion. While this method takes a little more care in preparation, it does not produce toxic waste.

Once reduced to the nitrite ion, some methods react sulfanilimide with the nitrite ion to produce a diazonium ion and then a subsequent reaction with NED (*N*-(1-naphthyl) ethylenediamine) to produce an azo dye. This highly conjugated molecule absorbs in the 540 nm region of the visible spectrum and makes the solution look pinkish. There are a variety of similar methods that use different derivatization agents, but they produce a derivative of the nitrite ion that absorbs in the visible region of the spectrum. This also means that the measurement can be made using relatively inexpensive spectrometers (operating in the visible instead of UV region) and sample cells (plastic or glass instead of quartz).

5.10.8.2 Ion-Selective Electrode An ISE for the nitrate ion is available, and it carries with it the same advantages and disadvantages as the DO ISE. It must be calibrated with a nitrate standard and suffers from interferences from the chloride and sulfate ions. A standard method for measuring the nitrate ion via an ISE can be found in *Standard Methods* 4500-NO3-D.

5.10.8.3 ***Ion Chromatography*** Ion chromatography is a method in which the ionic components of a sample are physically separated from each other, preventing each of the components from interfering with the detection of any other. A sample mixture is injected into a mobile phase containing water and the hydroxide ion. The sample is carried by the mobile phase into a column containing a resin that contains a charged surface. If the analyst wishes to detect anions, such as halides and nitrate, then a resin that contains quaternary amine groups ($-N(CH_3)_3^+$) is used. The anions interact to varying degrees with the positively charged resin, and in doing so, the anions that interact strongly slow down as they move through the resin. Anions that interact poorly with the resin move through the resin nearly as fast as the mobile phase. One by one, the anions leave the resin and are detected by either UV absorbance or by conductivity.

Ion chromatography is featured in EPA Methods 300.0 and 300.1 for the analysis of the nitrate ion along with standard methods sanctioned by other agencies such as the USGS (methods I-2057 and I-2058) and *Standard Methods* 4110B.

5.10.9 The Nitrite Ion

Detection of the nitrite ion is related to the detection of the nitrate ion. Since it was necessary to convert the nitrate ion into the nitrite ion in order to make chromophoric derivatization possible, detection of the nitrite ion proceeds, as described in Section 5.10.8 above, without the reduction step. See EPA Method 354.1, *Standard Methods* 4500-NO2-, and USGS-NWQL method I-2540-90 for more details. The nitrite ion can also be detected via ion chromatography as described in Section 5.10.8.3.

5.10.10 Ammoniacal Nitrogen

The detection of ammoniacal nitrogen is accomplished by the same general methods as were the nitrate and nitrite ions. Spectroscopic measurements require derivatization to an indophenol, which absorbs near 625 nm, such as in EPA Method 350.1.

A separate spectroscopic method, EPA Method 350.2 and Hach Method 8038, involves the use of Nessler tubes or Nessler's reagent (K_2HgI_4). The ammonium ion reacts with Nessler's reagent at low concentrations to form a yellow complex that absorbs at 425 nm.

$$NH_4^+(aq) + 2HgI_4^{2-}(aq) + 4OH^-(aq) \longrightarrow HgO{\cdot}Hg(NH_2)I(aq) + 7I^-(aq) + 3H_2O(l) \qquad (R5.38)$$

Similar to the cadmium reduction for the analysis of the nitrate ion, methods involving Nessler's reagent produce toxic mercury waste that must be disposed of properly. Ammonia determination can also be accomplished using EPA Method 350.2 (titrimetry) and *Standard Methods* 4500-NH3-E (ISE).

5.10.11 The Phosphate Ion

The phosphate ion can be found in several forms in the environment. The simplest form, orthophosphate carries from zero to three hydrogen ions depending on the pH of the sample. Polyphosphates are two or more orthophosphates bonded together in a chain or rings. The phosphate ion can also be bound to large organic molecules, both soluble and insoluble. Most standard methods derivatize the orthophosphate ion and provide for the hydrolysis or oxidation of the other forms of phosphate for inclusion in this measurement. Thus, the measurement of orthophosphate can be separated from the measurement of dissolved hydrolyzable phosphate (orthophosphate plus soluble polyphosphates), dissolved organic phosphate, and total phosphate.

The orthophosphate ion is derivatized with ammonium molybdate ($(NH_4)_2MoO_4$, which also forms a hydrate) to form a phosphomolybdate complex. When this complex is reduced, it forms a molybdenum blue-colored complex ($PMo_{12}O_{40}^{7-}$) that absorbs around 690 nm. The EPA Method 365.1 (and USGS I-2602) uses ascorbic acid ($C_6H_8O_6$) as the reducing agent and antimony potassium tartrate as a catalyst to speed up the derivatization. Another method (*Standard Methods* 4500-P) uses stannous chloride ($SnCl_2$) as the reducing agent.

A determination of just the orthophosphate ion is accomplished by producing the molybdenum blue complex described earlier. Only the soluble orthophosphate ions will produce the complex. Polyphosphates can be measured by hydrolysis with the addition of sulfuric acid and boiling for several minutes. Forming the molybdenum blue complex at this stage results in a signal that is the sum of the soluble ortho- and polyphosphates. Organic phosphorus is measured by persulfate digestion ($S_2O_8^{2-}$) in sulfuric acid, which frees the phosphate from the organic compounds and hydrolyzes all polyphosphates. The resulting measurement is the sum of the concentrations of the three phosphate forms. The phosphate ion can also be quantitated using EPA Method 300.1 (ion chromatography) and *Standard Methods* 4500-P (absorption spectrophotometry).

☞ *Gravimetric* methods of analysis use the mass of a sample to determine its concentration in solution. Typically, the analyte can be precipitated out of solution or the solvent can be evaporated to leave a residue containing the analyte. The mass of the dried solid can be stoichiometrically related to the concentration of the analyte in solution.

5.10.12 The Sulfate Ion

The sulfate ion can be measured using spectroscopic and *gravimetric* methods. Spectroscopic methods use the chloranilate ion (EPA Method 375.1) or the methylthymol blue ion (EPA Method 375.2) as the chromophore. Barium chloranilate is slightly soluble in an acidified 50:50 water/ethanol solution but more soluble than barium sulfate. When barium chloranilate is added to a solution containing the sulfate ion, the chloranilate ion is liberated in a single replacement reaction.

$$BaC_6O_4Cl_2(s) + SO_4^{2-}(aq) \longrightarrow BaSO_4(s) + C_6O_4Cl_2^{2-}(aq) \qquad (R5.39)$$

As long as the two barium salts remain insoluble, the concentration of the chloranilate ion in solution is proportional to the amount of sulfate ion that was in solution. This method is best utilized when the sulfate ion concentration is low (10 to 400 mg/L SO_4^{2-}). A related spectroscopic method of analysis uses solution *turbidity* for samples with sulfate ion concentrations between 1 and 40 mg/L (EPA Method 375.4 and *Standard Methods* 2130).

☞ *Turbidity* is a measure of the cloudiness of a solution. Suspended particles, such as silt or $BaSO_4$ (in the case of sulfate analysis), scatter light. Standard absorbance spectroscopes can measure turbidity by a decrease in the light that reaches a detector. A more sensitive instrument called a *nephelometer* measures the intensity of the scattered light at 90° from the light source. With either method, the absorbance or intensity of light scattered is proportional to the sulfate ion concentration. See Section F.1.1 on page 296.

For sulfate ion concentrations above 400 mg/L, such as in marine samples (around 2700 mg/L), the gravimetric method (EPA Method 375.3) is the most accurate. Barium chloride ($BaCl_2$) is added to an acidified sample to precipitate barium sulfate ($BaSO_4$). The precipitate is filtered and washed with deionized water to remove any soluble chloride salts. The rinse can be tested with silver nitrate ($AgNO_3$) – if a silver chloride precipitate forms, then more rinses are required. The mass of the dried barium sulfate is stoichiometrically related to the amount of sulfate in the sample.

5.11 IMPORTANT TERMS

- ★ open system
- ★ closed system
- ★ water stratification
  - · epilimnion
  - · hypolimnion
  - · metalimnion/thermocline
  - · seasonal overturn
- ★ thermohaline circulation
- ★ detritus
- ★ solvent, solute
- ★ homogeneous solution
- ★ heterogeneous solution
- ★ salinity
- ★ permil (‰)
- ★ estuary
- ★ dissolved oxygen
- ★ Henry's law
- ★ thermal pollution
- ★ ionic strength
- ★ activity coefficients
- ★ carbon cycle
- ★ nitrogen cycle
  - · fixation
  - · diazotrophic bacteria

- assimilation
- ammonification
- ammoniacal nitrogen
- nitrification
- denitrification
- Haber synthesis

★ phosphorus cycle
★ eutrophication
- natural
- cultural
- oligotrophic
- eutrophic
- dead zones

★ sulfur cycle
★ acid mine drainage
★ smelting
★ water quality
- alkalinity
- BOD/COD
- DO
- nitrogen
- phosphorus
- fecal coliform
- hardness
- metals
- turbidity
- temperature
- pH
- total dissolved solids

★ wastewater treatment
- primary
- secondary
- anaerobic digestion
- tertiary
- filtration
- disinfection

★ biosolids
★ standard method
★ conductivity
★ TIC & TOC & TC
★ sparging
★ DOC & POC
★ Winkler Method
- azide modification

★ standardization
★ ion-selective electrode
★ *in situ*, *in vitro*
★ derivatization
★ chromophore
★ fluorophore
★ spectroscopic
★ gravimetric
★ turbidity
★ nephelometry

EXERCISES

1. What is the coldest temperature you might find at the bottom of a lake (ignoring deviations from extreme pressures)? What happens when the temperature of the water drops below this?
2. If the Earth is bombarded by meteors on a daily basis, is it acting as an open or a closed system?
3. Describe the difference between the percent unit and the permil unit.
4. How is the density of water affected by salinity and temperature?
5. Describe the phenomenon of thermal pollution. Where might you find it?
6. Estimate the DO levels at two points in a freshwater heat reservoir (for an industrial facility) compared to DO levels in the headwaters, which maintain a summer temperature of 57 °F (14 °C). At the discharge point (point 1), the water temperature is 93 °F (34 °C), and 0.5 km away from the discharge point where the water temperature drops to 75 °F (24 °C)
7. Explain what happens to the DO levels in an estuary when high tides cause an increase in the salinity of the water (assuming that there is no temperature change).
8. What is the salinity of a 24.7842 g seawater sample if a preweighed 250-mL beaker (58.7254 g) containing the sample was boiled to dryness, baked in a drying oven overnight at 180 °C, cooled in a desiccator, and reweighed giving a mass of 59.5173 g?

Aliquot volume (mL)	D_1 $\left(\frac{\text{mg } O_2}{\text{L}}\right)$	D_2 $\left(\frac{\text{mg } O_2}{\text{L}}\right)$
5.00	7.9	5.4
10.00	7.8	2.9
15.00	8.0	0.6
20.00	8.0	0.0
Blank	7.8	7.7

Table 5.8 The results of a simulated BOD experiment.

Dissolved carbon (mg/L C)	Signal (AU × cm^{-1})
6.01	268.6
9.02	517.5
12.03	543.0
15.04	712.8
18.05	905.8
21.07	1003.5
24.07	1059.9
27.08	1271.7
30.09	1387.4

Table 5.9 AU × cm^{-1} (AU = Absorbance Units) is the integrated peak area, from 3800 to 3400 cm^{-1}, taken from the absorbance spectrum of CO_2.

9. Determine the oxidation state of N in hydrazine (N_2H_4), which is used as rocket fuel.
10. Given the BOD experimental data in Table 5.8, determine the BOD_5 for the sample if 300 mL BOD incubation bottles were used.
11. List the most common inorganic forms of carbon in the atmosphere, lithosphere, and hydrosphere (these are commonly called reservoirs).
12. Why is carbon essential to life?
13. List the five major reaction processes involved in the nitrogen cycle, describing the reactant and product form of nitrogen in each.
14. Why is nitrogen essential to life?
15. How did the work of Fritz Haber revolutionize agriculture and affect human population?
16. Describe the two major biotic and abiotic reservoirs of phosphorus.
17. Why is phosphorus essential to life?
18. Why was phosphate used in detergents?
19. Why is sulfur essential to life?
20. Sulfur exists in what inorganic form in its hydrospheric reservoir?
21. What essential water quality information do the BOD and COD methods provide? Describe the difference between the two methods.
22. A water quality analyst made calibration standards for a Total Inorganic Carbon (TIC) and Total Organic Carbon (TOC) analysis using a certified Na_2CO_3 standard (see Table 5.9). The TIC was measured first and gave a signal of 1229.5 AU×cm^{-1}. After the inorganic carbon was removed and the persulfate digestion was complete, the sample produced a TOC signal of 317.2 AU×cm^{-1}. Determine the TIC and TOC of the sample in units of mg/L carbon using linear regression.
23. How does the primary wastewater treatment process contribute to water quality?
24. How does the secondary wastewater treatment process contribute to water quality?
25. How does the tertiary wastewater treatment process contribute to water quality? Describe some of the implementations of the tertiary process through chemical and biological means.
26. Estimate the pH of a marine sample (salinity of 30.4‰) using Eq. (5.10) if, after the cresol red indicator is added, the measured absorbance at 573 nm is 0.126 and the absorbance at 433 nm is 0.027.
27. What is the difference between a turbidity spectrometer and a nephelometer?
28. The nitrate and nitrite ions both absorb UV radiation below 300 nm. Why must these two ions be derivatized in order to be quantitatively detected?
29. If 25.4 mg of $BaSO_4$ precipitated from a 200 mL sample of water from an estuary, what was the concentration of sulfate ion in the sample (in mg/L SO_4^{2-})?

30. In the Calculation section of EPA Method 375.3 for gravimetric sulfate analysis, the authors provide the following conversion for each 1 mL sample analyzed.

$$\text{mg/L } SO_4^{2-} = \frac{\text{mg } BaSO_4}{\text{1 mL sample}} \times 411.5 \tag{5.14}$$

Show how the 411.5 number was derived using a stoichiometric conversion between $BaSO_4$ and SO_4^{2-}.

31. In the Winkler titration, initially all of the DO reacts with Mn^{2+} to form the Mn precipitate, which happens under basic conditions. After adding the reagents in the next step, the solution is "fixed" and will not react with oxygen if exposed to air. What reaction condition keeps this solution fixed and prevents it from reacting with oxygen?

32. What is the titrant in the Winkler method?

33. What chemical species, indirectly produced from DO, is being titrated (is the limiting reactant and proportional to the DO)?

34. Why does the modified Winkler method contain sodium azide?

35. Why might an environmental chemistry lab want to invest in the purchase of an ion chromatograph to quantitate water quality parameters such as NO_3^-, NO_2^-, SO_4^{2-}, and PO_4^{3-} instead of using each of the specific standard methods used for the analytes?

BIBLIOGRAPHY

Ashley, K.; Cordell, D.; Mavinic, D. *Chemosphere* **2011**, *84*, 737–746.

Austin, S. A.; Humphreys, D. R. The Sea's Missing Salt: A Dilemma for Evolutionists, Proceedings of the Second International Conference or Creationism pp. 17–33, 1990.

Benson, B. B.; Krause, D. *Limnology and Oceanography* **1980**, *25(4)*, 662–671.

Benson, B. B.; Krause, D. *Limnology and Oceanography* **1984**, *29(3)*, 620–632.

Ben-Yaakov, S. *Limnology and Oceanography* **1970**, *15*, 326–328.

Bergquist, L. Wisconsin Study Says Untreated Drinking Water Has More Risk of Illness. Electronic, 2014; http://www.jsonline.com/news/health/wisconsin-study-says-untreated-drinking-water-has-more-risk-of-illness-b99214493z1-249347101.html.

Broecker, W. S. *Science* **1997**, *278*, 1582–1588.

Byrne, R. H.; Breland, J. A. *Deep Sea Research Part A. Oceanographic Research Papers* **1989**, *36(2)*, 803–810.

Capone, D. G.; Bronk, D. A.; Mulholland, M. R.; Carpenter, E. J., Eds. *Nitrogen in the Marine Environment*; Elsevier: San Diego, CA, 2008; Chapter 1 The Marine Nitrogen Cycle: Overview and Challenges, pp. 1–43.

Clark, P. U.; Pisias, N. G.; Stocker, T. F.; Weaver, A. J. *Nature* **2002**, *415*, 863–869.

Clesceri, L. S.; Greenberg, A. E.; Eaton, A. D., Eds. *Standard Methods for the Examination of Water and Wastewater*; American Public Health Association: Washington, DC, 1998.

Cross, D. W.; Carton, R. J. *International Journal of Occupational and Environmental Health* **2003**, *9(1)*, 24–29.

DelValls, T.; Dickson, A. *Deep Sea Research Part I: Oceanographic Research Papers* **1998**, *45*, 1541–1554.

Easley, R. A.; Byrne, R. H. *Environmental Science and Technology* **2012**, *46*, 5018–5024.

Emsley, J. *The Elements*; Claredon Press, Oxford, 1998.

Falkowski, P.; Scholes, R. J.; Boyle, E.; Canadell, J.; Canfield, D.; Elser, J.; Gruber, N.; Hibbard, K.; Högberg, P.; Linder, S.; Mackenzie, F. T.; Moore, B. III; Pedersen, T.; Rosenthal, Y.; Seitzinger, S.; Smetacek, V.; Steffen, W. *Science* **2000**, *290* (5490), 291–296.

Federal Energy Management Program, E. Wastewater, Digester Gas Can Produce High Quality Methane Fuel for Federal Facilities. Electronic, 2005; http://www1.eere.energy.gov/femp/news/news_detail.html?news_id=8961.

Filippelli, G. M. *Chemosphere* **2011**, *84*, 759–766.

Fischer, H. B.; List, E. J.; Koh, R. C. Y.; Imberger, J.; Brooks, N. H. *Mixing in Inland and Coastal Waters*; Academic Press: San Diego, CA, 1979; pp. 229–245.

Glindemann, D.; Stottmeister, U.; Bergmann, A. *Environmental Science and Pollution Research* **1996**, *3*, 17–19.

Goodman, S. Fewer Regulations for Bottled Water Than Tap, GAO Says. Electronic, 2009; http://www.nytimes.com/gwire/2009/07/09/09greenwire-fewer-regulations-for-bottled-water-than-tap-g-33331.html.

Hammond, A. *Science* **1971**, *172*, 361–363.

Haynes, W. *CRC Handbook of Chemistry and Physics*, 93rd ed.; *CRC Handbook of Chemistry and Physics*; Taylor & Francis Group: Boca Raton, FL, 2012.

Hem, J. D. *Study and Interpretation of the Chemical Characteristics of Natural Water*; 1985.

Jankowski, J. J.; Kieber, D. J.; Mopper, K. *Photochemistry and Photobiology* **1999**, *70(3)*, 319–326.

Kartal, B.; Kuypers, M. M. M.; Lavik, G.; Schalk, J.; den Camp, H. J. M. O.; Jetten, M. S. M.; Strous, M. *Environmental Microbiology* **2007**, *9(3)*, 635–642.

Kim, J.; Rees, D. C. *Biochemistry* **1994**, *33(2)*, 389–397.

Leman, L.; Orgel, L.; Ghadiri, M. R. *Science* **2004**, *306(5694)*, 283–286.

Lewis, E. R.; Schwartz, S. E. *Sea Salt Aerosol Production: Mechanisms, Methods, Measurements, and Models*; Geophysical Monograph Series; American Geophysical Union: Washington, DC, 2004; Vol. 152.

Livingstone, D. A. In *Data of Geochemistry*, 6th ed.; Fleischer, M., Ed.; USGS: Washington, DC, 1963.

Macknick, J.; Newmark, R.; Heath, G.; Hallett, K. C. *Environmental Research Letters* **2012**, *7*, 1–10.

Maly, J.; Fadrus, H. *Journal Water Pollution Control Federation* **1971**, *43(4)*, 641–650.

Migon, C.; Sandroni, V.; Bethoux, J. P. *Marine Environmental Research* **2001**, *52*, 413–426.

Mullen, J. *British Dental Journal* **2005**, *199*, 1–4.

Nelson, D. O. Fresh Water, Natural Composition of. Electronic, 2013; http://www.waterencyclopedia.com/En-Ge/Fresh-Water-Natural-Composition-of.html#ixzz2jbArYi5X.

Nichols, M. U.N.'s Ban Launches Bid to Stamp Out Cholera in Haiti. Electronic, 2012; http://articles.chicagotribune.com/2012-12-11/lifestyle/sns-rt-us-haiti-cholera-unbre8ba1fk-20121211_1_oral-cholera-vaccine-cholera-outbreak-dehydration-and-death.

Ogilvy, G. *The Alchemist's Kitchen: Extraordinary Potions and Curious Notions*; Wooden Books, UK, 2006; p. 58.

Olson, E. D. Bottled Water: Pure Drink or Pure Hype? Electronic, 2013; http://www.nrdc.org/water/drinking/bw/exesum.asp.

Piedrahita, R. H. *Aquacultural Engineering* **1995**, *14(4)*, 331–346.

Quere, C. L.; Raupach, M. R.; Canadell, J. G.; Marland, G. *Nature Geoscience* **2009**, *2*, 831–836.

Reeve, R. N. *Introduction to Environmental Analysis*; John Wiley & Sons, Ltd.: Chichester, 2002.

Schindler, D. W. *Science* **1977**, *195*, 260–262.

Schmidt, U. *Tellus* **1979**, *31*, 68–74.

Somasundaran, P.; Amankonah, J. O.; Ananthapadmabhan, K. P. *Colloids and Surfaces* **1985**, *15*, 309–333.

Staff, N. The Story of Fluoridation. Electronic, 2014; http://www.nidcr.nih.gov/oralhealth/topics/fluoride/thestoryoffluoridation.htm.

Stumm, W.; Morgan, J. J. *Aquatic Chemistry*, 3rd ed.; John Wiley & Sons, Inc.: New York, 1996.

Survey, U. G. Utah's Great Salt Lake and Ancient Bonneville. Electronic, 2013; http://geology.utah.gov/online_html/pi/pi-39/pi39pg9.htm.

Toggweiler, J. R.; Key, R. M. In *Enclyclopedia of Ocean Sciences*; Thorpe, S. A., Turekian, K. K., Eds.; J. H. Steele, 2001; Chapter Ocean Circulation: Thermohaline Circulation, pp. 2941–2947.

Vanotti, M.; Szogi, A. *Technology for Recovery of Phosphorus from Animal Wastewater Through Calcium Phosphate Precipitation*; USDA-ARS: Florence, SC, 2009.

Weiss, R. F. *Deep Sea Research and Oceanographic Abstracts* **1970**, *17(4)*, 721–735.

Wiesmann, U.; Choi, I. S.; Dombrowski, E.-M. *Fundamentals of Biological Wastewater Treatment*; Wiley-VCH Verlag GmbH & Co. KGaA: Weinheim, 2006; Chapter 1, pp. 1–23.

CHAPTER 1 REVIEW EXAMPLES AND END-OF-CHAPTER EXERCISES

A.1 SOLUTIONS TO IN-CHAPTER REVIEW EXAMPLES

Solution to Example 1.1 on page 3

1. Use Eq. (1.1) and substitute in the temperature.

$$\lambda_{max} = \frac{2.8977685 \times 10^{-3}\,\frac{\text{m}}{\text{K}}}{3000\ \text{K}}$$
$$= 9.659 \times 10^{-7}\ \text{m} = 966\ \text{nm}$$

   This is a wavelength in the infrared just below the red end of visible radiation (750 nm).

2. Use the same procedure for finding the infrared emission from the Earth. Make sure that the temperature is converted to units of Kelvin.

$$\lambda_{max} = \frac{2.8977685 \times 10^{-3}\,\frac{\text{m}}{\text{K}}}{273.15 + (60\ °\text{F} - 32) \times \frac{5}{9}}$$
$$= \frac{2.8977685 \times 10^{-3}\,\frac{\text{m}}{\text{K}}}{288.7\ \text{K}}$$
$$= 1.0 \times 10^{-5}\ \text{m} = 10\ \mu\text{m}$$

3. Any wavelength in the 10^{-5} m range is in the infrared region of the spectrum.

Solution to Review Example 1.1 on page 22

1. From the shortest to longest wavelength, the order is as follows: X-rays, ultraviolet, visible, infrared, radio waves.

2. Using Eq. (1.3), the frequency can be calculated. Make sure the metric prefix is converted into scientific notation.

$$\nu = \frac{c}{\lambda}$$
$$= \frac{2.99792458 \times 10^{8}\ \text{m/s}}{604 \times 10^{-9}\ \text{m}}$$
$$= 4.96 \times 10^{14}\ \text{s}^{-1} = 4.96 \times 10^{14}\ \text{Hz}$$

Environmental Chemistry: An Analytical Approach, First Edition. Kenneth S. Overway.

Companion website: www.wiley.com/go/overway/environmental_chemistry

3. Using Eq. (1.4), the energy of a yellow photon can be calculated. Either the wavelength or the frequency can be used. Make sure that the metric prefix is converted into scientific notation.

$$\begin{aligned} E &= \frac{hc}{\lambda} \\ &= \frac{(6.62606957 \times 10^{-34}\ \text{J} \cdot \text{s})(2.99792458 \times 10^{8}\ \text{m/s})}{604 \times 10^{-9}\ \text{m}} \\ &= 3.29 \times 10^{-19}\ \text{J} \end{aligned}$$

4. Using Eq. (1.3) the wavelength can be calculated. Make sure that the metric prefix is converted into scientific notation.

$$\begin{aligned} \lambda &= \frac{c}{\nu} \\ &= \frac{2.99792458 \times 10^{8}\ \text{m/s}}{90.7 \times 10^{6}\ \text{Hz}} \\ &= 3.31\ \text{m} \end{aligned}$$

Solution to Review Example 1.2 on page 23

1. Using Eqs (1.5)–(1.7), absorbance can be calculated.

$$\begin{aligned} \%T &= 15\% \rightarrow T = 0.15 \\ Abs &= -\log T = -\log(0.15) = 0.82 \end{aligned} \tag{A.1}$$

2. Equations (1.7) and (1.6) can be used to convert absorbance to %T.

$$\begin{aligned} Abs &= -\log T \\ T &= 10^{-Abs} \\ T &= 10^{-0.157} = 0.697 \\ \%T &= 69.7\% \end{aligned}$$

3. The solution to this exercise requires solving Eq. (1.8) for the concentration.

$$\begin{aligned} Abs &= \varepsilon b C \\ C &= \frac{Abs}{\varepsilon b} \\ &= \frac{0.088}{\left(54.2\ \text{M}^{-1}\text{cm}^{-1}\right)(1\ \text{cm})} = 1.6 \times 10^{-3}\ \text{M} = 1.6\ \text{mM} \end{aligned}$$

Solution to Review Example 1.3 on page 24

1.

$$1.05\ \text{mm} \times \frac{1\ \text{m}}{1000\ \text{mm}} \times \frac{10^{6}\ \mu\text{m}}{1\ \text{m}} = 1050\ \mu\text{m}$$

2.

$$46.2\ \frac{\text{ng}}{\mu\text{L}} \times \frac{1\ \text{g}}{10^{9}\ \text{ng}}\ \frac{10^{6}\ \mu\text{L}}{1\ \text{L}} = 0.0462\ \text{g/L}$$

3.

$$0.154 \times 10^{-6}\ \text{L} \times \frac{10^{9}\ \text{nL}}{1\ \text{L}} = 154\ \text{nL}$$

Solution to Review Example 1.4 on page 24

Complete Table A.1 using your knowledge of the structure of an atom.

Symbol	# of protons	# of neutrons	# of electrons
$\mathbf{^{80}_{35}}Br^-$	35	45	36
$^{23}_{11}Na^{+1}$	**11**	**12**	**10**

The bold items are the answers to the exercises.

Table A.1 Atomic structure example with answers.

In the first row, the correct symbol is $^{80}_{35}Br^-$ since the number of protons (atomic number) is 35. Summing the atomic number and the number of neutrons gives a mass number (A) of 80. Given that the element has 35 protons and 36 electrons, it must have a net −1 charge. For the second row, the lower number on the left gives an atomic number of 11. Subtracting 11 from the mass number of 23 gives 12 neutrons. Since sodium has a net +1 charge, there must be one less electron than the number of protons, so the number of electrons is 10.

Solution to Review Example 1.5 on page 25

1. Balance the following nuclear reactions and identify the unknown reactant or product.
   (a) $^{12}_{6}C + ^{12}_{6}C \rightarrow \mathbf{^{16}_{8}O} + 2^{4}_{2}He$
   (b) $^{12}_{6}C + \mathbf{^{44}_{20}Ca} \rightarrow ^{56}_{26}Fe$
   (c) $^{201}_{80}Hg + ^{0}_{-1}\beta \rightarrow \mathbf{^{201}_{79}Au}$
   (d) $^{19}_{9}F \rightarrow \mathbf{^{0}_{+1}}\beta^+ + ^{19}_{8}O$
2. The correct labels are as follows:
   (a) alpha decay
   (b) fusion
   (c) beta capture
   (d) positron emission
3. Start with Ni-56 and end with Fe-56.
   (a) $^{56}_{28}Ni + ^{0}_{-1}\beta \rightarrow ^{56}_{27}Co$
   (b) $^{56}_{27}Co + ^{0}_{-1}\beta \rightarrow ^{56}_{26}Fe$

Solution to Review Example 1.6 on page 25

1. Ni-56 has a first-order rate constant (k) of 0.115 day^{-1}. Determine its half-life.

$$t_{1/2} = \frac{\ln\left(\frac{1}{2}\right)}{-k} = \frac{\ln\left(\frac{1}{2}\right)}{-0.115 \text{ day}^{-1}} = 6.03 \text{ days}$$

2. Determine the second-order half-life of NO in the following reaction if it has a rate constant of 245 $atm^{-1}s^{-1}$ when the NO concentration (partial pressure in this case) is 0.00172 atm.

$$t_{1/2} = \frac{1}{k[A]_0}$$
$$= \frac{1}{(245\ atm^{-1}s^{-1})(0.00172\ atm)}$$
$$= 2.37\ s$$

Solution to Review Example 1.7 on page 26

Label each of the following compounds as soluble (S) or insoluble (I) in water.

I	$CaCO_3$	S	Na_2CO_3	S	$(NH_4)_3PO_4$
S	$MgSO_4$	I	$HgCl_2$	S	$HgNO_3$
S	Na_2CO_3	S	NaCl	S	$Pb(CH_3COO)_2$
S	K_3PO_4	S	NaOH	I	CuCl
I	$CaSO_4$	I	$Mg(OH)_2$	S	$CuCl_2$

Solution to Review Example 1.8 on page 27

1. Complete Table A.2.

Formula	Name
CoO	**Cobalt(II) oxide**
Co_2O_3	**Cobalt(III) oxide**
$Mg(NO_3)_2$	Magnesium nitrate
$Na_2SO_4 \cdot 6H_2O$	Sodium sulfate hexahydrate

The bold items are the answers to the exercises.

Table A.2 Name the following ionic compounds.

Solution to Review Example 1.9 on page 27

1. Complete Table A.3.

Formula	Name
SO_2	**Sulfur dioxide**
SO_3	Sulfur trioxide
P_2O_4	**Diphosphorus tetroxide**
CCl_4	Carbon tetrachloride

The bold items are the answers to the exercises.

Table A.3 Name the following covalent compounds.

Solution to Review Example 1.10 on page 28

1. Calculate the maximum photon wavelength that would break the N≡N triple bond. The bond energy, from Table I.7, is 946 kJ/mol.

$$\text{total energy per photon} = (946\ \text{kJ/mol})\left(\frac{1\ \text{mol}}{6.02214129 \times 10^{23}\ \text{photons}}\right)\left(\frac{1000\ \text{J}}{1\ \text{kJ}}\right)$$
$$= 1.5709 \times 10^{-18}\ \text{J/photon} \qquad \text{(A.2)}$$

$$\lambda = \frac{(6.62606957 \times 10^{-34}\ \text{J} \cdot \text{s})(2.99792458 \times 10^{8}\ \text{m/s})}{1.5709 \times 10^{-18}\ \text{J/photon}}$$
$$= 1.2646 \times 10^{-7}\ \text{m} = 126\ \text{nm} \qquad \text{(A.3)}$$

Solution to Review Example 1.11 on page 29

1. Since gases are more soluble in colder liquids, placing an open soft drink container in a refrigerator would keep it carbonated (at least compared to the warm can) longer.
2. When pressure decreases, the solubility of gases in a liquid also decreases. At high altitudes, the partial pressure of oxygen is lower than at sea level, so the solubility of oxygen in the lungs and blood is lower. This is why many hikers must take oxygen tanks on high-altitude climbs.
3. Dissolved oxygen is higher in cold water because gases are more soluble in colder liquids. This is why industrial facilities that use surface water to cool equipment must be careful not to warm the water too much and cause thermal pollution – dangerously reduced dissolved oxygen levels in aquatic environments.

Solution to Review Example 1.12 on page 29

1. Balance the following reactions.

   (a) $2C_4H_{10} + 13O_2 \rightarrow 10H_2O + 8CO_2$
   (b) $Na_2CO_3 + 2HCl \rightarrow CO_2 + 2NaCl + H_2O$
   (c) $2Al + 6HCl \rightarrow 2AlCl_3 + 3H_2$

Solution to Review Example 1.13 on page 30

Exercises

1. Determine the oxidation state of S in each of the following species.

   (a) S = <u>+2</u> in $S_2O_3^{2-}$
   (b) S = <u>+6</u> in SO_3
   (c) S = <u>−2</u> in H_2S
   (d) S = <u>0</u> in S_8

2. Which of the following reactions is a redox reaction?
   (b) $C_3H_8 + 5O_2 \rightarrow 4H_2O + 3CO_2$
   carbon is oxidized and oxygen is reduced
   (d) $2Al + 6HCl \rightarrow 2AlCl_3 + 3H_2$
   Al is oxidized and H is reduced

Solution to Review Example 1.14 on page 32

1. Calculate the free energy change for the following reactions (assume 25 °C). For each reaction, also state whether it is spontaneous or nonspontaneous.

(a) $N_2(g) + O_2(g) \rightarrow 2NO(g)$

$$\begin{aligned}\Delta G^\circ_{rxn} &= ((2\ \text{mol})(87.6\ \text{kJ/mol})) \\ &\quad - ((1\ \text{mol})(0\ \text{kJ/mol}) + (1\ \text{mol})(0\ \text{kJ/mol})) \\ &= 175.2\ \text{kJ}\end{aligned}$$

(b) $2C_4H_{10}(g) + 13O_2(g) \rightarrow 8CO_2(g) + 10H_2O(g)$

$$\begin{aligned}\Delta G^\circ_{rxn} &= ((8\ \text{mol})(-394.4\ \text{kJ/mol}) + (10\ \text{mol})(-228.6\ \text{kJ/mol})) \\ &\quad - ((2\ \text{mol})(-218.0\ \text{kJ/mol}) + (13\ \text{mol})(0\ \text{kJ/mol})) \\ &= -5005.2\ \text{kJ}\end{aligned}$$

(c) $2HCl(aq) + Fe(s) \rightarrow H_2(g) + Fe^{2+}(aq) + 2Cl^-(aq)$

$$\begin{aligned}\Delta G^\circ_{rxn} &= ((1\ \text{mol})(0\ \text{kJ/mol}) + (1\ \text{mol})(-78.9\ \text{kJ/mol}) + (2\ \text{mol})(-131.2\ \text{kJ/mol})) \\ &\quad - ((2\ \text{mol})(-131.2\ \text{kJ/mol}) + (1\ \text{mol})(0\ \text{kJ/mol})) \\ &= -78.9\ \text{kJ}\end{aligned}$$

2. Calculate the free energy change for the following reaction at 415 °C.

$$N_2(g) + 3H_2(g) \rightarrow 2NH_3(g) \qquad \text{(RA.1)}$$

From the calculations in the review example on page 32, where ΔH°_{rxn}=−91.8 kJ, ΔS°_{rxn}=−198.1 J/K, these values can be plugged into the following equation.

$$\begin{aligned}\Delta G_{sys} &= \Delta H_{sys} - T\Delta S_{sys} \\ &= -91.8\ \text{kJ} - (273.15 + 415\ \text{K})\left(-198.1\ \text{J/K} \times \frac{1\ \text{kJ}}{1000\ \text{J}}\right) \\ &= 44.5\ \text{kJ}\end{aligned}$$

3. Determine the temperature at which the following reaction becomes spontaneous.

$$N_2(g) + 3H_2(g) \rightarrow 2NH_3(g) \qquad \text{(RA.2)}$$

Using the temperature-dependent free energy equation, solve for the temperature using the ΔH°_{rxn} and ΔS°_{rxn} when the reaction reaches equilibrium (ΔG°_{rxn}=0) when it is just turning spontaneous.

$$\begin{aligned}\Delta G_{sys} &= \Delta H_{sys} - T\Delta S_{sys} \\ 0 &= \Delta H_{sys} - T\Delta S_{sys} \\ \Delta H_{sys} &= T\Delta S_{sys} \\ \frac{\Delta H_{sys}}{\Delta S_{sys}} &= T = \frac{-91.8\ \text{kJ}}{-0.1981\ \text{kJ/K}} \\ T &= 463\ \text{K}\end{aligned}$$

The reaction becomes spontaneous at 463 K, and since the enthalpy and entropy are negative, the reaction is only spontaneous *below* this temperature.

4. Which is the most stable form of sulfur in the hydrosphere, SO_3^{2-} or SO_4^{2-}?
Given that the ΔG°_f for SO_3^{2-} is −486.5 kJ/mol and the ΔG°_f for SO_4^{2-} is −744.53 kJ/mol, the sulfate ion has a lower free energy of formation and is the most stable ion in the hydrosphere.

Solution to Review Example 1.15 on page 33

1. Determine the molar mass of the following compounds.

   (a) Co_2O_3

$$MM = 58.9332\ g/mol \times 2 + 15.9994\ g/mol \times 3 = 165.8646\ g/mol$$

   (b) $CaSO_4 \cdot 4\,H_2O$

$$MM = 40.078 + 32.066 + 15.9994 \times 4 + 4 \times (1.0079 \times 2 + 15.9994)$$
$$= 208.2024\ g/mol$$

   (c) sodium dichromate ($Na_2Cr_2O_7$)

$$MM = 22.9898 \times 2 + 51.9961 \times 2 + 15.9994 \times 7 = 261.9676\ g/mol$$

   (d) C_8H_{18}

$$MM = 12.0107 \times 8 + 1.00794 \times 18 = 114.2285\ g/mol$$

2. How many moles are in 2.19 g of C_8H_{18}?

$$2.19\ g \times \frac{1\ mol}{114.2285\ g} = 0.0192\ mol$$

Solution to Review Example 1.16 on page 34

1. Balance the combustion reaction for octane (C_8H_{18}).

$$2C_8H_{18} + 25O_2 \rightarrow 16CO_2 + 18H_2O \qquad (RA.3)$$

   (a) If the reaction is started with 10 mol of octane and 10 mol of oxygen, which chemical would be the limiting reactant (LR)?

$$10\ mol\ C_8H_{18} \times \frac{16\ mol\ CO_2}{2\ mol\ C_8H_{18}} = 80\ mol\ CO_2$$
$$10\ mol\ O_2 \times \frac{16\ mol\ CO_2}{25\ mol\ O_2} = 6.4\ mol\ CO_2$$

   Since 10 mol of oxygen would produce less carbon dioxide than 10 mol of octane, oxygen is the LR.

   (b) If the reaction is started with 10.0 g of octane and 10.0 g of oxygen, which chemical would be the LR?

$$10.0\ g\ C_8H_{18} \times \frac{1\ mol\ C_8H_{18}}{114.229\ g\ C_8H_{18}} \times \frac{16\ mol\ CO_2}{2\ mol\ C_8H_{18}} = 0.70035\ mol\ CO_2$$
$$10.0\ g\ O_2 \times \frac{1\ mol\ O_2}{31.9988\ g\ O_2} \times \frac{16\ mol\ CO_2}{25\ mol\ O_2} = 0.2000\ mol\ CO_2$$

   Since 10.0 g of oxygen would produce less carbon dioxide than 10.0 g of octane, oxygen is the LR.

   (c) What mass of carbon dioxide would be made from 10.0 g of octane and 10.0 g of oxygen with 100% efficiency?

$$0.2000\ mol\ CO_2 \times \frac{44.0095\ g\ CO_2}{1\ mol\ CO_2} = 8.80\ g\ CO_2$$

Solution to Review Example 1.17 on page 35

Exercises

1. Write the Rate Law for the following steps.

   (a) (an overall reaction) $H_2 + Cl_2 \rightarrow 2HCl$

   $$Rate = k[H_2]^x[Cl_2]^y$$

   (b) (a slow step in a mechanism) $SO_2 + O_2 \rightarrow SO_3 + O$

   $$Rate = k[SO_2]^1[O_2]^1$$

   (c) (a slow step in a mechanism) $2NO \rightarrow N + NO_2$

   $$Rate = k[NO]^2$$

2. Propose two mechanisms for the production of NO from molecular oxygen and nitrogen. Assume that the first step is the slow step.

   $$N_2(g) + O_2(g) \rightleftharpoons 2NO(g) \qquad \text{(RA.4)}$$

   Write out the steps of each mechanism, and derive the correct Rate Law for each mechanism. These are just two possible examples.

   Mechanism I

   $$\text{step 1}: \; N_2 \rightarrow 2N \text{ (slow)} \qquad \text{(RA.5)}$$

   $$\text{step 2}: \; O_2 + N \rightarrow NO_2 \text{ (fast)} \qquad \text{(RA.6)}$$

   $$\text{step 3}: \; NO_2 + N \rightarrow 2NO \text{ (fast)} \qquad \text{(RA.7)}$$

   $$Rate = k[N_2]$$

   Mechanism II

   $$\text{step 1}: \; N_2 + O_2 \rightarrow N_2O + O \text{ (slow)} \qquad \text{(RA.8)}$$

   $$\text{step 2}: \; N_2O \rightarrow NO + N \text{ (fast)} \qquad \text{(RA.9)}$$

   $$\text{step 3}: \; N + O \rightarrow NO \text{ (fast)} \qquad \text{(RA.10)}$$

   $$Rate = k[N_2][O_2]$$

Solution to Review Example 1.18 on page 36

Determine the Lewis Dot Structure for the following species.

1. F_2

   - the number of valence electrons the structure has (h): $h = 7 + 7 = 14$
   - the number of valance electrons the structure needs (n): $n = 8 + 8 = 16$
   - the number of electrons that must be shared (s): $s = 16 - 14 = 2$
   - the minimum number of bonds in the structure (b): $b = 2 \div 2 = 1$

   Thus, the structure is

2. CO_2
    - the number of valence electrons the structure has (h): $h = 4 + 2 \times 6 = 16$
    - the number of valance electrons the structure needs (n): $n = 3 \times 8 = 24$
    - the number of electrons that must be shared (s): $s = 24 - 16 = 8$
    - the minimum number of bonds in the structure (b): $b = 8 \div 2 = 4$

    Thus, the structure is

    O=C=O

3. H_2O
    - the number of valence electrons the structure has (h): $h = 2 \times 1 + 6 = 8$
    - the number of valance electrons the structure needs (n): $n = 2 \times 2 + 8 = 12$
    - the number of electrons that must be shared (s): $s = 12 - 8 = 4$
    - the minimum number of bonds in the structure (b): $b = 4 \div 2 = 2$

    Thus, the structure is

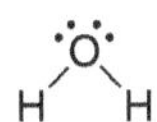

Solution to Review Example 1.19 on page 37

1. Circle the element in each pair that is more electronegative.
    (a) O
    (b) C
    (c) Br
2. Fluorine is the most electronegative element.
3. Circle the molecular compounds that have polar covalent bonds.
    a) HF b) ~~Br_2~~ c) ~~CS~~ - very small d) CO
    e) ~~N_2~~ f) ~~O_2~~ g) NO h) ~~F_2~~

Solution to Review Example 1.20 on page 38

1. Draw the Lewis Dot Structure for the following species. Make sure that you specify the h, n, s, b numbers for each structure. Calculate the formal charges for the elements, and find the most stable structure when considering formal charges and hypervalency.
    (a) SO_4^{2-}
    (b) CNO^- (atoms in that order: C–N–O)
    (c) CO_3^{2-}

The Lewis Structure of SO_4^{2-}

- the number of valence electrons the structure has: $h = 6 + 6 \times 4 + 2 = 32$
- the number of valence electrons the structure needs: $n = 8 + 8 \times 4 = 40$
- the number of electrons that must be shared: $s = 40 - 32 = 8$
- the minimum number of bonds in the structure: $b = 8 \div 2 = 4$

Thus, the initial structure is

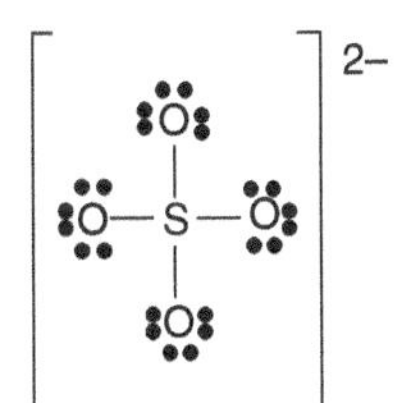

Formal Charges

- $FC_S = 6 - 0 - 4 = +2$
- $FC_{O_{\text{single bond}}} = 6 - 6 - 1 = -1$
- $\Delta FC = (+2) - (-1) = 3$

Since S can exhibit hypervalency, we can increase the number of bonds to the central atom in order to spread out the highly negative charge on the oxygen atoms. So, let us add two double bonds (each one decreases FC_S by one) to see what happens to the formal charges.

Formal Charges

- $FC_S = 6 - 0 - 6 = 0$
- $FC_{O_{\text{single bond}}} = 6 - 6 - 1 = -1$
- $FC_{O_{\text{double bond}}} = 6 - 4 - 2 = 0$
- $\Delta FC = (0) - (-1) = 1$

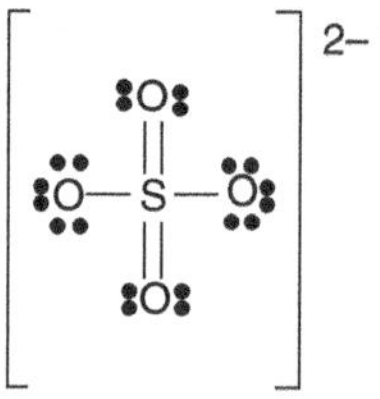

Let us see if adding another double bond is better.

Formal Charges

- $FC_S = 6 - 0 - 7 = -1$
- $FC_{O_{\text{single bond}}} = 6 - 6 - 1 = -1$
- $FC_{O_{\text{double bond}}} = 6 - 4 - 2 = 0$
- $\Delta FC = (0) - (-1) = 1$

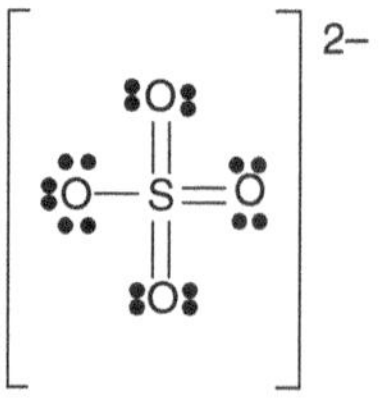

The ΔFC did not change, so nothing was gained, and the structure violates electronegativity rules since the least EN element (S) has a more negative charge than the most EN element (O). The previous structure with two double bonds is most stable (note that there would be a total of six resonance structures).

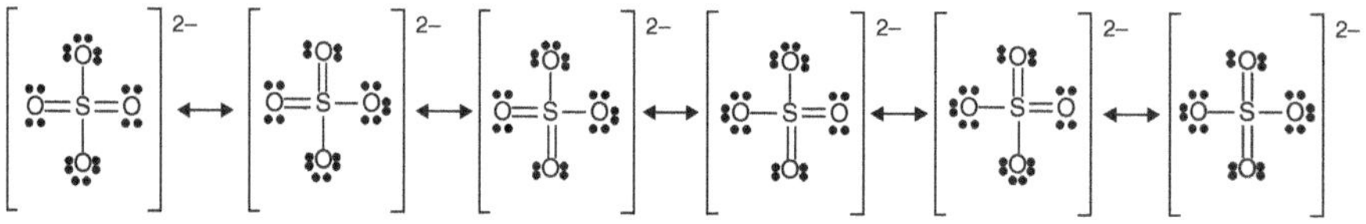

The Lewis Structure of CNO^-

- the number of valence electrons the structure has: h = 4 + 5 + 6 + 1 = 16
- the number of valence electrons the structure needs: n = 8 × 3 = 24
- the number of electrons that must be shared: s = 24 − 16 = 8
- the minimum number of bonds in the structure: b = 8 ÷ 2 = 4

With four bonds and three atoms, there are three nonequivalent structures possible: two with a single bond and triple bonds and one with two double bonds. None of the atoms can exhibit hypervalency. Formal charges will help us determine the most stable structure.

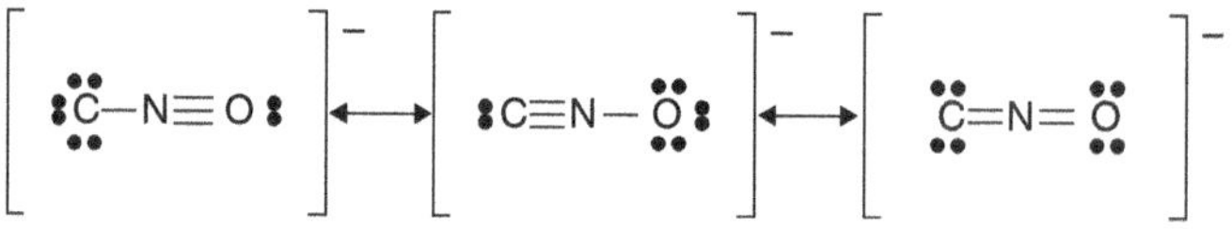

Formal Charges

- $FC_C = 4 - 6 - 1 = -3$
- $FC_N = 5 - 0 - 4 = +1$
- $FC_O = 6 - 2 - 3 = +1$
- $\Delta FC = (+1) - (-3) = 4$

- $FC_C = 4 - 2 - 3 = -1$
- $FC_N = 5 - 0 - 4 = +1$
- $FC_O = 6 - 6 - 1 = -1$
- $\Delta FC = (+1) - (-1) = 2$

- $FC_C = 4 - 4 - 2 = -2$
- $FC_N = 5 - 0 - 4 = +1$
- $FC_O = 6 - 4 - 2 = 0$
- $\Delta FC = (+1) - (-2) = 3$

The middle structure has the lowest ΔFC and gives C and O the most equitable FC given that O is more electronegative than C. The middle structure is the most stable.

The Lewis Structure of CO_3^{2-}

- the number of valence electrons the structure has: h = 4 + 6 × 3 + 2 = 24
- the number of valence electrons the structure needs: n = 8 × 4 = 32
- the number of electrons that must be shared: s = 32 − 24 = 8
- the minimum number of bonds in the structure: b = 8 ÷ 2 = 4

With four bonds and four atoms, there must be one double bond. None of the atoms can exhibit hypervalency, so there is no other possible combination. There are three resonance structures.

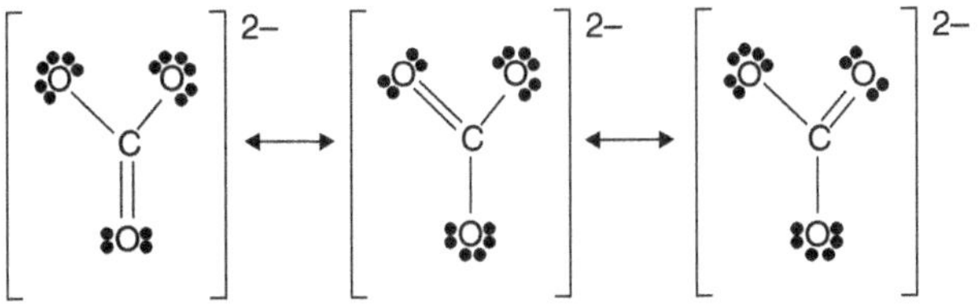

Formal Charges

- $FC_C = 4 - 0 - 4 = 0$
- $FC_{O_{single\ bond}} = 6 - 6 - 1 = -1$
- $FC_{O_{double\ bond}} = 6 - 4 - 2 = 0$
- $\Delta FC = (0) - (-1) = 1$

Solution to Review Example 1.21 on page 40

Since these problems involve log functions, remember the significant figure rules described in Section 2.2.2 on page 58.

1. Determine the pH of the following situations.

   (a) $[H_3O^+] = 1.0 \times 10^{-4}$ M

   $$pH = -\log([H_3O^+]) = -\log(1.0 \times 10^{-4}) = 4.00$$

   (b) $[H_3O^+] = 4.52 \times 10^{-5}$ M

   $$pH = -\log(4.52 \times 10^{-5}) = 4.345$$

   (c) $[OH^-] = 6.13 \times 10^{-6}$ M

   $$pOH = -\log([OH^-]) = -\log(6.13 \times 10^{-6}) = 5.212539$$
   $$pH = 14 - pOH = 14 - 5.2125 = 8.787$$

2. Determine the $[H_3O^+]$ for each of the following conditions.

   (a) pH = 9.84

   $$[H_3O^+] = 10^{-pH} = 10^{-9.84} = 1.4 \times 10^{-10}$$

   (b) pOH = 5.39

   $$[H_3O^+] = 10^{-pH} = 10^{-(14-pOH)} = 10^{-14+5.39} = 2.5 \times 10^{-9}$$

   (c) $[OH^-] = 4.37 \times 10^{-3}$ M

   $$[H_3O^+] = \frac{K_w}{[OH^-]} = \frac{1.0 \times 10^{-14}}{4.37 \times 10^{-3}} = 2.29 \times 10^{-12}$$

3. Label each of the following as a strong acid (SA), weak acid (WA), strong base (SB), weak base (WB), or an electrolyte (E).

   (a) WA HNO_2
   (b) SA HNO_3
   (c) E KCl

(d) WB NH_3
(e) SB NaOH
(f) WB NaCN

4. In each of the following comparisons, circle the acid or base that is the strongest of the pair.

   (a) CH_3COOH or $\boxed{HNO_2}$ – compare the K_a of each
   (b) $\boxed{NH_3}$ or CN^- – compare the K_b of each; NH_3 is slightly stronger
   (c) H_3PO_4 or $\boxed{H_2SO_4}$ – H_2SO_4 is a strong acid
   (d) $\boxed{NaOH}$ or NH_3 – NaOH is a strong base

5. Write the weak acid equilibrium reaction for $HClO_2$.

$$HClO_2(aq) \rightleftharpoons H^+(aq) + ClO_2^-(aq) \qquad \text{(RA.11)}$$

6. Write the weak base equilibrium reaction for ClO_2^-.

$$ClO_2^-(aq) + H_2O(l) \rightleftharpoons OH^-(aq) + HClO_2(aq) \qquad \text{(RA.12)}$$

Solution to Review Example 1.22 on page 42

1. Write the equilibrium expression for the following reactions.

   (a) $H_2SO_4(aq) + H_2O(l) \rightleftharpoons H_3O^+(aq) + HSO_4^-(aq)$
      i. $K_a = \dfrac{[H_3O^+][HSO_4^-]}{[H_2SO_4]}$
   (b) $CaSO_4(s) \rightleftharpoons Ca^{2+}(aq) + CO_3^{2-}(aq)$
      i. $K_{sp} = [Ca^{2+}][CO_3^{2-}]$
   (c) $Hg^{2+}(aq) + 4Cl^-(aq) \rightleftharpoons HgCl_4^{2-}(aq)$
      i. $K_f = \dfrac{[HgCl_4^{2-}]}{[Hg^{2+}][Cl^-]^4}$

2. Write the reverse of Reaction R1.61.

$$2NO(g) \rightleftharpoons N_2(g) + O_2(g) \qquad K_p = \frac{(P_{N_2})(P_{O_2})}{(P_{NO})^2} \qquad \text{(RA.13)}$$

   The K_p expression for the aforementioned reaction is the reciprocal of Reaction R1.61. If the forward reaction has a $K_p = 7.2 \times 10^{-5}$, then the reverse reaction has a $K_p = 1.4 \times 10^4$.

3. Complete Table A.4, and predict which way the reaction will shift in each case.

$$2NO_2(g) \rightleftharpoons N_2O_4(g) \qquad \text{(RA.14)}$$

   The equilibrium expression for the aforementioned reaction is $K_p = \frac{[N_2O_4]}{[NO_2]^2}$ because the following units are specified as Molar (otherwise, the concentrations would be expressed as pressures).

Exp. #	Initial conc.(M) $[NO_2]$	$[N_2O_4]$	Q	Equilibrium Conc.(M) $[NO_2]$	$[N_2O_4]$	K_{eq}	The reaction will shift ...
1	0.0200	0.0000	**0**	0.0172	0.00139	**4.70**	→
2	0.0300	0.0300	**33.3**	0.0244	0.00280	**4.70**	←
3	0.0400	0.0000	**0**	0.0310	0.00452	**4.70**	→
4	0.0000	0.0200	**∞**	0.0310	0.00452	**4.70**	←

The bold items are the answers to the exercises.

Table A.4 Simulated equilibrium experiments and answers.

4. Calculate Q and K_{eq}, and enter these values in the aforementioned table.
5. Which way will the reaction shift? Enter ← or → in the last column.

Solution to Review Example 1.23 on page 44

1. Determine the pH of a 0.0100 M solution of HCN.

	HCN(aq)	⇌	H^+(aq)	+	CN^-(aq)
I	0.0100		0		0
C	$-x$		$+x$		$+x$
E	$0.0100 - x$		x		x

$$K_a = \frac{[H^+][CN^-]}{[HCN]}$$

$$6.2 \times 10^{-10} = \frac{(x)(x)}{(0.0100 - x)}$$

Assume $x \ll 0.0100$ M.

$$6.2 \times 10^{-10} = \frac{(x)(x)}{(0.0100 - \cancel{x}_0)}$$

$$6.2 \times 10^{-10} = \frac{x^2}{0.0100}$$

$$(6.2 \times 10^{-10})(0.0100) = x^2$$

$$\sqrt{(6.2 \times 10^{-10})(0.0100)} = x$$

$$2.490 \times 10^{-6} = x$$

Check the 5% rule assumption to see if it was valid.

$$\%\text{err} = \frac{x}{0.0100} \times 100\% = \frac{2.490 \times 10^{-6}}{0.0100} \times 100\% = 0.025\% \text{ (valid!!)}$$

From the E line of the ICE table, $[H^+] = x = 2.490 \times 10^{-6}$ M so

$$\text{pH} = -\log[H^+] = -\log[2.490 \times 10^{-6}] = 5.60$$

2. Determine the pH of a 0.0100 M solution of NH_3.

	$NH_3(aq)$	$+ H_2O(l)$	$\rightleftharpoons$	$OH^-(aq)$	+	$NH_4^+(aq)$
I	0.0100	–		0		0
C	$-x$			$+x$		$+x$
E	$0.0100 - x$			x		x

$$K_b = \frac{[OH^-][NH_4^+]}{[NH_3]}$$
$$1.8 \times 10^{-5} = \frac{(x)(x)}{(0.0100 - x)}$$

Assume $x \ll 0.0100$ M.

$$1.8 \times 10^{-5} = \frac{(x)(x)}{(0.0100 - \cancel{x}^{0})}$$
$$1.8 \times 10^{-5} = \frac{x^2}{0.0100}$$
$$(1.8 \times 10^{-5})(0.0100) = x^2$$
$$\sqrt{(1.8 \times 10^{-5})(0.0100)} = x$$
$$4.243 \times 10^{-4} = x$$

Check the 5% rule assumption to see if it was valid.

$$\%err = \frac{x}{0.0100} \times 100\%$$
$$= \frac{4.243 \times 10^{-4}}{0.0100} \times 100\%$$
$$= 4.24\% \text{ (close, but valid)}$$

From the E line of the ICE table, $[OH^-] = x = 4.243 \times 10^{-4}$ M so

$$pOH = -\log[OH^-] = -\log[4.243 \times 10^{-4}]$$
$$= 3.3723$$
$$pH = 14 - pOH = 14 - 3.3723 = 10.63$$

3. (Challenge) Determine the pH of a 0.0100 M solution of HNO_2 (nitrous acid, not nitric acid!!).

	$HNO_2(aq)$	$\rightleftharpoons$	$H^+(aq)$	+	$NO_2^-(aq)$
I	0.0100		0		0
C	$-x$		$+x$		$+x$
E	$0.0100 - x$		x		x

$$K_a = \frac{[H^+][NO_2^-]}{[HNO_2]}$$
$$5.6 \times 10^{-4} = \frac{(x)(x)}{(0.0100 - x)}$$

Assume $x \ll 0.0100$ M.

$$5.6 \times 10^{-4} = \frac{(x)(x)}{(0.0100 - \cancel{x}^{0})}$$

$$5.6 \times 10^{-4} = \frac{x^2}{0.0100}$$
$$(5.6 \times 10^{-4})(0.0100) = x^2$$
$$\sqrt{(5.6 \times 10^{-4})(0.0100)} = x$$
$$2.366 \times 10^{-3} = x$$

Check the 5% rule assumption to see if it was valid.

$$\%\text{err} = \frac{x}{0.0100} \times 100\%$$
$$= \frac{2.366 \times 10^{-3}}{0.0100} \times 100\%$$
$$= 24\% \text{ (not valid!!)}$$

We need to abandon the 5% rule and use successive approximations.

$$x_2 = \sqrt{(5.6 \times 10^{-4})(0.0100 - x)}$$
$$= \sqrt{(5.6 \times 10^{-4})(0.0100 - 2.366 \times 10^{-3})}$$
$$= 2.068 \times 10^{-3}$$
$$x_3 = \sqrt{(5.6 \times 10^{-4})(0.0100 - 2.068 \times 10^{-3})}$$
$$= 2.108 \times 10^{-3}$$
$$x_4 = \sqrt{(5.6 \times 10^{-4})(0.0100 - 2.108 \times 10^{-3})}$$
$$= 2.102 \times 10^{-3}$$

From the E line of the ICE table, $[H^+] = x = 2.102 \times 10^{-3}$ M so

$$\text{pH} = -\log[H^+] = -\log[2.102 \times 10^{-3}]$$
$$= 2.68$$

4. (Challenge) Determine the pH of a 0.100 M solution of H_2SO_4 (the first H comes off as a strong acid, but the second H is part of a weak acid).

Since H_2SO_4 is a strong acid, we need to assume that 100% of the first protons will fall off the acid.

	$H_2SO_4(aq)$	→	$H^+(aq)$	+	$HSO_4^-(aq)$
I	0.100		0		0
C	−0.100		+0.100		+0.100
E	0		0.100		0.100

The next acid dissociation reaction starts from here.

	$HSO_4^-(aq)$	⇌	$H^+(aq)$	+	$SO_4^{2-}(aq)$
I	0.100		0.100		0
C	$-x$		$+x$		$+x$
E	$0.100 - x$		$0.100 + x$		x

$$K_{a2} = \frac{[H^+][SO_4^{2-}]}{[HSO_4^-]}$$
$$1.0 \times 10^{-2} = \frac{(0.100 + x)(x)}{(0.100 - x)}$$

Assume $x \ll 0.100$ M.

$$1.0 \times 10^{-2} = \frac{(0.100 + \cancel{x}_0)(x)}{(0.100 - \cancel{x}_0)}$$
$$1.0 \times 10^{-2} = \frac{(0.100)(x)}{0.100}$$
$$1.0 \times 10^{-2} = \frac{\cancel{(0.100)}(x)}{\cancel{0.100}}$$
$$1.0 \times 10^{-2} = x$$

Check the 5% rule assumption to see if it was valid.

$$\begin{aligned}\%\text{err} &= \frac{0.010}{0.100} \times 100\% \\ &= 10\% \text{ (not valid!!)}\end{aligned}$$

We need to abandon the 5% rule and use successive approximations.

$$\begin{aligned}x_2 &= \frac{(0.100 - x)(1.0 \times 10^{-2})}{(0.100 + x)} \\ &= \frac{(0.100 - 0.010)(1.0 \times 10^{-2})}{(0.100 + 0.010)} \\ &= 0.0081818\end{aligned}$$
$$\begin{aligned}x_3 &= \frac{(0.100 - 0.0081818)(1.0 \times 10^{-2})}{(0.100 + 0.0081818)} \\ &= 0.0084874\end{aligned}$$
$$\begin{aligned}x_4 &= \frac{(0.100 - 0.0084874)(1.0 \times 10^{-2})}{(0.100 + 0.0084874)} \\ &= 0.0084353\end{aligned}$$
$$\begin{aligned}x_4 &= \frac{(0.100 - 0.0084353)(1.0 \times 10^{-2})}{(0.100 + 0.0084353)} \\ &= 0.008444\end{aligned}$$

From the E line of the ICE table, $[H^+] = 0.100 + x = 0.100 + 0.008444 = 0.108444$ M so

$$\begin{aligned}pH &= -\log[H^+] = -\log[0.108444] \\ &= 0.96\end{aligned}$$

Solution to Review Example 1.24 on page 47

Exercises

1. Balance the following redox reactions (states of matter can be ignored).
   (a) $Cr_2O_7^{2-} + NO_2^- \rightarrow Cr^{3+} + NO_3^-$ (acidic)
   (b) $Cu(s) + HNO_3(aq) \rightarrow Cu^{2+}(aq) + NO(g)$ (basic)
   (c) $Cr(OH)_3(s) + ClO_3^-(aq) \rightarrow CrO_4^{2-}(aq) + Cl^-(aq)$ (basic)
   (d) $Pb(s) + PbO_2(s) + HSO_4^-(aq) \rightarrow PbSO_4(s)$ (acidic)

$Cr_2O_7^{2-}(aq) + NO_2^-(aq) \rightarrow Cr^{3+}(aq) + NO_3^-(aq)$ (acidic)

(step 1 – half-reactions)

$Cr_2O_7^{2-}(aq) \rightarrow Cr^{3+}(aq)$ $\qquad$ $NO_2^-(aq) \rightarrow NO_3^-(aq)$

(step 2 – balance for all elements except H, O)

$Cr_2O_7^{2-}(aq) \rightarrow \mathbf{2}Cr^{3+}(aq)$ $\qquad$ $NO_2^-(aq) \rightarrow NO_3^-(aq)$

(step 3 – balance for O by adding H_2O)

$Cr_2O_7^{2-}(aq) \rightarrow 2Cr^{3+}(aq) + \mathbf{7H_2O(l)}$ $\qquad$ $\mathbf{H_2O(l)} + NO_2^-(aq) \rightarrow NO_3^-(aq)$

(step 4 – balance for H by adding H^+)

$\mathbf{14H^+(aq)} + Cr_2O_7^{2-}(aq) \rightarrow 2Cr^{3+}(aq) + 7H_2O(l)$ $\qquad$ $H_2O(l) + NO_2^-(aq) \rightarrow NO_3^-(aq) + \mathbf{2H^+(aq)}$

(step 5 – if the reaction is under basic conditions... the reaction is acidic – nothing to do)

(step 6 – balance each half-reaction for charge by adding electrons)

$\mathbf{6e^-} + 14H^+(aq) + Cr_2O_7^{2-}(aq) \rightarrow 2Cr^{3+}(aq) + 7H_2O(l)$ $\qquad$ $H_2O(l) + NO_2^-(aq) \rightarrow NO_3^-(aq) + 2H^+(aq) + \mathbf{2e^-}$

(step 7 – find the least common multiple of the electrons in the two half-reactions and equate them)

$\mathbf{1}\times(6e^- + 14H^+(aq) + Cr_2O_7^{2-}(aq) \rightarrow 2Cr^{3+}(aq) + 7H_2O(l))$ $\qquad$ $\mathbf{3}\times(H_2O(l) + NO_2^-(aq) \rightarrow NO_3^-(aq) + 2H^+(aq) + \mathbf{2e^-})$

(step 8 – add the half-reactions and cancel the like terms)

$6e^- + 14H^+(aq) + Cr_2O_7^{2-}(aq) + 3H_2O(l) + 3NO_2^-(aq) \rightarrow 2Cr^{3+}(aq) + 7H_2O(l) + 3NO_3^-(aq) + 6H^+(aq) + 6e^-$

$\cancel{6e^-} + \cancel{14}_8H^+(aq) + Cr_2O_7^{2-}(aq) + \cancel{3H_2O}(l) + 3NO_2^-(aq) \rightarrow 2Cr^{3+}(aq) + \cancel{7}_4\,H_2O(l) + 3NO_3^-(aq) + \cancel{6H^+}(aq) + \cancel{6e^-}$

$8H^+(aq) + Cr_2O_7^{2-}(aq) + 3NO_2^-(aq) \rightarrow 2Cr^{3+}(aq) + 4H_2O(l) + 3NO_3^-(aq)$

$Cu(s) + HNO_3(aq) \rightarrow Cu^{2+}(aq) + NO(g)$ (basic)

(step 1 – half-reactions)

$Cu(s) \rightarrow Cu^{2+}(aq)$ $\qquad$ $HNO_3(aq) \rightarrow NO(g)$

(step 2 – balance for all elements except H, O) - nothing to do

(step 3 – balance for O by adding $H_2O(l)$)

$Cu(s) \rightarrow Cu^{2+}(aq)$ $\qquad$ $HNO_3(aq) \rightarrow NO(g) + \mathbf{2H_2O(l)}$

(step 4 – balance for H by adding $H^+(aq)$)

$Cu(s) \rightarrow Cu^{2+}(aq)$ $\qquad$ $\mathbf{3H^+(aq)} + HNO_3(aq) \rightarrow NO(g) + 2H_2O(l)$

(step 5 – if the reaction is under basic conditions, add enough $OH^-(aq)$ to ***both sides*** to neutralize all $H^+(aq)$)

$Cu(s) \rightarrow Cu^{2+}(aq)$ $\qquad$ $\mathbf{3OH^-(aq)} + 3H^+(aq) + HNO_3(aq) \rightarrow NO(g) + 2H_2O(l) + \mathbf{3OH^-(aq)}$

$Cu(s) \rightarrow Cu^{2+}(aq)$ $\qquad$ $\mathbf{3H_2O(l)} + HNO_3(aq) \rightarrow NO(g) + 2H_2O(l) + 3OH^-(aq)$

(step 6 – balance each half-reaction for charge by adding electrons)

$Cu(s) \rightarrow Cu^{2+}(aq) + \mathbf{2e^-}$ $\qquad$ $\mathbf{3e^-} + 3H_2O(l) + HNO_3(aq) \rightarrow NO(g) + 2H_2O(l) + 3OH^-(aq)$

(step 7 – find the least common multiple of the electrons in the two half-reactions and equate them)

$\mathbf{3}\times(Cu(s) \rightarrow Cu^{2+}(aq) + 2e^-)$ $\qquad$ $\mathbf{2}\times(3e^- + 3H_2O(l) + HNO_3(aq) \rightarrow NO(g) + 2H_2O(l) + 3OH^-(aq))$

(step 8 – add the half-reactions and cancel the like terms)

$3Cu(s) + 6e^- + 6H_2O(l) + 2HNO_3(aq) \rightarrow 3Cu^{2+}(aq) + 6e^- + 2NO(g) + 4H_2O(l) + 6OH^-(aq)$

$3Cu(s) + \cancel{6e^-} + \cancel{6}_2\,H_2O(l) + 2HNO_3(aq) \rightarrow 3Cu^{2+}(aq) + \cancel{6e^-} + 2NO(g) + \cancel{4H_2O(l)} + 6OH^-(aq)$

$3Cu(s) + 2H_2O(l) + 2HNO_3(aq) \rightarrow 3Cu^{2+}(aq) + 2NO(g) + 6OH^-(aq)$

$Cr(OH)_3(s) + ClO_3^-(aq) \rightarrow CrO_4^{2-}(aq) + Cl^-(aq)$ (basic)

(step 1 – half-reactions)

$Cr(OH)_3(s) \rightarrow CrO_4^{2-}(aq)$

$ClO_3^-(aq) \rightarrow Cl^-(aq)$

(step 2 – balance for all elements except H, O) – nothing to do

(step 3 – balance for O by adding H_2O)

$Cr(OH)_3(s) + \mathbf{1H_2O(l)} \rightarrow CrO_4^{2-}(aq)$

$ClO_3^-(aq) \rightarrow Cl^-(aq) + \mathbf{3H_2O(l)}$

(step 4 – balance for H by adding $H^+(aq)$)

$Cr(OH)_3(s) + H_2O(l) \rightarrow CrO_4^{2-}(aq) + \mathbf{5H^+(aq)}$

$\mathbf{6H^+(aq)} + ClO_3^-(aq) \rightarrow Cl^-(aq) + 3H_2O(l)$

(step 5 – if the reaction is under basic conditions, add enough $OH^-(aq)$ to ***both sides*** to neutralize all $H^+(aq)$)

$\mathbf{5OH^-(aq)} + Cr(OH)_3(s) + H_2O(l) \rightarrow CrO_4^{2-}(aq) + \mathbf{5OH^-(aq)} + 5H^+(aq)$

$\mathbf{6OH^-(aq)} + 6H^+(aq) + ClO_3^-(aq) \rightarrow Cl^-(aq) + 3H_2O(l) + \mathbf{6OH^-(aq)}$

$5OH^-(aq) + Cr(OH)_3(s) + H_2O(l) \rightarrow CrO_4^{2-}(aq) + \mathbf{5H_2O(l)}$

$\mathbf{6H_2O(l)} + ClO_3^-(aq) \rightarrow Cl^-(aq) + 3H_2O(l) + 6OH^-(aq)$

(step 6 – balance each half-reaction for charge by adding electrons)

$5OH^-(aq) + Cr(OH)_3(s) + H_2O(l) \rightarrow CrO_4^{2-}(aq) + 5H_2O(l) + \mathbf{3e^-}$

$\mathbf{6e^-} + 6H_2O(l) + ClO_3^-(aq) \rightarrow Cl^-(aq) + 3H_2O(l) + 6OH^-(aq)$

(step 7 – find the least common multiple of the electrons in the two half-reactions and equate them)

$\mathbf{2}\times(5OH^-(aq) + Cr(OH)_3(s) + H_2O(l) \rightarrow CrO_4^{2-}(aq) + 5H_2O(l) + 3e^-)$

$\mathbf{1}\times(6e^- + 6H_2O(l) + ClO_3^-(aq) \rightarrow Cl^-(aq) + 3H_2O(l) + 6OH^-(aq))$

(step 8 – add the half-reactions and cancel the like terms)

$10OH^-(aq) + 2Cr(OH)_3(s) + 8H_2O(l) + 6e^- + ClO_3^-(aq) \rightarrow 2CrO_4^{2-}(aq) + 13H_2O(l) + 6e^- + Cl^-(aq) + 6OH^-(aq)$

$\cancel{10}\,4OH^-(aq) + 2Cr(OH)_3(s) + \cancel{6e^-} + \cancel{8H_2O(l)} + ClO_3^-(aq) \rightarrow 2CrO_4^{2-}(aq) + \cancel{13}\,5\,H_2O(l) + \cancel{6e^-} + Cl^-(aq) + \cancel{6OH^-(aq)}$

$4OH^-(aq) + 2Cr(OH)_3(s) + ClO_3^-(aq) \rightarrow 2CrO_4^{2-}(aq) + 5H_2O(l) + Cl^-(aq)$

$Pb(s) + PbO_2(s) + HSO_4^-(aq) \rightarrow PbSO_4(s)$ (acidic)

(step 1 – half-reactions)

$Pb(s) + HSO_4^-(aq) \rightarrow PbSO_4(s)$

$PbO_2(s) + HSO_4^-(aq) \rightarrow PbSO_4(s)$

(step 2 – balance for all elements except H, O)

$Pb(s) + HSO_4^-(aq) \rightarrow PbSO_4(s)$

$PbO_2(s) + HSO_4^-(aq) \rightarrow PbSO_4(s)$

(step 3 – balance for O by adding $H_2O(l)$)

$Pb(s) + HSO_4^-(aq) \rightarrow PbSO_4(s)$

$PbO_2(s) + HSO_4^-(aq) \rightarrow PbSO_4(s) + \mathbf{2H_2O(l)}$

(step 4 – balance for H by adding $H^+(aq)$)

$Pb(s) + HSO_4^-(aq) \rightarrow PbSO_4 + \mathbf{H^+(aq)}$

$\mathbf{3H^+(aq)} + PbO_2(s) + HSO_4^-(aq) \rightarrow PbSO_4(s) + 2H_2O(l)$

(step 5 – if the reaction is under basic conditions... the reaction is acidic – nothing to do)

(step 6 – balance each half-reaction for charge by adding electrons)

$Pb(s) + HSO_4^-(aq) \rightarrow PbSO_4(s) + H^+(aq) + \mathbf{2e^-}$

$\mathbf{2e^-} + 3H^+(aq) + PbO_2(s) + HSO_4^-(aq) \rightarrow PbSO_4(s) + 2H_2O(l)$

(step 7 – find the least common multiple of the electrons in the two half-reactions and equate them)

$\mathbf{1}\times(Pb(s) + HSO_4^-(aq) \rightarrow PbSO_4(s) + H^+(aq) + 2e^-)$

$\mathbf{1}\times(2e^- + 3H^+(aq) + PbO_2(s) + HSO_4^-(aq) \rightarrow PbSO_4(s) + 2H_2O(l))$

(step 8 – add the half-reactions and cancel the like terms)

$Pb(s) + 2HSO_4^-(aq) + 2e^- + 3H^+(aq) + PbO_2(s) \rightarrow 2PbSO_4(s) + H^+(aq) + 2e^- + 2H_2O(l)$

$Pb(s) + 2HSO_4^-(aq) + \cancel{2e^-} + \cancel{3}\,2\,H^+(aq) + PbO_2(s) \rightarrow 2PbSO_4(s) + \cancel{H^+} + \cancel{2e^-} + 2H_2O(l)$

$Pb(s) + 2HSO_4^-(aq) + 2H^+(aq) + PbO_2(s) \rightarrow 2PbSO_4(s) + 2H_2O(l)$

A.2 QUESTIONS ABOUT THE BIG BANG, SOLAR NEBULAR MODEL, AND THE FORMATION OF THE EARTH

The Big Bang and Nucleosynthesis

1. **How old is the universe?** 13.8 billion years
2. **How old is our solar system?** 4.6 billion years
3. **What are the two major pieces of evidence for the Big Bang theory mentioned in this chapter?**
   - The microwave background measurement showing the temperature of the universe following the plasma era was 3000 K, red-shifted by the expansion of the universe to approximately 3 K (which is observed empirically).
   - The hydrogen/helium ratio that is observed empirically reflects theoretical calculations.
4. **What elements are predominantly the result of the Big Bang?** Hydrogen and helium, with traces of lithium and beryllium
5. **What elements are predominantly the result of the nuclear fusion in massive stars?** Elements from lithium to iron
6. **What elements are predominantly the result of supernova explosions of massive stars?** From nickel to uranium
7. **What are the two most abundant elements in the Sun?** Hydrogen and helium
8. **Which stars have longer lives – low-mass or high-mass stars?** Low-mass stars live longer because the temperatures and pressures are lower and therefore the nuclear reactions proceed slower than in massive stars
9. **Which element has the most stable nucleus?** Fe-56
10. **What element is the main product of hydrogen fusion?** Helium
11. **What element is the main product of helium fusion?** Carbon
12. **List a few of the elements produced from product of carbon fusion?** Mg, Al, O
13. **What element is formed as a result of oxygen fusion and is the primary reactant in the final fusion process in a massive stars?** Silicon
14. **Why can't a massive star use iron and heavier elements as a source of nuclear fusion?** Since Fe-56 is the most stable nucleus and has the highest binding energy, any fusion process involving Fe-56 will require the addition of energy instead of a release of energy.
15. **What happens to a massive star once all of the nuclear fuel has been used up?** The star undergoes a supernova explosion.
16. **What two general categories of nuclear reactions are predominantly responsible for the synthesis of the elements heavier than iron?** Neutron capture and beta emission
17. **Our solar system was** (correct answers in *italics*):
   (a) made from the remains of one or more small to medium star(s).
   (b) made from the remains of one or more massive star(s).
   (c) at the beginning or shortly after the beginning of the universe.
18. **Describe the main evidence for your choice earlier.** Since our planet contains significant amounts of elements from lithium through uranium, our solar system must have formed from the remains of at least one or more generations of massive stars. Since massive stars have short lifetimes and given the gap of 9 Gyr between the Big Bang and the birth of our solar system, there was plenty of time for several generations of massive stars to form and explode in these 9 Gyrs.

The Solar Nebular Model

1. **What is the solar wind and what did it do to the early solar system?** The solar wind is a stream of charged particles that flow away from the Sun as a result of high-energy nuclear reactions. These particles act as a wind in that they pushed gas, dust, and debris (from the nebular cloud that gave birth to our solar system) out toward the edges of the solar system. Most of the light gases within the frost zone were pushed out and absorbed by the giant gas planets. The first atmosphere of the Earth, which contained hydrogen and helium (which contributes most of the mass of the entire solar system) was blown off the forming Earth by the solar wind.
2. **List two pieces of evidence stemming from the orbits and spins of the planets that corroborate the solar nebular model.** All of the planets circle the Sun counterclockwise and in the same plane. Most planets also spin in the same direction as the Sun (CCW) except Venus (CW) and Uranus (90 ° tilt).
3. **Why is the solar system mostly a two-dimensional structure?** The law of the conservation of angular momentum caused the swirling nebular cloud to flatten as gravity began to pull gases and debris into the center of the solar system.
4. **What is the region of the solar system called that separates the terrestrial planets from the gas giants?** The frost zone separates the gas giants in the outer portion of the solar system from the rocky planets in the inner part of the solar system.
5. **If you were to classify elements as volatile and nonvolatile, how were these two categories of elements distributed between the two regions of the solar system separated by the asteroid belt?** The heavier, less volatile elements and compounds remained in the inner part of the solar system, and the light, more volatile components were pushed to the outer portion of the solar system.
6. **What two planets are exceptions to the normal planetary axis of rotation?** Venus, which rotates clockwise, and Uranus, which rotates with a 90 ° tilt with respect to the planetary plane.

The Ages of the Earth

1. **Very briefly, list the major changes to the Earth during the Hadean eon.** Accretion, massive bombardment, differentiation, formation of the Moon, the second atmosphere formed as a result of vulcanism, planetary cooling, formation of the oceans, and the transition from a heavy greenhouse atmosphere to a moderate greenhouse
2. **Specify the ages of the Earth covered by the Hadean eon.** 4.6–4.0 Gyr
3. **Very briefly, list the major changes to the Earth during the Archean eon.** Further cooling, simple life forms spread, photosynthetic microbes produce significant levels of atmospheric oxygen (up to 1 ppm at the end), oceans transition from reducing and acidic to alkaline, cooling climate.
4. **Specify the ages of the Earth covered by the Archean eon.** 4.0–2.5 Gyr
5. **Very briefly, list the major changes to the Earth during the Proterozoic eon.** Great Oxidation Events, an increase in atmospheric oxygen levels to 10%, oxidative atmosphere, global ice ages, ozone layer begins to form, simple life begins to colonize the land.
6. **Specify the ages of the Earth covered by the Proterozoic eon.** 2.5–0.5 Gyr
7. **Very briefly, list the major changes to the Earth during the Phanerozoic eon.** the Cambrian explosion and the proliferation of life, climate stabilizes, oxygen levels rise to current levels.
8. **Specify the ages of the Earth covered by the Phanerozoic eon.** 500 Myrs to present
9. **How did the Moon form?** Probably from the collision between the proto-Earth and a planetoid in the same orbit. The Moon seems to be formed from the mantle of the proto-Earth. The 23 ° axis tilt and the current rotational speed were probably set by this collision.

10. **The first atmosphere of the Earth was made of what gases? How did the Earth lose this first atmosphere?** It was likely composed of molecular hydrogen and helium gases, which comprise the bulk mass of the Sun. The relentless solar wind blew off these light gases since the Earth's mass is not large enough to hold onto them.
11. **Where did the second atmosphere come from?** Vulcanism – dissolved gases became less soluble as the molten Earth cooled.
12. **What gases were the major components of the second atmosphere?** CO_2, H_2O, N_2
13. **Why does the Earth have a core? Why isn't the Earth homogeneous in composition? What is the core composed of?** The process of differentiation describes the separation of compounds in the molten Earth from the interaction of gravity and density. The less dense compounds, such as silicates, floated to the surface while the dense iron and nickel sunk to the core.
14. **What elements compose most of the Earth's crust? Why are these elements so abundant?** Silicates and aluminosilicates contain silicon, oxygen, aluminum, magnesium, and iron. These elements are so abundant because they are the stable end-products of various stages of nuclear fusion in the core of massive stars.
15. **How are the age of rocks measured?** The age of rocks are measured by the radioactive decay of various isotopes, such as uranium.
16. **Where did most of the CO_2 from the atmosphere go?** Most of the carbon dioxide came out of the early atmosphere when the water condensed and precipitated. Much of this carbon dioxide is locked in geologic formations such as limestone.
17. **The early atmosphere of the Earth was believed to be lacking what gas that is tremendously important to most life forms? The current atmosphere contains what percentage of this gas now?** Free oxygen (O_2) was originally absent from the atmosphere. The current atmosphere contains about 21% of O_2.
18. **Describe the geological evidence that is the result of the growing concentration of this gas.** Banded Iron Formations (BIFs) tell the geologic story of the rise of atmospheric molecular oxygen, both from marine and terrestrial sources.

B

CHAPTER 2 EXAMPLES AND END-OF-CHAPTER EXERCISES

B.1 SOLUTIONS TO IN-CHAPTER EXAMPLES

Solution to Example 2.1 on page 57

Use a spreadsheet to determine the average ($\bar{x}$), standard deviation (s), relative standard deviation (RSD), maximum value, minimum value, range, and number of data points of the numbers in Table B.1.

7.05	6.00	4.77	6.69	7.10	9.56	4.66	5.12
8.16	7.25	9.46	12.09	8.95	3.91	7.03	10.29
6.62	7.61	6.27	8.77	4.16	8.66	8.20	11.04

Table B.1 A set of numbers in Gaussian distribution

Figure B.1 shows formulas used to obtain the results. The average is obtained by using the function *AVERAGE( )*. The range of the cells comprises the input parameters. The standard deviation is given by the function *STDEV( )* (or *STDEV.S( )* in newer versions of Excel). The RSD is found by dividing the standard deviation by the average to give a relative measure of the precision of the data set. The maximum and minimum values are given by the *MAX( )* and *MIN( )* functions, the range is the difference between the maximum and minimum values, and the count of the data points is given by the *COUNT( )* function. While you may have been able to determine these values with a calculator or by a short examination of the data set, you should take this opportunity to practice with spreadsheet functions. They are quick to use on small and large data sets.

	A	B	C	D	E	F	G	H
1	7.05	6.00	4.77	6.69	7.10	9.56	4.66	5.12
2	8.16	7.25	9.46	12.09	8.95	3.91	7.03	10.29
3	6.62	7.61	6.27	8.77	4.16	8.66	8.20	11.04
4								
5	average =AVERAGE(A1:H3)					average	7.5	
6	std dev =STDEV(A1:H3)					std dev	2.2	
7	RSD =B6/B5*100					RSD	29%	
8	max =MAX(A1:H3)					max	12.09	
9	min =MIN(A1:H3)					min	3.91	
10	range =B8-B9					range	8.18	
11	count =COUNT(A1:H3)					count	24	

Figure B.1 Spreadsheet analysis using general statistics.

Solution to Example 2.2 on page 57

Calculate the baseline, noise, signal level, and signal-to-noise ratio (SNR) for two different samples (Sample A and Sample B) using the data given in Table B.2. The following questions will get you through the process.

Environmental Chemistry: An Analytical Approach, First Edition. Kenneth S. Overway.

Companion website: www.wiley.com/go/overway/environmental_chemistry

Blank	Sample A	Sample B
1.1844	1.9831	9.4616
0.5324	1.2627	9.1638
0.5516	1.9993	8.5835
1.1929	1.1270	9.5922
0.6368	1.4045	9.3784
0.8920	1.1045	9.1483
0.7263	1.8622	9.1141
0.3387	1.5284	8.8063
0.7667	1.3332	9.2327

Table B.2 A set of simulated data for the measurement of a blank and two samples.

Solution:

1. **What is the baseline value of the blank?** Using the AVERAGE( ) function, the baseline of noise is 0.75798.
2. **What is the noise level calculated from the measurement of the blank?** Using the STDEV( ) function, the noise of the blank is 0.29026.
3. **What is the signal level for Sample A?** Signal for Sample A is $1.51166 - 0.75798 = 0.75862$.
4. **What is the noise level in Sample A?** The noise for Sample A is 0.35395.
5. **Calculate the SNR for Sample A.** SNR $= 0.75862 \div 0.35395 = 2.14$.
6. **Calculate the noise, signal, and SNR for Sample B.** SNR $= (9.1645 - 0.75798) \div 0.314299 = 26.75$.
7. **Given the SNR for Sample A, is Sample A not detectable (ND), only above the Limit of Detection (>LOD) or above the Limit of Quantitation (>LOQ)?** Given that the SNR = 2.14, Sample A is ND.
8. **Given the SNR for Sample B, is Sample B ND, only above the Limit of Detection (>LOD) or above the Limit of Quantitation (>LOQ)?** Given that the SNR = 26.75, Sample B is >LOQ.

Solution to Example 2.4 on page 59

For each of the following examples, the result of a data analysis is given as *result* ± *uncertainty*. In each case, format this pair of numbers with the proper significant figures (sig figs). Assume the two sig fig rule on the uncertainty.

1. 624.694 ± 228.357
2. 3840.2 ± 4739.17
3. 470.254 ± 42.5641

Solution: One correct and several incorrect options are provided for each example. An explanation is given for each incorrect option.

1. 624.694 ± 228.357
   (a) 625 ± 228 (This option is incorrect because the uncertainty has three sig figs, not two.)
   (b) 620 ± 230 (This is the correctly formatted option.)
   (c) 625 ± 230 (This option is incorrect because the result has more precision than the uncertainty.)

(d) 600 ± 200 (This option is only correct if a one sig fig rule is followed. Both the result and uncertainty have only one sig fig.)

(e) 1000 ± 0 (This item presents the result as a gross overestimate and the uncertainty as an underestimate.)

2. 3840.2 ± 4739.17

   (a) 0 ± 0 (This conveys the detectability of the result but eliminates all of the information of the measurement.)

   (b) 3800 ± 4700 (This is the correctly formatted option.)

   (c) 3840.0 ± 4740.0 (This option is incorrect because the uncertainty has five sig figs, not two.)

   (d) 3800 ± 5000 (This option is incorrect because the result and uncertainty have different levels of precision and the uncertainty has only one sig fig and not two.)

   (e) 4000 ± 5000 (This option is only correct if a one sig fig rule is followed. Both the result and uncertainty have only one sig fig.)

3. 470.254 ± 42.5641

   (a) 500 ± 0 (This option is incorrect because the result is a gross overestimate and the uncertainty is a gross underestimate.)

   (b) 470 ± 40 (This option is incorrect because the uncertainty has one sig fig, not two.)

   (c) 470 ± 43 (This is the correctly formatted option. The result, however, unfortunately rounds to a zero in the units place and gives a number with vague precision. It would be best to format this set of numbers in scientific notation as $(4.70 \pm 0.43) \times 10^2$).

   (d) 470.3 ± 43 (This option is incorrect because the result and uncertainty have different levels of precision.)

   (e) 470.3 ± 42.6 (This option is incorrect because the uncertainty has three sig figs, not two.)

Examine the correctly formatted results, and label each as ND, *detectable*, or *quantitatively detectable*. Assume that the number of trials was well over 20 in each case.

1. 0.1422 ± 0.0017
2. 640 ± 170
3. 4.7 ± 1.8
4. 8.3 ± 1.3

Solution: Remember that analytically undetectable measurements are those with an SNR <3. Qualitatively detectable measurements are those where the SNR ≥ 3 and SNR <10. Quantitatively detectable measurements are those with an SNR ≥ 10.

1. $0.1422 \pm 0.0017 \rightarrow \text{SNR} = 0.1422 \div 0.0017 = 84$, so this result is quantitatively detectable.
2. $640 \pm 170 \rightarrow \text{SNR} = 640 \div 170 = 3.8$, so this result is detectable.
3. $4.7 \pm 1.8 \rightarrow \text{SNR} = 4.7 \div 1.8 = 2.6$, so this result is ND.
4. $8.3 \pm 1.3 \rightarrow \text{SNR} = 8.3 \div 1.3 = 6.4$, this result is detectable.

Solution to Example 2.5 on page 59

A calcium carbonate primary standard has a certified purity of 99.97%. A stock standard was made by adding 0.1764 g of the primary standard to a 250-mL volumetric flask. After dissolving the calcium carbonate with a few milliliters of HCl, the solution was diluted to volume with deionized water. This standard was used to produce a calibration curve according to the volumes in Table B.3.

Std #	Vol. of stock (mL)	Total vol. (mL)
0	0	25.00
1	0.100	25.00
2	0.250	25.00
3	0.500	25.00
4	1.000	25.00
5	1.500	25.00

Table B.3 A table of volumes used to obtain a set of calibration standards.

1. What is the concentration, in units of mg Ca/L, of the stock solution?
2. Calculate the concentration of each of the calibration standards.

Solution: This problem requires the use of formula weights, a solution, and dilution calculations.

1. **What is the concentration, in units of mg Ca/L, of the stock solution?** First, the purity of the calcium carbonate standard must be taken into account. A purity of 99.97% implies that out of every 100 g of the primary standard, there is 99.97 g of $CaCO_3$. Next, since the units of mg Ca/L were specified, the mass of $CaCO_3$ must be converted into a mass of Ca using a formula weight ratio (FW_{Ca} = 40.078 g/mol and FW_{CaCO_3} = 100.09 g/mol). Next, convert grams to milligrams. Finally, divide it by the total volume of the solution.

$$0.1764 \text{ g CaCO}_3 \text{ std} \left(\frac{99.97 \text{ g CaCO}_3}{100 \text{ g CaCO}_3\text{std}}\right)\left(\frac{40.078 \text{ g Ca}}{100.09 \text{ g CaCO}_3}\right)\left(\frac{1000 \text{ mg}}{1 \text{ g}}\right) \div 0.250 \text{ L} \quad \text{(B.1)}$$

$$= 282.451 = 282.5 \frac{\text{mg Ca}}{\text{L}}$$

2. **Calculate the concentration of each of the calibration standards.** The concentration of each calibration standard is found by using the dilution equation (Eq. (2.14)). The stock solution calculation C_1 = 282.56 mg Ca/L and the total volume of calibration standard V_2 = 25 mL in the third column of the table. The stock volume (V_1 found in the second column) changes for each standard. The diluted concentration (C_2) is the target variable. Here is the calculation for the first two standards.

$$C_2 = \frac{C_1 V_1}{V_2}$$

$$C_2 = \frac{\left(282.451 \frac{\text{mg Ca}}{\text{L}}\right)(0 \text{ mL})}{25.00 \text{ mL}} = 0 \frac{\text{mg Ca}}{\text{L}}$$

$$C_2 = \frac{\left(282.451 \frac{\text{mg Ca}}{\text{L}}\right)(0.100 \text{ mL})}{25.00 \text{ mL}} = 1.130 \frac{\text{mg Ca}}{\text{L}}$$

The other standards are as follows: std3 = 2.825, std4 = 5.649, std5 = 11.30, & std6 = 16.95 $\frac{\text{mg Ca}}{\text{L}}$

Solution to Example 2.9 on page 68

A set of nitrite ion calibration standards were obtained using volumetric glassware. The results in Table B.4 were obtained for the standards.

Conc. (μM)	Absorbance
2.00	0.065
6.00	0.205
10.00	0.338
14.00	0.474
18.00	0.598
25.00	0.821

Table B.4 Simulated calibration curve results from using the method of external standards.

A single unknown sample was prepared in the same way and also measured with the spectrophotometer, resulting in an absorbance of 0.658.

1. Determine the slope of the best-fit line using a linear regression.
2. Determine the y intercept of the best-fit line using a linear regression.
3. What is the concentration of nitrite ion in the unknown?
4. What is uncertainty in the unknown concentration?

Solution: If you have the external standards regression template already prepared, then answering these questions will be very simple. You may want to read Section 2.8 before tackling this problem.

1. The slope and y intercept of the regression are available in the Regression Analysis portion of the external standards regression spreadsheet after the concentrations are added to the Concentration column (A) and the absorbance values into the Signal column (B). If you have not completed this, you could determine the slope and y intercept by plotting the absorbance as the y values and concentration as the x values, then add a trendline to the graph showing the equation of best fit.
   (a) The slope is $3.28 \times 10^{-2}\ \mu M^{-1}$.
2. The y intercept is 6.67×10^{-3} (no units here because absorbance is a unitless measure).
3. To determine the concentration of the unknown sample, you must use the signal from the unknown as the y value in the equation of best fit and solve for the x value. If you are using the regression template, enter the unknown absorbance into one of the yellow signal cells in the "non-replicates" section of the spreadsheet. The concentration should be calculated.
   (a) The unknown nitrite ion concentration is 19.85 μM. This should make sense since its absorbance (0.658) is between the absorbance signals of the 18 and 25 μM standards.
4. Calculating the uncertainty is very easy once the regression template has been constructed. In order to do it without the template, you need to use Eq. (2.26) (see Section 2.8 on page 84).
   (a) The uncertainty in the unknown concentration is 0.23 μM, yielding a final unknown result of 19.85 ± 0.23 μM.

B.2 SOLUTIONS TO END-OF-CHAPTER EXERCISES

1. **Describe the difference between qualitative and quantitative measurements?** Qualitative measurements produce a binary result, such as indicating if the analyte is present or not. Quantitative measurements result in numerical concentrations of the analyte.
2. **Describe random noise, how it can be reduced or eliminated from measurements, whether it is associated with accuracy or precision, and which parameter of a Gaussian distribution is it associated with.** Random noise is characterized by random deviations from the baseline or signal, approximately half of which are above the baseline and half below the baseline. Signal averaging can reduce the noise since the positive and negative deviations tend to partially cancel out, but random noise can never be completely eliminated. Random noise is associated with precision because the deviations are above and below the baseline and do not shift the baseline or signal level to any significant degree. In a Gaussian distribution, the standard deviation is a measure of the random noise.
3. **Describe systematic errors, how they can be reduced or eliminated from measurements, whether they are associated with accuracy or precision, and which parameter of a Gaussian distribution are they associated with.** Systematic errors offset the signal or baseline positively or negatively, but never both in the same measurement. Since they offset the signal or baseline, they affect the accuracy of the measurement. Systematic errors affect the average or mean of a sample distribution. They can be eliminated in some cases.

4. **What mathematical distribution describes most measurements? Why?** A repetitive measurement is a snapshot of sample distribution, often an average of the signal over a fixed period of time. The Central Limit Theorem states that the measurement of the average of any distribution will be normally distributed, so the Gaussian function is a good mathematical model for most measurements.

5. **Under what conditions can a signal be first identified from noise?** A signal can be classified as such when its amplitude is at least three times larger than the standard deviation of the noise.

6. **How is a SNR calculated?** In a typical measurement that results in an average and standard deviation, the SNR is calculated by dividing the average by the standard deviation.

7. **Describe the Limit of Detection (LOD).** The LOD is reached when the SNR of a measurement drops down to 3. Any measurement with an SNR between 3 and 10 is qualitatively detectable, and any measurement with an SNR below 3 is considered noise.

8. **Describe the Limit of Quantitation (LOQ).** The LOQ is reached when the SNR of a measurement drops down to 10. Any measurement with an SNR above 10 is quantitatively detectable.

9. **In two steps, summarize the rule for specifying significant figures in the results of a measurement.** First, round the standard deviation or uncertainty to two significant figures. Second, round the average measurement so that it has the same precision as the uncertainty.

10. **What are the differences between a primary and secondary standard?** A primary standard is usually highly pure and has a certified concentration. A secondary standard is used when the primary standard is unstable or prohibitively expensive. The secondary standard is standardized with the primary standard and then used in its place.

11. **Why is a standard necessary in most measurements?** Most instruments measure analyte signals relative to the environmental conditions (temperature, humidity, etc.) and the instrument design. The relationship between the signal and the analyte must therefore be determined experimentally. A standard provides the link between the relative signal and the value of the parameter measured, such as concentration.

12. **Describe the differences between a population and a sample distribution. What statistical variables are used in each?** A population distribution is the set of all possible measurements of a given parameter. The population is characterized by a mean (μ) and a standard deviation (σ). A sample distribution is a subset of the population, usually characterized by a set of less than 20 measurements. The sample distribution is characterized by a mean ($\bar{x}$) and a standard deviation (s).

13. **What is the purpose of a pooled standard deviation?** When several data sets must be combined, a pooled standard deviation provides a weighted average of the standard deviations, representing the standard deviation of the combined set.

14. **What is the purpose of performing a hypothesis test?** A hypothesis test provides a nonarbitrary way to determine the statistical significance of a result. It relies on the parameters of the measurement, such as the mean and standard deviation, and not on the bias of the technician.

15. **List three different scenarios where a hypothesis test would be useful in an environmental measurement.** (1) When calibrating an instrument, the measurement result must be statistically indistinguishable from the primary standard's certified value. (2) When comparing the analyte level from two different samples, it is important to know if the measurement means are statistically the same or different. (3) When comparing two different instruments for an important measurement, the average result or noise level can be compared to see if the instruments perform the same or not.

16. **List the advantages and disadvantages of each of the following methods of quantitation.**

    (a) **the method of external standards** This method has the advantage of high sample throughput, since a single set of calibration standards can be used to quantitate many samples. One disadvantage is that if there is a matrix mismatch, then a systematic error could be present. Another disadvantage is that without an internal standard, variations in the volume of the standards and samples used will lead to a low precision.

(b) **The method of multiple standard additions** The advantage of this method is that if a matrix mismatch causes a proportional error in the measurement, the method of multiple standard additions will be able to eliminate the systematic error. The disadvantage of this method is that an entire set of standards must be obtained in order to quantitate one unknown sample.

(c) **The method of single-point internal standard** The advantage of this method is that it is very quick, requiring only the measurement of two solutions - a standard spiked with the IS and the unknown spiked with the IS. The disadvantage of this method is that it does not yield an uncertainty for the concentration of the unknown, and with only two data points, it is difficult to detect an error in the preparation of one of the samples, which would lead to a significant systematic error.

(d) **The method of multipoint internal standard** The advantage of this method carries with it the high-throughput advantages of the method of external standards plus the added advantage of the elimination of the sample-to-sample variations that would lead to a loss of precision. The disadvantage of this method is that it requires the preparation and measurement of the IS.

17. **What method is being employed if a sample or standard is spiked with a stock standard solution of the analyte?** This implies that one of the methods of multiple standard additions is being employed.

18. **What method is being employed if a sample or standard is spiked with a stock standard solution of a chemical similar to, but different from, the analyte?** This implies that an internal standard is being used in one of the methods.

19. **Why is the variable-volume method of multiple standard additions convenient to use with a micropipettor?** In this method, each additional measurement only requires an added spike, mixing the sample, and remeasurement.

20. **How does the method of multiple standard additions achieve a set of standards with the same matrix?** Each standard contains the same volume of the sample, so the dilution of the standard results in equal concentration of the interference in the matrix.

21. **Why must objects be at room temperature when placed on an electronic balance?** Objects warmer than the surroundings could heat the air inside the balance chamber and generate a buoyancy (with air convection) that would bias the result. Object colder than room temperature could also disturb air currents or condense moisture from the atmosphere, resulting in a positive bias.

22. **What physical difference between a 100 mL graduated cylinder and a 100 mL volumetric flask results in the difference in the respective accuracy of the glassware?** The very narrow neck of the volumetric flask is more sensitive to a change in volume and provides for a more defined meniscus. Both of these results make for a more accurate measurement.

23. **List the advantages and disadvantages of glass pipettes compared to micropipettors?** Glass volumetric pipettes are very useful for large volumes over 1 mL. They must be cleaned and rinsed for every sample. They are usually calibrated at 20 °C. Micropipettors are very useful for volume of 5 mL and less. The disposable tips make switching between samples easy. They require regular calibration and must be disassembled in order to be cleaned.

24. **If an analyte had a strong absorbance at 240 nm, which cuvette materials would be most appropriate?** Only quartz or special UV plastic can be used this far into the UV. Other materials absorb this radiation, making the measurement impossible.

25. **If glass cuvettes are useful only in the visible radiation range and are more expensive than the various plastic cuvettes, why would anyone ever use glass?** Glass, while limited in its spectral range, can be used with a wide range of organic solvents, unlike the plastic cuvettes.

26. **Why is it useful to go through all of the trouble of a full regression analysis of unknown and calibration standards?** A regression analysis provides not only the concentration of the unknown but also its level of uncertainty, which is essential for determining the detectability of the analyte. Without the uncertainty and the SNR, the reliability of the measurement is unknown.

C

CHAPTER 3 EXAMPLES AND END-OF-CHAPTER EXERCISES

C.1 SOLUTIONS TO IN-CHAPTER EXAMPLES

Solution to Example 3.3 on page 101
Calculate the standard enthalpy of Reaction R3.18, and then determine the wavelength of the photon with the minimal energy to break this bond.

Using the standard enthalpies of the species involved and the equation to calculate enthalpy (Eq. (3.1)), the enthalpy of this reaction is given as follows.

$$\begin{aligned}\Delta H_{rxn} &= \Sigma\Delta H^{\circ}_{f\ products} - \Sigma\Delta H^{\circ}_{f\ reactants}\\ &= (249.18 + 249.18) - (0)\ \text{kJ/mol}\\ &= 498.36\ \text{kJ}\end{aligned}$$

This enthalpy needs to be converted to energy per molecule in order to derive the energy of a single photon.

$$(498.36\ \text{kJ/mol})\left(\frac{1\ \text{mol}}{6.02214129 \times 10^{23}\ \text{mcs}}\right) = 8.27546 \times 10^{-22}\ \text{kJ/mcs}$$

Now apply the photon energy equation.

$$\begin{aligned}\lambda_{max} &= \frac{hc}{E}\\ &= \frac{(6.62606957 \times 10^{-34}\ \text{J}\cdot\text{s})(2.99792458 \times 10^{8}\ \text{m/s})}{(8.27546 \times 10^{-22}\ \text{kJ/mcs})\left(\frac{1000\ \text{J}}{1\ \text{kJ}}\right)\left(\frac{1\ \text{mcs}}{1\ \text{photon}}\right)}\left(\frac{10^{9}\ \text{nm}}{1\ \text{m}}\right)\\ &= 240.04\ \text{nm}\end{aligned}$$

We know from the previous examples that using thermodynamic data gives us a low estimate of the photon energy since it does not consider the actual activation energy. Consequently, we can assume that UV radiation with a $\lambda < 240$ nm would be responsible for initiating this reaction.

Solution to Example 3.4 on page 102

What are the rate laws for Reactions R3.18 and R3.21?

Solution: In Chapter 1, you reviewed chemical kinetics and rate laws. Writing the rate law typically involves the rate constant for a reaction and the concentration of the reactants. One uncertainty in

Environmental Chemistry: An Analytical Approach, First Edition. Kenneth S. Overway.

Companion website: www.wiley.com/go/overway/environmental_chemistry

writing rate laws is knowing whether the reaction represents a step in a mechanism or an overall reaction. In rate laws for overall reactions, each reactant gets raised to an unknown exponent because it is unknown what the actual coefficients are for each reactant in each step (there may be some reactions that add together and cancel species). In mechanistic steps, the coefficients are known, so each reactant gets raised to the power of its reaction coefficient.

$$Rate_1 = k_1[O_2]^1 \tag{C.1}$$

Reaction R3.21 is an overall reaction, so its rate law is indeterminate.

$$Rate_{overall} = k_2[O]^x[O_2]^y[M]^z \tag{C.2}$$

We must drill down into the two steps in the mechanism to obtain more concrete rate laws. In Reaction R3.19, there is a fast equilibrium, implying that the forward (f) rate and the reverse (r) rate are equal.

$$Rate_{2a_f} = k_f[O]^1[O_2]^1 = Rate_{2a_r} = k_r[O_3^*]^1 \tag{C.3}$$

Then for Reaction R3.20, the rate law is

$$Rate_{2b} = k_{2b}[O_3^*]^1[M]^1 \tag{C.4}$$

The rate law for Reaction R3.20 contains a reaction intermediate (O_3^*) and can be simplified by replacing its concentration with the rate law from Reaction R3.19. We simply need to solve for $[O_3^*]$.

$$[O_3^*] = \frac{k_f}{k_r}[O][O_2] \tag{RC.1}$$

Now substitute $[O_3^*]$ into the rate law for the second step.

$$\begin{aligned} Rate_{2b} &= k_{2b}\left(\frac{k_f}{k_r}[O][O_2]\right)[M] \\ &= k_2[O][O_2][M] \end{aligned} \tag{C.5}$$

The rate constant k_2 represents a combination of k_{2b}, k_f, and k_r. Since the constants are not important individually, they can just be combined. The rate law for Reaction R3.21 turns out to be the same as if we were to assume that the reaction was a mechanistic step with $x = 1$ and $y = 1$. This is because the only the intermediate was canceled in the combination of the reactions.

Solution to Example 3.5 on page 102

Estimate the wavelength of the photons necessary in order to photolyze ozone according to Reaction R3.22 at standard temperature assuming all combinations of ground- and excited-state atomic oxygen and molecular oxygen. Once again we need to estimate photon wavelength by using thermodynamic data, knowing full well that this will be an underestimate of the actual photon energy (thermodynamic ΔH vs. E_a).

$$\underline{O_3 + h\nu \xrightarrow{k_3} O^* + O_2^*}$$

$$\begin{aligned} \Delta H_{rxn} &= \Sigma\Delta H^\circ_{f\,products} - \Sigma\Delta H^\circ_{f\,reactants} \\ &= (438.05 + 94.29) - (142.7)\ \text{kJ/mol} \\ &= 389.64\ \text{kJ} \end{aligned}$$

$$\begin{aligned} \lambda_{max} &= \frac{hc}{E} \\ &= \frac{(6.62606957 \times 10^{-34}\ \text{J}\cdot\text{s})(2.99792458 \times 10^8\ \text{m/s})}{(389.64\ \text{kJ/mol})\left(\frac{1000\ \text{J}}{1\ \text{kJ}}\right)\left(\frac{1\ \text{mol}}{6.02214129\times10^{23}\ \text{mcs}}\right)}\left(\frac{10^9\ \text{nm}}{1\ \text{m}}\right) \\ &= 307.0\ \text{nm} \end{aligned}$$

$$O_3 + h\nu \xrightarrow{k_3} O^* + O_2^{**}$$

$$\begin{aligned}\Delta H_{rxn} &= \Sigma\Delta H^\circ_{f\,products} - \Sigma\Delta H^\circ_{f\,reactants}\\ &= (438.05 + 156.96) - (142.7)\ \text{kJ/mol}\\ &= 452.31\ \text{kJ}\end{aligned}$$

$$\begin{aligned}\lambda_{max} &= \frac{hc}{E}\\ &= \frac{(6.62606957 \times 10^{-34}\ \text{J}\cdot\text{s})(2.99792458 \times 10^{8}\ \text{m/s})}{(452.31\ \text{kJ/mol})\left(\frac{1000\ \text{J}}{1\ \text{kJ}}\right)\left(\frac{1\ \text{mol}}{6.02214129\times 10^{23}\ \text{mcs}}\right)}\left(\frac{10^9\ \text{nm}}{1\ \text{m}}\right)\\ &= 264.5\ \text{nm}\end{aligned}$$

The other combinations result in the following:

$O_3 + h\nu \xrightarrow{k_3} O + O_2$: $\Delta H_{rxn} = 106.5$ kJ and $\lambda_{max} = 1123$ nm

$O_3 + h\nu \xrightarrow{k_3} O + O_2^{**}$: $\Delta H_{rxn} = 263.4$ kJ and $\lambda_{max} = 454.1$ nm

$O_3 + h\nu \xrightarrow{k_3} O + O_2^{*}$: $\Delta H_{rxn} = 200.8$ kJ and $\lambda_{max} = 595.8$ nm

$O_3 + h\nu \xrightarrow{k_3} O^* + O_2$: $\Delta H_{rxn} = 295.4$ kJ and $\lambda_{max} = 405.0$ nm

Figure 3.19 on page 121 shows the absorption spectrum of ozone with these reactions placed in the proper context.

Solution to Example 3.9 on page 107

If an extrasolar planet at an orbit of 234×10⁶ km from HD 4308 has an albedo of 0.44 and a planetary radius of 5931 km, what is the expected planetary temperature (ignoring any greenhouse effect)?

- Distance from the Earth: 72 light years
- Solar mass: 0.833 × the mass of our Sun
- Radius: 0.916 × the radius of our Sun
- Surface temperature: 5597 K
- Age: 7.072 Gyr
- Our Sun's radius is 6.96342×10^5 km.
- Our Sun's mass = 1.9891×10^{30} kg.

Solution: In order to answer this question, we need to calculate the energy given off by the HD 4308 star and the solar constant it would impart to the planet of interest. We know from Eq. (3.6) how to calculate the total energy of the output of the star, requiring the surface temperature of the star. We also know that Eq. (3.7) allows us to calculate the solar constant using the results of Eq. (3.6), the radius of the star, and the orbit radius. Combining these two equations, we get the solar constant.

$$\begin{aligned}E_{sc} &= (E_{star})\left(\frac{r^2_{star}}{r^2_{orbit}}\right)\\ &= (\sigma_{SB}T^4)\left(\frac{r^2_{star}}{r^2_{orbit}}\right)\\ &= ((5.67053 \times 10^{-8}\ \text{W/(m}^2\ \text{K}^4))(5597\ \text{K})^4)\left(\frac{[(0.916)(6.96342 \times 10^5\ \text{km})]^2}{(234 \times 10^6\ \text{km})^2}\right)\\ &= 413.48\ \text{W/m}^2\end{aligned}$$

Next, we need to account for the size of the planet and its albedo. Equation (3.8) shows us that the radius of the planet does not matter because the input energy is given on a per m^2 basis but that the curvature of the planet does matter – it reduces the solar constant by a factor of 4. Finally, we can reuse Planck's Law, as provided by Eqs (3.9) and (3.10).

$$\begin{aligned} T &= \left[\left(\frac{E_{sc}}{4}\right)(1 - \text{albedo})\left(\frac{1}{\sigma_{SB}}\right)\right]^{\frac{1}{4}} \\ &= \left[\left(\frac{413.48\ \text{W/m}^2}{4}\right)(1 - 0.44)\left(\frac{1}{5.67054 \times 10^{-8}\ \text{W/(m}^2\ \text{K}^4)}\right)\right]^{\frac{1}{4}} \\ &= 178.7 = 180\ \text{K} \end{aligned} \tag{C.6}$$

In summary, in order to determine the expected temperature of a planet you need to know (1) the surface temperature of the star, (2) the radius of the star, (3) the orbit distance of the planet, (4) the albedo of the planet, and access to various physical constants.

Solution to Example 3.10 on page 109

Which of the following molecules would you expect to have IR-active vibrational modes and thus be categorized as a greenhouse gas? Draw each molecule using the Lewis structure methodology, and then imagine the molecules bending and vibrating.

$$SO_3, H_2, Ar, CH_4.$$

Solution:

1. SO_3: Let us draw the Lewis structure for this gas to see how it can vibrate.

electron accounting system

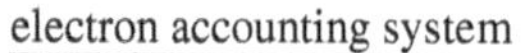

(h): S → 6 plus 3 × O → 6 = 24 e^-
(n): Four atoms (in the p-block) × 8 = 32 e^-
(s): s = 32 − 24 = 8 e^-
(b): b = $\frac{1}{2}$ × 8 = 4 bonds

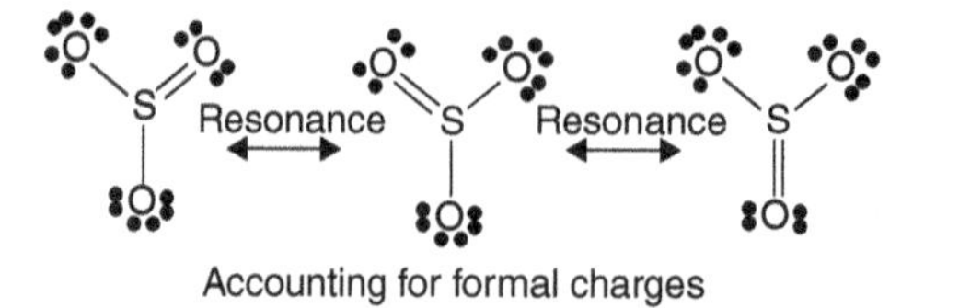

$FC_S = 6 - 0 - 4 = +2$
$FC_{O_{\text{single bond}}} = 6 - 6 - 41 = -1$
$FC_{O_{\text{double bond}}} = 6 - 4 - 2 = 0$
$\Delta FC = (+2) - (-1) = 3$

No net dipole

$FC_S = 6 - 0 - 46 = 0$
$FC_{O_{\text{double bond}}} = 6 - 4 - 2 = 0$
$\Delta FC = (0) - (0) = 0$

The structure with three double bonds is the most stable.

Sulfur trioxide appears to be a trigonal planar molecule that is nonpolar. Its symmetrical stretch vibration would maintain the equal and opposite bond dipole and would be IR-inactive. You can hopefully imagine SO_3 vibrating such that the sulfur moves out of the plane of the oxygen atoms in a flapping mode, causing an oscillating dipole moment (see the figure). This would result in an IR-active band because the three bond dipoles would not be equal and opposite. Sulfur trioxide has a total of six vibrational modes, but you do not need to know all of them to label it a greenhouse gas. It has one IR-active vibrational mode, and that is enough information.

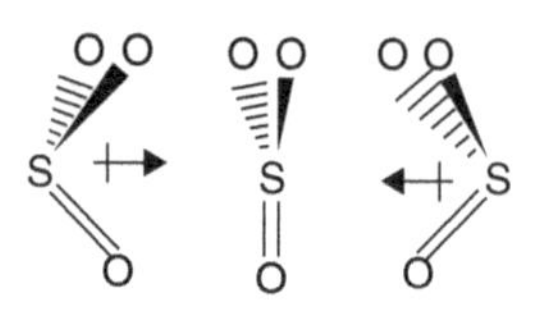

2. H_2: Molecular hydrogen is similar to molecular oxygen – no IR-active vibrational modes. It is not a greenhouse gas.
3. Ar: Argon has no bond through which it can vibrate. It is not a greenhouse gas.
4. CH_4: Methane has a tetrahedral shape similar to CCl_2F_2, except that it has four identical substituents on the central carbon. There is no permanent dipole. With a little imagination, you could have predicted that some asymmetric stretching or bending would have resulted in an oscillating dipole moment. Sometimes it does get difficult to imagine when there are several atoms. The IR spectrum in Figure C.1 shows two IR-active vibrational modes.

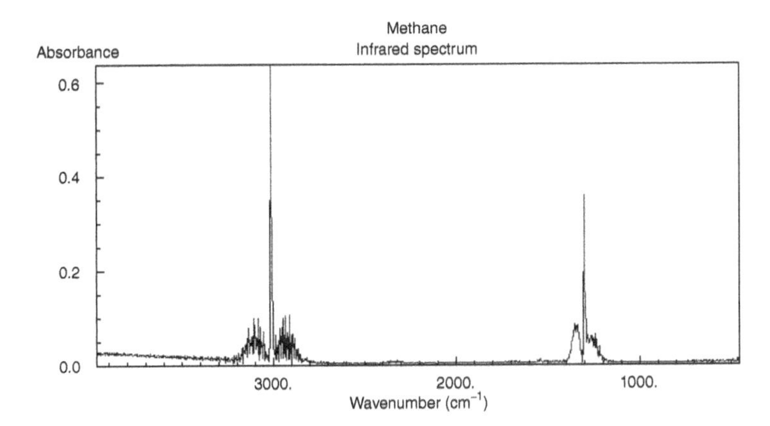

Figure C.1 The IR spectrum of methane.

Solution to Example 3.18 on page 126
Calculate the lifetime of propene against oxidation from the hydroxyl radical at 10 °C, assuming $[OH] = 1.5 \times 10^6 \frac{mcs}{cm^3}$.

Solution: The reaction in question can be abbreviated by the following.

$$C_3H_6 + OH \rightarrow \text{products}$$

The rate law is given by $Rate = k[C_3H_6][OH]$. Using Eq. (3.3) and Table I.9 on page 327, we get the following.

$$\tau = \frac{1}{\left(2.88 \times 10^{-11} \frac{cm^3}{mcs \cdot s}\right)\left(1.5 \times 10^6 \frac{mcs}{cm^3}\right)} = 2.3148 \times 10^4 \text{ s}$$

If this τ is converted to hours, then it is $\tau = 6.4$ h.

Solution to Example 3.23 on page 139
Draw the Lewis structure for nitrous oxide and determine the formal charge of each atom. Can you rationalize why half of the time Reaction R3.94 occurs and Reaction R3.95 occurs the other half of the time?

Solution: First, set up the electron accounting scheme, then draw the structure, and then use formal charges to determine the most stable form.

Four bonds spread over three atoms produce a few possibilities.

Step 1
$(h) = N + N + O = 5 + 5 + 6 = 16$
$(n) = 3 \times 8 = 24$
$(s) = 24 - 16 = 8$
$(b) = 8 \div 2 = 4$

:N̈=N=Ö:	:N≡N–Ö:	:N̈–N≡O:
$FC_{N1} = 5 - 4 - 2 = -1$	$FC_{N1} = 5 - 2 - 3 = 0$	$FC_{N1} = 5 - 6 - 1 = -2$
$FC_{N2} = 5 - 0 - 4 = +1$	$FC_{N2} = 5 - 0 - 4 = +1$	$FC_{N2} = 5 - 0 - 4 = +1$
$FC_O = 6 - 4 - 2 = 0$	$FC_O = 6 - 6 - 1 = -1$	$FC_O = 6 - 2 - 3 = +1$
$\Delta FC = +1 - (-1) = 2$	$\Delta FC = +1 - (-1) = 2$	$\Delta FC = +1 - (-2) = 3$

The first two structures have the lowest ΔFC, so the third structure can be ruled out. The real structure is something between the first and second, but closer to the second structure because O is more electronegative than N and the second structure gives the negative charge to O.

When atomic oxygen collides with N_2O, approximately half of the time it will collide with the right side of N_2O where it will hit the oxygen. This will result in a bond between the two oxygen atoms. When the N–O bond breaks, O_2 and N_2 will result. The other half of the time the oxygen atom will strike the left side of N_2O where it will hit a nitrogen atom. This will form an N–O bond and will cause the N=N bond to break, resulting in two NO molecules.

C.2 SOLUTIONS TO END-OF-CHAPTER EXERCISES

1. **Why does the thermosphere have the temperature trend that it has? What range of electromagnetic radiation and what gases are responsible for this trend?** The thermosphere is hot at the top and cold at the bottom as a result of absorbing the highest energy radiation coming from the Sun, roughly less than 100 nm in wavelength. Less of this UV light is available for absorption as altitude decreases, so the temperature drops.

2. **Why does the mesosphere have the temperature trend that it has? What range of electromagnetic radiation and what gases are responsible for this trend?** The mesosphere is cold at the top and warm at the bottom as a result of the absorption of the 100–230 nm radiation as the gas pressure increases with decreasing altitude.

3. **Why does the stratosphere have the temperature trend that it has? What range of electromagnetic radiation and what gases are responsible for this trend?** The stratosphere is warm at the top and cold at the bottom because most of the 220–300 nm radiation is absorbed by the ozone in the middle and top of the layer.

4. **Why does the troposphere have the temperature trend that it has? What range of electromagnetic radiation and what gases are responsible for this trend?** The troposphere is cold at the top and warm at the bottom because the UV and visible radiation that is not absorbed by the atmosphere gets absorbed by the surface (except what is reflected). The surface becomes warmer as a result, which warms the troposphere at the bottom.

5. **Describe the role that a chaperone molecules plays in atmospheric reactions. Given an example.** Chaperone molecules collide with excited atoms and molecules, and in the process, some of the vibrational energy is transferred. The result is a more stable atom or molecule. A specific example is the second step in the Chapman cycle.

$$O_3^* + M \longrightarrow O_3 + M \qquad \text{(RC.2)}$$

6. **Summarize the key reason why ozone persists at significant levels in the stratosphere and not in the mesosphere or thermosphere.** A kinetic analysis of the Chapman cycle shows that the concentration of gases above the stratosphere is so small that the lifetime of the oxygen atom, produced in the first step of the Chapman cycle, is very high because the probability of a collision is small. This results in an improbable formation of ozone at higher altitudes.

7. **Describe, in your own words, what a species lifetime (represented by τ) says about a species or related reaction.** The lifetime of a species is the time that the species persists until it collides and reacts with the other reactants.

8. **Write the four reactions of the Chapman Cycle.**

$$O_2 + h\nu \xrightarrow{k_1} O + O$$
$$O + O_2 + M \xrightarrow{k_2} O_3 + M$$
$$O_3 + h\nu \xrightarrow{k_3} O^* + O_2^*$$
$$O + O_3 \xrightarrow{k_4} 2O_2$$

9. **Describe the reactions in the Chapman Cycle. What are the two main species that are formed and destroyed?** The first step of the Chapman is the photolysis of molecular oxygen, producing atomic oxygen. The atomic oxygen collides with molecular oxygen (and a chaperone) in the second step to form ozone. The third step is the photolysis of ozone, and the fourth step is the collision of atomic oxygen with ozone to reform molecular oxygen. The two species that interconvert in this cycle are molecular oxygen and ozone.

10. **Describe the usefulness and limitation of using the enthalpy (ΔH) of an endothermic reaction to estimate the photon energy necessary to initiate a reaction.** The enthalpy calculation in a photolysis gives a minimal estimate of the energy required to break molecular bonds. The actual energy is often higher because of the existence of the activation energy required to complete the reaction.

11. **Derive the lifetime formulas for the following reactions.**

    (a) the lifetime of NO in $NO + O_3 \xrightarrow{k} NO_2 + O_2$

$$\tau_{NO} = \frac{[NO]}{Rate} = \frac{1}{k[O_3]}$$

    (b) the lifetime of O_3 in $NO_2 + O_3 \xrightarrow{k} NO_3 + O_2$

$$\tau_{O_3} = \frac{[O_3]}{Rate} = \frac{1}{k[NO_2]}$$

    (c) the lifetime of NO_2 in $NO_2 + OH + M \xrightarrow{k} HNO_3 + M$

$$\tau_{NO_2} = \frac{[NO_2]}{Rate} = \frac{1}{k[OH][M]}$$

12. **Calculate the lifetime of atomic oxygen in Reaction R3.21 at an altitude of 90 km. Refer to the values in Table 3.6.** This problem is similar to Example 3.6 on page 103.

$$\tau_O = \frac{1}{k_2[O_2][M]}$$
$$= \frac{1}{\left(1.84 \times 10^{-33} \frac{cm^6}{mcs^2\,sec}\right)\left(6.63 \times 10^{11} \frac{mcs}{cm^3}\right)\left(3.16 \times 10^{12} \frac{mcs}{cm^3}\right)}$$
$$= 2.59 \times 10^8 \text{ s or } 8.22 \text{ years}$$

13. **Calculate the lifetime of formaldehyde against OH oxidation at 15 °C assuming that the concentration of OH is 3.67×10^6 $\frac{mcs}{cm^3}$. Convert the lifetime to days or years.** Given the temperature, the correct kinetic rate constant for formaldehyde must be obtained from Table I.9 on page 327. This rate constant and the concentration of OH can be used in the lifetime equation for formaldehyde (H_2CO).

$$H_2CO + OH \xrightarrow{k} \text{products}$$
$$Rate = k[H_2CO][OH]$$
$$\tau_{H_2CO} = \frac{[H_2CO]}{Rate} = \frac{1}{k[OH]} = \frac{1}{\left(9.36 \times 10^{-12} \frac{cm^3}{mcs \cdot s}\right)\left(3.67 \times 10^6 \frac{mcs}{cm^3}\right)}$$
$$\tau_{H_2CO} = 2.91 \times 10^3 \text{ s} = 8.09 \text{ hours}$$

14. **What does the albedo describe?** The albedo of a planet or landscape is a measure of its reflectivity. An albedo of 0.85 means that 85% of the solar radiation reaching this surface is reflected back into space.

15. **If the Earth were covered entirely by deciduous trees (and no clouds), would the albedo increase or decrease? Would this lead to an increase or a decrease in the Earth's surface temperature?** The trees would absorb more light compared to snow or ice during the growing season, which would cause the albedo to decrease. This would result in the retention of more heat and an increase in the surface temperature, assuming that all other factors are constant.

16. **If the Earth were covered entirely by snow (and no clouds), would the albedo increase or decrease? Would this lead to an increase or decrease in the Earth's surface temperature? Calculate the expected temperature of the Earth using the albedo of snow from Table 3.7.** Snow would reflect light more than most other surface conditions, increasing the albedo and lowering the surface temperature. Given the Earth's solar constant, the temperature of a snow-covered Earth can be found.

$$T = \left[\left(\frac{1360\ \frac{\text{W}}{\text{m}^2}}{4}\right)(1-0.85)\left(\frac{1}{5.67053\times10^{-8}\ \frac{\text{W}}{(\text{m}^2\cdot\text{K}^4)}}\right)\right]^{\frac{1}{4}} \tag{C.7}$$

This results in a surface temperature of 173 K.

17. **Ignoring any greenhouse effect, calculate the maximum distance from our Sun that a planet can orbit while barely maintaining a global surface layer of snow. Compare this distance to the orbits of the inner planets, and draw a conclusion about the necessity of a greenhouse effect.** Assuming that the maximum temperature at sea level for snow to exist is 273 K, the orbit can be calculated with the albedo for snow.

$$r_{orbit} = \sqrt{\frac{(1-albedo)(E_{\text{Sun}})(r^2_{\text{Sun}})}{(\sigma_{SB})(T^4)(4)}}$$
$$= \sqrt{\frac{(1-0.85)(6.3203\times10^7\ \text{W/m}^2)((6.96342\times10^5\ \text{km})^2)}{(5.67053\times10^{-8}\ \text{W/(m}^2\cdot\text{K}^4))((273\ \text{K})^4)(4)}} = 6.04\times10^7\ \text{km}$$

Without any greenhouse effect, about 60 million km is the maximum distance. Any orbit smaller than that would not allow an ice ball planet. This is about the orbit of Mercury, so even the planet closest to the Sun could be an ice ball without an atmosphere with a greenhouse effect if it had the albedo of snow.

18. **Would the solar constant (E_{sc}) be larger or smaller if the Earth's orbit was closer to the Sun? If the orbit decreased by a factor of 2, how would the solar constant change?** If the Earth's orbit decreased, then E_{sc} would increase. If it were halved, then E_{sc} would increase by a factor of 4 according to Eq. (3.7).

19. **If our Sun were replaced by a small, white dwarf star (smaller diameter, all else being the same) and the Earth had the same orbit, how would it affect the climate on the Earth?** A smaller diameter star would yield a smaller E_{sc}, so the Earth's climate would be much colder.

20. **Why do two-stroke and four-stroke gasoline engines differ on the (a) HC and (b) NO_x emissions?**

    (a) HC: Two-stroke engines use the new fuel:air mixture to flush out the exhaust from a previous cycle and mix a high-molecular-weight lubricant in the gasoline, both of which lead to much higher HC emissions compared to four-stroke engines.

    (b) NO_x: Four-stroke gasoline and diesel engines run hotter than two-stroke engines and therefore produce more NO_x emissions.

21. **Why do four-stroke gasoline and diesel engines differ on the (a) PM and (b) NO_x emissions?**

    (a) PM: Diesel engines require a higher molecular weight fuel (mostly cetane) compared to gasoline engines, which is less flammable and volatile, and diesel engines inject the fuel into the cylinder at the very last moment before combustion. This leads to poor dispersal and mixing of the fuel with the air and leads to soot formation.

(b) NO_x: Diesel engines run at very high compression ratios and hotter temperatures compared to gasoline engines, which leads to higher NO_x emissions.

22. **Why do fuel additives, such as ethanol and MTBE, and biodiesel reduce CO and PM emissions?** Petroleum-based fuels, such as gasoline and diesel fuel, are entirely hydrocarbons and contain no oxidizing atoms within the fuel's chemical structure. Additives such as MTBE, ethanol, and biodiesel contain oxygen atoms within their chemical structures. These oxygen atoms are in an oxidized form and can become oxidizing agents within the fuel during combustion, leading to a more complete combustion of the fuel and less production of CO.

23. **How could you design a car with an internal combustion engine, without worrying about expense, so that it would have zero NO_x emissions?** One option is to substitute pure oxygen for the air that is mixed with the fuel by using a tank of oxygen. If ambient air is avoided, then no nitrogen gets in the combustion chamber.

24. **Methane and nitrous oxide emissions from anthropogenic sources are on the increase. Explain how each affects ozone chemistry and the greenhouse effect.** Tropospheric ozone is generated during daylight hours by nitrogen oxides and hydrocarbons, so more N_2O and CH_4 lead to increased levels of ground-level ozone. Any N_2O and CH_4 left would absorb IR radiation from the Earth's surface and contribute to greenhouse warming.

25. **Describe the anthropogenic pollutants that ozone destruction and the enhanced greenhouse effect have in common. Describe how the chemicals act in each context.** CFCs and HCFCs are anthropogenic compounds that are troublesome gases in each environmental issue. In the troposphere, they absorb in an open spectral window of the IR spectrum and contribute to greenhouse warming. When CFCs escape into the stratosphere, they are photolyzed by UV radiation producing catalytic chlorine atoms that contribute to the destruction of stratospheric ozone.

26. **Knowing that the following reaction occurs**

$$H_2O + N_2O_5 \rightarrow 2\ HNO_3$$

draw the Lewis structure for N_2O_5 next to a water molecule in such a way that you can clearly circle the two HNO_3 molecules that result from the collision. HINT: there is a bridging oxygen atom.

- the number of valence electrons the structure has (h): $h = 5 \times 2 + 6 \times 5 = 40$
- the number of valance electrons the structure needs (n): $n = 8 \times 7 = 56$
- the number of electrons that must be shared (s): $s = 56 - 40 = 16$
- the minimum number of bonds in the structure (b): $b = 16 \div 2 = 8$

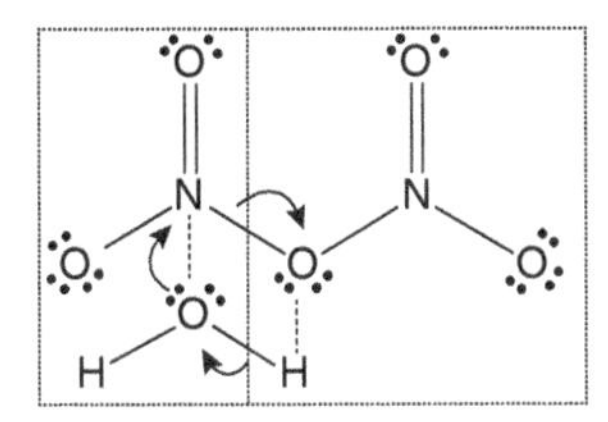

Figure C.2 The formation of nitric acid from N_2O_5 and H_2O.

Figure C.2 shows the Lewis structures of N_2O_5 and water. The oxygen of water attacks one of the nitrogen atoms and causes a rearrangement in the bonds to form two molecules of HNO_3.

27. **Knowing that the following reaction occurs**

$$N_2O_4 \rightleftharpoons 2\ NO_2$$

draw the Lewis structure for N_2O_4.

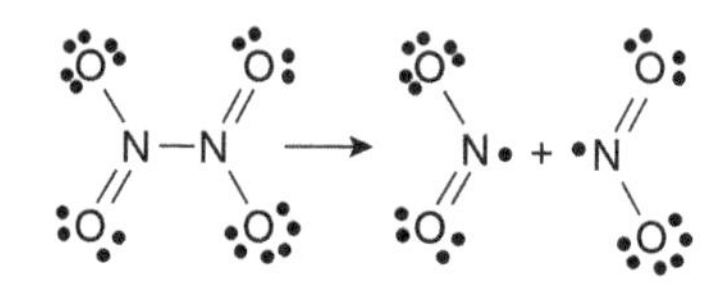

Figure C.3 The decomposition of N_2O_4.

Figure C.3 shows the Lewis structures of N_2O_4 and NO_2 as N_2O_4 decomposes.

28. **Which biogenic species is the least stable against oxidation by the hydroxyl radical at 20 °C – isoprene or α-pinene?** According to Table I.9 on page 327, the OH rate constant for isoprene is $1.03 \times 10^{-10}\ \frac{cm^3}{mcs \cdot s}$, whereas the rate constant for α-pinene is $5.50 \times 10^{-11}\ \frac{cm^3}{mcs \cdot s}$. This would predict that isoprene would have a shorter lifetime against OH attack and therefore would be less stable.

29. **Describe the conditions that can overcome kinetic barriers and drive a species to its thermodynamic fate.** Kinetic barriers are overcome by the addition of kinetic and electronic energy to reacting molecules. Thus, the presence of heat and/or light would increase the reactivity of species. The addition of a catalysis would reduce the activation energy and achieve a similar result.

30. **Propose a mechanism for Reaction R3.33 that uses NO_3 as an intermediate.**

$$NO + O_2 \rightarrow NO_3 \qquad \text{(RC.3)}$$

$$NO_3 \rightarrow NO_2 + O \qquad \text{(RC.4)}$$

$$O + NO \rightarrow NO_2 \qquad \text{(RC.5)}$$

31. **While nitric acid is highly soluble in water, in the gas phase, it can undergo the following reaction.**

$$HNO_3 \rightarrow H + NO_3 \qquad \text{(RC.6)}$$

Is this reaction likely to occur during the day or night? Give a reason for your answer and explain what reactant is missing from the reaction. The free energy for this reaction is +392.7 kJ (enthalpy is +425.6 kJ), which means that the reaction must occur during the day when solar radiation is available. The reactant that is missing from the aforementioned reaction is a photon, with a wavelength no longer than 280 nm.

32. **Describe the difference between photochemical smog and classical smog. Where can they be found, what are the constituents, and what human activities can cause each?**

    (a) Photochemical smog is produced from the primary pollutants of hydrocarbons or volatile organic compounds, NO_x, and carbon monoxide. This mixture produces ground-level ozone, aldehydes, peroxyacetyl nitrates (PANs), nitric acid, and CO_2. It is associated with the combustion of petroleum fuels.

    (b) Classical smog consists of SO_x emissions, soot, and smoke. The sulfur compounds eventually produce sulfuric acid. It is often associated with the combustion of wood and coal.

33. **What is the difference between O and O^* or O_2, O_2^*, and O_2^{**}? What can happen to these species?** The starred species are in excited states. These species can emit radiation when they fall back to their ground electronic states, as in the case of night glow, or the extra energy can lead to broken bonds and molecular decomposition, as in the second step of the Chapman cycle.

34. **Which photolysis reaction will require a photon with the smallest wavelength? Explain.**

    (a) $O_3 + h\nu \rightarrow O_2^{**} + O$

(b) $O_3 + h\nu \rightarrow O_2 + O$

By calculating the enthalpy of each reaction, the answer becomes apparent.
(a) $\Delta H_{rxn} = 156.96 + 249.18 - 142.7 = 263.44$ kJ
(b) $\Delta H_{rxn} = 0 + 249.18 - 142.7 = 106.48$ kJ
The reaction in (a) requires more energy to complete, so the energy of the photon would have to be higher and the wavelength shorter than the photon needed for the reaction in (b). This energy difference is equal to the energy difference between O_2^{**} and O_2.

35. **Describe a photolytic reaction. How is it different than a radical reaction?** A photolytic reaction requires the absorption of a photon that leads to broken chemical bond and a chemical reaction. This could result in the formation of a radical, but not necessarily. A radical reaction may or may not involve light, but it is driven by one or more reactants that have unpaired electrons.

36. **How can radical reactions degrade stable species in the absence of light, whereas photolytic degradation of the same stable species requires light?** Radicals are high-energy and highly reactive species. Whereas photons provide the activation energy in photolytic reactions, radicals have enough inherent energy that they can overcome kinetic barriers.

37. **Write the radical reaction mechanism between butane and the hydroxyl radical, which yields carbon dioxide (stop after the first CO_2 formed). How many molecules of ozone would be formed if all of the carbons in butane were to be converted to CO_2?**

$$CH_3CH_2CH_2CH_3 + \dot{O}H \rightarrow CH_3CH_2CH_2\dot{C}H_2 + H_2O$$
$$CH_3CH_2CH_2\dot{C}H_2 + O_2 \rightarrow CH_3CH_2CH_2CH_2O\dot{O}$$
$$CH_3CH_2CH_2CH_2O\dot{O} + \dot{N}O \rightarrow CH_3CH_2CH_2CH_2\dot{O} + \dot{N}O_2$$
$$CH_3CH_2CH_2CH_2\dot{O} + O_2 \rightarrow CH_3CH_2CH_2COH + HO\dot{O}$$
$$CH_3CH_2CH_2COH + \dot{O}H \rightarrow CH_3CH_2CH_2\dot{C}O + H_2O$$
$$CH_3CH_2CH_2\dot{C}O + O_2 \rightarrow CH_3CH_2CH_2COO\dot{O}$$
$$CH_3CH_2CH_2COO\dot{O} + \dot{N}O \rightarrow CH_3CH_2CH_2CO\dot{O} + \dot{N}O_2$$
$$CH_3CH_2CH_2CO\dot{O} \rightarrow CH_3CH_2\dot{C}H_2 + CO_2$$

In this first set of eight reactions, two NO_2 and one $HO\dot{O}$ are formed and will eventually react with O_2 to form three molecules of O_3. This would occur for each carbon atom in butane, resulting in a total yield of 12 molecules of ozone.

38. **What is the difference between a biogenic, an anthropogenic, and a geologic source?** A biogenic source is one that is produced by living organisms. An anthropogenic source is one that results from human activity. A geologic source is one that comes from inorganic activity of the Earth's crust and mantle.

39. **Name one anthropogenic, one geologic, and one biogenic source for sulfur emissions and the specific form of sulfur that is the primary pollutant.**

(a) Anthropogenic source: Combustion of coal or smelting – both result in SO_2 emissions
(b) Geologic: Volcanoes emit SOx
(c) Biogenic: Swamp gas contains H_2S and phytoplankton release $(CH_3)_2S$ (DMS)

40. **Name one anthropogenic and one biogenic source for nitrogen emissions and the specific form of nitrogen that is the primary pollutant.**

(a) Anthropogenic: Combustion of any fuel produces NO; agricultural activities use significant amounts of NH_3 and NO_3^- fertilizers; microbial degradation of the fertilizer releases N_2O
(b) Biogenic: Decay of protein (NH_3).

41. **Name one anthropogenic and one biogenic source for HC or VOC and the specific chemical formula of the primary pollutant.**

    (a) Anthropogenic: All internal combustion engines emit unburned fuel (C_8H_{18}, for example); natural gas drilling rigs emit HCs, such as CH_4 and C_2H_6

    (b) Biogenic: Vegetation emit terpenes, such as isoprene

42. **What is the pH of a solution that contains 1.43×10^{-5} M HNO_3?** Since nitric acid is a strong acid, it completely dissociates to produce $[H_3O^+] = 1.43 \times 10^{-5}$ M.

$$pH = -\log[H_3O^+] = -\log(1.43 \times 10^{-5}) = 4.845 \qquad (C.8)$$

43. **If a sample of acid rain had a pH of 4.81, what is the hydronium ion concentration?**

$$[H_3O^+] = 10^{-pH} = 10^{-4.81} = 1.5 \times 10^{-5}\ M \qquad (C.9)$$

44. **Determine the pH of natural rain when the average concentration of CO_2 reaches 550 ppmv or 5.50×10^{-4} atm, the level expected in the year 2050 if the industrialized countries follow the "business as usual" model and do not enact a carbon tax or any CO_2 sequestration initiative.** This calculation would be identical to the one used in Example 3.21 on page 134, except that the partial pressure of CO_2 is different.

$$x = [H^+] = \sqrt{\left(3.5 \times 10^{-2}\ \frac{M}{atm}\right)(4.5 \times 10^{-7})(5.50 \times 10^{-4}\ atm)}$$
$$= 2.94 \times 10^{-6}\ M$$

Given this hydronium ion concentration, the pH would be simple to calculate.

$$pH = -\log[H^+] = -\log(2.94 \times 10^{-6}\ M) = 5.53$$

45. **Calculate the pH of the rain droplets in a cloud that is exposed to a SO_2 concentration of 2.0 ppb (at a total pressure of 1.0 atm).** We need to write a reaction where SO_2 dissolves in water (Henry's Law) and participates in the first acid dissociation.

$$SO_2(g) \rightleftharpoons SO_2(aq) \qquad K_H = 1.2\ M/atm \qquad (RC.7)$$

Sulfur dioxide reacts with water in the aqueous phase to form sulfurous acid, which is a weak acid and ionizes in solution to provide a hydronium ion. Because it is a diprotic acid, there are two ionization steps.

$$SO_2(aq) + 2\,H_2O(l) \rightleftharpoons H_2SO_3(aq) + H_2O(l) \rightleftharpoons H_3O^+(aq) + HSO_3^-(aq) \qquad (RC.8)$$

$$HSO_3^-(aq) + H_2O(l) \rightleftharpoons H_3O^+(aq) + SO_3^{2-}(aq) \qquad (RC.9)$$

The equilibrium constants are $K_{a1} = 1.4 \times 10^{-2}$ and $K_{a2} = 6.3 \times 10^{-8}$. Since the K_{a2} is very small, let us ignore it for now and use only the first K_{a1}. Adding up Reactions RC.7 and RC.8 results in the following.

$$SO_2(g) \rightleftharpoons \cancel{SO_2(aq)} \qquad (RC.10)$$
$$\cancel{SO_2(aq)} + 2\,H_2O(l) \rightleftharpoons H_3O^+(aq) + HSO_3^-(aq) \qquad (RC.11)$$
$$SO_2(g) + 2\,H_2O(l) \rightleftharpoons H_3O^+(aq) + HSO_3^-(aq) \qquad (RC.12)$$

Using Hess's Law, the sum of a series of reactions implies the multiplication of each reaction's equilibrium constant.

$$K_{SO_2\ rain} = K_H \times K_{a1} \qquad (C.10)$$

We need to take Reaction RC.12 and set up an ICE table with it. We will need the gaseous concentration of SO_2 which, at 2.0 ppb, is

$$P_{SO_2} = (2.0 \times 10^{-9})(1.0\ atm) = 2.0 \times 10^{-9}\ atm \qquad (C.11)$$

We need to also remember that the units of P_{SO_2} are atm and the units of x are molar.

	$SO_2(g)$	+	$2\ H_2O(l)$	$\rightleftharpoons$	$H_3O^+(aq)$	+	$HSO_3^-(aq)$
I (atm)	2.0×10^{-9}		- - -		0		0
C	$-x(RT)$		- - -		$+x$		$+x$
E	$2.0 \times 10^{-9} - x(RT)$		- - -		x		x

We now need to plug in the equilibrium values into the $K_{SO_2\ rain}$ expression.

$$\frac{[H_3O^+][HSO_3^-]}{P_{SO_2}} = K_{SO_2\ rain} = K_H \times K_{a1}$$

$$\frac{(x)(x)}{2.0 \times 10^{-9} - x(RT)} = K_H \times K_{a1}$$

To simplify the math, we can use the 5% rule and then solve for x. Assume $x \ll 2.0 \times 10^{-9}$ atm. Once again, we should consider the volume of the atmosphere to the volume of the rain droplets. The amount of SO_2 absorbed by the water is insignificant compared to the amount in the atmosphere, so let us assume that P_{SO_2} is constant and unaffected by x.

$$\frac{x^2}{2.0 \times 10^{-9} - \cancel{x}(RT)_{\ 0}} = K_H \times K_{a1}$$

$$x^2 = K_H \times K_{a1}\left(2.0 \times 10^{-9}\right)$$

$$x = \sqrt{K_H \times K_{a1}\,(2.0 \times 10^{-9})}$$

$$x = \sqrt{(1.2)\,(1.4 \times 10^{-2})\,(2.0 \times 10^{-9})}$$

$$x = 5.80 \times 10^{-6}\ \text{M}$$

Checking the assumption, we find that it failed terribly, but we are assuming that P_{SO_2} is constant.

Looking back at the equilibrium ICE table, x represents the concentration of hydronium ion, so finding the pH is easy.

$$\text{pH} = -\log[H_3O^+] = -\log(5.8 \times 10^{-6}\ \text{M}) = 5.24$$

The pH of SO_2 in rain water, saturated with dissolved SO_2 at a mixing ratio of 2.0 ppb, is 5.24. Recall the example of natural rain, where P_{CO_2} was 397 ppm or 200,000 times more concentrated than P_{SO_2} at 2.0 ppb. Sulfurous acid is less weak than carbonic acid and thus can depress the pH of rain at a concentration of SO_2 10^5 smaller than CO_2.

Now let us check to see if the second acid dissociation (K_{a2}) would have contributed to the pH. We will use x as initial values for $[HSO_3^-]$ and $[H_3O^+]$.

	$HSO_3^-(aq)$	+	$H_2O(l)$	$\rightleftharpoons$	$H_3O^+(aq)$	+	$SO_3^{2-}(aq)$
I (M)	5.8×10^{-6}		- - -		5.8×10^{-6}		0
C	$-x$		- - -		$+x$		$+x$
E	$5.8 \times 10^{-6} - x$		- - -		$5.8 \times 10^{-6} + x$		x

We now need to plug in the equilibrium values into the K_{a2} expression.

$$\frac{[H_3O^+][SO_3^{2-}]}{[HSO_3^-]} = K_{a2}$$

$$\frac{(5.8 \times 10^{-6} + x)(x)}{5.8 \times 10^{-6} - x} = K_{a2}$$

To simplify the math, we can use the 5% rule and then solve for x. Assume $x \ll 5.8 \times 10^{-6}$ M.

$$\frac{(5.8 \times 10^{-6} + \cancel{x}_0)(x)}{5.8 \times 10^{-6} - \cancel{x}_0} = K_{a2}$$
$$\frac{(5.8 \times 10^{-6})(x)}{5.8 \times 10^{-6}} = K_{a2}$$
$$\frac{\cancel{(5.8 \times 10^{-6})}(x)}{\cancel{5.8 \times 10^{-6}}} = K_{a2}$$
$$(x) = K_{a2} = 6.3 \times 10^{-8} \tag{C.12}$$

Checking the assumption, we find that the assumption was good (1%). The second dissociation of sulfurous acid only contributed about 1% to the total acid concentration $[H_3O^+]$.

46. **How can you recognize a catalyst in the reactions of a mechanism?** A catalyst is a reactant in an early step and a product in a later step, such that the catalyst is regenerated and drops out of the overall reaction.

47. **What does the catalyst do to the rate of the reaction? How does the catalyst manage to do this (discuss the answer in terms of energy)?** Catalysts lower the activation energy of a reaction, resulting in a much faster reaction.

48. **What are the three major factors that lead to significant ozone loss over the Antarctic pole? Describe their synergy. How is this different from the Arctic pole?**

    (a) Polar vortex: The South Pole is exceptionally cold during the winter and develops a weather pattern that isolates the gases above the pole, not allowing gases from other parts of the globe to mix and raise the temperature. When the ozone destruction starts, the ozone from the stratosphere in other parts of the globe cannot mix into the polar air and raise the concentration.

    (b) Polar stratospheric clouds (PSCs): The exceptionally cold temperatures allow clouds made of ice crystals to form. These crystals convert relatively inactive forms of chlorine into catalytically active forms during the winter months.

    (c) Dark winter: The absence of light allows the chlorine species to accumulate since no photolysis of other stable molecules is possible.

    (d) The synergistic combination of these three features means that when light begins to reach the South Pole, the ozone above the pole is trapped and mixed thoroughly with catalytically active chlorine. The Chapman cycle is unable to replace the ozone loss, and ozone from the rest of the stratosphere is unable to mix in and replace the lost ozone.

49. **Greenhouse gases trap heat in the troposphere and result in colder temperatures in the tropopause and lower stratosphere. How might increases in GHG emissions affect ozone destruction over the Arctic pole?** As the top of the troposphere gets colder, this will aid in the formation of PSCs, possibly even above the North Pole, leading to an increased loss of ozone.

50. **Describe one possible positive feedback cycle and one negative feedback cycle related to Climate Change.**

    (a) Positive feedback: As the CO_2 concentration in the atmosphere increases, the troposphere and oceans will warm; warmer oceans mean gases are less soluble, which means that the oceans will no longer be able to remove the 20–30% of the anthropogenically emitted CO_2, and CO_2 levels will rise even faster.

    (b) Negative feedback: As the planet warms, the thermohaline circulation, which delivers heat to northern Europe and Iceland, may slow down and deliver less heat, which means that the land masses in the northern latitudes of the Atlantic Ocean will develop a colder climate, which would promote more snow cover for longer periods, which would increase the albedo of the planet, which would make the planet colder.

51. **Determine the concentration of Ba^{2+} in the human stomach given that the HCl concentration present is approximately 0.1 M. Consider the K_{sp} of barium sulfate and the K_b of the sulfate and hydrogen sulfate ions (or the K_a values of sulfuric acid and hydrogen sulfate ion). Also, remember that the stomach continually produces acid, so the concentration of acid remains relatively constant despite any reaction that may occur as a result of the barium sulfate.** We need the solubility product reaction for barium sulfate.

$$BaSO_4(s) \rightleftharpoons Ba^{2+}(aq) + SO_4^{2-}(aq) \qquad K_{sp} \qquad \text{(RC.13)}$$

Since the sulfate ion is a weak base, it reacts with the acid in the stomach.

$$SO_4^{2-}(aq) + H_3O^+(aq) \rightleftharpoons HSO_4^-(aq) + H_2O(l) \qquad 1/K_{a2} \qquad \text{(RC.14)}$$

Adding these reactions yields the following.

$$BaSO_4(s) + H_3O^+(aq) \rightleftharpoons Ba^{2+}(aq) + HSO_4^-(aq) + H_2O(l) \quad K = K_{sp}/K_{a2} \qquad \text{(RC.15)}$$

We need to set up an ICE table.

	$BaSO_4(s)$	+	$H_3O^+(aq)$	$\rightleftharpoons$	$Ba^{2+}(aq)$	+	$HSO_4^-(aq)$
I (M)	- - -		0.10		0		0
C	- - -		$-x$		$+x$		$+x$
E			$0.10-x$		x		x

We now need to plug in the equilibrium values into the K expression.

$$\frac{[Ba^{2+}][HSO_4^-]}{[H_3O^+]} = \frac{K_{sp}}{K_{a2}}$$

$$\frac{(x)(x)}{0.10 - x} = \frac{K_{sp}}{K_{a2}}$$

To simplify the math, we can use the 5% rule and then solve for x. Assume $x \ll 1.0$ M.

$$\frac{(x)(x)}{0.10 - \cancel{x}_0} = \frac{K_{sp}}{K_{a2}}$$

$$\frac{x^2}{0.10} = \frac{K_{sp}}{K_{a2}}$$

$$x = \sqrt{\frac{(0.10)\left(K_{sp}\right)}{K_{a2}}}$$

$$x = \sqrt{\frac{(0.10)\,(1.08 \times 10^{-10})}{1.0 \times 10^{-2}}}$$

$$x = 3.3 \times 10^{-5}\ \text{M}$$

Only 0.033 mM Ba^{2+} dissolves in the stomach, where the pH is the lowest.

Alternative calculation: If you used the K_{b1} of the sulfate ion instead of using Reaction 51, then you would need the following combination.

$$SO_4^{2-}(aq) + \cancel{H_2O(l)} \rightleftharpoons HSO_4^-(aq) + \cancel{OH^-(aq)} \qquad K_{b1} \qquad \text{(RC.16)}$$

$$H_3O^+(aq) + \cancel{OH^-(aq)} \rightleftharpoons \cancel{2}\,H_2O(l) \qquad 1/K_w \qquad \text{(RC.17)}$$

$$SO_4^{2-}(aq) + H_3O^+(aq) \rightleftharpoons HSO_4^-(aq) + H_2O(l) \qquad K = \frac{K_{b1}}{K_w} \qquad \text{(RC.18)}$$

The resulting reaction is the same as Reaction 51, and the equilibrium constant is also the same since

$$K_{a2} \times K_{b1} = K_w$$

You can look up these values from Appendix I to confirm this.

D

CHAPTER 4 EXAMPLES AND END-OF-CHAPTER EXERCISES

D.1 SOLUTIONS TO IN-CHAPTER EXAMPLES

Solution to Review Example 4.1 on page 167

Identify the oxidation states of the elements that were oxidized or reduced in Reaction R4.2, and label the oxidizing agent and reducing agent.

$$4FeO \cdot Fe_2O_3(s) + O_2(g) \rightarrow 6Fe_2O_3(s) \quad \text{(RD.1)}$$

Solution: Magnetite has a formula that is a bit unusual since it contains two forms of iron in two different oxidation states. We can approach this by remembering how to assign oxidation states, as described in Review 1.13 on p. 30 of Chapter 1.

Total	+2	−2		+6	−6		0		+6	−6
Individual	+2	−2		+3	−2		0		+3	−2
	Fe	O	·	Fe_2	O_3	+	O_2	→	Fe_2	O_3

The "individual" line represents the oxidation state of each atom, and the "total" line represents the oxidation value when the subscript is considered. Thus, it appears that one of the iron atoms in magnetite is oxidized (+2 → +3) and that O_2 is reduced (0 → −2). Since oxidizing agents cause another compound to oxidize, O_2 is the oxidizing agent and magnetite is the reducing agent.

Solution to Review Example 4.2 on page 172

Determine the oxidation state of carbon before and after the following reaction.

$$HCO_3^-(aq) + 4H_2(g) + H^+(aq) \rightarrow CH_4(g) + 3H_2O(l) \quad \text{(RD.2)}$$

Solution:

- HCO_3^-: Oxygen is the most electronegative element, so it obtains a −2 oxidation state for all three Os. Hydrogen is the least electronegative, so it obtains the positive oxidation state, which must be +1. The species has an overall charge of −1, so

$$-1 = (+1 \text{ from H}) + (3 \times -2 \text{ for O's}) + (x \text{ for the C}) \quad \text{(D.1)}$$

  The value of x must be +4.
- CH_4: Carbon is slightly more electronegative compared to hydrogen, so hydrogen obtains a +1 oxidation state. With four Hs at +1, each carbon must have a −4 oxidation state to result in an overall charge of zero.

Environmental Chemistry: An Analytical Approach, First Edition. Kenneth S. Overway.

Companion website: www.wiley.com/go/overway/environmental_chemistry

Solution to Example 4.1 on page 173

Using the three categories listed earlier, describe the following organisms by food source, respiration, and temperature preference.

1. Humans
2. Microbes that produce methane from carbohydrates in the absence of oxygen, find oxygen toxic, and prefer temperatures around 55 °C
3. Algae that thrive near the surface of the ocean at temperatures near 25 °C
4. Archaebacteria that thrive in the isolated lakes below the ice sheets of Antarctica and feed on reduced sulfur compounds

Solution:

1. Humans are heterotrophic obligate aerobic mesophiles.
   (a) Humans eat other organisms, use oxygen for metabolism, cannot survive without oxygen, and have a body temperature of 37 °C.
2. These microbes are heterotrophic obligate anaerobic thermophiles.
   (a) Methanogenic microbes consume carbohydrates produced by other organisms, thrive in anoxic conditions only, and prefer temperatures above 45 °C.
3. These algae are photoautotrophic aerobic mesophiles.
   (a) These algae make their own food from solar radiation, thrive near the surface of the ocean where oxygen is plentiful, and prefer temperatures in the mesophilic range.
4. These archaebacteria are chemoautotrophic anaerobic psychrophiles.
   (a) These archaebacteria metabolize inorganic chemicals (reduced sulfur) and live under thousands of feet of ice where temperatures are near freezing and far away from atmospheric oxygen. Not enough information was given to determine if the archaebacteria are facultative or obligate anaerobes.

Solution to Example 4.2 on page 174

1. Determine the standard free energy for the oxidation of glucose at 298 K, given by the following reaction.

$$C_6H_{12}O_6(s) + 6O_2(g) \rightarrow 6CO_2(g) + 6H_2O(l) \qquad \text{(RD.3)}$$

2. Now find the free energy for Reaction R4.21 under reasonable environmental conditions of $P_{O_2} = 0.2095$ atm, $P_{CO_2} = 3.97 \times 10^{-4}$ atm, and $T = 16.5$ °C.
3. Determine the standard-state cell potential as well as the cell potential under the environmental conditions of $P_{O_2} = 0.2095$ atm, $P_{CO_2} = 3.97\times10^{-4}$ atm, and $T = 16.5$ °C for the combination of the following half-reactions.

$$6CO_2(g) + 24H^+(aq) + 24e^- \rightarrow C_6H_{12}O_6(s) + 6H_2O(l) \qquad \text{(RD.4)}$$

$$O_2(g) + 4H^+(aq) + 4e^- \rightarrow 2H_2O(l) \qquad \text{(RD.5)}$$

Solution:

1. The free energy of any reaction can be calculated by using tabular thermodynamic values found in Table I.2 on pages 318 to 321.

$$\Delta G^{\circ}_{rxn} = \Sigma \Delta G^{\circ}_{f\ products} - \Sigma \Delta G^{\circ}_{f\ reactants}$$

$$\begin{aligned}\Delta G^{\circ}_{rxn} &= (6 \text{ mol } CO_2)(-394.4 \text{ kJ/mol}) + (6 \text{ mol } H_2O)(-237.1 \text{ kJ/mol}) \\ &\quad -(6 \text{ mol } O_2)(0 \text{ kJ/mol}) - (1 \text{ mol } C_6H_{12}O_6)(-910.4 \text{ kJ/mol}) \\ &= -2878.6 \text{ kJ}\end{aligned}$$

2. In order to find the free energy under nonstandard conditions, we need to use Eq. (4.2), which contains the temperature and the reaction quotient. The reaction quotient requires the balance reaction.

$$\begin{aligned}\Delta G &= \Delta G^{\circ} + RT \ln Q \\ &= \Delta G^{\circ} + RT \ln \left(\frac{(P_{CO_2})^6}{(P_{O_2})^6} \right) \\ &= -2878.6 \text{ kJ} + \left(8.3145 \times 10^{-3} \frac{\text{kJ}}{\text{mol}\cdot\text{K}} \right) (273.15 + 16.5 \text{ K}) \ln \left(\frac{(3.97\times10^{-4} \text{ atm})^6}{(0.2095 \text{ atm})^6} \right) \\ &= -2878.6 \text{ kJ} + (2.40829 \text{ kJ})(-37.611256) \\ &= -2969.2 \text{ kJ}\end{aligned}$$

You can see that reaction concentrations and temperatures affect the free energy of a reaction and, in this case, make the combustion of glucose more spontaneous.

3. When using half-reactions to assemble a complete redox reaction, we need to find the least common multiple (LCM). The glucose reaction consumes 24 electrons and the oxygen reaction consumes 4 electrons, so the LCM is 24. This means that the oxygen reaction must be multiplied by 6 before the reactions are added. When adding reactions, all electrons must cancel, so we need to flip one of the reactions so that the electrons are on the product side (making the reaction an oxidation instead of a reduction). Since the glucose reaction has the smallest reduction potential, flipping the glucose reaction and adding it to the oxygen reaction will result in an overall positive cell potential.

$$\begin{array}{rl} C_6H_{12}O_6(s) + \cancel{6H_2O(l)} \rightarrow 6CO_2(g) + \cancel{24H^+(aq)} + \cancel{24e^-} & 0.014 \text{ V} \\ 6O_2(g) + \cancel{24H^+(aq)} + \cancel{24e^-} \rightarrow \cancel{12}\,6\ H_2O(l) & 1.229 \text{ V} \\ \hline C_6H_{12}O_6(s) + 6O_2(g) \rightarrow 6CO_2(g) + 6H_2O(l) & 1.243 \text{ V} \end{array}$$

Remember that the cell potential is an intensive property, so when the oxygen half-reaction is multiplied by 6, the potential does not change.

4. To find the cell potential using nonstandard conditions, we need to use the Nernst equation with the concentrations and temperature.

$$\begin{aligned}E &= E^{\circ} - \frac{RT}{nF} \ln Q \\ &= E^{\circ} - \frac{RT}{nF} \ln \left(\frac{(P_{CO_2})^6}{(P_{O_2})^6} \right) \\ &= 1.243 \text{ V} - \frac{\left(8.3145 \frac{\text{J}}{\text{mol}\cdot\text{K}} \right)(273.15 + 16.5 \text{ K})}{(24 \text{ mol } e^-)(96,485.309 \text{ C/mol})} \ln \left(\frac{(3.97\times10^{-4} \text{ atm})^6}{(0.2095 \text{ atm})^6} \right) \\ &= 1.243 \text{ V} - (1.04 \times 10^{-3} \text{ V})(-37.611256) \\ &= 1.282 \text{ V}\end{aligned}$$

Solution to Review Example 4.3 on page 176

What are the oxidation states of the two carbon atoms in the acetate ion and how do they change when they become HCO_3^- and CH_4 according to Reaction R4.18? How many electrons are exchanged in this reaction and which direction do the electrons flow?

$$CH_3COO^-(aq) + H_2O(l) \rightarrow CH_4(g) + HCO_3^-(aq) \qquad \text{(RD.6)}$$

Referring back to Review Example 4.2 (page 172), we know that the oxidation state for carbon in CH_4 is −4 and it is +4 in HCO_3^-. The trickier oxidation states for carbon are in the acetate ion. The methyl carbon in the acetate ion has the same oxidation state as the carbon in ethane (C_2H_6). Since carbon is the most electronegative element, the hydrogens obtain a +1 oxidation state. The H total is +6, so the two carbon atoms must have a combined −6 oxidation state to balance the Hs. Since each carbon is identical, both Cs must have the same oxidation state of −3. If the methyl carbon in the acetate ion has an oxidation state of −3, then the carboxylate carbon must have an oxidation state of +3.

$$(-1 \text{ total charge}) = (+3 \text{ from H's}) + (-3 \text{ from methyl C}) + x + (2 \times -2 \text{ from O's})$$

The value of x must be +3. Thus, the methyl carbon in the reaction above goes from a −3 oxidation state to a −4 oxidation state in CH_4. It has been reduced. The carboxylate carbon goes from an oxidation state of +3 to +4 in the HCO_3^- species. It has been oxidized. Thus, one electron has been transferred from the carboxylate carbon to the methyl carbon in this disproportionation reaction.

Solution to Review Example 4.4 on page 178

1. What is the electron configuration of Zn(II)?
   $1s^2\ 2s^2\ 2p^6\ 3s^2\ 3p^6\ 3d^{10}$ or $[Ar]\ 3d^{10}$
2. What is the electron configuration of Cu(II)?
   $1s^2\ 2s^2\ 2p^6\ 3s^2\ 3p^6\ 3d^9$ or $[Ar]\ 3d^9$
3. What is the electron configuration of Sn(II)?
   $1s^2\ 2s^2\ 2p^6\ 3s^2\ 3p^6\ 4s^2\ 3d^{10}\ 4p^6\ 5s^2\ 4d^{10}$ or $[Kr]\ 5s^2\ 4d^{10}$
4. What is the electron configuration of Sn(IV)?
   $1s^2\ 2s^2\ 2p^6\ 3s^2\ 3p^6\ 4s^2\ 3d^{10}\ 4p^6\ 4d^{10}$ or $[Kr]\ 4d^{10}$
5. What is the electron configuration of Ca(II)?
   [Ar]

D.2 SOLUTIONS TO END-OF-CHAPTER EXERCISES

1. **Describe the purpose of dividing soil into horizons. Why are they useful?** Soil horizons are useful because they represent visible transitions between layers of soil that have different compositions and explain the trends for organic and inorganic content.
2. **Which soil horizon is likely to be the most effective (short term) at buffering soil: Horizon B (containing small particle sizes and high surface area) or Horizon C (containing some small particles and some larger particles with less surface area)?** Horizon B would be a more effective soil buffer since the finer particles and higher surface area would promote faster kinetic reaction with acid precipitation.

3. **Separate the items in the following list into chemical and physical weathering processes: hydration, abrasion, freeze–thaw, dissolution, oxidation/reduction, hydrolysis, expansion/contraction**

   (a) Chemical – Hydration, dissolution, oxidation/reduction, hydrolysis

   (b) Physical – Abrasion, freeze-thaw, expansion/contraction

4. **Which type of mineral is the most abundant on the Earth's surface? Why? Consider what you know about the formation of the Earth.** Silicate and aluminosilicate minerals are most abundant on the Earth's surface. This is because while the Earth was molten, the process of differentiation caused the less dense material to float to the surface of the globe and the more dense material to sink to the core. Since the nucleosynthesis process of massive stars favors elements such as silicon, iron, and oxygen, the silicate minerals were very abundant and were less dense compared to other abundant elements such as iron.

5. **What role does water play in the formation of biopolymers (saccharides, proteins, and fats)?** Water is the molecule involved in the linkage between monomers. When two monomers link, a hydroxide from one monomer combines with a hydrogen from the other monomer and forms a water molecule when the linage is made in a process call condensation. When the polymer breaks up through the process of hydrolysis, water splits into hydroxide and hydrogen to cap off the ends of the former linkage.

6. **How are lignins made hydrophilic as a result of leaf senescence?** The final stages of photosynthesis cause the formation of superoxides and other oxidizing agents, which attack the lignins and cause them to become more hydrophilic.

7. **Organic matter is further degraded into what chemical forms by what processes?** Organic molecules are oxidized in a process called acidogenesis, where these organic molecules form carboxylic acids. Some processes yield acetic acid, in a process called acetogenesis.

8. **Classify the microbes that are used in the tertiary treatment stage of wastewater, by their metabolic pathway, temperature preference, and food source. These "bugs" thrive in the tanks with alternating stages of vigorous mixing with air and stages where dissolved oxygen drops to near zero, and they consume the organic nutrients at temperatures between 0 °C and 40 °C.** These microbes would be classified as heterotropic facultative anaerobes, since they can survive with and without oxygen, and psychropilic or mesophilic depending on the actual temperature that they prefer.

9. **List the important chemical species that replace O_2 in anaerobic respiration reactions in the order of their energetic potential under environmental conditions. What chemicals do these species (C, N, S elements) produce in the reactions?** The nitrate has the highest energy yield of all of the anaerobic reactions, followed by the manganese oxide, sulfate ion, iron oxide hydroxide, another variant of the sulfate ion reaction, and finally the disproportionation of carbon in methanogenesis. The nitrate ion produces molecular nitrogen. Manganese and iron oxides become magnese and iron carbonates. The sulfate ion becomes hydrogen sulfide. The carbon disproportionation results in the formation of carbon dioxide and methane.

10. **In the limestone buffer system involving minerals such as $CaCO_3$ and $MgCO_3$, which ionic species does the actual buffering and how does the other species make a difference in the buffer?** The limestone buffering system is really a carbonate buffering system. It is the carbonate ion that is involved in the acid/base

chemistry. The cation, such as Ca^{2+} or Mg^{2+}, only affects the buffer in its solubility in water. The more soluble the mineral, the more carbonate ion is mobilized to act as a buffer.

11. **Write a representative reaction for each inorganic buffering mechanism, listing the pH range where each buffers the soil.**

(a) The limestone buffer maintains a soil pH range of around 7.5–9.3, depending on the mineral present in the soil. A representative reaction is

$$CaCO_3(s) + 2H^+(aq) \rightleftharpoons Ca^{2+}(aq) + H_2O(l) + CO_2(g)$$
$$\text{or}\quad CO_2(g) + CaCO_3(s) + H_2O(l) \rightleftharpoons 2HCO_3^-(aq) + Ca^{2+}(aq) \tag{RD.7}$$

(b) The cation-exchange buffer maintains a soil pH range of 4.2–6.2, with a representative reaction of

$$\text{soil} - Mg(s) + 2H^+(aq) \rightleftharpoons \text{soil} - H_2(s) + Mg^{2+}(aq) \tag{RD.8}$$

(c) The aluminum buffer maintains a soil pH range of 4.3–5.3, depending on the aluminosilicate mineral present in the soil. A representative reaction could be

$$Al(OH)_3(s) + 3H^+(aq) \rightleftharpoons Al^{3+}(aq) + 3H_2O(l) \tag{RD.9}$$

12. **Explain which of the soils in Table D.1 would maintain the highest pH.**

Soil 1		Soil 2		Soil 3	
Compound	Percent composition	Compound	Percent composition	Compound	Percent composition
SiO_2	43.38	$MgCO_3$	37.68	SiO_2	57.29
$NaAlSi_3O_8$	28.91	SiO_2	32.17	Al_2O_3	27.35
Fe_2SiO_4	14.55	Al_2O_3	26.84	Fe_2O_3	10.32
Al_2O_3	13.16	humus	3.31	TiO_2	5.04

Table D.1 Simulated compositional analysis of some soil samples.

Soil 2 would maintain the highest pH because of the presence of $MgCO_3$, which would drive the soil pH to near 9. The other soils would be buffered at no higher than about pH 6.

13. **Which class of metal would have a stronger affinity to humic acids with only carboxylate groups?** Carboxylate groups are rich in anionic oxygen (high electronegativity), which would tend to form stronger interactions with Class A metals.

14. **Which class of metal would be more likely to bond with proteins containing sulfur?** Sulfur represents an anionic ligand that has a low electronegativity value compared to anionic oxygen and would thus form covalent bonds with Class B metals.

15. **Given the data in Table D.2, write a short narrative describing the level and type of microbial activities (identified by respiration preferences) that are thriving at each stage of the measurement.**

Days	O_2(aq) (mg/L)	CO_2(aq) (mg/L)	CH_4(aq) (mg/L)	H_2S(aq) (mg/L)
1	6.2	0.6	–	–
2	1.5	1.3	–	–
5	0.3	1.8	–	–
10	–	2.0	0.6	0.01
15	–	2.1	0.8	0.16

Table D.2 Simulated water analysis after contamination of organic waste.

Following the timeline experiment, initially the water is oxic and supports aerobic microbial activity. Methane and H_2S are not thermodynamically favorable in these conditions. As the dissolved oxygen is used up, the metabolism shifts from aerobic to anaerobic. Anaerobic and methanogenic microbes begin to flourish, and the reduced forms of carbon and sulfur become thermodynamically favorable.

16. **What deleterious effects does soluble aluminum have on plants?** Cationic aluminum forms insoluble minerals with nutrients such as phosphate and sulfate, making them unavailable to the plant roots. It can also bind to DNA and enzymes and disrupt important cell functions, such as water and nutrient absorption in roots.

17. **List two major instruments that are used to measure the elemental analysis of metal in various samples.** An atomic absorption spectrometer measures metals dissolved in a liquid solution or solid sample through a process of atomization and light absorption using a hollow cathode lamp. Another very common instrument used in metal analysis is an inductively coupled plasma spectrometer. This instrument atomizes and electronically excites the atoms inside a plasma flame, and the resulting ions and atoms are quantitated by either their emission spectrum or their mass using a mass spectrometer.

E

CHAPTER 5 EXAMPLES

E.1 SOLUTIONS TO IN-CHAPTER EXAMPLES

Solution to Review Example 5.1 on page 192

1. The dissolution of KCl in water can be represented in the following reaction.

$$KCl(s) \longrightarrow K^{+}(aq) + Cl^{-}(aq) \qquad \text{(RE.1)}$$

At 25 °C the enthalpy, entropy, and free energy can be calculated using the thermodynamic data found in Table I.2 on pages 318–321.

$$\begin{aligned}\Delta G^{\circ}_{rxn} &= \Sigma \Delta G^{\circ}_{f\ products} - \Sigma \Delta G^{\circ}_{f\ reactants}\\ \Delta G^{\circ}_{rxn} &= (1\ \text{mol}\ K^{+})(-283.3\ \text{kJ/mol}) + (1\ \text{mol}\ Cl^{-})(-131.2\ \text{kJ/mol})\\ &\quad -(1\ \text{mol KCl})(-408.5\ \text{kJ/mol})\\ &= -6.0\ \text{kJ}\end{aligned}$$

The same can be done for enthalpy and entropy.

$$\begin{aligned}\Delta H^{\circ}_{rxn} &= \Sigma \Delta H^{\circ}_{f\ products} - \Sigma \Delta H^{\circ}_{f\ reactants}\\ \Delta H^{\circ}_{rxn} &= (1\ \text{mol}\ K^{+})(-252.4\ \text{kJ/mol}) + (1\ \text{mol}\ Cl^{-})(-167.2\ \text{kJ/mol})\\ &\quad -(1\ \text{mol KCl})(-436.5\ \text{kJ/mol})\\ &= 16.9\ \text{kJ}\end{aligned}$$

$$\begin{aligned}\Delta S^{\circ}_{rxn} &= \Sigma S^{\circ}_{f\ products} - \Sigma S^{\circ}_{f\ reactants}\\ \Delta S^{\circ}_{rxn} &= (1\ \text{mol}\ K^{+})(102.5\ \text{J/(K}\cdot\text{mol)}) + (1\ \text{mol}\ Cl^{-})(56.5\ \text{J/(K}\cdot\text{mol)})\\ &\quad -(1\ \text{mol KCl})(82.6\ \text{J/(K}\cdot\text{mol)})\\ &= 76.4\ \text{J/K}\end{aligned}$$

Notice that in this case the change in enthalpy is positive, indicating that this is an endothermic reaction and the solution should drop in temperature as the salt dissolves. The entropy change is so large for this endothermic dissolution that it drives the reaction to be spontaneous, indicated by the free energy change.

2. The thermodynamics of the dissolution of methanol is solved in a similar manner.

$$CH_3OH(l) \longrightarrow CH_3OH(aq) \qquad \text{(RE.2)}$$

Environmental Chemistry: An Analytical Approach, First Edition. Kenneth S. Overway.

Companion website: www.wiley.com/go/overway/environmental_chemistry

$$\Delta G^{\circ}_{rxn} = (1 \text{ mol } CH_3OH(aq))(-175.4 \text{ kJ/mol}) - (1 \text{ mol } CH_3OH(l))(-166.6 \text{ kJ/mol})$$
$$= -8.8 \text{ kJ}$$
$$\Delta H^{\circ}_{rxn} = (1 \text{ mol } CH_3OH(aq))(-245.9 \text{ kJ/mol}) - (1 \text{ mol } CH_3OH(l))(-239.2 \text{ kJ/mol})$$
$$= -6.7 \text{ kJ}$$
$$\Delta S^{\circ}_{rxn} = (1 \text{ mol } CH_3OH(aq))(133 \text{ J/(K} \cdot \text{mol)}) - (1 \text{ mol } CH_3OH(l))$$
$$(126.8 \text{ J/(K} \cdot \text{mol)}) = 6.2 \text{ J/K}$$

In this case the change in enthalpy is negative, indicating that this is an exothermic reaction and the solution should increase in temperature as the methanol dissolves. The entropy change is also large for the dissolution of methanol. With both enthalpy and entropy favorable for a spontaneous reaction, a negative free energy was inevitable.

Solution to Example 5.3 on page 197

Using Eq. (5.5), plot the Henry's Law solubility curve for molecular oxygen from 5 to 45 °C in intervals of 5 °C, then calculate the dissolved oxygen (DO) level in units of mg/L O_2 for this range, and plot this data set in the same graph as K_H. Assume an atmosphere with 1 atm of pressure and 21% oxygen.

The solution to this question involves the use of a spreadsheet to calculate the values for K_H and the DO levels. The first two columns in the spreadsheet should list the temperatures, from 5 to 45 °C in steps of 5 °C. The third column should employ Eq. (5.5) using K°_H and δ_H for O_2 as found in Table I.3 on page 321 and T° = 298.15 K. Since the units for K_H are M/atm, multiplying the calculated K_H by the ambient partial pressure of O_2 (1.0 atm × 0.21%) would yield the DO level in units of molar. Converting a molar concentration to mg/L requires multiplying the molar concentration by formula weight of O_2 in units of mg/mol (2 × 15.9994 × 1000) (displayed in the fourth column). All of these values are included in Table E.1.

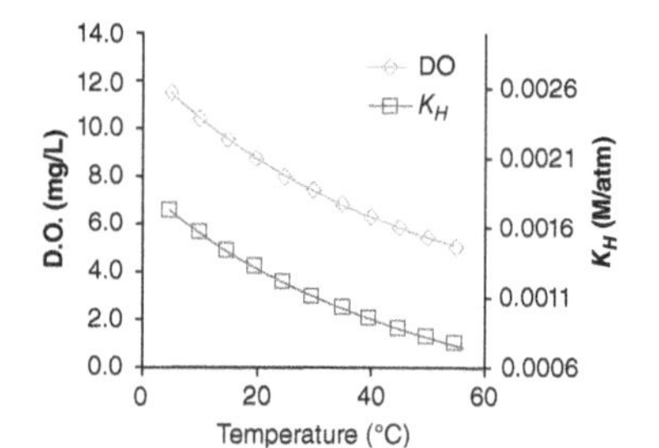

Figure E.1 A graphical depiction of the data found in Table E.1.

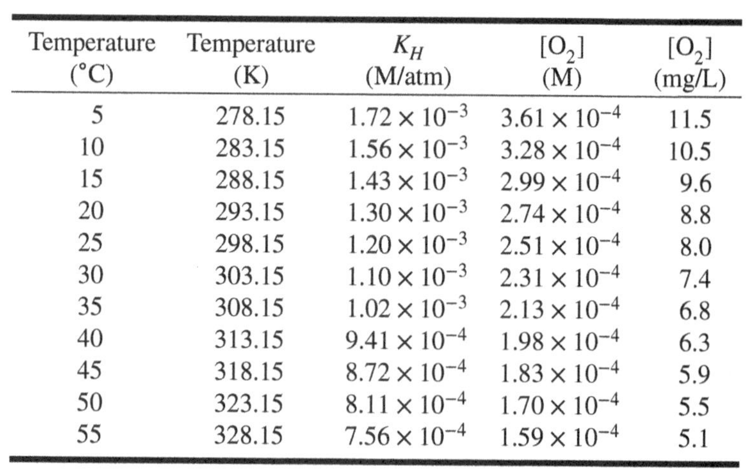

Temperature (°C)	Temperature (K)	K_H (M/atm)	$[O_2]$ (M)	$[O_2]$ (mg/L)
5	278.15	1.72×10^{-3}	3.61×10^{-4}	11.5
10	283.15	1.56×10^{-3}	3.28×10^{-4}	10.5
15	288.15	1.43×10^{-3}	2.99×10^{-4}	9.6
20	293.15	1.30×10^{-3}	2.74×10^{-4}	8.8
25	298.15	1.20×10^{-3}	2.51×10^{-4}	8.0
30	303.15	1.10×10^{-3}	2.31×10^{-4}	7.4
35	308.15	1.02×10^{-3}	2.13×10^{-4}	6.8
40	313.15	9.41×10^{-4}	1.98×10^{-4}	6.3
45	318.15	8.72×10^{-4}	1.83×10^{-4}	5.9
50	323.15	8.11×10^{-4}	1.70×10^{-4}	5.5
55	328.15	7.56×10^{-4}	1.59×10^{-4}	5.1

See Figure E.1 for a graphical depiction.

Table E.1 A theoretical prediction of the value of K_H and DO levels as a function of temperature.

Solution to Review Example 5.2 on page 199

Use the USGS DO table website under Section B Oxygen Solubility Tables to plot the DO levels for a freshwater system (salinity = 0‰), where the water temperature may range from 5 to 40 °C in intervals of 5 °C, and then retrieve similar data for a seawater system (salinity = 34‰, much like the Antarctic ocean) where the water temperature may range from 5 to

40 °C in intervals of 5 °C. Plot both of these data sets on the same graph to demonstrate the effect temperature and solubility have on DO. Assume that the atmospheric pressure ranges from 1.0 to 1.1 atm in increments of 0.1 atm. Plot only the data sets for the atmospheric pressure condition of 1.0 atm.

The data generated by the USGS website produces Table E.2 when the 1.1 atm data set is removed. This data was retrieved from a comma-separated text file (csv) to make it easier to import into a spreadsheet.

	DO levels (mg/L)	
Temp. (°C)	Salinity 0‰	Salinity 34‰
5	12.77	10.18
10	11.29	9.08
15	10.08	8.19
20	9.09	7.44
25	8.26	6.81
30	7.56	6.27
35	6.95	5.80
40	6.41	5.38

See Figure E.2 for a graphical depiction.

Table E.2 A theoretical prediction of the DO levels of a freshwater and seawater system as a function of temperature.

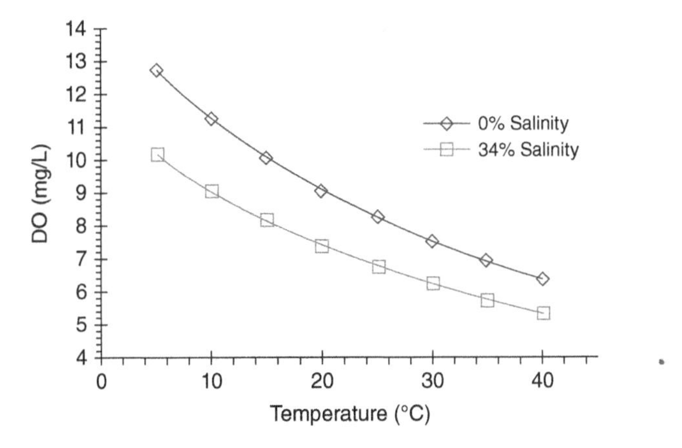

Figure E.2 A graphical depiction of the data found in Table E.2.

Solution to Review Example 5.3 on page 201

Remember that determination of oxidation states starts with the most electronegative element. Refer to the rules set out in the Review box 1.13 in Chapter 1 on page 30.

1. N_2
   (a) N oxidation state = 0
2. NH_3/NH_4^+
   (a) N oxidation state = −3
   (b) H oxidation state = +1

3. NO_2^-
   (a) O oxidation state = −2
   (b) N oxidation state = +3
4. NO_3^-
   (a) O oxidation state = −2
   (b) N oxidation state = +5

Fixing nitrogen requires a good bit of energy and raises the oxidation state of nitrogen from 0 to −3. This step requires 16 units of adenosine triphosphate (ATP) as described in Reaction R5.6 (on page 202). This necessary, yet costly, step provides an energy-rich molecule (NH_3) from which other organisms can benefit. It should be obvious that once nitrogen is fixed, there is a stepwise oxidation from N(−3) in NH_3 to N(+3) in NO_2^- to N(+5) in NO_3^-.

Solution to Review Example 5.4 on page 203

1. Calculate the change in free energy for Reaction R5.8.

$$2\,NH_4^+(aq) + 3\,O_2(g) \longrightarrow 2\,NO_2^-(aq) + 2\,H_2O(l) + 4\,H^+(aq) \qquad (RE.3)$$

$$\begin{aligned}\Delta G^\circ_{rxn} &= (2\text{ mol } NO_2^-(aq))(-32.2\text{ kJ/mol}) + (2\text{ mol } H_2O(l))(-237.1\text{ kJ/mol})\\ &\quad + (4\text{ mol } H^+(aq))(0.0\text{ kJ/mol})\\ &\quad - (2\text{ mol } NH_4^+(aq))(-79.31\text{ kJ/mol}) - (3\text{ mol } O_2(g))(0.0\text{ kJ/mol})\\ &= -379.98\text{ kJ}\end{aligned}$$

2. Calculate the change in free energy for Reaction R5.9.

$$2\,NO_2^-(aq) + O_2(aq) \longrightarrow 2\,NO_3^-(aq) \qquad (RE.4)$$

$$\begin{aligned}\Delta G^\circ_{rxn} &= (2\text{ mol } NO_3^-(aq))(-108.7\text{ kJ/mol})\\ &\quad - (2\text{ mol } NO_2^-(aq))(-32.2\text{ kJ/mol}) - (1\text{ mol } O_2(g))(0.0\text{ kJ/mol})\\ &= -153.0\text{ kJ}\end{aligned}$$

Solution to Review Example 5.5 on page 206
What are the oxidation states for the following forms of sulfur? SO_4^{2-}, S, $(CH_3)_2S$, COS, and H_2S. Refer to the rules set out in the Review box 1.13 in Chapter 1 on page 30.

1. SO_4^{2-}
   (a) S oxidation state = +6
   (b) O oxidation state = −2
2. S
   (a) S oxidation state = 0

3. $(CH_3)_2S$
   (a) H oxidation state = +1
   (b) S oxidation state = −2
   (c) C oxidation state = −2
4. COS
   (a) O oxidation state = −2
   (b) S oxidation state = −2
   (c) C oxidation state = +4
5. H_2S
   (a) S oxidation state = −2
   (b) H oxidation state = +1

E.2 SOLUTIONS TO END-OF-CHAPTER EXERCISES

1. **What is the coldest temperature that you might find at the bottom of a lake (ignoring deviations from extreme pressures)? What happens when the temperature of the water drops below this?** Ignoring pressure effects, the coldest temperature in deep water would be around 4 °C. If the water were to get any colder, the density would decrease and the water would rise to the surface of the body of water.
2. **If the Earth is bombarded by meteors on a daily basis, is it acting as an open or a closed system?** The Earth is an open system since it is constantly receiving debris from our solar system and losing some gases to space.
3. **Describe the difference between the percent unit and the permil unit.** The percent unit represent parts per one hundred and uses the symbol %, whereas the permil unit expresses parts per thousand and uses the ‰ unit.
4. **How is the density of water affected by salinity and temperature?** Density is proportional to salinity – as salinity increases, so does the density. Density is inversely proportional to temperature in the liquid state of water between 4 and 100 °C. As the temperature of the water increases from 4 °C, the density decreases.
5. **Describe the phenomenon of thermal pollution. Where might you find it?** Thermal pollution is the anthropogenic process of using surface water to cool industrial machinery. As the temperature of the water increases, the solubility of oxygen decreases, which can have severe affects on the local ecology. Thermal pollution is likely to occur when industrial facilities are built next to bodies of water, assuming that some of the water is to be used for cooling purposes.
6. **Estimate the DO levels at two points in a freshwater heat reservoir (for an industrial facility) compared to DO levels in the headwaters, which maintain a summer temperature of 57 °F (14 °C). At the discharge point (point 1), the water temperature is 93 °F (34 °C) and 0.5 km away from the discharge point where the water temperature drops to 75 °F (24 °C).** This problem deals with the temperature dependence of the Henry's Law constant. Using the given temperatures and Eq. (5.5), given on page 197, each of the DO values can be calculated.

$$K_H = K_H^\circ \cdot e^{\left(\delta_H\left(\frac{1}{T^\circ}-\frac{1}{T}\right)\right)}$$

$$= \left(1.2 \times 10^{-3}\ \frac{\text{M}}{\text{atm}}\right) e^{\left(-1800\left(\frac{1}{298.15\ \text{K}} - \frac{1}{14 + 273.15\ \text{K}}\right)\right)}$$

$$= 1.512 \times 10^{-3}\ \frac{\text{M}}{\text{atm}}$$

With Henry's Law constant determined for a given temperature, what remains is to calculate the DO from the K_H equilibrium expression.

$$[\text{O}_2] = K_H \cdot P_{\text{O}_2}$$

$$= \left(1.512 \times 10^{-3}\ \frac{\text{M}}{\text{atm}}\right)[(21\%)(1.0\ \text{atm})]$$

$$= 3.1752 \times 10^{-4}\ \text{M}$$

$$= \left(3.1752 \times 10^{-4}\ \text{M}\right)\left(\frac{2 \times 15.9994\ \frac{\text{g}}{\text{mol}}}{1\ \text{mol}}\right)\left(\frac{1000\ \text{mg}}{1\ \text{g}}\right)$$

$$= 10.\ \frac{\text{mg}}{\text{L}}\text{O}_2$$

At 34 °C (93.2 °F), $K_H = 1.005 \times 10^{-3}\ \frac{\text{M}}{\text{atm}}$ and DO = 6.7 $\frac{\text{mg}}{\text{L}}\text{O}_2$. At 24 °C (75.2 °F), $K_H = 1.225 \times 10^{-3}\ \frac{\text{M}}{\text{atm}}$ and DO = 8.2 $\frac{\text{mg}}{\text{L}}\text{O}_2$. It is obvious that the colder headwaters maintain a higher DO compared to the others and that the DO improves as the water temperature drops farther from the discharge point.

7. **Explain what happens to the DO levels in an estuary when high tides cause an increase in the salinity of the water (assuming that there is no temperature change).** As salinity increases, the solubility of O_2 decreases, so the estuary would see DO levels drop.

8. **What is the salinity of a 24.7842 g seawater sample if a preweighed 250-mL beaker (58.7254 g) containing the sample was boiled to dryness, baked in a drying oven overnight at 180 °C, cooled in a desiccator, and reweighed giving a mass of 59.5173 g?** salinity = 0.7919 g (net mass) ÷ 24.7842 g (total mass) × 1000‰ = 31.95‰

9. **Determine the oxidation state of N in hydrazine (N_2H_4), which is used as rocket fuel.** In N_2H_4, nitrogen is the most electronegative element, so H must have an oxidation state of +1, which produces a +4 charge to be balanced by the two N. Thus, the oxidation state of N is −2.

10. **Given the biochemical oxygen demand (BOD) experimental data in Table E.3, determine the BOD_5 for the sample if 300 mL BOD incubation bottles were used.** Only the first two dilutions are useful because they still contain 30% of the original DO.

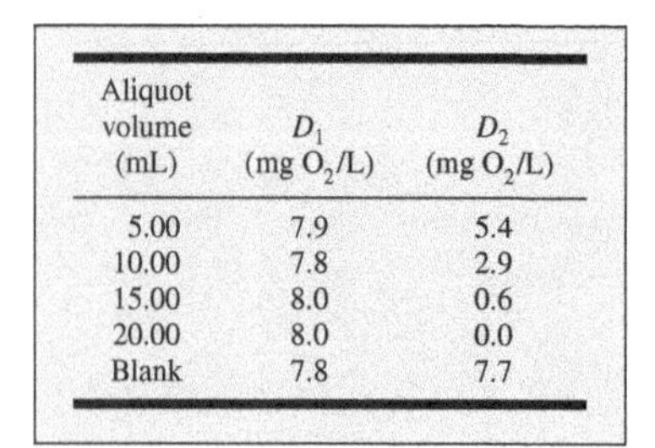

Aliquot volume (mL)	D_1 (mg O_2/L)	D_2 (mg O_2/L)
5.00	7.9	5.4
10.00	7.8	2.9
15.00	8.0	0.6
20.00	8.0	0.0
Blank	7.8	7.7

Table E.3 A simulated BOD_5 analysis.

$$\text{BOD}_5\ (\text{mg O}_2/\text{L}) = \frac{D_1 - D_2}{P}$$

$$= \frac{7.9 - 5.4}{\frac{5\ \text{mL}}{300\ \text{mL}}}$$

$$= 150\ \frac{\text{mg}}{\text{L}}\text{O}_2$$

The BOD_5 for the second trial is 150 mg/L O_2.

11. **List the most common inorganic forms of carbon in the atmosphere, lithosphere, and hydrosphere (these are commonly called reservoirs).** The atmospheric reservoir of inorganic carbon is in the form of CO_2. The lithospheric reservoir of inorganic carbon is in the form of carbonate minerals. The hydrospheric reservoir of inorganic carbon is in the form of HCO_3^-.

12. **Why is carbon essential to life?** Carbon forms the backbone of virtually all of the biomolecules used by organisms.

13. **List the five major reaction processes involved in the nitrogen cycle, describing the reactant and product form of nitrogen in each.**

    (a) Fixation: $N_2 \rightarrow$ ammoniacal nitrogen (NH_3/NH_4^+)

    (b) Assimilation: Ammoniacal nitrogen $\rightarrow$ amino acids

    (c) Ammonification: Amino acids $\rightarrow$ ammoniacal nitrogen

    (d) Nitrification: Ammoniacal nitrogen $\rightarrow$ nitrite ion $\rightarrow$ nitrate ion

    (e) Denitrification: Nitrate ion $\rightarrow N_2$, nitrite ion + ammonium $\rightarrow N_2$

14. **Why is nitrogen essential to life?** Nitrogen is an essential element in amino acids, which form all of the proteins that make life function.

15. **How did the work of Fritz Haber revolutionize agriculture and affect human population?** Fritz Haber discovered a reaction to synthesize ammonia from molecular nitrogen and hydrogen gas, allowing humans to make artificial fertilizer. Prior to this discovery, nitrogen fertilizer was limited to manure, and the limited process of fixation brought about by some common agricultural plants. Synthetic fertilizer could be produced in much larger quantities and allowed agricultural production to increase dramatically. With a much larger food supply, human population boomed.

16. **Describe the two major biotic and abiotic reservoirs of phosphorus.** Biotic phosphorus can be found in the form of phosphate minerals as part of bird guano. Several islands and other sites along the avian migratory routes host large deposits of guano. Abiotic phosphate minerals can be found in a few places in the world, such as Morocco and Florida.

17. **Why is phosphorus essential to life?** Phosphorus, in the form of phosphate, is essential for DNA, RNA, ATP, and the phospholipids that form cell walls. It is also integral to vitamins B6 and B12.

18. **Why was phosphate used in detergents?** Forms of sodium phosphate were added as builders to detergents. The phosphate ion scavenged all of the cations that would bind to the detergent molecule and form a precipitate so that the detergent was free to form micelles.

19. **Why is sulfur essential to life?** Sulfur is integral to two essential amino acids, methionine and cysteine, and is an element in vitamins B1 and B7.

20. **Sulfur exists in what inorganic form in its hydrospheric reservoir?** The sulfate ion is one of the most abundant components of seawater.

21. **What essential water quality information do the BOD and chemical oxygen demand (COD) methods provide? Describe the difference between the two methods.** Biochemical oxygen demand and chemical oxygen demand provide a measure of the nutrient load in a water sample. The lower the BOD or COD, the lower the nutrient load and the cleaner the water sample is. BOD requires a 5-day incubation, where a sample's DO is measured before and after the incubation. The loss of DO is proportional to the BOD. COD is a 2-h test in which the sample is mixed with a dichromate ion solution. The dichromate ion oxidizes all of the nutrients and nonnutrients. A spectroscopic measurement of the remaining dichromate ion or the Cr^{3+} ion is proportional to the COD. The COD usually provides an overestimate of the BOD since it oxidizes nearly everything in the solution.

Dissolved carbon (mg/L C)	Signal (AU × cm^{-1})
6.01	268.6
9.02	517.5
12.03	543.0
15.04	712.8
18.05	905.8
21.07	1003.5
24.07	1059.9
27.08	1271.7
30.09	1387.4

Table E.4 AU × cm^{-1} (AU = Absorbance Units) is the integrated peak area, from 3800 to 3400 cm^{-1}, taken from the absorbance spectrum of CO_2.

22. **A water quality analyst obtained calibration standards for a Total Inorganic Carbon (TIC) and Total Organic Carbon (TOC) analysis using a certified Na_2CO_3 standard (see Table E.4).**

    The TIC was measured first and gave a signal of 1229.5 AU × cm^{-1}. After the inorganic carbon was removed and the persulfate digestion was complete, the sample produced a TOC signal of 317.2 AU × cm^{-1}. Determine the TIC and TOC of the sample in units of mg/L carbon using linear regression. Using the method of external standards, the regression spreadsheet yields a TIC result of 26.5 ± 1.2 mg/L C and a TOC result of 6.1 ± 1.2 mg/L C.

23. **How does the primary wastewater treatment process contribute to water quality?** Primary treatment removes the large floating debris and most of the suspended inorganic and organic solids through settling tanks. This improves the clarity of the water and removes the BOD associated with the settled organic solids.

24. **How does the secondary wastewater treatment process contribute to water quality?** In secondary treatment, the wastewater is treated by an aerobic oxidation process with a host of microbes. The microbes consume much of the nutrient load and convert it into sludge. The resulting water has had most of its carbon nutrient load removed.

25. **How does the tertiary wastewater treatment process contribute to water quality? Describe some of the implementations of the tertiary process through chemical and biological means.** Tertiary treatment focuses on removing nitrogen and phosphorus nutrients. Chemical removal of nitrogen involves a process called ammonia stripping, where the pH of the solution is raised by lime, resulting in the conversion of ammoniacal nitrogen into ammonia. The ammonia is much less soluble and can be sparged out of the solution. By adding lime, the phosphate ion is precipitated out of the solution by the presence of calcium cations. Biological removal of nitrogen usually is an alteration of the secondary treatment stage, where the oxic zones are alternated with anoxic zones. In the anoxic zone, microbes consume the nitrate ion in a denitrifying reaction to produce N_2. In the oxic zones, any form of nitrogen is nitrified, forming the nitrate ion, which is consumed in the anoxic zones. The secondary process can be seeded with phosphate absorbing organisms, which consume inordinate amounts of phosphate ion and precipitate as sludge.

26. **Estimate the pH of a marine sample (salinity of 30.4‰) using Eq. (5.10) if, after the cresol red indicator is added, the measured absorbance at 573 nm is 0.126 and the absorbance at 433 nm is 0.027.**

$$\mathrm{pH} = 7.8164 + 0.004(35 - S) + \log\left(\frac{\frac{A_{573}}{A_{433}} - 0.0286}{2.7985 - 0.09025 \cdot \frac{A_{573}}{A_{433}}}\right)$$

$$= 7.8164 + 0.004(35 - 30.4) + \log\left(\frac{\frac{0.126}{0.027} - 0.0286}{2.7985 - 0.09025 \cdot \frac{0.126}{0.027}}\right)$$

$$= 8.13$$

27. **What is the difference between a turbidity spectrometer and a nephelometer?** While both instruments measure turbidity, the turbidity spectrometer uses a linear optical path and measures the amount of light that is not scattered or absorbed and infers the turbidity from the absorbance. A nephelometer measures turbidity by measuring the intensity of the scattered light at a 90° optical path.

28. **Both the nitrate and nitrite ions absorb UV radiation below 300 nm. Why must these two ions be derivatized in order to be quantitatively detected?** Most environmental solutions will contain some dissolved organic compounds, which also

absorb UV radiation. These DOCs would lead to an overestimate of the nitrate and nitrite ion concentrations. When derivatized to chromophores absorbing visible radiation, the measurement will avoid the systematic error since there are fewer contaminants that absorb in the visible region of the spectrum.

29. **If 25.4 mg of $BaSO_4$ precipitated from a 200 mL sample of water from an estuary, what was the concentration of sulfate ion in the sample (in mg/L SO_4^{2-})?** This is a simple stoichiometric conversion.

$$\text{mg/L } SO_4^{2-} = (25.4 \text{ mg } BaSO_4)\left(\frac{96.0636 \text{ mg } SO_4^{2-}}{233.3906 \text{ mg } BaSO_4}\right) \div (0.2 \text{ L})$$

This results in a concentration of 52.3 mg/L SO_4^{2-}.

30. **In the Calculation section of EPA Method 375.3 for gravimetric sulfate analysis, the authors provide the following conversion for each 1 mL sample analyzed.**

$$\text{mg/L } SO_4^{2-} = \frac{?\text{ mg } BaSO_4}{1 \text{ mL sample}} \times 411.5$$

Show how the 411.5 number was derived using a stoichiometric conversion between $BaSO_4$ and SO_4^{2-}. The number comes from the ratio of the formula weights of SO_4^{2-} and $BaSO_4$.

$$\left(\frac{?\text{ mg } BaSO_4}{1 \text{ mL sample}}\right)\left(\frac{1000 \text{ mL}}{1 \text{ L}}\right)\left(\frac{96.0636 \text{ g } SO_4^{2-}}{233.3906 \text{ g } BaSO_4}\right)$$

Taking $(1000 \times 96.0636) \div 233.3906 = 411.6$, which is very close to the value in the EPA method.

31. **In the Winkler titration, initially all of the DO reacts with Mn^{2+} to form the Mn precipitate, which happens under basic conditions. After adding the reagents in the next step, the solution is "fixed" and will not react with oxygen if exposed to air. What reaction condition keeps this solution fixed and prevents it from reacting with oxygen?** After adding the sulfamic acid, the pH of the solution makes it impossible for the oxygen to react (this requires basic conditions).

32. **What is the titrant in the Winkler method?** The titrant is the thiosulfate ion ($S_2O_3^{2-}$).

33. **What chemical species, indirectly produced from DO, is being titrated (is the limiting reactant and proportional to the DO)?** Iodine, in the form of I_2, is the limiting reactant that results from the DO.

34. **Why does the modified Winkler method contain sodium azide?** The nitrite ion can cause a systematic error if present in the sample. The nitrite ion reacts with the azide ion, removing the nitrite ion and eliminating the error.

35. **Why might an environmental chemistry lab want to invest in the purchase of an ion chromatograph to quantitate water quality parameters such as NO_3^-, NO_2^-, SO_4^{2-}, and PO_4^{3-} instead of using each of the specific standard methods used for the analytes?** The individual standard methods require separate derivatization methods, whereas an IC could separate and detect a mixture of all of these analytes in one injection since they are all anions. This would save a significant amount of time.

F

COMMON CHEMICAL INSTRUMENTATION

Most of the information in this instrumentation primer was gleaned from Skoog, Holler, and Crouch (2007). You are encouraged to consult this source for more detailed information.

F.1 UV-VIS SPECTROPHOTOMETERS

Spectrophotometers that operate in the range of the ultraviolet (UV) and visible (Vis) portions of the electromagnetic (EM) spectrum are ubiquitous in undergraduate laboratories. It is highly likely that you have already used such an instrument in your general chemistry or biology laboratories. The common instantiation in these settings is either a Spectronic-20 or a smaller, less sensitive instrument (see Figure F.1). While these instruments are quite common, a more sensitive and higher quality instrument is needed for the analytical measurements required for analytes that can be found in the environment at low concentrations. An example of such an instrument can be seen in Figure F.2. Most chemistry departments have at least one of these instruments, and they are usually more sensitive than the instruments shown in Figure F.1.

Figure F.1 The Spectronic-20 (top, larger spectrometer) has been the preferred instrument in general biology and chemistry laboratories for decades. In many cases, they are now being replaced by smaller spectroscopes, such as the Vernier Colorimeter (bottom, smaller spectrometer) due to significantly lower cost. Source: Courtesy Ken Overway.

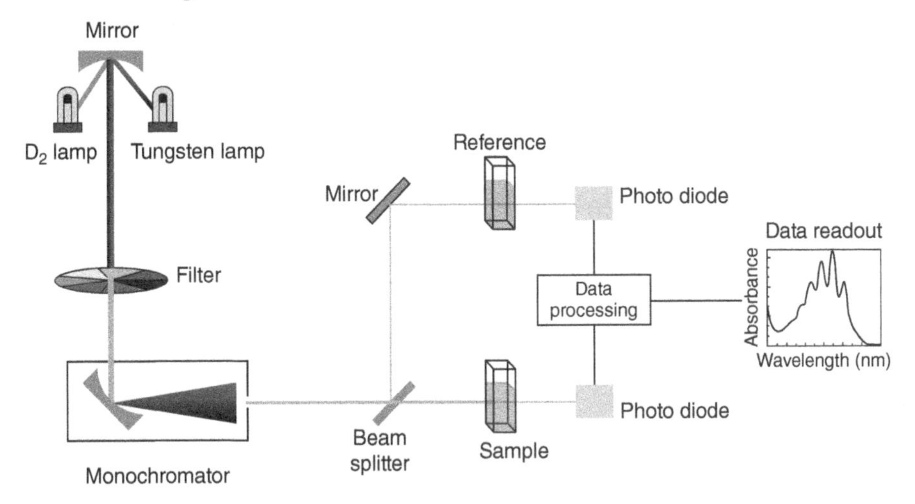

Figure F.2 UV-Vis spectrophotometers are common in biology and chemistry departments, and in upper-level lab courses, students are more likely to use more sophisticated instruments compared to the Spectronic 20. A schematic example of one such instrument uses a combination of lamps to deliver UV-vis radiation to the sample. A monochromator narrows the radiation to a specific band of wavelengths, chosen by the analyst, and thereby increases the instrument sensitivity and selectivity. While mirrors and other optics turn the light, the essential absorption instrument has a linear path from source to detector. Source: Courtesy Sobarwiki.

Environmental Chemistry: An Analytical Approach, First Edition. Kenneth S. Overway.

Companion website: www.wiley.com/go/overway/environmental_chemistry

Tungsten/halogen lamp: A low-intensity, continuum light source that covers some low-energy UV wavelengths, the entire visible spectrum, and some of the near-infrared (IR) spectrum

Deuterium lamp: A low-intensity, continuum light source that covers the UV region of the spectrum

Light-emitting diode (LED): A low-intensity line source that individually emits a narrow band of radiation; several LEDs can cover the visible region of the spectrum

Xenon arc lamp: A high-intensity, continuum light source that covers the UV and visible regions of the spectrum

Continuum sources emit a broad spectrum of polychromatic radiation centered in some region of the EM spectrum. Line sources emit a very narrow band of monochromatic radiation.

Table F.1 Typical light sources found in common absorption and fluorescence spectrophotometers.

☞ *Sensitivity* is a measure of the ability of an instrument or method to distinguish between small changes in the concentration of the analyte. It is manifested as the slope of the calibration curve. *Selectivity* is a measure of the ability of an instrument or method to detect only a set of analytes while ignoring other components of the sample matrix. Both sensitivity and selectivity are *Figures of Merit*, factors that make one instrument or method of analysis better than another.

A *scanning* UV-vis spectrophotometer uses the absorption of light by an analyte as a means of quantitation. In a scanning spectrophotometer, the light from a radiation source (see Table F.1) is sent into a monochromator (a device that disperses light and selects one wavelength) and then is focused into the sample container (usually a cuvette). The light that is not absorbed is directed toward a photonic detector, which measures the intensity of the light. A sample blank is measured first, and the intensity is designated at I_0. This represents the intensity of light when the analyte is not present. When the sample is measured, the intensity is assigned to I. The resulting absorbance is a logarithmic calculation based on the ratio of the two intensities, as described with Beer's Law in Chapter 1 in Eqs (1.7) and (1.8) on page 23.

$$Abs = -\log\left(\frac{I}{I_0}\right) \tag{F.1}$$

$$Abs = \varepsilon bC \tag{F.2}$$

Equation (F.1) converts the blank intensity and the sample intensity to an absorbance value, which is proportional to the relatively low concentrations of the analyte (described in Eq. (F.2) in the form of Beer's Law). If calibration standards are used (review Sections 2.3 and 2.6 in Chapter 2) to establish the value of εb (the molar absorptivity times the path length), which is the slope of the calibration curve, then samples containing an unknown concentration of the analyte can be quantitated.

UV-vis spectrophotometers can be found in other variants. Some instruments do not use a monochromator to disperse the light but send the full spectrum of light from the source through the sample. The transmitted light is then dispersed into its individual colors, and this dispersed spectrum is detected by a multielement detector called a *photodiode array*. The "elements" in this case are not from of the periodic table, but they are individual photodiodes that measure the light intensity. Since the dispersion of the transmitted light spreads it out in a single direction, the photodiode on one end of the array detects the blue light and the photodiode on the other end of the array detects the red light. Thus, the entire absorption spectrum of an analyte can be measured simultaneously. This contributes to a speed and convenience advantage, but usually not a sensitivity advantage over a scanning instrument (described earlier) because photodiodes are not as sensitive as photomultiplier tubes (the common detector for scanning instruments). There are other instruments that use many fiber optics to simultaneously measure the absorbance of several samples when they are placed in a plate containing many sample wells. The well plates, described in Section 2.7.5.3 on page 83, can significantly decrease the measurement time if several samples are to be measured. These spectrophotometers are often called *plate readers* and use roughly the same principle as scanning and diode array instruments.

F.1.1 Turbidity

UV-vis spectrophotometers can also be used to measure the turbidity of a solution, which is a quantification of the suspended solids. Insoluble particles scatter light that is passed through the sample. This is an important measure of water quality and is also used commercially to determine the clarity of beverages such as drinking water, beer, and wine. In a linear optical path instrument, such as a UV-vis spectrophotometer, light that is scattered by the suspended solids does not reach the detector. When compared to a blank, this apparent absorbance is proportional to the density of the suspended solids. Turbidity measured in this absorbance mode is only appropriate for solutions with relatively high turbidity since the detector noise levels improve when the blank and sample are significantly different. Low-turbidity solutions should be measured using a nephelometer, which is described in Section F.2.1.

As with most other analytical methods, turbidity measurements require a primary standard to calibrate the instrument. Turbidity standard is usually a solution of formazin, a polymer produced when hydrazine sulfate and hexamethylenetetramine are mixed. The turbidity unit is the TU, equivalent to the NTU (nephelometric turbidity unit) when measured in nephelometry. A turbidity measurement can be related to suspended clay particles in water, where 1 NTU is equivalent to 1 mg/L of suspended kaolin clay. The formazin standard can be purchased at different concentrations and can be diluted to make calibration standards to be used in one of the methods of quantitation described in Chapter 2.

F.1.2 Quantitation

When using a UV-vis absorption spectrophotometer to quantitate environmentally significant analytes, it is important to first determine the spectral location where the species absorbs radiation. This is accomplished by running a spectral scan on one of the calibration standards. Usually, each analyte will have its own characteristic *absorption spectrum*, such as the derivatized nitrite ion from EPA Method 354.1. The absorbance spectrum is measured by placing a cuvette containing a dilute solution of derivatized sample. The peak absorbance, which will result in the most sensitive measurement, is used in one of the methods of quantitation described in Chapter 2 using Beer's Law.

Many environmental methods of analysis make use of absorption spectrophotometry as the method of detection, such as EPA Methods 354.1 (nitrite ion) and 352.1 (nitrate ion), and *Standard Methods* 4500-NO3-E (nitrate ion) and 4500-P-E (phosphate ion).

☞ Absorption spectrophotometers come in several forms. *Scanning* instruments have *monochromators* and step through a range of light to measure the *absorption spectrum* of an analyte and quantitate its concentration. These instruments are usually very sensitive. *Photodiode array* instruments can instantaneously measure an absorption spectrum of an analyte and quantitate its concentration. *Plate readers* can quantitate the concentration of many samples once the correct absorption wavelength is known. Simple spectrophotometers, as seen in Figure F.1, are a cost-effective, albeit less useful and sensitive, way to introduce students to spectrophotometry.

F.2 FLUOROMETERS

Fluorometers harness additional degrees of selectivity compared to UV-vis spectrophotometers. Some molecules that absorb a photon hold on to the energy for a few nanoseconds and then release it as photon. This process is called *fluorescence*, and it is a form of *luminescence* similar to *phosphorescence*, *chemiluminescence*, and *bioluminescence*. In brief, the analyte is excited with a photon, there is some loss of energy through nonradiative mechanisms, and then the analyte emits a photon that is usually red-shifted from the excitation wavelength. The red shift, known as a *Stokes shift*, is an important source of selectivity that allows the use of very powerful light sources while preventing the excitation photons from reaching the detector. Another source of selectivity comes from the rarity of fluorescence itself – since most molecules do not have the ability to fluoresce, the number of contaminants in the sample and standards that can contribute to the background noise is very small. The combination of these important sources of selectivity often allows fluorescence measurements to have lower detection limits compared to absorption measurements.

The rarity of fluorescence also becomes an important limitation of this method since it is very likely that the environmental chemical of interest will not have a fluorescence signal. This can be remedied by derivatization, where a fluorescent tag can be attached to the analyte, similar to the derivatization of some of the common water quality methods.

☞ An *absorption spectrum* is a graphical depiction of the range of EM radiation that an analyte absorbs. Its maximum absorbance is often used to quantitate the amount of analyte present.

☞ *Luminescence* is a category of processes that result in an atom or molecule emitting a photon. *Fluorescence* and *phosphorescence* are caused by photonic excitation. The only practical difference between them is their lifetime (fluorescence is over in nanoseconds after the excitation, whereas phosphorescence can last a few seconds), and phosphorescence is exceedingly rare. *Chemiluminescence* is caused by a chemical reaction, which results in a product that starts in an excited electronic state. The product eventually emits a photon as the electron falls to a ground-state orbital. *Bioluminescence* is the same process that occurs in biomolecules. You will observe fluorescence and phosphorescence at a bowling alley during a kid's night when the black lights are used. Chemiluminescence is observed in glowsticks.

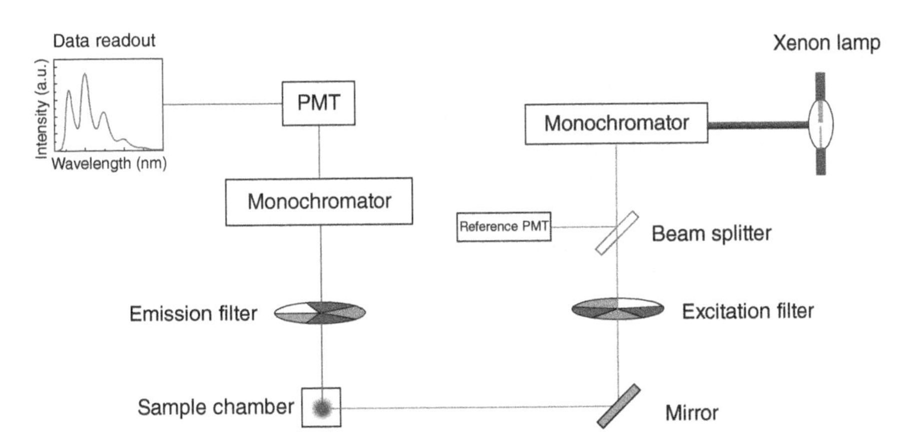

Figure F.3 A fluorescence spectrophotometer or fluorometer. Source: Courtesy Sobarwiki.

Figure F.3 shows the major components of a typical fluorometer. You will notice that many of the components are similar to a UV-vis spectrophotometer. In this figure, the light source is a xenon arc lamp, which emits both UV and visible radiation. Fluorometers benefit from intense light sources because the modified Beer's Law equation that links fluorescence signal to analyte concentration contains the source intensity term (P_0).

$$F = P_0 \Phi \varepsilon b C \tag{F.3}$$

Intense light sources increase the sensitivity of the measurements. The other terms should be familiar to you from Beer's Law except the Φ term, which represents the quantum yield of the analyte – a fractional measure of how many excited analyte molecules actually emit radiation. The two key features that distinguish a fluorometer from an absorbance spectrophotometer are the presence of two monochromators and a 90° optical path. Since fluorescence is an emission of radiation, the detector needs to look at the analyte emission and not the light source. Thus, the detector usually looks at the sample cuvette at an angle 90° from the incident excitation beam. This allows the emission monochromator to remove nearly all of the excitation radiation and allows for the use of very sensitive detectors compared to those in absorption instruments. Ideally, the fluorescence signal blank is the result of the detector seeing zero photons and any photons reaching the detector are from fluorescence. In an absorption instrument, the blank signal results from the detector seeing the full intensity of the source, so the detector cannot be very sensitive or it will produce excessive current for the blank and could be damaged or produce nonlinear results.

F.2.1 Nephelometry

A turbidity measurement made with the detector at a right angle to the line of the light path is called nephelometry. In this mode, the amount of scattered light detected is proportional to the turbidity of the solution. A fluorometer can be used to make nephelometric measurements because they share the same design. Nephelometers, specifically designed for measuring turbidity, are simplified fluorometers. Turbidity measured in the nephelometric mode is only appropriate for solutions with low turbidity, else the relationship between the scattered light and the density of suspended solids becomes nonlinear. Highly turbid solutions should be measured using a turbidometer, which is described in Section F.1.1.

☞ *Fluorometers* measure the emission of an analyte as set by λ_{ex} and λ_{em}. The difference between these wavelengths is the *Stokes shift*, which results from energy losses in the excited state of the analyte before emission and improves the measurement by allowing the excitation wavelength to be removed. High-intensity sources, such as lasers, make fluorometers very sensitive. The selectivity enhancements from the Stokes shift, monochromators, and the reduction in interferences (fluorescence is rare among potential interferents) usually result in a lower detection limit compared to absorption measurements.

F.2.2 Quantitation

Equation (F.3) shows that fluorescence signal is proportional to the analyte concentration, so quantitation in fluorescence methods is ideally done using linear regression. Since fluorescence is an emission, the characteristic excitation and emission wavelengths of the analyte must be determined. These can be looked up in a table if the analyte or fluorescent tag has a known λ_{ex} and λ_{em}. For a complete unknown, this is accomplished by running a blind emission scan (setting the excitation monochromator to 0 nm and very narrow slits) to determine the approximate λ_{em}, then running an excitation scan (parking the emission monochromator at λ_{em}) to determine the λ_{ex}, and then rerunning the emission scan (with the excitation monochromator parked at λ_{ex}) to determine a more precise λ_{em}. Once these are determined, the monochromators of the fluorometer are set to λ_{ex} and λ_{em}, respectively, and the fluorescence signal is measured for the calibration standards and unknowns.

Fluorescence detection is used in several methods, usually in conjunction with liquid chromatographic separation, extraction, or with postcolumn derivatization of the analyte. Some of these methods include EPA Method 547 (glyphosate herbicide) and EPA Methods 500 and 550.1 (oil contamination of water).

F.3 ATOMIC ABSORPTION SPECTROPHOTOMETERS

Atomic absorption spectrophotometers (AAS) work on the same principle as UV-vis spectrophotometers. An AAS uses the absorption of atoms (instead of molecular analytes in UV-vis spectrophotometers) and Beer's Law to quantitate analytes in a sample. The instrument begins with a *hollow cathode lamp* (HCL) as the light source. This light source passes through a sample container (see the following subsections). The transmitted light passes into a monochromator, which selects a specific wavelength, and then onto a detector. If the choice of the HCL matches the analyte present in the flame, then an absorption occurs ($I \neq I_0$) and Beer's Law can be used. A simplified diagram of this instrument can be seen in Figure F.4.

☞ *Hollow cathode lamps* (HCL) are low-intensity line sources that emit in the UV and visible regions of the EM spectrum. Each lamp is usually designed for a specific element, so each analyte requires its own lamp. The cathode in the lamp contains the metal of interest, and electronic excitation of the metal produces radiation that is characteristic of and uniquely absorbed by the metal.

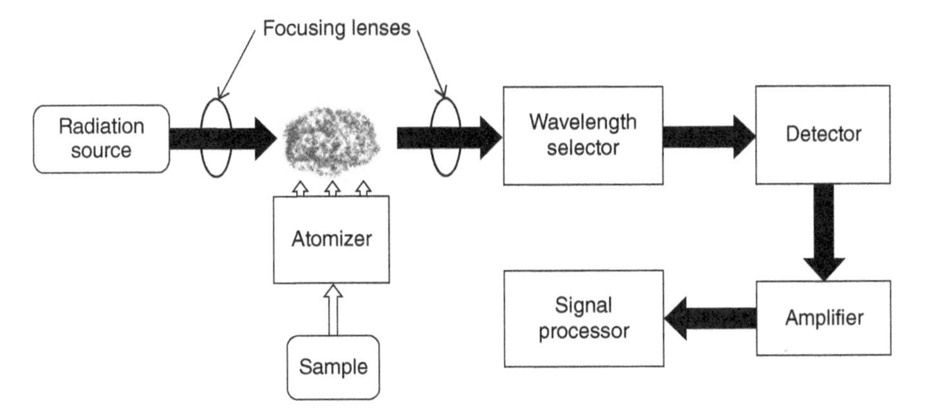

Figure F.4 A simplified block diagram of an atomic absorbance spectrometer. Source: Courtesy K05en01.

☞ A *nebulizer* is a device that draws the liquid sample into a chamber where it is sprayed as a mist and mixed with the flame's fuel and oxidant gases. A series of baffles inside of the nebulizer prevent all but the finest mist from passing through the chamber and into the flame.

F.3.1 Flame Atomization

One way an AAS sample can be prepared is by means of flame atomization. In this method, the sample "container" is the flame and the sample, which is in the solution phase, is first dispersed into a very fine mist with a *nebulizer* and then atomized in a flame. The flame heats the sample to the point that all molecular bonds are broken and only neutral atoms remain (ideally). Once in its neutral, atomic form, the analyte can absorb the light that is produced by the HCL. Different combinations of fuel and oxidant provide different flame temperatures, which is important for striking a balance between enough heat to atomize the sample but not enough heat to ionize the sample.

The advantage that flame atomization conveys to the measurement is one of reproducibility. The relative standard deviation of the analyte signal is usually below 1%, giving the measurement relatively high precision. The disadvantage is one of sensitivity. Most of the sample does not end up in the flame since the baffles in the nebulizer block the larger and medium droplets from getting to the flame. This portion of the sample drains out of the nebulizer and into the waste container; thus, the sample needs to be relatively high in concentration since only a fraction of it will be detected.

F.3.2 Electrothermal Atomization

A second common way AAS delivers the analyte to the sample chamber is by means of a *graphite furnace* by which *electrothermal atomization* is accomplished. In this method, the sample (usually in the solution phase but sometimes in the solid phase) is placed in a graphite tube. After a brief time of heating to evaporate the solvent, the graphite tube is heated up to 3000 °C by running a very high current through the tube (thus the term *electrothermal atomization*). At this temperature, the sample is vaporized and atomized. The light from the HCL is directed through the graphite tube and then into the monochromator and detector.

The advantage of this method of atomization is one of sensitivity. Nearly all of the analyte is vaporized in the graphite tube and available for absorption. Further, the vapor remains in the tube for up to a few seconds, giving the detector and electronics time to perform signal averaging and increase the signal-to-noise ratio. The disadvantages of electrothermal atomization are low reproducibility (5–10% RSD) compared to flame atomization, which can be corrected by using an internal standard. Since the graphite tube must be cooled between samples, the time for analysis is also longer.

F.3.3 Summary

Thus, the two common means of atomization are complementary in their advantages and disadvantages. Flame atomization provides a precise and quick method of sample delivery, whereas electrothermal vaporization provides a high sensitivity, low precision, and moderately slow means of sample delivery.

☞ A *method* is a set of experimental conditions that optimize a measurement. These conditions can be stored in a digital file and reloaded each time the same measurement is made in order to provide consistency and the best possible measurement conditions. Methods include items such as the wavelength of the source, temperature of the sample, spectral scanning range, gas pressures, alarms when certain parameters are out of compliance, and many other parameters that are specific for the analyte and the instrumentation. Method development is a highly valued skill, employing industrial scientists to develop methods for optimizing the measurement or the production of pharmaceuticals, food additives, environmental contaminants, and many other commercial products.

There are many methods of analysis that employ AAS instruments, such as EPA Method 7473: Mercury in Solids and Solutions by Thermal Decomposition, and *Standard Methods* 3111: Metals by Flame Atomic Absorption Spectroscopy Methods.

F.3.4 Quantitation

Quantitation methods in the context of AAS instruments rely, either directly or indirectly, on the Beer's Law relationship. When using flame atomization or electrothermal vaporization, the primary measurement made is absorbance. This measurement starts with the selection of an analyte and the HCL that is specific to the element. The instrument software will allow the analyst to choose among one or more characteristic wavelengths that the analyte will absorb. Often each of these wavelengths offers a different sensitivity since the analyte absorbs some of these wavelengths very strongly and some wavelengths weakly. Further complications include the choice of the optimal flame chemistry (air:acetylene, nitrous oxide:acetylene, and others) and the avoidance of certain interfering contaminants. Fear not, for most of these instrumental parameters have been optimized and described in a manual that usually accompanies the instrument or can be found in a search of the Internet. When setting up an instrument for a measurement, these parameters will be selected and stored in an instrumental *method* and will be set by the computer software that communicates with the instrument (or can be set by the analyst in the case of older instrumentation). Once these parameters are set, the sample must be prepared and delivered to the instrument according to the method. The signal in AAS is an absorbance, and it is used in Beer's Law through the chosen method of quantitation.

F.4 INDUCTIVELY COUPLED PLASMA INSTRUMENT

While atomic absorption instruments rely on flame and electrothermal atomization to convert the analyte into elements, inductively coupled plasma (ICP) instruments use plasma flames to atomize analytes. Conventional flame temperatures, used in AAS, range from around 2000 to over 3000 °C (depending on the flame chemistry). As mentioned earlier, these temperatures can be too hot such that elements are put into an excited electronic state or ionized, or too cold such that compounds found in the analyte are not sufficiently atomized. Either case leads to a reduction in signal, but the "goldilocks" zone for each element is achievable. An ICP torch reaches temperatures near 10,000 K, making atomization of even the most stubborn compound inevitable. Temperatures in the plasma torch are so hot that the atoms and ions of the analyte are promoted into excited electronic states. From here, the elements are analyzed using their emission (ICP-OES) or by their mass (ICP-MS).

Instead of relying on the absorption of radiation as AAS does to quantitate, ICP optical emission spectrometry (ICP-OES) relies on the emission of the excited atoms and ions. The radiation emitted by these species is highly characteristic of the element, spectrally narrow, and each element has multiple emissions. Because of these characteristics, ICP-OES instruments can isolate and quantitate the concentration of around 70 elements simultaneously. This is the first major advantage of ICP instruments over AAS instruments.

ICP-OES instruments have comparable detection limits to AAS-electrothermal instruments, both of which have usually 10 to 100 times lower detection limits compared to AAS-flame instruments. The major disadvantage of ICP-OES instruments over AAS instruments is cost – ICP-OES instruments are much more expensive.

Instead of relying on the optical emission of the analyte (as in ICP-OES), attaching a mass spectrometer (MS) to the ICP torch allows the analyst to quantitate the analyte based on the mass of each atom. The ICP-MS instrument injects the sample, after it has passed through the ICP torch into a vacuum chamber where the elements are sorted based on their masses (actually mass-to-charge ratio). Mass spectrometers have inherently lower detection limits compared to optical instruments and isotopes of elements can be identified, giving ICP-MS an advantage over ICP-OES. Mass spectrometers, however, require high vacuum conditions and various pumps to maintain the vacuum, so ICP-MS instruments are usually more expensive compared to ICP-OES instruments.

F.4.1 Summary

ICP instruments, compared to AAS instruments, carry the main advantage of simultaneous detection of multiple elements. This equates to a large decrease in the analysis time when several different elements are the target of the analysis. Second, ICP instruments usually have the same detection limits as AAS-electrothermal or better.

There are many methods of analysis that employ ICP instruments, such as EPA Method 200.7: Determination of Metals and Trace Elements in Water and Wastes by ICP-OES, and *Standard Methods* 3120: Metals in Water by Plasma Emission Spectroscopy.

F.4.2 Quantitation

The process for using an ICP instrument for quantitation is very similar to quantitation in AAS. The sample is processed according to the needs of the quantitation method and instrument. The instrumental method is developed to optimize the measurement parameters. In an ICP-OES measurement, the signal measured is an emission signal from the analyte. This signal is measured by selecting the characteristic wavelength for the analyte by using a monochromator and then converting the light into an electrical signal with a detector. In an ICP-MS measurement, the signal is either a total ion current (TIC) or a reconstructed ion current (RIC) – both are proportional to the amount of analyte present in the sample. Each of these signals is collected, as defined in the instrumental method, and used in a modified Beer's Law equation that can be summarized as

$$S = mC + S_{bl} \tag{F.4}$$

☞ In a *mass spectrometer*, the analyte is ionized and, depending on the ionization process, fragmentation often results. All of the ion fragments strike an ion detector after they have been sorted by mass. When each ion strikes the detector, a small current is generated. The sum of all of these currents becomes the TIC. This signal is proportional to the amount of analyte. In the presence of interferences, the analyst can choose to measure only a single ion fragment that is unique to the analyte. This signal, called the reconstructed ion current (RIC), often improves the measurement if interferences are present.

where S is the signal received from the instrument, m is the slope of the measurement (as found on the calibration curve), C is the concentration of the analyte, and S_{bl} is the signal of the blank. This linear equation is functionally the same as the version of Beer's Law used in any of the various methods of quantitation described in Chapter 2.

F.5 CHROMATOGRAPHY

☞ In most chromatography techniques, there is a *stationary phase* and a *mobile phase*. The mobile phase is the liquid or gas solvent into which the sample is injected. The sample and mobile phase move through the stationary phase under the pressure of a pump, high gas pressure, or via gravity. As the mobile phase *elutes*, the components of the sample mixture interact with the stationary phase. Strong interactions lead to a slow-moving component, and weak interactions lead to a fast-moving component. By the time the mixture elutes from the *column*, the components have been separated in time and the faster moving components come out first. The graphical record of this separation is called a *chromatogram*.

Almost all of the samples that you may want to collect and analyze from the environment will contain a mixture of various compounds. Some of the instruments mentioned previously, such as AAS and ICP instruments, have the ability to select only the analyte out of the mixture for measurement and ignore most or all of the other contaminants. These methods are very selective, but they achieve this selectivity by completely destroying the sample. This is not a problem if only an elemental analysis is required, but it does not allow an analyst to distinguish between different compounds. If an analyst wanted to determine whether chlordane ($C_{10}H_6Cl_8$), dieldrin ($C_{12}H_8Cl_6O$), or dichlorodiphenyltrichloroethane (also known as DDT, $C_{14}H_9Cl_5$)[1] was contaminating a local stream, these compounds would yield similar results if analyzed using AAS and ICP methods since they are all made of carbon, hydrogen, chlorine. In order to identify the offending insecticide, the mixture must be separated from other interfering components and other potential insecticides (maybe the sample contains more than one of these compounds). This is where chromatography can be used.

Chromatography, literally "color writing," was first developed for separating dyes that were observable with the unaided eye. Now it is applied to any separation of a mixture of compounds that have different molecular weights, polarities, sizes, and chirality. In a chromatograph, a sample mixture is injected into a long tube (called a *column*) and, based on the strength of the interaction of the mixture components and the *stationary phase*, a separation of the components is achieved. This enables each component of the sample mixture to be selectively detected and quantitated without the interference of the other components.

The three major types of chromatographic instruments you are likely to find in a typical undergraduate laboratory. Gas chromatographs (GCs) require that the sample and solvent be volatile and stable at temperatures below 300 °C (above this most stationary phases will begin to bleed off from the column). The separation occurs in a capillary column that is a typically 10–100 m long glass tube with the stationary phase forming a thin coating on the inner walls and the mobile phase as an inert carrier gas such as helium or nitrogen, as seen in Figure F.5. Another common column is a packed column, which consists of metal or glass tube packed with inert beads that are coated with the stationary phase. The sample is injected with a syringe into an injector port where it is vaporized and swept into the column with the carrier gas. The sample components are separated based on their boiling points and interaction with the stationary phase and elute from the column and through a detector. There is a wide array of detectors that can be added to the output of a GC to detect a variety of analytes. Electron-capture detectors (ECDs) are selective for organohalide compounds (such as the insecticides mentioned before) and nitrogen–phosphorus detectors (NPDs) are selective for compounds containing nitrogen and phosphorus (many pesticides, such as parathion and atrazine). More generalized GC detectors include a flame ionization detector (FID), which detects most organic compounds, or the truly universal detector known as a thermal conductivity dectector (TCD).

High performance liquid chromatographs (HPLCs) do not require volatile analytes, only that the sample be soluble in the mobile phase. The mobile phase can range from a

[1]Chlordane, dieldrin, and DDT are all organochlorine insecticides that were commonly used in the middle 20th century and now are all banned.

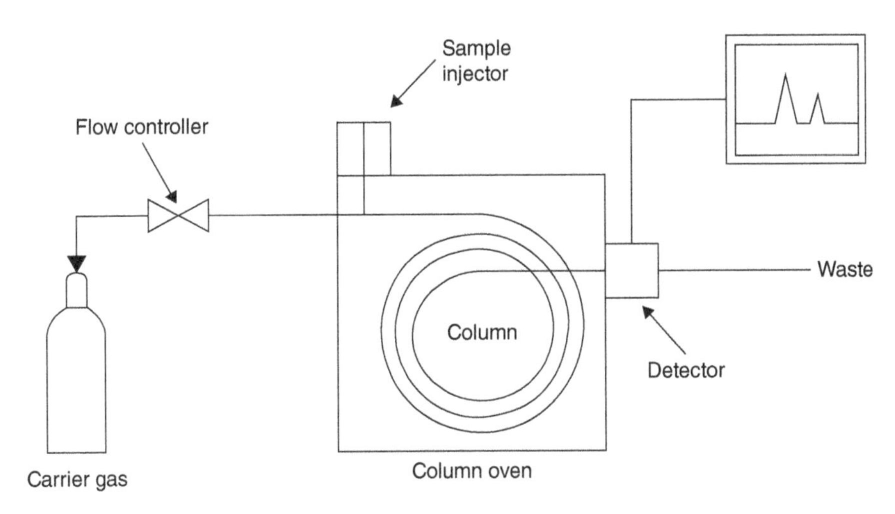

Figure F.5 A simplified diagram of a gas chromatograph. Source: Courtesy Offnfopt.

polar aqueous solution to a nonpolar hexane solution. The stationary phase is a coating on microbeads that are packed into a steel tube that is typically 10 cm to 25 cm long. The sample is injected with a syringe into the column, and the individual compounds are separated based on their differential affinities for the mobile and stationary phases. HPLCs have a limited range of detectors, and you are only likely to find an HPLC with a UV absorbance detector – sufficient for most organic compounds.

A third common chromatograph is a ion chromatograph (IC), which specializes in ionic analytes, such as nitrate, nitrite, phosphate, halides, and most metal cations. The stationary phase is a cationic or anionic resin packed into a steel or plastic column typically 25 cm long, and the mobile phase is an aqueous solution with a variable ionic strength. The components are separated based on the differential affinities for the mobile and stationary phases. Anionic analytes require a cationic stationary phase resin and vice versa. ICs typically employ a conductivity detector.

Some HPLCs and many GCs are sold with a type of mass spectrometer as the detector. The type of mass spectrometer can vary in design, but all of them add an additional dimension of information compared to the detectors described earlier. In addition to a chromatogram with retention times, mass spectrometers record a mass spectrum for each analyte as it elutes and is detected. Mass spectrometers ionize the analyte, which often causes the analyte to fragment. The fragmentation pattern (see Figure F.6) is similar to a fingerprint that can be compared to a database of mass spectra, which provides the potential identity of the analyte if the two spectra match. Even without fragmentation, the analyte's mass is measured and can provide some information about the analyte if it is unknown.

F.5.1 Quantitation

There are a wide variety of chromatographic instrument designs, but each contains some sort of detector that produces an electrical signal that is proportional to the concentration of the analyte. This signal is plotted as a function of time to produce a *chromatogram*, as seen in Figure F.6a. To quantitate the sample, the analyst chooses either the peak height or the peak area and then uses Eq. (F.4) and a quantitation method to determine the concentration of a single or several analytes. Most modern chromatography instruments have software that will automatically measure the peak height or integrate the peak area after the chromatogram has been collected.

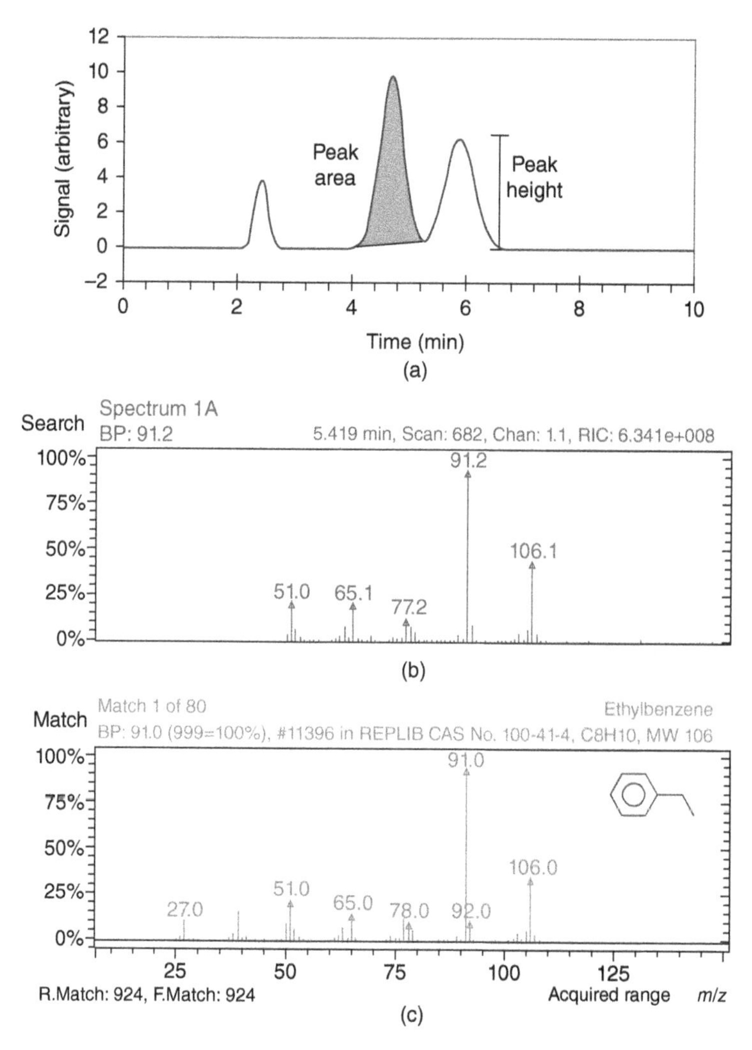

Figure F.6 A *chromatogram* (a) shows the analyte signal as a function of time since the sample mixture is separated by the chromatography method and detected as each analyte elutes from the column. The peak height and peak area are proportional to the concentration of the analyte. When a mass spectrometer is used as the detector, ionization of the analyte weakens the molecular bonds and causes a reproducible fragmentation pattern. This *mass spectrum* can be compared to a library of patterns previously recorded from known samples. Computer software allows the analyst to click on a peak in the chromatogram to display the mass spectrum of the analyte (b). A search of the library can provide a probable identification of the compound (c).

Standardized chromatographic methods of analysis are wide-ranging, including EPA Methods 505 (GC-ECD for pesticides), 525.2 (GC-MS for chlordane), 8015B (GC-FID for nonhalogenated organics), and 531.2 (HPLC for carbamate insecticides).

Bond type	Substituent class	Frequency absorbed (cm^{-1})
C–H	Alkane	2850–2970
O–H	Alcohol	3200–3650
N–H	Amine, amide	3300–3500
C=C	Alkene	1610–1680
C≡C	Alkyne	2100–2260
CO_2		3400–3850
CO		2000–2500
NO_2		1550–1900

Table F.2 A select list of some of the common organic functional groups and gases and their generalized absorption bands.

F.6 INFRARED SPECTROMETRY

If you have taken an organic chemistry course and/or laboratory, then you have probably learned about or used an infrared spectrometer. These instruments allow an analyst to determine several structural characteristics of an unknown organic compound. Mid-infrared radiation, in the range of 2–25 μm (about 400–4000 cm^{-1}), is sent into a gaseous, liquid, or solid sample and certain wave numbers are absorbed by certain organic functional groups. Table F.2 shows a brief list of some of the many characteristic absorption bands.

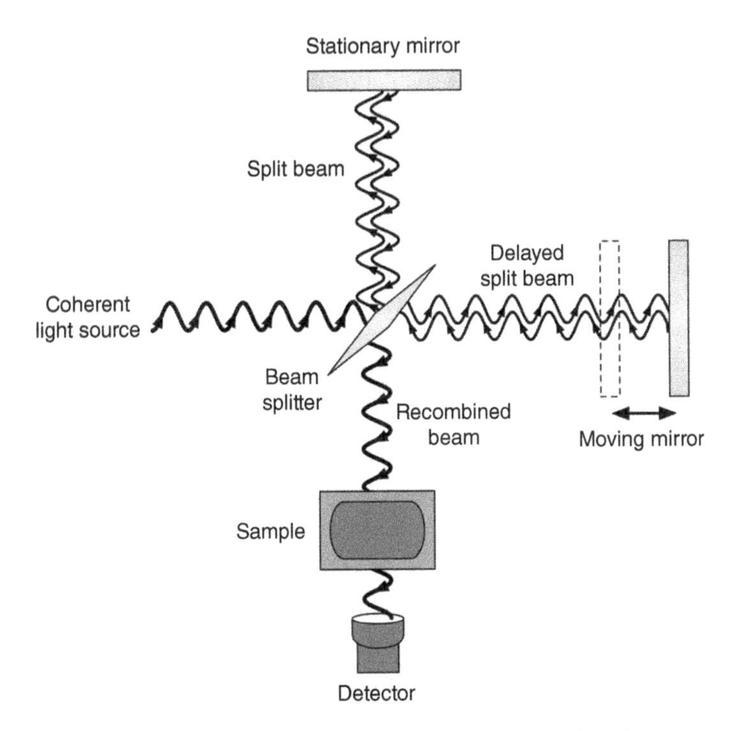

Figure F.7 A diagram of the interferometer of a Fourier Transform infrared spectrometer. Source: Courtesy Sanchonx

Infrared spectrometry is commonly performed using a Fourier transform infrared spectrometer (FTIR). These instruments work on the basis of absorbance but use a more complicated optical design, which provides faster measurements and improved resolution over scanning absorbance instruments working in the IR range. The interferometer (Figure F.7) at the heart of an FTIR splits the IR radiation from the source into two beams that travel along different paths, one of which reflects off a fixed mirror and the other reflects off a movable mirror. When the two beams of IR recombine, they experience constructive and destructive interference as the mirror moves. Imagine, first, that only a single wavelength of IR travels through the interferometer. When the movable mirror is at the same distance from the center as the fixed mirror, the light waves from the two paths are in phase, the constructive interference is maximized, and the detector registers a maximum signal. As the mirror moves, the constructive interference decreases, the destructive interference increases, and the detector registers less and less signal until the destructive interference is maximized when the waves are 180° out of phase. If the mirror continues to move in the same direction, the constructive interference will begin to increase until it maximizes again. Hopefully in your mind, you are imagining a sine wave representing the interference pattern, where the amplitude is the intensity of the radiation and the x-axis is the distance the mirror has moved. This wave is the reconstruction of the electromagnetic wavelength of the radiation, which is determined by the speed of light and the time-based frequency of the oscillating wave, to a new wave that is a function of the movement of the mirror. The detector electronics are not fast enough to reproduce the high-frequency electric field of an IR photon, but in an interferometer, the detector is fast enough to follow the new wave of the interference pattern based on the movement of the mirror, which is much slower by comparison. If the sample is placed in the interferometer and the sample absorbed that single IR wavelength, then the detector would register an absorbance signal. Now imagine the entire spectrum of IR photons racing through the interferometer – the interference pattern becomes very complex and the absorption of the sample, as a function of the IR wavelengths, is not obvious.

Applying a Fourier transform to the interferometer interference pattern converts the unintelligible signal into a spectrum of absorbance bands. Think of a Fourier transform

☞ *Wave number* ($\tilde{\nu}$), similarly to wavelength (λ), frequency (ν), and energy, can be used to describe properties of light. It is usually used for photons in the IR and microwave regions of the spectrum. Most IR spectra use wave number as the x axis. The wave number is the reciprocal of the wavelength and usually has units of cm^{-1}.

$$\tilde{\nu} = \frac{1}{\lambda} \tag{F.5}$$

For example, 973 nm is 10,300 cm^{-1} and can be calculated as follows.

$$\tilde{\nu} = \left(\frac{1}{973\ \text{nm}}\right) \times \left(\frac{10^9\ \text{nm}}{1\ \text{m}}\right)\left(\frac{1\ \text{m}}{100\ \text{cm}}\right) = 10277.49\ \text{cm}^{-1}$$

as the digital or analog display on a radio dial. When the display reads 90.7 MHz, you know that the actual radio wave is an oscillating electric field, but the display reduces the oscillating field to a single number that represents the oscillation. A Fourier transform of the interference pattern created by the moving mirror (measured in cm) generates a spectrum with the units of wave numbers. Thus, the entire IR spectrum from the source can be scanned very quickly to obtain absorbance information about the sample. In fact, this method of spectral scanning is much faster than the tradition scanning with a monochromator, such as in a UV-vis spectrophotometer, and conveys additional advantages of sensitivity and resolution. As such, most modern instruments use this method.

You may remember from the description of the greenhouse effect in Section 3.6.2 that greenhouse gases absorb infrared radiation. This makes infrared spectrometry a natural tool for determining the identity and quantitation of a greenhouse gas. Infrared spectrometry is also useful in helping to characterize the nature of soil organic material. Unfortunately, the absorption bands in infrared spectrometry are usually very wide compared to the very narrow bands observed in atomic absorption spectroscopy (mentioned in Section F.3). This limits the ability of infrared spectrometry to spectrally separate the components of a mixture as selectively as AAS or ICP methods, but it is possible.

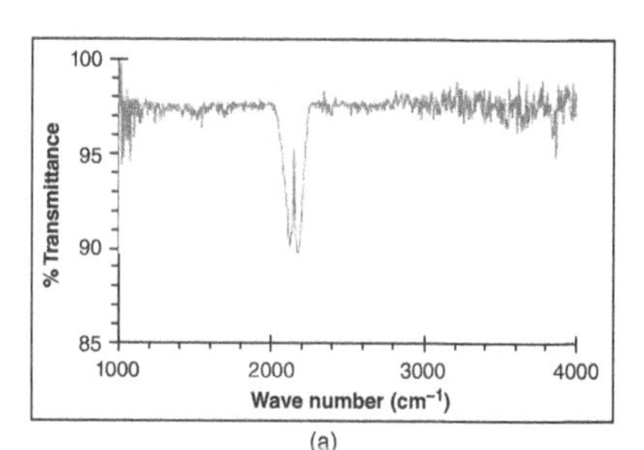

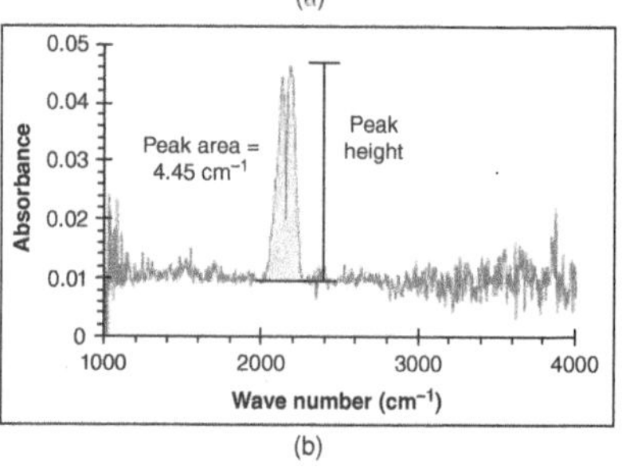

Figure F.8 (a) Infrared spectra are often recorded as % Transmittance versus Wave number. (b) A more useful graph for quantitation is one of Absorbance versus Wave number. The peak height and peak area are proportional to the analyte concentration, but peak area is often more useful when the absorption shows a complex structure, as is the case for this spectrum of carbon monoxide.

F.6.1 Quantitation

An infrared instrument will collect a spectrum of a sample that reveals the wavelengths where absorption occurs. These absorptions are either recorded as % transmittance or as absorbance. Since absorbance is proportional to the analyte concentration (Beer's Law), either the graph needs to be converted to an absorbance graph, as in Figure F.8, or the peak height (in transmittance) must be recalculated as absorbance (see Eq. (F.1)). Figure F.8 shows the difficulty in using peak height in the context of IR spectra. When an absorption has a complex structure, choosing which peak to measure is arbitrary, but it also provides a method of controlling sensitivity. Using a weak absorption peak allows the analyst to quantitate high-concentration samples (low sensitivity). A strong absorption peak provides high sensitivity. Figure 3.14 on page 110 shows an IR spectrum for CO_2, where the absorption near 3800 cm^{-1} is weak and the absorption at 2400 cm^{-1} is strong. Since sensitivity is defined as the slope of the calibration curve, a higher signal for the same concentration leads to greater sensitivity. For this reason, peak area is the preferred signal in the case of infrared spectra.

Most modern infrared instruments, mostly of the Fourier transform design (FTIR), are controlled by software that provides the functionality to integrate peaks and measure heights. Once samples are prepared and measured according to a chosen method of quantitation, the resulting infrared spectra are transformed into absorbance spectra, unoverlapped peaks associated with specific analytes are integrated (or their heights are measured), and the integrated peak areas are used according to Eq. (F.4) to determine the concentration of unknown samples.

Infrared spectroscopy is used in several methods, such as EPA Method 8440 (petroleum vapor).

EXERCISES

F.6.2 UV-Vis Spectrophotometry

1. What function does a monochromator perform in a spectrometer?
2. Explain how Beer's Law is connected to the interaction of photons and matter.
3. What is the difference between scanning and diode array spectrophotometers? What advantage does each provide?
4. How is a line source different from a continuum source?
5. What is the difference between a spectrophotometer that stores the sample in a cuvette compared to one that reads a well plate? What advantage does each provide? Review Section 2.7.5.3 on page 83.

F.6.3 Fluorometers

6. What is the difference between the instrument design of a fluorometer and an absorption spectrophotometer?
7. Why would a fluorometer signal contain less interferences compared to an absorbance signal? Provide two reasons.
8. What do λ_{ex} and λ_{em} represent in a fluorescence measurement?

F.6.4 Atomic Absorption Spectrophotometry (AAS) and ICP-MS/OES

9. What interaction between light and matter does atomic absorption spectrophotometery require in order to quantitate element concentrations?
10. What two methods of atomization are common for AAS?
11. What happens to the analyte signal in AAS using flame atomization when the flame is too hot or too cold?
12. What is the major advantage of flame atomization over electrothermal atomization? What is it about electrothermal atomization that causes its disadvantage?
13. What is the major advantage of electrothermal atomization over flame atomization? What is it about flame atomization that causes the disadvantage?
14. What is the main advantage that ICP instruments have over AAS instruments when analyzing an environmental sample?
15. What does an ICP-OES instrument actually detect in the sample that allows for quantitation?
16. What does an ICP-MS instrument actually detect in the sample that allows for quantitation?

F.6.5 Chromatography

17. What are the three common forms of chromatography you might see in an undergraduate lab?

18. Which GC detector would specialize in the analysis of the insecticide Aldrin ($C_{12}H_8Cl_6$)?
19. The boiling point of the insecticide Acifluorfen is 422 °C. Could you use GC or HPLC to separate and quantitate it?
20. What advantage does a mass spectrometer provide when coupled with a GC or HPLC?
21. If you wanted to quantitate the levels of nitrate and nitrite ions in a water sample, which type of IC stationary phase would you use – anionic or cationic?

F.6.6 FTIR Spectrometer

22. When quantitating analytes using IR spectrometry, what must be done to a spectrum measured in % transmittance?
23. In a sample of car exhaust, where in the IR spectrum would you look for nitrogen dioxide?
24. Is it possible to quantitate two different analytes with IR spectrometry if the absorption peaks are overlapped?

F.7 ANSWERS TO COMMON INSTRUMENTATION EXERCISES

F.7.1 UV-Vis Spectrophotometry

1. **What function does a monochromator perform in a spectrometer?** It disperses polychromatic radiation and outputs monochromatic radiation. The monochromator can be scanned to output any specific wavelength contained by the input radiation.
2. **Explain how Beer's Law is connected to the interaction of photons and matter.** The result of an analyte absorbing a photon is a reduction in I compared to I_0 and produces a nonzero absorbance when $-\log \frac{I}{I_0}$ is taken.
3. **What is the difference between scanning and diode array spectrophotometers? What advantage does each provide?** A scanning spectrophotometer contains a monochromator and takes longer to measure an absorption spectrum compared to a diode array spectrometer, but is more sensitive because of its photomultiplier tube detector.
4. **How is a line source different from a continuum source?** A line source emits a narrow range of radiation. A continuum source emits a broad range of radiation.
5. **What is the difference between a spectrophotometer that stores the sample in a cuvette compared to one that reads a well plate? What advantage does each provide?** A well-plate spectrophotometer is inherently less sensitive because of its small path length, compared to a 1-cm cuvette spectrophotometer. Well plates contain multiple samples, which can be read must faster than an equal number of cuvettes.

F.7.2 Fluorometers

6. **What is the difference between the instrument design of a fluorometer and an absorption spectrophotometer?** A fluorometer has a 90° detection angle, whereas an absorption instrument has a linear detection angle. The detector of an absorption instrument measures the full intensity of the source and measures the absorbance of the sample when it absorbs the light and reduces the intensity. A fluorometer's detector sees only the emission of the sample, and the emission monochromator can block the radiation of the source and any interferences.

7. **Why would a fluorometer signal contain less interferences compared to an absorbance signal? Provide two reasons.** (1) Fluorescence is rare, so most of the potential interferences in a sample will not fluorescence and infect the signal. (2) The excitation and emission monochromators are set for the specific excitation and emission wavelengths of the analyte, so only interferences that coincidentally have the same λ_{ex} and λ_{em} will infect the signal.
8. **What do λ_{ex} and λ_{em} represent in a fluorescence measurement?** The λ_{ex} represents the wavelength that the analyte absorbs, and λ_{em} represents the wavelength that the analyte emits; λ_{em} is the wavelength detected.

F.7.3 Atomic Absorption Spectrophotometry (AAS) and ICP-MS/OES

9. **What interaction between light and matter does atomic absorption spectrophotometery require in order to quantitate element concentrations?** Light must be absorbed by the analyte in its atomic state, not ionized state. This limitation is imposed by the HCL.
10. **What two methods of atomization are common for AAS?** Liquid samples can be dispersed with a nebulizer and atomized with a flame. Liquid and solid samples can be placed in an electrothermal atomizer.
11. **What happens to the analyte signal in AAS using flame atomization when the flame is too hot or too cold?** If the flame is too hot, then the analyte ionizes and the signal decreases. If the flame is too cold, then the sample may not atomize sufficiently and the signal decreases.
12. **What is the major advantage of flame atomization over electrothermal atomization? What is it about electrothermal atomization that causes its disadvantage?** Flame atomization with a nebulizer provides a very reproducible signal because the sample is delivered to the flame uniformly. Electrothermal atomization provides an imprecise measurement because of sample-to-sample variances in mass.
13. **What is the major advantage of electrothermal atomization over flame atomization? What is it about flame atomization that causes the disadvantage?** Electrothermal atomization provides a relatively sensitive measurement because nearly all of the sample is vaporized and analyzed. This is not true of flame atomization, where the nebulizer sends much of the sample into the waste.
14. **What is the main advantage that ICP instruments have over AAS instruments when analyzing an environmental sample?** An ICP instrument (either OES or MS detection) can analyze multiple elements at the same time, whereas an AAS can only analyze one element at a time.
15. **What does an ICP-OES instrument actually detect in the sample that allows for quantitation?** An ICP-OES detects the emission of elements that are atomized and excited in the plasma flame.
16. **What does an ICP-MS instrument actually detect in the sample that allows for quantitation?** The ICP-MS detects the sample based on the mass-to-charge ratio of the ionized analyte.

F.7.4 Chromatography

17. **What are the three common forms of chromatography you might see in an undergraduate lab?** Gas Chromatography (GC), high-performance liquid chromatography (HPLC), and ion chromatography (IC).
18. **Which GC detector would specialize in the analysis of the insecticide Aldrin ($C_{12}H_8Cl_6$)?** Aldrin is an organochlorine compound, so the electron capture detector would be an appropriate choice.

19. **The boiling point of the insecticide Acifluorfen is 422 °C. Could you use GC or HPLC to separate and quantitate it?** Most GC columns have a temperature limit near 300–350 °C, so analysis would have to be done using HPLC. There are some methods that derivatize the analyte and reduce its boiling point (esterification), allowing for GC analysis.
20. **What advantage does a mass spectrometer provide when coupled with a GC or HPLC?** Mass spectrometer can provide the analyst with the molar mass of the analyte or a probable identity of the analyte if the fragmentation pattern matches one found in a library.
21. **If you wanted to quantitate the levels of nitrate and nitrite ions in a water sample, which type of IC stationary phase would you use – anionic or cationic?** Anionic analytes require a cationic stationary phase.

F.7.5 FTIR Spectrometer

22. **When quantitating analytes using IR spectrometry, what must be done to a spectrum measured in % transmittance?** Beer's Law equates the absorbance to the concentration on the analyte. A spectrum measured in % transmission must be converted to absorbance.
23. **In a sample of car exhaust, where in the IR spectrum would you look for nitrogen dioxide?** A characteristic absorption of NO_2 is found between 1550 and 1900 cm^{-1}.
24. **Is it possible to quantitate two different analytes with IR spectrometry if the absorption peaks are overlapped?** Not easily. Overlapped peaks mix the signals of the two different absorptions so that both the peak height and the peak area are not proportional to a single analyte.

BIBLIOGRAPHY

Fosbol, P. L.; Thomsen, K.; Stenby, E. H. *Corrosion Engineering Science and Technology* **2010**, *45(2)*, 115–135.

Gurvich, L. V.; Veyts, I. V.; Alcock, C. B. *Thermodynamic Properties of Individual Substances*, 4th ed.; Hemisphere Publishing Corp.: New York, 1991; Vol. 2.

Haynes, W. *CRC Handbook of Chemistry and Physics*, 93rd ed.; Taylor & Francis: Boca Raton, FL, 2012.

Majzlan, J.; Grevel, K.-D.; Navrotsky, A. *American Mineralogist* **2003**, *88(5-6)*, 855–859.

Moore, C. E. *Atomic Energy Levels*; NSRDS: Washington, DC, 1971; Vol. 1.

Natrella, M. In *NIST/SEMATECH e-Handbook of Statistical Methods*; Croarkin, C., Tobias, P., Eds.; National Institute of Standards and Technology, 2013, http://www.nist.gov.

Roue, R. A.; Haselton, H. T. Jr.; Hemingway, B. S. *American Mineralogist* **1984**, *69*, 340–357.

Roussel, M. R. *A Life Scientist's Guide to Physical Chemistry*; Cambridge university Press: Cambridge, 2012.

Skoog, D. A.; Holler, F. J.; Crouch, S. R. *Principles of Instrumental Analysis*, 6th ed.; Thomson Brooks/Cole: Belmont, CA, 2007.

Stumm, W.; Morgan, J. J. *Aquatic Chemistry*, 3rd ed.; John Wiley & Sons, Inc.: New York, 1996.

Zen, E.-A.; Chernosky, J., Joseph V. *American Mineralogist* **1976**, *61*, 1156–1166.

G

DERIVATIONS

G.1 THE EQUAL VOLUME METHOD OF MULTIPLE STANDARD ADDITIONS FORMULA

Start with Eq. (2.8). The signal is due to the total concentration of analyte, which comes from the unknown and the standard.

$$\begin{aligned} S &= kC \\ &= k\left(\frac{C_{unk}V_{unk} + C_{std}V_{std}}{V_{tot}}\right) \\ &= \left(\frac{kC_{unk}V_{unk}}{V_{tot}}\right) + \left(\frac{kC_{std}}{V_{tot}}\right)V_{std} \end{aligned} \tag{G.1}$$

where C_{unk} is the concentration of the undiluted unknown, V_{unk} is the aliquot volume of the unknown sample, C_{std} is the concentration of the standard, V_{std} is the volume of the standard added, and V_{tot} is the total volume of the solution. Equation (G.1) has the form of a linear equation ($y = mx + b$) with y as S, b as $\frac{kC_{unk}V_{unk}}{V_{tot}}$, m as $\frac{kC_{std}}{V_{tot}}$, and x as V_{std} (this allows the usual graph of signal vs. volume of standard added). The *x-intercept*, which is the value of x when y=0, provides an indirect measure of the unknown concentration. To designate this, the V_{std} term becomes V_{x-int} to show that it is a distinct value of the independent variable.

$$\begin{aligned} S &= \left(\frac{kC_{unk}V_{unk}}{V_{tot}}\right) + \left(\frac{kC_{std}}{V_{tot}}\right)V_{std} \\ 0 &= \left(\frac{kC_{unk}V_{unk}}{V_{tot}}\right) + \left(\frac{kC_{std}}{V_{tot}}\right)V_{x-int} \\ -\left(\frac{kC_{unk}V_{unk}}{V_{tot}}\right) &= +\left(\frac{kC_{std}}{V_{tot}}\right)V_{x-int} \\ -C_{unk}V_{unk} &= +C_{std}V_{x-int} \\ C_{unk} &= -V_{x-int}\left(\frac{C_{std}}{V_{unk}}\right) \end{aligned} \tag{G.2}$$

If you examine Eq. (G.2), you will see the mathematical equivalent of the "probably incomprehensible" statement I made on page 73.

Environmental Chemistry: An Analytical Approach, First Edition. Kenneth S. Overway.

Companion website: www.wiley.com/go/overway/environmental_chemistry

G.2 TWO-POINT VARIABLE-VOLUME METHOD OF STANDARD ADDITION FORMULA

Start with Eq. (2.8) to relate the signal before the spike, and set up equations for the unknown before and after the addition (the spike)

$$S_{unk} = kC_{unk}$$
$$S_{spike} = k\left[C_{unk}\left(\frac{V_{unk}}{V_{unk}+V_{spike}}\right) + C_{std}\left(\frac{V_{spike}}{V_{unk}+V_{spike}}\right)\right] \tag{G.3}$$

where S_{unk} is the signal measured from the unknown aliquot, k is a constant of proportionality, C_{unk} is the concentration of the unknown, S_{spike} is the signal of the sample after it has been spiked, V_{unk} is the volume of the unknown aliquot, and V_{spike} is the volume of the spike of the standard.

Since the two k values are equal, solve for k in each equation, substitute, and solve for C_{unk}.

$$k = k$$
$$\frac{S_{unk}}{C_{unk}} = \frac{S_{spike}}{C_{unk}\left(\frac{V_{unk}}{V_{unk}+V_{spike}}\right) + C_{std}\left(\frac{V_{spike}}{V_{unk}+V_{spike}}\right)}$$
$$\left(\frac{S_{unk}}{S_{spike}}\right)\left[C_{unk}\left(\frac{V_{unk}}{V_{unk}+V_{spike}}\right) + C_{std}\left(\frac{V_{spike}}{V_{unk}+V_{spike}}\right)\right] = C_{unk}$$
$$\left(\frac{S_{unk}}{S_{spike}}\right)\left[C_{unk}\left(\frac{V_{unk}}{V_{unk}+V_{spike}}\right)\right] + \left(\frac{S_{unk}}{S_{spike}}\right)\left[C_{std}\left(\frac{V_{spike}}{V_{unk}+V_{spike}}\right)\right] = C_{unk}$$
$$\left(\frac{S_{unk}}{S_{spike}}\right)\left[C_{std}\left(\frac{V_{spike}}{V_{unk}+V_{spike}}\right)\right] = C_{unk} - \left(\frac{S_{unk}}{S_{spike}}\right)\left[C_{unk}\left(\frac{V_{unk}}{V_{unk}+V_{spike}}\right)\right]$$
$$\left(\frac{S_{unk}}{S_{spike}}\right)\left[C_{std}\left(\frac{V_{spike}}{V_{unk}+V_{spike}}\right)\right] = C_{unk}\left[1 - \left(\frac{S_{unk}}{S_{spike}}\right)\left(\frac{V_{unk}}{V_{unk}+V_{spike}}\right)\right]$$
$$\frac{\left(\frac{S_{unk}}{S_{spike}}\right)\left[C_{std}\left(\frac{V_{spike}}{V_{unk}+V_{spike}}\right)\right]}{\left[1 - \left(\frac{S_{unk}}{S_{spike}}\right)\left(\frac{V_{unk}}{V_{unk}+V_{spike}}\right)\right]} = C_{unk}$$
$$\left(\frac{\left(\frac{S_{spike}}{S_{unk}}\right)}{\left(\frac{S_{spike}}{S_{unk}}\right)}\right)\frac{\left(\frac{S_{unk}}{S_{spike}}\right)\left[C_{std}\left(\frac{V_{spike}}{V_{unk}+V_{spike}}\right)\right]}{\left[1 - \left(\frac{S_{unk}}{S_{spike}}\right)\left(\frac{V_{unk}}{V_{unk}+V_{spike}}\right)\right]} = C_{unk}$$
$$C_{unk} = \frac{\left[C_{std}\left(\frac{V_{spike}}{V_{unk}+V_{spike}}\right)\right]}{\left[\left(\frac{S_{spike}}{S_{unk}}\right) - \left(\frac{V_{unk}}{V_{unk}+V_{spike}}\right)\right]} \tag{G.4}$$

G.3 VARIABLE-VOLUME METHOD OF MULTIPLE STANDARD ADDITIONS FORMULA

Start with Eq. (2.8). The signal is due to the total concentration of analyte, which comes from the unknown and the standard.

$$S = k\left(\frac{C_{unk}V_{unk} + C_{std}NV_{spike}}{V_{unk} + NV_{spike}}\right) \tag{G.5}$$

where S is the signal measured for any solution, k is a constant of proportionality, C_{unk} is the concentration of the unknown, V_{unk} is the volume of the unknown aliquot, N is the number of cumulative spikes added, and V_{spike} is the repeated spike volume of the standard. This method assumes that the additions of the standard are repeated spikes of the same volume. For example, if V_{spike}=1.000 mL, then the first time the unknown is spiked, 1.000 mL would be added. The second time it is spiked, another 1.000 mL would be added, making the total cumulative spike volume of 2.000 mL or 2 × 1.000 mL or NV_{spike}.

The next step is to linearize the equation so that it can be plotted and fit to a linear equation ($y = mx + b$).

$$\left(V_{unk} + NV_{spike}\right) S = kC_{unk}V_{unk} + kC_{std}NV_{spike} \tag{G.6}$$

In this linear equation, $y = \left(V_{unk} + NV_{spike}\right) S$, $b = kC_{unk}V_{unk}$, $m = kC_{std}V_{spike}$, and $x = N$. Plotting $\left(V_{unk} + NV_{spike}\right) S$ versus N allows the slope and intercept to be calculated. The value of k is unknown, and it needs to be eliminated. To do this, divide the intercept by the slope and then substitute their values from Eq. (G.6).

$$\frac{b}{m} = \frac{kC_{unk}V_{unk}}{kC_{std}V_{spike}} \tag{G.7}$$

The k can be canceled. Solving for C_{unk} reveals that it can be calculated from a combination of initial data (C_{std}, V_{spike}, V_{unk}) and experimental data from the linear fit (b and m). Each of the other terms needs to be multiplied or divided in order to isolate C_{unk}.

$$\left(\frac{C_{std}V_{spike}}{V_{unk}}\right)\left(\frac{b}{m}\right) = C_{unk}$$

$$C_{unk} = \frac{bC_{std}V_{spike}}{mV_{unk}} \tag{G.8}$$

TABLES

H.1 STUDENT'S t TABLE

See Table H.1.

α (1 tail)	90%	95%	97.5%	99%	99.5%	99.9%
α (2 tail)	80%	90%	95%	98%	99%	99.8%
DOF						
1	3.078	6.314	12.706	31.821	63.657	318.313
2	1.886	2.920	4.303	6.965	9.925	22.327
3	1.638	2.353	3.182	4.541	5.841	10.215
4	1.533	2.132	2.776	3.747	4.604	7.173
5	1.476	2.015	2.571	3.365	4.032	5.893
6	1.440	1.943	2.447	3.143	3.707	5.208
7	1.415	1.895	2.365	2.998	3.499	4.782
8	1.397	1.860	2.306	2.896	3.355	4.499
9	1.383	1.833	2.262	2.821	3.250	4.296
10	1.372	1.812	2.228	2.764	3.169	4.143
11	1.363	1.796	2.201	2.718	3.106	4.024
12	1.356	1.782	2.179	2.681	3.055	3.929
13	1.350	1.771	2.160	2.650	3.012	3.852
14	1.345	1.761	2.145	2.624	2.977	3.787
15	1.341	1.753	2.131	2.602	2.947	3.733
16	1.337	1.746	2.120	2.583	2.921	3.686
17	1.333	1.740	2.110	2.567	2.898	3.646
18	1.330	1.734	2.101	2.552	2.878	3.610
19	1.328	1.729	2.093	2.539	2.861	3.579
20	1.325	1.725	2.086	2.528	2.845	3.552
25	1.316	1.708	2.060	2.485	2.787	3.450
30	1.310	1.697	2.042	2.457	2.750	3.385
35	1.306	1.690	2.030	2.438	2.724	3.340
40	1.303	1.684	2.021	2.423	2.704	3.307
45	1.301	1.679	2.014	2.412	2.690	3.281
50	1.299	1.676	2.009	2.403	2.678	3.261
60	1.296	1.671	2.000	2.390	2.660	3.232
70	1.294	1.667	1.994	2.381	2.648	3.211
80	1.292	1.664	1.990	2.374	2.639	3.195
90	1.291	1.662	1.987	2.368	2.632	3.183
100	1.290	1.660	1.984	2.364	2.626	3.174
∞	1.282	1.645	1.960	2.326	2.576	3.090

Source: Natrella (2013).

Table H.1 Student's t values for one-tailed and two-tailed comparisons.

Environmental Chemistry: An Analytical Approach, First Edition. Kenneth S. Overway.

Companion website: www.wiley.com/go/overway/environmental_chemistry

H.2 *F* TEST TABLE

See Table H.2.

***F* Distribution critical values for 90% confidence**

Denom. DOF	Numerator degrees of freedom (DOF) 1	2	3	4	5	7	10	20	30
1	39.864	49.500	53.593	55.833	57.240	58.906	60.195	61.740	62.265
2	8.5264	8.9999	9.1618	9.2434	9.2926	9.3491	9.3915	9.4413	9.4580
3	5.5384	5.4624	5.3907	5.3426	5.3092	5.2661	5.2304	5.1845	5.1681
4	4.5448	4.3245	4.1909	4.1073	4.0505	3.9790	3.9198	3.8443	3.8175
5	4.0605	3.7798	3.6194	3.5202	3.4530	3.3679	3.2974	3.2067	3.1740
7	3.5895	3.2575	3.0740	2.9605	2.8833	2.7850	2.7025	2.5947	2.5555
10	3.2850	2.9244	2.7277	2.6054	2.5216	2.4139	2.3226	2.2007	2.1554
20	2.9746	2.5893	2.3801	2.2490	2.1582	2.0397	1.9368	1.7939	1.7383
30	2.8808	2.4887	2.2761	2.1423	2.0493	1.9269	1.8195	1.6674	1.6064

***F* Distribution critical values for 95% confidence**

Denom. DOF	Numerator degrees of freedom (DOF) 1	2	3	4	5	7	10	20	30
1	161.45	199.50	215.71	224.58	230.16	236.77	241.88	248.01	250.10
2	18.513	19.000	19.164	19.247	19.296	19.353	19.396	19.446	19.462
3	10.128	9.5522	9.2766	9.1172	9.0135	8.8867	8.7855	8.6602	8.6165
4	7.7086	6.9443	6.5915	6.3882	6.2560	6.0942	5.9644	5.8026	5.7458
5	6.6078	5.7862	5.4095	5.1922	5.0504	4.8759	4.7351	4.5582	4.4958
7	5.5914	4.7375	4.3469	4.1202	3.9715	3.7871	3.6366	3.4445	3.3758
10	4.9645	4.1028	3.7082	3.4780	3.3259	3.1354	2.9782	2.7741	2.6996
20	4.3512	3.4928	3.0983	2.8660	2.7109	2.5140	2.3479	2.1241	2.0391
30	4.1709	3.3159	2.9223	2.6896	2.5336	2.3343	2.1646	1.9317	1.8408

***F* Distribution critical values for 99% confidence**

Denom. DOF	Numerator degrees of freedom (DOF) 1	2	3	4	5	7	10	20	30
1	4052.2	4999.5	5403.4	5624.6	5763.6	5928.4	6055.8	6208.7	6260.6
2	98.503	99.000	99.166	99.249	99.299	99.356	99.399	99.449	99.466
3	34.116	30.817	29.457	28.710	28.237	27.672	27.229	26.690	26.504
4	21.198	18.000	16.694	15.977	15.522	14.976	14.546	14.020	13.838
5	16.258	13.274	12.060	11.392	10.967	10.455	10.051	9.5526	9.3793
7	12.2460	9.5467	8.4513	7.8466	7.4605	6.9929	6.6201	6.1554	5.9920
10	10.0440	7.5594	6.5523	5.9944	5.6363	5.2001	4.8492	4.4055	4.2469
20	8.0960	5.8489	4.9382	4.4306	4.1027	3.6987	3.3682	2.9377	2.7785
30	7.5624	5.3903	4.5098	4.0179	3.6990	3.3046	2.9791	2.5486	2.3859

Table H.2 *F* Test data from MedCalc at http://www.medcalc.org/manual/f-table.php

I

CHEMICAL AND PHYSICAL CONSTANTS

I.1 PHYSICAL CONSTANTS

See Table I.1.

Constants	Symbols	Values
Avogadro's constant	N_A	$6.02214129 \times 10^{23}$ things/mol
Boltzmann's constant	k_B	$1.3806488 \times 10^{-23}$ J/K
Coulomb's constant	k_e	$8.9875517873681764 \times 10^{9}\ \mathrm{N \cdot m^2/C^2}$
Gas law constant	R	$8.3144621 \frac{\mathrm{J}}{\mathrm{K \cdot mol}}$
Gas law constant	R	$0.08205746 \frac{\mathrm{L \cdot atm}}{\mathrm{K \cdot mol}}$
Elementary charge	e	$1.602176565 \times 10^{-19}$ C
Electron mass	m_e	$9.10938291 \times 10^{-31}$ kg
Electron molar mass	$\mathcal{M}_e$	$5.4857990946 \times 10^{-7}$ kg/mol
Neutron mass	m_n	$1.674927351 \times 10^{-27}$ kg
Neutron molar mass	$\mathcal{M}_\mathrm{n}$	$1.00866491600 \times 10^{-3}$ kg/mol
Proton mass	m_p	$1.672621777 \times 10^{-27}$ kg
Proton molar mass	$\mathcal{M}_\mathrm{p}$	$1.007276466812 \times 10^{-3}$ kg/mol
Planck's constant	h	$6.62606957 \times 10^{-34}\ \mathrm{J \cdot s}$
Rydberg constant	R_∞	$2.179872171 \times 10^{-18}$ J
Stefan–Boltzmann constant	σ_{SB}	$5.670373 \times 10^{-8}\ \mathrm{W/(m^2\ K^4)}$
Velocity of light	c	2.99792458×10^{8} m/s
Wien's wavelength displacement law constant	b_Wien	$2.8977721 \times 10^{-3}\ \mathrm{m \cdot K}$

Table I.1 These values come from the National Institute of Standards and Technology Reference on Constants, Units, and Uncertainty, physics.nist.gov/cuu/index.html.

Environmental Chemistry: An Analytical Approach, First Edition. Kenneth S. Overway.

Companion website: www.wiley.com/go/overway/environmental_chemistry

I.2 STANDARD THERMOCHEMICAL PROPERTIES OF SELECTED SPECIES

See Table I.2.

Species	ΔH_f° (kJ/mol)	ΔG_f° (kJ/mol)	S_f° (J/(K mol))
- - - Selected inorganic species - - -			
Ag(s)	0.0		42.6
Ag^+(aq)	105.79	77.1	73.45
Ag_2S(s)	−32.6	−40.7	144.0
AgBr(s)	−100.4	−96.9	107.1
AgCl(s)	−127.0	−109.8	96.3
$AgNO_3$(s)	−124.4	−33.4	140.9
Al(s)	0.0	0.0	28.3
Al_2O_3(s)	−1675.7	−1582.3	50.9
Al_2S_3(s)	−724.0		116.9
Al^{3+}(aq)	−538.4	−485.	−321.7
As(s)	0.0	0.0	35.1
AsH_3(g)	66.4	68.9	222.8
Au(s)	0.0	0.0	47.4
B_2O_3(s)	−1273.5		
H_3BO_3(s)	−1094.3		
Ba(s)	0.0	0.0	62.5
Ba^{2+}(aq)	−537.6	−560.8	9.6
BaS(s)	−460.0	−456.0	78.2
$BaSO_4$(s)	−1473.2	−1362.2	132.2
Br^-(aq)	−121.4	−104.0	82.4
Br_2(l)	0.0	0.0	152.2
Br_2(g)	30.9	3.1	245.5
BrO_3^-(aq)	−67.07	18.6	161.71
Ca(s)	0.0	0.0	41.6
$Ca(OH)_2$(s)	−985.2	−897.5	83.4
Ca^{2+}(aq)	−542.8	−553.6	−53.1
$Ca_3(PO_4)_2$(s)	−4120.8	−3884.7	236.0
$CaCl_2$(s)	−795.4	−748.8	108.4
CaO(s)	−634.9	−603.3	38.1
CaS(s)	−482.4	−477.4	56.5
$CaSO_4$(s)	−1434.5	−1322.0	106.5
Cd(s)	0.0	0.0	51.8
Cd^{2+}(aq)	−75.9	−77.61	−73.2
CdS(s)	−161.9	−156.5	64.9
Cl(g)	121.3	105.3	165.2
Cl^-(aq)	−167.2	−131.2	56.5
Cl_2(g)	0.0	0.0	223.1
ClO_2(g)	102.5	120.5	256.8
ClO_4^-(aq)	−129.33	−8.52	182
$Cr_2O_7^{2-}$(aq)	−1490.3	−1301.1	261.9
CrO_4^{2-}(aq)	−881.1	−727.8	50.21
Cu(s)	0.0	0.0	33.2
$Cu(NO_3)_2$(s)	−302.9		
Cu^+(aq)	71.7	50.0	40.6
Cu^{2+}(aq)	64.8	65.5	−99.6
Cu_2S(s)	−79.5	−86.2	120.9
CuCl(s)	−137.2	−119.9	86.2
$CuCl_2$(s)	−220.1	−175.7	108.1
$Cu(NH_3)_4^{2+}$(aq)	−348.5	−111.07	273.6
CuS(s)	−53.1	−53.6	66.5
$CuSO_4$(s)	−771.4	−662.2	109.2
F(g)	79.4	62.3	158.8
F^-(aq)	−332.6	−278.8	−13.8
F_2(g)	0.0	0.0	202.8
Fe(s)	0.0	0.0	27.3
Fe^{2+}(aq)	−89.1	−78.9	−137.7
Fe^{3+}(aq)	−48.5	−4.7	−315.9
Fe_2O_3(s)	−824.2	−742.2	87.4
$Fe(OH)_3$(s)	−823.0	−696.5	106.7
$FeCl_2$(s)	−341.8	−302.3	118.0
$Fe(CN)_6^{3-}$(aq)	561.9	729.4	270.3
$Fe(CN)_6^{4-}$(aq)	455.6	695.08	95
$Fe(CNS)^{2+}$(aq)	23.4	71.1	−130
$FeCO_3$(s)[a]	−738.28	−665.16	95.47
FeS(s)	−100.0	−100.4	60.3
FeS_2(s)	−178.2	−166.9	52.9
$FeSO_4$(s)	−928.4	−820.8	107.5
H(g)	218.0	203.3	114.7
H^+(aq)	0.0	0.0	0
HCl(g)	−92.3	−95.3	186.9
HCl(aq)	−167.2	−131.2	56.5
H_2(g)	0.0	0.0	130.7
H_2O(l)	−285.8	−237.1	70.0
H_2O(g)	−241.8	−228.6	188.8
H_2O_2(l)	−187.8	−120.4	109.6
H_2O_2(g)	−136.3	−105.6	232.7
H_3PO_4(s)	−1284.4	−1124.3	110.5
H_3PO_4(l)	−1271.7	−1123.6	150.8
H_2S(g)	−20.6	−33.4	205.8
H_2SO_4(aq)	−909.3	−744.5	20.1
H_2SO_4(l)	−814.0	−690.0	156.9

Species	ΔH_f° (kJ/mol)	ΔG_f° (kJ/mol)	S_f° (J/(K mol))
$HNO_2(g)$	−79.5	−46.0	254.1
$HNO_3(l)$	−174.1	−80.7	155.6
$HNO_3(g)$	−133.9	−73.5	266.9
$HNO_3(aq)$	−207.4	−111.3	−146.4
$HO_2(g)$	10.5	22.6	229.0
$HOCl(g)$	−78.7	−66.1	236.7
$HS^-(aq)$	−17.6	12.08	62.8
$HSO_3^-(aq)$	−626.2	−527.7	139.7
$HSO_4^-(aq)$	−887.3	−755.9	131.8
$Hg(l)$	0.0	0.0	75.9
$Hg(g)$	61.4	31.8	175.0
$Hg_2^{2+}(aq)$	172.4	153.5	84.5
$Hg_2Cl_2(s)$	−265.4	−210.7	191.6
$Hg^{2+}(aq)$	171.1	164.4	−32.2
$Hg_2SO_4(s)$	−743.1	−625.8	200.7
$HgCl_2(s)$	−224.3	−178.6	146.0
$HgCl_4^{2-}(aq)$	−554	−446.8	293
$HgO(s)$	−90.8	−58.5	70.3
$HgS(s)$	−58.2	−50.6	82.4
$I(g)$	106.8	70.2	180.8
$I^-(aq)$	−55.2	−51.6	111.3
$I_2(g)$	62.4	19.3	260.7
$I_2(s)$	0.0	0.0	116.1
$I_3^-(aq)$	−51.5	−51.4	239.3
$IO_3^-(aq)$	−221.3	−128	118.4
$K^+(aq)$	−252.4	−283.3	102.5
$K(s)$	0.0	0.0	64.7
$K_2SO_4(s)$	−1437.8	−1321.4	175.6
$KBr(s)$	−393.8	−380.7	95.9
$KCl(s)$	−436.5	−408.5	82.6
$KH_2PO_4(s)$	−1568.3	−1415.9	134.9
$KI(s)$	−327.9	−324.9	106.3
$KMnO_4(s)$	−837.2	−737.6	171.7
$KNO_2(s)$	−369.8	−306.6	152.1
$KNO_3(s)$	−494.6	−394.9	133.1
$KOH(s)$	−424.6	−379.4	81.2
$Mg(s)$	0.0	0.0	32.7
$Mg(NO_3)_2(s)$	−790.7	−589.4	164.0
$Mg(OH)_2(s)$	−924.5	−833.5	63.2
$Mg^{2+}(aq)$	−466.9	−454.8	−138.1
$MgO(s)$	−601.6	−569.3	27.0

Species	ΔH_f° (kJ/mol)	ΔG_f° (kJ/mol)	S_f° (J/(K mol))
$MgSO_4(s)$	−1284.9	−1170.6	91.6
$Mn(s)$	0.0	0.0	32.0
$Mn^{2+}(aq)$	−220.8	−228.1	−73.6
$MnCl_2(s)$	−481.3	−440.5	118.2
$MnCO_3(s)^b$	−891.91	−818.13	98.03
$MnO_2(s)$	−520.0	−465.1	53.1
$MnO_4^-(aq)$	−541.4	−447.2	191.2
$N(g)$	472.7	455.5	153.3
$N_2(g)$	0.0	0.0	191.6
$N_2O(g)$	81.6	103.7	220.0
$N_2O_3(g)$	86.6	142.4	314.7
$N_2O_4(l)$	−19.5	97.5	209.2
$N_2O_4(g)$	11.1	99.8	304.4
$NH_3(aq)$	−80.3	−26.5	111.3
$NH_3(g)$	−45.9	−16.4	192.8
$NH_4^+(aq)$	−133.26	−79.31	113.4
$NH_4Cl(s)$	−314.4	−202.9	94.6
$NH_4OH(l)$	−361.2	−254.0	165.6
$NO(g)$	91.3	87.6	210.8
$NO_2(g)$	33.2	51.3	240.1
$NO_3(g)$	73.7	115.9	258.4
$NO_2^-(aq)$	−104.6	−32.2	146.4
$NO_3^-(aq)$	−207.4	−108.7	146.4
$N_2O_5(l)$	−43.1	113.9	178.2
$N_2O_5(g)$	13.3	117.1	355.7
$Na(s)$	0.0	0.0	51.3
$Na^+(aq)$	−240.1	−261.9	59.0
$Na_2HPO_4(s)$	−1748.1	−1608.2	150.5
$Na_2SO_4(s)$	−1387.1	−1270.2	149.6
$NaBr(s)$	−361.1	−349.0	86.8
$NaHSO_4(s)$	−1125.5	−992.8	113.0
$NaN_3(s)$	21.7	93.8	96.9
$NaNO_2(s)$	−358.7	−284.6	103.8
$NaNO_3(s)$	−467.9	−367.0	116.5
$NaOH(s)$	−425.8	−379.7	64.4
$O(g)$	249.18	231.7	161.1
$O^*(g)$	438.05		
$O^{**}(g)$	654.4		
$O_2(g)$	0.0	0.0	205.2
$O_2^*(g)$	94.29		205.2
$O_2^{**}(g)$	156.96		205.2

Species	ΔH_f° (kJ/mol)	ΔG_f° (kJ/mol)	S_f° (J/(K mol))
O_3(g)	142.7	163.2	238.9
OH(g)	39.0	34.2	183.7
OH^-(aq)	−230	−157.2	−10.75
P(s)(white)	0.0	0.0	41.1
P(s) (red)	−17.6		22.8
Pb(s)	0.0	0.0	64.8
Pb^{2+}(aq)	−1.7	−24.4	10.5
PbS(s)	−100.4	−98.7	91.2
$PbSO_4$(s)	−920.0	−813.0	148.5
PH_3(g)	5.4	13.5	210.2
PO_4^{3-}(aq)	−1277.4	−1018.7	−222
S(s)	0.0	0.0	32.1
S^{2-}(aq)	33.1	85.8	−14.6
$S_2O_3^{2-}$(aq)	−648.5	−522.5	67
$S_4O_6^{2-}$(aq)	−1224.2	−1040.4	257.3
Si(s)	0.0	0.0	18.8
SiO_2(s)	−910.7	−856.3	41.5
SO(g)	6.3	−19.9	222.0
SO_2(g)	−296.8	−300.1	248.2
SO_3(s)	−454.5	−374.2	70.7
SO_3(l)	−441.0	−373.8	113.8
SO_3(g)	−395.7	−371.1	256.8
SO_3^{2-}(aq)	−635.5	−486.5	−29.0
SO_4^{2-}(aq)	−909.3	−744.53	20.1
SO_2Cl_2(g)[c]	−364.0	−320.0	311.8
SO_2Cl_2(aq)[c]	−394.1	−314	207
Zn(s)	0.0	0.0	41.6
Zn^{2+}(aq)	−153.9	−147.1	−112.1
ZnO(s)	−350.5	−320.5	43.7
Zr(s)	0.0	0.0	39.0
$Zr(SO_4)_2$(s)	−2217.1		
ZrO_2(s)	−1100.6	−1042.8	50.4

Species	ΔH_f° (kJ/mol)	ΔG_f° (kJ/mol)	S_f° (J/(K mol))
- - - *Selected carbon-containing species* - - -			
$C_{graphite}$	0	0.0	5.7
$C_{diamond}$	1.9	2.9	2.4
CCl_4(g)	−95.7		
CCl_4(l)	−128.2		
CCl_2F_2(g)	−477.4	−439.4	300.8
$CHClF_2$(g)	−482.6		280.9
$CHCl_2F$(g)			293.1
CH_3CH_2OH(aq)	−277.6	−174.8	160.7
CH_3CH_2OH(l)	−277.6	−174.8	160.7
CH_3COO^-(aq)	−486	−369.3	−6.3
CH_3COOH(aq)	−484.5	−389.9	178.7
CH_3OH(aq)[d]	−245.9	−175.4	133
CH_3OH(l)	−239.2	−166.6	126.8
$(CH_3)_2S$(aq)	−37.5	−57.8	68.25
CH_4(g)	−74.6	−50.5	186.3
C_2H_2(g)	227.4	209.9	200.9
C_2H_4(g)	52.3	68.4	219.3
C_2H_6(g)	−84.0	−32.0	229.2
C_4H_{10}(g)	−125.6	−218.0	310.23
CN^-(aq)	150.6	172.4	94.1
CNS^-(aq)	76.44	92.71	144.3
CO(g)	−110.53	−137.2	197.7
CO_2(g)	−393.5	−394.4	213.8
CO_2(aq)[c]	−412.9	−386.2	121
COS(g)[d]	−137.2	−169.2	234.5
H_2CO_3(aq)	−699.65	−623.08	187.4
HCO_3^-(aq)	−692.0	−586.8	91.2
CO_3^{2-}(aq)	−677.1	−527.8	−56.9
$C_2O_4^{2-}$(aq)	−825.1	−673.9	45.6
$C_6H_{12}O_6$(s)[c]	−1273.3	−910.4	212.1
$C_6H_{12}O_6$(aq)[c]	−1263.06	−914.25	264.01
C_8H_{18}(g)	−208.5	16.40	466.7

Name	Formula	ΔH_f° (kJ/mol)	ΔG_f° (kJ/mol)	S_f° (J/(K mol))
	- - - *Selected minerals* - - -			
Brucite	$Mg(OH)_2$	−925.6	−834.53	−305.1
Clinochrysotile	$Mg_3Si_2O_5(OH)_4$	−4361.4	−4030.8	−1098
Clinoenstatite	$MgSiO_3$	−1549	−1462	−291.3
Enstatite	$MgSiO_3$	−1.547	−1460	
Forsterite	Mg_2SiO_4	−2170.2	−2051.7	−399.0
Goethite	FeO(OH)[e]	−561.9	−489.8	59.2
Talc	$Mg_3Si_4O_{10}(OH)_2$	−5916.2	−5536.6	−1273

Excited state atomic and molecular oxygen comes from Gurvich, Veyts, and Alcock (1991) and Moore (1971).
[a]From Fosbol, Thomsen, and Stenby (2010).
[b]From Roue, Haselton, and Hemingway (1984).
[c]Roussel (2012).
[d](Stumm and Morgan, 1996, p.1003).
[e]For goethite from Majzlan, Grevel, and Navrotsky (2003).

Table I.2 The values come from *CRC Handbook of Chemistry and Physics*, 89th ed.; Lide, D.R., Ed.; CRC Press: Boca Raton, FL, 2008; Section 5.

I.3 HENRY'S LAW CONSTANTS

See Table I.3.

Species	K_H° (M/atm)	Temperature correction (δ_H)
Ar	1.4×10^{-3}	−1500
Br_2	0.80	−3900
CCl_4	3.4×10^{-2}	−4200
$(CH_3)_2S$	0.56	−3700
CH_4	1.4×10^{-3}	
$(CH_3)_2S$	0.48	−3100
Cl_2	9.5×10^{-2}	−2100
$ClNO_2$	2.4×10^{-2}	
CO	9.9×10^{-4}	−1300
CO_2	3.5×10^{-2}	−2400
OCS (or COS)	2.2×10^{-2}	−2100
H_2	7.8×10^{-4}	−490
H_2O_2	$8.3 \times 10^{+4}$	−7400
H_2S	0.10	−2100
He	3.8×10^{-4}	−92
HNO_2	49	−4800
HNO_3	$2.1 \times 10^{+5}$	−8700
HO_2	$5.7 \times 10^{+3}$	−4800
HOCl	$6.6 \times 10^{+2}$	−5900
I_2	3.0	−4400
N_2	6.5×10^{-4}	−1300
N_2O	2.5×10^{-2}	−2600
NH_3	76	−3400
NO	1.9×10^{-3}	−1400
NO_2	1.2×10^{-2}	−2500
NO_3	2.0	−2000
O_2	1.2×10^{-3}	−1800
O_3	1.3×10^{-2}	−2000
OH	29	−3100
H_2S	8.7×10^{-2}	−2100
SO_2	1.2	−3100
SO_3	∞	

The temperature correction factor (δ_H) represents the $\frac{\Delta H^\circ}{R}$ found in eq 5.5, and is negative because of the definition I have chosen for Henry's Law $\left(K_H = \frac{[O_2]}{P_{O_2}}\right)$. Other sources that list it as a positive value define the equilibrium as $K_H = \frac{P_{O_2}}{[O_2]}$. To calculate the temperature effect, use $K_H = K_H^\circ \cdot e^{\left(\delta_H\left(\frac{1}{T^\circ}-\frac{1}{T}\right)\right)}$
Source: Max Planck Institute.

Table I.3 Henry's law equilibrium constants.

I.4 SOLUBILITY PRODUCT CONSTANTS

See Table I.4.

Compound	Formula	K_{sp}	pK_{sp}
Aluminum phosphate	$AlPO_4$	9.84×10^{-21}	20.007
Barium carbonate	$BaCO_3$	2.58×10^{-9}	8.588
Barium fluoride	BaF_2	1.84×10^{-7}	6.735
Barium hydroxide octahydrate	$Ba(OH)_2 \cdot 8H_2O$	2.55×10^{-4}	3.593
Barium nitrate	$Ba(NO_3)_2$	4.64×10^{-3}	2.333
Barium sulfate	$BaSO_4$	1.08×10^{-10}	9.967
Barium sulfite	$BaSO_3$	5×10^{-10}	9.301
Cadmium carbonate	$CdCO_3$	1×10^{-12}	12.000
Cadmium fluoride	CdF_2	6.44×10^{-3}	2.191
Cadmium hydroxide	$Cd(OH)_2$	7.2×10^{-15}	14.143
Cadmium phosphate	$Cd_3(PO_4)_2$	2.53×10^{-33}	32.597
Calcium carbonate (calcite)	$CaCO_3$	3.36×10^{-9}	8.474
Calcium fluoride	CaF_2	3.45×10^{-11}	10.462
Calcium hydroxide	$Ca(OH)_2$	5.02×10^{-6}	5.299
Calcium oxalate monohydrate	$CaC_2O_4 \cdot H_2O$	2.32×10^{-9}	8.635
Calcium phosphate	$Ca_3(PO_4)_2$	2.07×10^{-33}	32.684
Calcium sulfate	$CaSO_4$	4.93×10^{-5}	4.307
Calcium sulfate dihydrate	$CaSO_4 \cdot 2H_2O$	3.14×10^{-5}	4.503
Copper(II) oxalate	CuC_2O_4	4.43×10^{-10}	9.354
Copper(II) phosphate	$Cu_3(PO_4)_2$	1.4×10^{-37}	36.854
Iron(II) carbonate	$FeCO_3$	3.13×10^{-11}	10.504
Iron(II) fluoride	FeF_2	2.36×10^{-6}	5.627
Iron(II) hydroxide	$Fe(OH)_2$	4.87×10^{-17}	16.312
Iron(III) hydroxide	$Fe(OH)_3$	2.79×10^{-39}	38.554
Iron(III) phosphate dihydrate	$FePO_4 \cdot 2H_2O$	9.91×10^{-16}	15.004
Magnesium carbonate	$MgCO_3$	6.82×10^{-6}	5.166
Magnesium carbonate trihydrate	$MgCO_3 \cdot 3H_2O$	2.38×10^{-6}	5.623
Magnesium carbonate pentahydrate	$MgCO_3 \cdot 5H_2O$	3.79×10^{-6}	5.421
Magnesium fluoride	MgF_2	5.16×10^{-11}	10.287
Magnesium hydroxide	$Mg(OH)_2$	5.61×10^{-12}	11.251
Magnesium phosphate	$Mg_3(PO_4)_2$	1.04×10^{-24}	23.983
Manganese(II) carbonate	$MnCO_3$	2.24×10^{-11}	10.650
Mercury(I) carbonate	Hg_2CO_3	3.6×10^{-17}	16.444
Mercury(I) chloride	Hg_2Cl_2	1.43×10^{-18}	17.845
Mercury(I) sulfate	Hg_2SO_4	6.5×10^{-7}	6.187
Silver(I) acetate	$AgCH_3COO$	1.94×10^{-3}	2.712
Silver(I) carbonate	Ag_2CO_3	8.46×10^{-12}	11.073
Silver(I) chloride	$AgCl$	1.77×10^{-10}	9.752
Silver(I) phosphate	Ag_3PO_4	8.89×10^{-17}	16.051
Silver(I) sulfate	Ag_2SO_4	1.2×10^{-5}	4.921
Silver(I) sulfite	Ag_2SO_3	1.5×10^{-14}	13.824
Strontium carbonate	$SrCO_3$	5.6×10^{-10}	9.252
Tin(II) hydroxide	$Sn(OH)_2$	5.45×10^{-27}	26.264
Zinc carbonate	$ZnCO_3$	1.46×10^{-10}	9.836
Zinc carbonate monohydrate	$ZnCO_3 \cdot H_2O$	5.42×10^{-11}	10.266
Zinc hydroxide	$Zn(OH)_2$	3×10^{-17}	16.523

Source: CRC Handbook, 87th ed.

Table I.4 Solubility product constants.

I.5 ACID DISSOCIATION CONSTANTS

See Table I.5.

Acid	Formula	K_a	pK_a
Acetic	CH_3COOH	1.75×10^{-5}	4.756
Ammonium ion	NH_4^+	5.6×10^{-10}	9.25
Carbonic	H_2CO_3	4.5×10^{-7}	6.35
	HCO_3^-	4.7×10^{-11}	10.33
Chlorous	$HClO_2$	1.1×10^{-2}	1.94
Chromic	H_2CrO_4	1.8×10^{-1}	0.74
	$HCrO_4^-$	3.2×10^{-7}	6.49
Hydrofluoric	HF	6.3×10^{-4}	3.20
Hydrogen sulfide	H_2S	8.9×10^{-8}	7.05
	HS^-	1.0×10^{-19}	19
Hypochlorous	$HClO$	4.0×10^{-8}	7.40
Hydrocyanic	HCN	6.2×10^{-10}	9.21
Nitrous	HNO_2	5.6×10^{-4}	3.25
Nitric	HNO_3	∞	–
Phenol	C_6H_5OH	1.0×10^{-10}	9.99
Phosphoric	H_3PO_4	6.9×10^{-3}	2.16
	$H_2PO_4^-$	6.2×10^{-8}	7.21
	HPO_4^{2-}	4.8×10^{-13}	12.32
Sulfuric	H_2SO_4	∞	–
	HSO_4^-	1.0×10^{-2}	1.99
Sulfurous	H_2SO_3	1.4×10^{-2}	1.85
	HSO_3^-	6.3×10^{-8}	7.2
Silicic	H_4SiO_4	1.3×10^{-10}	9.9
	$H_3SiO_4^-$	1.6×10^{-12}	11.8
	$H_2SiO_4^{2-}$	1.0×10^{-12}	12
	$HSiO_4^{3-}$	1.0×10^{-12}	12
Water	H_2O	1.01×10^{-14}	13.995

Source: Haynes (2012).

Table I.5 Acid dissociation constants.

I.6 BASE DISSOCIATION CONSTANTS

See Table I.6.

Base	Formula	K_b	pK_b
Acetate	CH_3COO^-	5.70×10^{-10}	9.244
Ammonia	NH_3	1.8×10^{-5}	4.75
Carbonate	CO_3^{2-}	2.1×10^{-4}	3.67
	HCO_3^-	2.2×10^{-8}	7.65
Chlorite	ClO_2^-	8.7×10^{-13}	12.06
Chromate	CrO_4^{2-}	5.5×10^{-14}	13.26
	$HCrO_4^-$	3.1×10^{-8}	7.51
Fluoride	F^-	1.6×10^{-11}	10.80
Hypochlorite	ClO^-	2.5×10^{-7}	6.60
Cyanide	CN^-	1.6×10^{-5}	4.79
Nitrite	NO_2^-	1.8×10^{-11}	10.75
Phenoxide	$C_6H_5O^-$	9.8×10^{-5}	4.01
Phosphate	PO_4^{3-}	2.1×10^{-2}	1.68
	HPO_4^{2-}	1.6×10^{-7}	6.79
	$H_2PO_4^-$	1.4×10^{-12}	11.84
Sulfate	SO_4^{2-}	9.8×10^{-13}	12.01
Sulfite	SO_3^{2-}	1.6×10^{-7}	6.8
	HSO_3^-	7.1×10^{-13}	12.15

Source: Haynes (2012).

Table I.6 Base dissociation constants.

I.7 BOND ENERGIES

See Table I.7.

Bond	Enthalpy (H°) (kJ/mol)	Bond	Enthalpy (H°) (kJ/mol)	Bond	Enthalpy (H°) (kJ/mol)
H–H	436	N–CO	360	C=C	611
H–B	377	N–N	161	N=N	456
H–C	414	N–O	230	O=O	498
H–N	389	O–CO	460	C=N	615
H–O	464	O–O	146	C=O (CO_2)	803
H–F	565	O–F	190	C=O (aldehyde)	741
H–Si	314	O–Cl	203	C=O (ketone)	745
H–P	322	O–Br	201	C=O (ester)	749
H–S	339	Si–O	460	C=O (amide)	749
H–Cl	431	Si–F	565	C=O (halide)	741
H–Br	366	Si–Si	218	C=S (CS_2)	577
H–I	297	Si–Cl	377	N=O (HONO)	598
C–B	377	S–O	364	P=O ($POCl_3$)	460
C–C	347	S–S	226	P=S ($PSCl_3$)	293
C–N	305	P–O	377	S=O (SO_2)	536
C–O	358	P–P	209	S=O (DMSO)	389
C–F	485	P–Cl	331	P=P	351
C–Si	318	P–Br	272	P≡P	490
C–P	293	F–F	153	C≡O	1080
C–S	272	Cl–Cl	243	C≡C	837
C–Cl	339	Br–Br	192	N≡N	946
C–Br	285	I–I	151	C≡N	891
C–I	213				

Source: Sanderson; *Polar Covalence*, 1983 and *Chemical Bonds and Bond Energy*, 1976.

Table I.7 Average bond energies (enthalpies) for some common single, double, and triple bonds.

I.8 STANDARD REDUCTION POTENTIALS

See Table I.8.

Reduction half-reaction	$E°$ (V)
$2H_2O(l) + 2e^- \longrightarrow H_2(g) + 2OH^-(aq)$	−0.8277
$S(s) + 2e^- \longrightarrow S^{2-}(aq)$	−0.47627
$N_2(g) + 8H^+(aq) + 6e^- \longrightarrow 2NH_4^+(aq)$	−0.28
$6CO_2(g) + 24H^+(aq) + 24e^- \longrightarrow C_6H_{12}O_6(s) + 6H_2O(l)$	−0.014[a]
$2H^+(aq) + 2e^- \longrightarrow H_2(g)$	0.00
$NO_3^-(aq) + H_2O(l) + 2e^- \longrightarrow NO_2^-(aq) + 2OH^-(aq)$	0.01
$S_4O_6^{2-}(aq) + 2e^- \longrightarrow 2S_2O_3^{2-}(aq)$	0.08
$6CO_2(aq) + 24H^+(aq) + 24e^- \longrightarrow C_6H_{12}O_6(aq) + 6H_2O(l)$	0.032[a]
$N_2(g) + 2H_2O(l) + 6H^+(aq) + 6e^- \longrightarrow 2NH_4OH(aq)$	0.092
$2\,NO_2^-(aq) + 3H_2O(l) + 4e^- \longrightarrow N_2O(g) + 6OH^-(aq)$	0.15
$AgCl(s) + e^- \longrightarrow Ag(s) + Cl^-(aq)$	0.22233
$SO_4^{2-}(aq) + 10H^+(aq) + 8e^- \longrightarrow H_2S(g) + 4H_2O(l)$	0.248
$I_2(aq) + 2e^- \longrightarrow 2I^-(aq)$	0.5355
$I_3^-(aq) + 2e^- \longrightarrow 3I^-(aq)$	0.536
$FeO(OH)(s) + HCO_3^-(aq) + 2H^+(aq) + e^- \longrightarrow FeCO_3(s) + 2H_2O(l)$	0.648[a]
$O_2(g) + 2H^+(aq) + 2e^- \longrightarrow H_2O_2(l)$	0.695
$Fe^{3+}(aq) + e^- \longrightarrow Fe^{2+}(aq)$	0.771
$ClO^-(aq) + H_2O(l) + 2e^- \longrightarrow Cl^-(aq) + 2OH^-(aq)$	0.841
$NO_3^-(aq) + 3H^+(aq) + 2e^- \longrightarrow HNO_2(aq) + H_2O(l)$	0.934
$NO_3^-(aq) + 4H^+(aq) + 3e^- \longrightarrow NO(g) + 2H_2O(l)$	0.957
$O_2(g) + 4H^+(aq) + 4e^- \longrightarrow 2H_2O(l)$	1.229
$MnO_2(s) + HCO_3^-(aq) + 3H^+(aq) + 2e^- \longrightarrow MnCO_3(s) + 2H_2O(l)$	1.245[a]
$MnO_2(s) + 4H^+(aq) + 2e^- \longrightarrow Mn^{2+}(aq) + 2H_2O(l)$	1.224
$O_2(aq) + 4H^+(aq) + 4e^- \longrightarrow 2H_2O(l)$	1.271[a]
$Cl_2(g) + 2e^- \longrightarrow 2Cl^-(aq)$	1.35827
$Cr_2O_7^{2-}(aq) + 14H^+(aq) + 6e^- \longrightarrow 2Cr^{3+}(aq) + 7H_2O(l)$	1.36
$MnO_4^-(aq) + 8H^+(aq) + 5e^- \longrightarrow 4H_2O(l) + Mn^{2+}(aq)$	1.507
$HClO(aq) + H^+(aq) + e^- \longrightarrow 1/2Cl_2(g) + H_2O(aq)$	1.611
$MnO_4^-(aq) + 4H^+(aq) + 3e^- \longrightarrow MnO_2(s) + 2H_2O(l)$	1.679
$H_2O_2(l) + 2H^+(aq) + 2e^- \longrightarrow 2H_2O(l)$	1.776
$OH(g) + e^- \longrightarrow OH^-(aq)$	2.02
$O_3(g) + 2H^+(aq) + 2e^- \longrightarrow O_2(g) + H_2O(l)$	2.076

[a]Calculated from standard free energy values at 298 K and 1 M concentrations.

Table I.8 Environmentally useful reduction potentials from the 92nd CRC.

I.9 OH OXIDATION RATE CONSTANTS VALUES

See Table I.9.

Temp. (°C)	Ethane	Propane	Propene	Formaldehyde	Acetaldehyde	Isoprene	α-Pinene
0	1.83×10^{-13}	9.25×10^{-13}	3.07×10^{-11}	9.37×10^{-12}	1.73×10^{-11}	1.14×10^{-10}	6.15×10^{-11}
5	1.96×10^{-13}	9.63×10^{-13}	2.97×10^{-11}	9.37×10^{-12}	1.70×10^{-11}	1.11×10^{-10}	5.97×10^{-11}
10	2.10×10^{-13}	1.00×10^{-12}	2.88×10^{-11}	9.36×10^{-12}	1.66×10^{-11}	1.08×10^{-10}	5.80×10^{-11}
15	2.24×10^{-13}	1.04×10^{-12}	2.79×10^{-11}	9.36×10^{-12}	1.63×10^{-11}	1.05×10^{-10}	5.65×10^{-11}
20	2.39×10^{-13}	1.08×10^{-12}	2.71×10^{-11}	9.36×10^{-12}	1.60×10^{-11}	1.03×10^{-10}	5.50×10^{-11}
25	2.54×10^{-13}	1.12×10^{-12}	2.63×10^{-11}	9.37×10^{-12}	1.58×10^{-11}	1.00×10^{-10}	5.36×10^{-11}
30	2.70×10^{-13}	1.16×10^{-12}	2.56×10^{-11}	9.38×10^{-12}	1.55×10^{-11}	9.82×10^{-11}	5.23×10^{-11}
35	2.87×10^{-13}	1.21×10^{-12}	2.49×10^{-11}	9.38×10^{-12}	1.52×10^{-11}	9.61×10^{-11}	5.11×10^{-11}

All values in the aforementioned table have units of $\frac{cm^3}{mcs \cdot sec}$.

Table I.9 Kinetic rate constants for VOC oxidation against the hydroxyl radical.

BIBLIOGRAPHY

Fosbol, P. L.; Thomsen, K.; Stenby, E. H. *Corrosion Engineering Science and Technology* **2010**, 45*(2)*, 115–135.

Gurvich, L. V.; Veyts, I. V.; Alcock, C. B. *Thermodynamic Properties of Individual Substances*, 4th ed.; Hemisphere Publishing Corp.: New York, 1991; Vol. 2.

Haynes, W. *CRC Handbook of Chemistry and Physics*, 93rd ed.; Taylor & Francis: Boca Raton, FL, 2012.

Majzlan, J.; Grevel, K.-D.; Navrotsky, A. *American Mineralogist* **2003**, 88*(5-6)*, 855–859.

Moore, C. E. *Atomic Energy Levels*; NSRDS: Washington, DC, 1971; Vol. 1.

Natrella, M. In *NIST/SEMATECH e-Handbook of Statistical Methods*; Croarkin, C., Tobias, P., Eds.; National Institute of Standards and Technology, 2013, http://www.nist.gov.

Roue, R. A.; Haselton, H. T. Jr.; Hemingway, B. S. *American Mineralogist* **1984**, 69, 340–357.

Roussel, M. R. *A Life Scientist's Guide to Physical Chemistry*; Cambridge University Press: Cambridge, 2012.

Skoog, D. A.; Holler, F. J.; Crouch, S. R. *Principles of Instrumental Analysis*, 6th ed.; Thomson Brooks/Cole: Belmont, CA, 2007.

Stumm, W.; Morgan, J. J. *Aquatic Chemistry*, 3rd ed.; John Wiley & Sons, Inc.: New York, 1996.

Zen, E.-A.; Chernosky, J., Joseph V. *American Mineralogist* **1976**, 61, 1156–1166.

Index

(Note: Entries in *italics* refer to figures, entries in **bold** refer to tables)

Environmental Chemistry: An Analytical Approach, First Edition. Kenneth S. Overway.

Companion website: www.wiley.com/go/overway/environmental_chemistry